EVOLUTIONARY THEORY AND PROCESSES: MODERN HORIZONS

Papers in Honour of Eviatar Nevo

EVOLUTIONARY THEORY AND PROCESSES: MODERN HORIZONS

Papers in Honour of Eviatar Nevo

Edited by

Solomon P. Wasser
Institute of Evolution, University of Haifa, Haifa, Israel;
N.G. Kholodny Institute of Botany, NASU, Kiev, Ukraine

SPRINGER-SCIENCE+BUSINESS MEDIA, B.V.

A C.I.P. Catalogue record for this book is available from the Library of Congress.

ISBN 978-90-481-6457-8   ISBN 978-94-017-0443-4 (eBook)
DOI 10.1007/978-94-017-0443-4

Printed on acid-free paper

Contents

Part One: Evolution of Life and Evolutionary Theory

Part Two: Genome Evolution

Part Three: Phylogeography and Phylogeny

Part Four: Human Evolution and Ecology

Professor Eviatar Nevo

Preface

Evolution is the most profound of human ideas integrating all natural phenomena: cosmic, biological, and cultural into a continuous universal change. This volume deals with evolutionary observations, experiments, and theories contributing to a deeper understanding of the evolutionary process, honoring the 75th birthday of Eviatar (Eibi) Nevo. I first met Eibi in 1966 when he was a Fellow in the Museum of Comparative Zoology at Harvard University and working mostly on cricket frog vocalization and speciation in the United States. His unique discovery of pipid fossil frogs in the Israeli Early Cretaceous, central Negev, is possibly the largest world collection of ancient fossil frogs. Our acquaintance developed into mutual friendship and admiration. Since then our long-lasting friendship has included a visit to Israel, enabling me to follow Eibi's major scientific achievements, in particular, his founding of the Institute of Evolution in the University of Haifa and now the pending establishment of the International Graduate School of Evolution.

The research program of Eibi Nevo, in collaboration with numerous colleagues and students in Israel and across the world, encompasses diverse perspectives of evolutionary biology and biodiversity of genes, populations, species, and ecosystems integrating modern and classical evolutionary approaches, molecular and organismal. They deal with model organisms in all forms from bacteria through plants, fungi, animals, and humans conducted over local, regional, and global scales. His research program is interdisciplinary integrating ecology, molecular biology, cytogenetics, genetics, genomics, morphology, physiology, and behavior in an evolutionary context shedding new light on organism-environment relationships and on the evolution of biodiversity primarily on the patterns and dynamics of adaptation and speciation. He advanced the environmental theory of genetic diversity by highlighting the interface between ecology and genetics across diverse organisms from bacteria to mammals. He substantially contributed to the ecological and genetic theory of speciation and adaptation at both the molecular and organismal levels. His classical studies on the blind subterranean mammals led to the discovery of many new species of the family Spalacidae than were previously known, based on chromosomal and allozymic variation. He already showed in the early 1970s that active speciation can proceed in mole rats with relatively minor genetic changes and unraveled the evolutionary driving forces leading to similar regressive, progressive, and convergent evolution of subterranean mammals across the planet. Furthermore, he and his colleagues showed how mole rats could dramatically contribute to human gene therapy potentially overcoming human ischemia by their powerful angiogenic potentials that evolved as an adaptive complex to cope with underground hypoxia stress. Likewise, they elucidated how a blind mammal with a subcutaneous degenerated eye can sense photoperiodic perception, perform circadian rhythms underground, and tell night from day in total darkness by means of three retinal pigments and a battery of clock genes. Eibi advanced the science of ecological genomics by elucidating genome structure, dynamics, and evolution at the protein, DNA, and diverse organismal levels.

He demonstrated the sweep of his competency by applying his findings to crop improvements primarily of wheat and barley. Eibi substantiated the idea that wild progenitors of cereals and other cultivars harbor rich, yet untapped genetic resources that should be conserved *in situ* and *ex situ* and utilized in breeding elite crops, thereby advancing the second genetic green revolution and stabilizing world food production. He revealed the sensitivity of allozyme systems to marine pollution proposing their use as genetic monitoring of pollution in an attempt to safeguard the quality of the marine environment. His population genetic and ecology studies across the geographic ranges of many species revealed the geographic patterning of genetic diversity providing rigorous guidelines for preservation of wild genes, populations, and species.

In the "Evolution Canyon" project in Mount Carmel, 2500 species have been identified from bacteria to mammals providing "gold mines" of evolutionary studies at a microscale. Four such "Evolution Canyons" are now investigated in the Carmel, Galilee, Negev, and Golan mountains. His now classical studies of the "Evolution Canyons" model in Israel highlighted biodiversity evolution across life, from bacteria to mammals, elucidating adaptation and speciation at a microscale. The local studies paralleled the regional and global ones singling out natural selection as a major evolutionary driving force interacting with all other forces such as mutation, migration, and drift but eventually orienting molecular and organismal evolution. His contributions, along with those of his colleagues in Israel and the Ukraine, in elucidating fungal life in the Dead Sea, enriched our knowledge of biology at the hypersaline edge of life. Currently, he and his colleagues cloned salt and drought resistance genes from these Dead Sea fungi and similar genes plus disease resistance genes from wild cereals for crop improvement. A recent human study unraveled a very high mutation rate in the offspring of Chernobyl victims born after the accident, but not in their siblings born before the exposure to the atomic radiation of the parents.

A major contribution to evolutionary research and teaching was Eibi's foundation and direction of the Institute of Evolution at the University of Haifa since 1976. The Institute, unique in Israel and the world, started modestly with two labs but now consists of 25 research labs, 80 Institute members, 25 academic staff, 8 full Professors covering diverse past, present, and future perspectives of evolutionary biology from paleoecology and paleobotany through biodiversity of genes and species, genomics, proteomics, bioinformatics, of viruses, bacteria, plants, fungi, animals, and humans while developing nature conservation conceptually and practically. The Institute of Evolution combines fieldwork in Israel, the Near East, and across the globe and integrates observations, experiments, and theoretical studies. The absorption of leading scientists as immigrants from the former Soviet Union provided a quantum jump to the current research programs of the Institute introducing biodiversity studies of algae, lichens and fungi, recombination, modeling, genetic mapping, molecular plant and animal cytogenetics, genomics and proteomics, and recently paleoecology and paleobotany. The Institute investigates several major model organisms belonging to cyanobacteria, bacteria, algae, and fungi including yeast, lichens, medicinal mushrooms, wild cereals, dicot plants, *Drosophila,* insects, reptiles, subterranean mammals, and humans. The research done on the evolution of wild emmer wheat and wild barley, the progenitors of wheat and barley, became classic. The Institute annually publishes more than 200 scientific papers and around five books in evolutionary biology. It collaborates with 400 labs in 40 countries. The Institute harbors the world's largest genetically and agronomically evaluated gene banks of wild wheat and barley, wild lettuce, subterranean mole rats, and the largest Israeli gene bank of fungi. The teaching program of the Institute comprises 40 PhD students and 10 postdocs, 30 M.Sc. students and an interdisciplinary division of evolutionary biology of 1300 students from the faculties of social sciences and humanities exposed to evolutionary studies.

The crown of Eibi's developmental plans is the founding (already inaugurated unofficially) of an International Graduate School of Evolution (IGSE) intimately intertwined with the Institute of Evolution and networked to all Israeli and top world universities, opening world scientific facilities, resources, and know-how to graduate students from all over the world. The scores of graduate students of evolutionary biology will be international, conducting research programs in all the aforementioned fields investigated in the Institute. The IGSE will benefit from the rich biodiversity of Israel and the fact that it is the cradle of Old World domestication of plants and animals. Likewise, it will integrate interdisciplinarily molecular biology, genomics, and phenomic studies linking classical and modern perspectives and advancing research programs of biodiversity, bioinformatics, and biotechnology guaranteeing the education of a new cohort of evolutionary biology leaders that will make a constructive contribution toward solving the predicaments of biodiversity and of the planetary environment.

Eibi Nevo has shown how persistence, resilience, and energy of a vision could yield a blessed harvest of scientific discoveries. His contributions to evolutionary biology research, teaching, and leadership are exemplary. In a land soaked with conflicts and human suffering, he contributed grains of hope, creativity, and innovation aiming at better understanding the evolutionary process including that of human evolution.

On his 75th birthday I wish my dear friend, Eibi, continues in the best of health and full success in developing his exciting plans. These plans are based on an understanding of the most important concept we humans have ever conceived, evolution, and a successful coping with our future depends on such an understanding.

Ernst Mayr
Museum of Comparative Zoology
Harvard University
26 Oxford St.
Cambridge, MA 02138, USA

Foreword

This volume consists of papers written by friends and colleagues in Honour of Professor Eviatar (Eibi) Nevo on the occasion of his 75th anniversary. The papers are written by evolutionary, molecular and organismal biologists, systematists, paleontologists, ecologists, geneticists, but also by physicists and mathematicians who are fascinated by the major thread that holds all natural phenomena together, namely, the evolutionary process - cosmic, biological, and cultural. The scope of papers embraces many of the fields to which Eibi Nevo, his many students, colleagues and collaborators contributed in the past and actively contribute at present. His widespread influence and activity as a leading evolutionary biologist in Israel and worldwide is reflected in the Institute of Evolution he directs since he founded it in 1976, in his current proposal to establish an International Graduate School of Evolution linked to the Institute of Evolution and networked to all Israeli and top world universities, and in his scientific output in papers and books embracing major evolutionary problems.

The research interests of Professor Eviatar Nevo involve evolution of biological diversity in nature, at all levels including genes, genomes, phenomes, populations, species, and ecosystems of bacteria, fungi, plants, animals, and humans, always focusing on the twin evolutionary processes of adaptation and speciation. He conducted research programs at local, regional, and global scales in ways which linked interdisciplinarily ecology, cytogenetics, genetics, genomics, and phenomics (and including morphology, physiology, behavior, and which bridged genotypes and phenotypes, structure and function, micro- and macroevolution). This work also often integrated molecular with organ organismal biology, shedding light on relationships between organism and environment. In application, he contributed to the genetic monitoring of the quality of the marine environment; but primarily reinforced the idea that wild progenitors of cereals and other cultivars harbor rich genetic resources that should be conserved *in situ* and *ex situ* to preserve the genetic resources for crop improvement as the indispensable foundation of successful domestication. He founded and currently directs the Institute of Evolution at the University of Haifa, a center of excellence conducting active integrative research in evolutionary biology and linking field, laboratory, and theoretical research programs. Eibi has worked on many model organisms, from microorganisms to plants, animals, and humans. He has discovered many species that are new in Israel, and several that are new to science more generally. Perhaps his most remarkable achievement is the evolutionary studies in the "Evolution Canyons" model in Israel which reflects microcosms of life's evolution and demonstrates the evolution of biological diversity at microsites on many different levels.

The unifying thread running through all the chapters in this volume is evolutionary biology. A good deal of the material is new. In particular, evolutionary theory has continued to expand since Darwin's 1859 *Origin of Species*, but particularly so since the beginning of the genomic era.

Many fundamental problems remain unsolved. The origin of life is still a major challenge and so is the language of genomic sequences. Particularly challenging is the understanding of the regulatory functions of the noncoding genomes and their evolution under diverse ecological stresses. Even the origin of species leaves much to be understood, as does elucidation of the magnitude and meaning of past and present biological diversity: how many species exist, how many have existed since the origin of life, and why these numbers not others. Why is it that biological diversity appears to have increased, on average, over the sweep of life's history, despite massive past extinctions?

How are complex systems like the tropical rain forest and the human brain organized and function? Do they follow global or rather local optimality theory, or are they largely accidental assemblies? What is the relative importance of random versus nonrandom factors in driving evolution? What are the most promising future lines of research in evolutionary biology? Can structure and function both be linked to environmental stresses? Ultimately how to develop a truly integrated biology, linking genomes, proteomes, phenomes, and biomes? How can theory, observation, and experiment best interact in advancing our understanding of the workings of evolution?

The papers in this volume contribute notably to sharpening the definition of these questions. They also carry us someway down the road to beginning to answer some of the questions. And, most appropriately, they represent some of the questions and problems that have inspired Eibi's lifetime of work.

Robert M. May
Department of Zoology
University of Oxford
Oxford OX1 3PS, UK

EVIATAR NEVO

In Honour of His 75th Birthday

On February 2nd, 2004, Professor Eviatar (Eibi) Nevo will be seventy-five years old. His name occupies a prominent place in world Evolutionary Biology because of his outstanding talent, original style of work, and significant scientific achievements. Nevo's wide circle of interests, his clarity of scientific vision, enormous energy, enthusiasm, and honesty have always attracted young researchers and encouraged them to imitate many traits of this charming man.

Nevo is the son of the late Lea Goldis-Levitas and David Levitas. He was born on February 2nd, 1929, in Tel Aviv, as Eviatar Levitas and changed his name to Nevo in the early 1950s. Nevo has a daughter Orit, a dancer and trapeze artist. His son Tal died in the Yom Kippur War (October, 1973).

Nevo studied biology at the Hebrew University of Jerusalem receiving the M.Sc. degree with special distinction in 1958 and a Ph.D. summa cum laude in 1964. In 1956 Nevo began his career as a Lecturer in Biology at the Kibbutz Teachers College on the Oranim Campus. He served as Visiting Professor in Zoology at the University of Texas from 1964 to 1965, and was a Fellow in Biology at Harvard University a year later. In 1967 he accepted the position of Research Associate in the Department of Genetics at the Hebrew University. In 1968 he was promoted to Lecturer and in 1970 to Senior Lecturer. In 1971 he established the Department of Biology at the University of Haifa, a basis of biological studies in the Medical School that moved to the Technion. He then established the Institute of Evolution, at the University of Haifa, and the Teaching Division of Evolutionary Biology. From 1972 to 1973, he was a Research Associate at the Museum of Vertebrate Zoology in the University of California at Berkeley and a Senior Postdoctoral Research Fellow in the Department of Biology at the University of Chicago. Returning to Israel in 1973, Nevo was appointed Associate Professor of Biology at the University of Haifa. Since 1975 he has been a Professor of Biology and, since 1977, the founder and Director of the Institute of Evolution, at the University of Haifa. Since 1984 he has been the incumbent of the Chair in Evolutionary Biology.

Professor Nevo is a leading international evolutionary biologist in the following fields: (i) Biodiversity of plants and animals; (ii) Genetic diversity in natural populations of plants and animals at the protein, DNA, and chromosomal levels; (iii) Speciation; (iv) Natural selection and adaptation; (v) Genetic resources in wild germplasms for crop improvement; (vi)

Environmental quality; (vii) Aggression evolution; (viii) Molecular ecological-genetics and ecological genomics; and (ix) Genetic mapping sequencing and cloning stress genes in fungi, plants, and animals.

Nevo's scientific work includes the following aspects:

1. Geology and Paleontology (1950 - 1960)

Nevo's work as a student in the 1950s is marked by geological mapping of the Israeli Negev and discoveries of ore (kaolin, copper, iron). He discovered one of the world's largest collections of fossil frogs from the Early Cretaceous and published it in the Museum of Comparative Zoology, Harvard University as his Ph.D. dissertation describing two new genera and three species and highlighting some basic aspects of frog origins, phylogeny, and evolution. These paleontological studies culminated in discovering Early Cretaceous fossil salamanders involving a new genus and species in the Early Cretaceous of the Negev and pipid tadpoles representing a new genus and species from the Early Cretaceous of Samaria.

2. Evolutionary Dynamics of Frogs (1960 - 1990)

In the 1960s Nevo's work shifted into evolutionary dynamics of living frogs covering molecular biology, genetics, morphology, physiology, and behavioral perspectives primarily vocal communication as the basis of speciation. These studies were conducted across the USA and the Middle East, but primarily in Israel.

3. Origin of Species and Adaptations (1950 - continuing actively today)

Starting in the early 1950s, Nevo initiated a long-term project, which is ongoing, on the speciation and adaptation of subterranean mammals. This project focuses on the mole rats, S*palax*, primarily in the Near East and Israel, but then extended regionally into Africa and Asia Minor, East Europe, and other subterranean mammals in North America and across the planet. During these studies, he described about 30 new biological species of subterranean mammals, primarily spalacids, based on chromosomal and genic diversity. Using multidisciplinary means, he studied their climatic adaptations involving diverse molecular (protein and DNA) and organismal (morphology, physiology, and behavior) strategies. Nevo showed that the classical species, *Spalax ehrenbergi,* is a complex of at least 12 biological allospecies, some in the final stages of species formation. The studies in Israel were reviewed in a monograph describing *Spalax* as a uniquely productive and rich evolutionary model of speciation and adaptation at the molecular and organismal levels, and in two additional monographs: the first describing the mosaic evolution that included the regression, progression, and convergent evolution of subterranean mammals across the planet, with more than 250 species in 11 families and three mammalian orders (Nevo, 1999. Mosaic Evolution of Subterranean Mammals. Oxford

University Press). The second monograph reviews and analyzes the Pleistocene active speciation and adaptive radiation of four species of the *S. ehrenbergi* superspecies, each adapted to a different climatic regime in Israel [*S. galili* (2n=52), *S. golani* (2n=54), *S. carmeli* (2n=58) and *S. judaei* (2n=60)] (Nevo et al., 2001. Adaptive Radiation of Blind Subterranean Mole Rats. Backhuys Publ.). Reductionist and integrationist aspects were fruitfully combined to display evolution in action. *A major discovery in the early 1970s demonstrated that speciation could occur with only minor genomic changes*. The role of chromosomal ecologic speciation has been highlighted. Extensive research programs have been conducted and currently continue on brain architecture reorganization, photoperiodic perception, and respiratory adaptations focusing on hypoxia tolerance. The *Spalax* Genome Project has recently advanced by FISH, genetic libraries, genetic mapping, and the cloning of respiratory, eye, circadian rhythm, and clock genes.

4. *Genetic Diversity in Nature: Patterns and Theory (1969 - continuing actively today)*

Starting in the late 1960s and still continuing vigorously, Nevo embarked upon another long-term research program aiming at unraveling patterns and processes of genetic and genomic diversity and evolution in natural populations of bacteria, plants, fungi, and animals *locally*, *regionally*, and *globally*, i.e., both micro- and macrogeographically, and at the protein, DNA, and chromosomal levels. The main thrust involved the interface between genetics and ecology. It substantiated diverse selection theories at the level of proteins (allozymes) in single and multilocus structures, DNA (RFLP, DNA fingerprinting of mini- and microsatellites, RAPD PCR, AFLP, and sequence polymorphism), and chromosomes as the best explanatory models for genetic diversity and divergence in nature. These studies rejected the neutral theory of molecular evolution as a major theory of molecular evolution, *and substantiated natural selection as a primary driving force of molecular evolution, as is also true for organismal evolution.*

5. *Discovery of new species*

Besides demonstrating that natural selection orients evolution at both the molecular and organismal levels and highlighting the nature of allozyme polymorphisms in nature, Nevo has described new biological species on the basis of molecular biology including mole crickets (subterranean insects) in Israel, frogs in the USA and the Middle East, particularly Israel, and rodent species living above and under ground in Israel, Turkey, and northern, eastern, and southern Africa. In collaboration with other scientists in Israel and the Ukraine, Nevo discovered many new species of algae and fungi in Israel.

6. *Quality of Marine Environment (1976 - 1993)*

In the mid-1970s, Nevo started a long-term project on pollution biology in an attempt to develop a biological monitor to record the quality of the marine environment. The program included controlled field and laboratory experiments involving thermal and chemical (inorganic and organic pollutants) stresses on a wide range of marine organisms, i.e., crustaceans and mollusks. It was demonstrated that allozyme diversity is sensitive to environmental pollution stresses. Nevo proposed using allozyme and DNA variation, at both the single and multilocus levels, as genetic monitoring systems to alert conservationists about environmental deterioration before pollution causes species extinction.

7. *Genetic Resources in Wild Wheat and Barley for Crop Improvement (1977 - continuing actively today)*

In the mid-1970s, Nevo started a long-term research program on the exploration, sampling, evaluation, conservation, and utilization of wild germplasms, primarily of wild cereals (wheat, barley, and oats) and later wild lettuce for crop improvement. In this program more than 10,000 genotypes have been tested for allozyme variation of 30-50 enzymatic genes and partly for DNA variations. In addition, a battery of agronomic traits including abiotic and biotic resistances has been evaluated in these genotypes (disease resistance to fungi and viruses; storage proteins and amino acid quality and quantity; photosynthetic yield, herbicide resistance, drought and salt tolerances). Resistant and elite genotypes have been bred into hexaploid bread wheat and cultivated barley *generating improved cultivars now grown extensively in the USA and Europe*. Currently, the wild germplasm is utilized to generate new cultivars in Israel. These studies advanced the idea that the best hope for crop improvement resides in the highly adapted and variable wild germplasms that can be utilized by classical breeding and modern biotechnology for crop improvement. Salt, herbicide, drought, and rust resistant genes in wild barley and wheat have been chromosomally mapped and are currently in the process of active cloning and genetic transformation. Furthermore, additional progenitors of cultivated plants are being added to the wild cereal project involving artichoke, lettuce, almonds, figs, olives, and vines. These studies are currently being extended into the examination of extensive mapping and ancient DNA in an attempt to unravel the molecular genetics and evolution of domestication.

8. *Evolution of Behavior: Communication, Aggression, and Pacifism in Animals (1975 - continuing actively today)*

Nevo started a long-term project in the mid-1960s in animal communication underlying speciation and adaptation (vocalization, olfaction, and vibration) in mole crickets, toads, frogs, and mole rats

(*Spalax*). Another behavioral program begun in the mid-1970s involved the genetic evolution of aggression and pacifism in *Spalax*. This program culminated in a verbal model of aggression evolution predicting and then leading to the discovery of a totally pacifistic mole rat species at the northern margin of the Sahara Desert. This species originated from Israeli polymorphic populations of mole rats involving behavioral phenotypes for aggression with militant, intermediate, and pacifist types. A genetic model based on a few major genes underlies this aggression evolution, which may be relevant to humans. Currently, efforts are being made to sequence and clone aggression genes from *Spalax* genetic libraries and to sequence them in an attempt to unravel the molecular genetic basis of aggression in mole rats.

9. *Chernobyl Project: Revealing Genetic Mutations in Victims (1994 - continued actively today)*

Recently, a new collaborative research project was started on detecting radiation mutations in Jewish immigrants to Israel, a total of 300,000 people from the Chernobyl disaster area. The genetic research is being conducted primarily on 200 liquidators and their families. Families are tested with children born before and after the atomic disaster detecting *de novo* mutations caused by the radiation that are transferred to the offspring via the germline, which affect health conditions of victims.

10. *Genetic Research in Amazonian Flood Forests (1992 - 1994)*

The study of genetic structure and divergence of plants and animals in the Amazonian flooded rainforest was conducted from 1993 to 1995 in collaboration with the late Professor W.D. Hamilton of Oxford University, UK. The goals were to unravel the genetic basis of adaptation and speciation of certain plant species including the rubber tree and wild rice. The research yielded interesting preliminary results, but was terminated due to lack of funds.

11. *Microcosm of Life's Evolution: "Evolution Canyon" I Project on Mount Carmel (1993 - continuing actively today) in "Evolution Canyon" II (Upper Galilee) and "Evolution Canyon" III (Negev). All three canyons are located in Israel.*

"Evolution Canyon" model (EC) on Mount Carmel, Israel, is a microscale ecological theater examined in the Mountains of Carmel ("EC" I), the Galilee ("EC" II), and the Negev ("EC" III) where the African biota meet the European biota and *local* tests are conducted in an attempt to discover *global* biodiversity and genomic diversity patterns. In particular, the effects of abiotic and biotic stress factors on evolution in action, including the twin evolutionary processes of adaptation and speciation, are being extensively and intensively studied. These include biodiversity studies across phylogeny from microorganisms, cyanobacteria, and soil bacteria to mammals and the evaluation of the relative importance of the forces driving evolution (mutation, recombination, migration, genetic drift, and natural selection). To date (August 2003), about 2500 species have been identified in "Evolution Canyon" I, more than 1000 species in "Evolution Canyon" II, and several hundred species in "Evolution Canyon" III including many species new to science for Israel. Model organisms are currently under in-depth study to evaluate the effects of stress on the evolutionary process. The levels and interslope divergence in natural populations of genetic polymorphisms (at the protein and DNA levels), mutations, recombination, gene conversion, adaptive complexes, stress genes, and incipient sympatric speciation across life are actively investigated.

12. *Founding and Directing the Institute of Evolution, University of Haifa 1971 - continuing actively today)*

In 1971 Nevo established the Department of Biology, which in 1977 was renamed the Institute of Evolution at the University of Haifa. The Institute is unique in structure and function. It consists of 25 active laboratories conducting field, laboratory, and theoretical research in cytogenetics, genetics, genomics, proteomics, ecology, behaviour, systematics, and biodiversity evolution of microorganisms, fungi, plants, animals, and humans focusing on the interfaces of these disciplines and highlighting interdisciplinarity as a major structure of the advancement of science. Molecular evolution of proteins and DNA including gene mapping and gene cloning are under active study. Molecular sequence comparisons of proteins and DNA are conducted from ecological perspectives focusing on ecological genomics. Currently, the Institute of Evolution collaborates with 400 labs in 40 countries across the globe. The annual publication record includes over 200 articles in Evolutionary Biology and about 5 books. As of 1991, the Institute absorbed 30 immigrant scientists from the former USSR, the USA, and returning Israelis and established ten of its laboratories. The Institute has an extensive undergraduate teaching program (see below). Today the Institute conducts field, experimental, and theoretical studies from viruses to humans including microorganisms, plants, fungi, animals, and humans. The Institute hosts the world's largest genetically and agronomically evaluated gene banks of wild cereals (wheat and barley), wild lettuce, and blind mole rats and the largest gene bank in Israel of soil and medicinal mushrooms.

13. *Teaching of Evolutionary Biology (1977 - continuing actively today)*

Nevo established the Division of Evolutionary Biology at the University of Haifa, which was a liberal arts college without a Faculty of Natural Sciences until 2000. This is the largest division in Israel and the world that offers an interdisciplinary teaching program, now including 2200 participant students from all 30 departments of the University. The program is open to non-biologists and includes courses in evolutionary biology involving ecology, genetics, behavior, systematics, and evolution as well as human evolution. The number of enrolled students rose from 200 (in 1977) to 2200 (in 1998).

14. Honors B.A. Program

In 1978 Nevo established the Honors B.A. program in the University, which continues very successfully today.

15. Graduate Program in Evolutionary Biology

In 1994 Nevo established the graduate program in Evolutionary Biology within the framework of the Authority of Advanced Studies at the University of Haifa. It now has forty students. He opened the M.Sc. program in Evolutionary and Environmental Biology in collaboration with the Department of Biology at the Oranim campus of the University of Haifa. Nevo proposed the establishment of a National School of Evolutionary Biology embracing all seven Israeli universities. This plan expanded into an International Graduate School of Evolutionary Biology enthusiastically endorsed by all seven Israeli universities and some 40 top world universities. The plan to initiate the school at the University of Haifa is underway.

Professor Nevo is the author of more than 900 papers in various fields of evolutionary biology including genetics, ecology, physiology, morphology, biodiversity, and behavior involving plants, fungi, and animals. He also is the author and co-editor of 18 books.

Nevo is a member of many scientific societies including the Society of the Study of Evolution, the International Society of Molecular Evolution, Fellow of the American Association for the Advancement of Science, the American Society of Naturalists, the Genetics Society of America, the Genetics Society of Israel, the Zoological Society of Israel, and the Geological Society of Israel. He is Foreign Associate of the National Academy of Sciences, USA; Foreign Member of the Linnean Society, London; Foreign Member of the National Academy of Sciences of Ukraine in the category Evolutionary Biology and Genetics; a Member of the New York Academy of Sciences; an Honorary Member of the Ukrainian Botanical Society; an Honorary Member of the American Society of Mammalogists; an inaugural Member of the Charles Darwin Associates of the New York Academy of Sciences; and Vice President of the Society of the Study of Evolution in 1978. He is also a Member of the Human Genome Organization and a Member of the Directing Council of the Association Iberoameciana de Biologia Evolutiva.

In recognition of his academic and career accomplishments, Eviatar Nevo has received many awards. In 1964 he was named a Fulbright Fellow; he was Fellow of the Guggenheim Foundation from 1978 to 1980; and also a

Fellow of the Explorers Club, the American Association of the Advancement of Science. He received the Honourary Cultural Doctorate from the World University Roundtable in 1990; the International Cultural Diploma of Honour from the American Biographical Institute (ABI) and the Presidential Seal of Honour from ABI in 1997. In 1996 Nevo received a degree of merit from the International Biographical Centre (IBC) in Cambridge, UK. He was awarded the IBC's certificate of inclusion in the International Leaders of Achievements, was appointed Deputy Director of the IBC in Asia in 1997, and was nominated a Life Fellow of the International Biographical Association in 1997. He was elected the Man of the Year by ABI, USA, for 1996, 1997, and 2003, and by IBC Cambridge, UK, for 1998 for his outstanding scientific achievements in Evolutionary Biology.

The diversity of Nevo's research programs, scientific, organizational, and pedagogical activities and his significant achievements in various scientific fields honor him as a scientist of a wide scope of interests, a brilliant science organizer, a world scientific leader of the school of evolutionary biology. His achievements and accomplishments have been recognized in world science.

Eviatar Nevo is an extraordinary and vital man in the broadest sense of the word. He attracts people like a magnet by his charm, wit, and erudition. From the first encounter with him you sense the scale of his personality, and his outgoing, unconventional nature. Nevo is not only a creative and gifted scientist, but also an intellectual of wide education with a deep interest in the fine arts and classical music. He celebrates his birthday at the peak of his creative activity. We, the contributors to this *Festschrift*, friends, colleagues, collaborators, and students wish him good health, happiness in all his activities, new creative achievements, good fortune, and wise, organizational decisions to strengthen the positions of his brainchild, the Institute of Evolution, at the University of Haifa, and the forthcoming linked International Graduate School of Evolution.

Solomon P. Wasser
Haifa-Kiev
August, 2003

List of Selected Publications of Eviatar Nevo

Nevo E. 1956. Fossil frogs from a Lower Cretaceous bed in southern Israel (Central Negev). Nature 178: 1191-1192.

Nevo E. 1961. Observations on Israeli populations of the mole rat, *Spalax ehrenbergi* Nehring 1898. Mammalia 25: 127-144.

Nevo E. and Amir E. 1961. Biological observations on the forest doormouse, *Dryomys nitedula* Pallas, in Israel (Rodentia, Muscardinidae). Bull. Res. Council Israel Zool. 9B: 200-201.

Nevo E. 1963. The Jurassic strata of Makhtesh Ramon. Isr. J. Earth Sci. 12 (2).

Nevo E. 1963. Fossil urodeles in early Lower Cretaceous deposits of Makhtesh Ramon, Israel. Nature 201: 415-416.

Nevo E. and Amir E. 1963. Geographic variation in reproduction and hibernation patterns of the forest dormouse. J. Mamm. 45: 69-87.

Nevo E. 1963. Population studies of anurans from the Lower Cretaceous of Makhtesh Ramon, Israel. Ph.D. Thesis, Hebrew University, Jerusalem. 132 pp. (unpublished).

Nevo E. 1968. Pipid frogs from the early Cretaceous of Israel and pipid evolution. Bull. Museum Comp. Zool. Harvard University 136: 255-318.

Nevo E. 1969. Mole rat *Spalax ehrenbergi*: Mating behavior and its evolutionary significance. Science 163: 484-486.

Wahrman J., Goitein R. and Nevo E. 1969. Mole rat *Spalax*: Evolutionary significance of chromosome variation. Science 164: 82-84.

Wahrman J., Goitein R. and Nevo E. 1969. Geographic variation of chromosome forms in *Spalax*, a subterranean mammal of restricted mobility. In: *Comparative Mammalian Cytogenetics*. K. Benirschke ed., Springer Verlag, N.Y., pp. 30-48.

Costa M. and Nevo E. 1969. Nidicolous arthropods associated with different chromosomal types of *Spalax ehrenbergi* Nehring. Zool. J. Linn. Soc. 48: 199-215.

Nevo E. 1969. Mole rat *Spalax*: A model of prolific speciation. In: *NATO Advanced Study Institute Vertebrate Evolution: Mechanism and Process*. Proc. of Meeting, Istanbul, Turkey, August 4-15, 1969, Hebrew University, Jerusalem, pp. 1-2.

Nevo E. 1969. Discussion on "The systematic significance of isolating mechanisms." In: *Systematic Biology*, Proc. Intl. Conf., Natl. Acad. Sci. USA, Washington, D.C., pp. 485-489.

Dessauer H. and Nevo E. 1969. Geographic variation of blood and liver proteins in cricket frogs. Biochem. Genet. 3: 171-188.

Salthe S.N. and Nevo E. 1969. Geographic variation of lactate dehydrogenase in the cricket frog, *Acris crepitans*. Biochem. Genet. 3: 335-341.

Nur U. and **Nevo E.** 1969. Supernumerary chromosomes in the cricket frog, *Acris crepitans*. Caryologia 22: 97-102.

Nevo E. 1969. Early Cretaceous pipid frogs from Israel and pipid evolution. *Scienza and Tecnica* 1969. Yearbook Enciclopedia della Scienza e della Tecnica.

Nevo E. 1969. *Ramonellus longispinus*, an early Cretaceous salamander from Israel. Copeia 3: 540-547.

Wahrman J., Goitein R. and **Nevo E.** 1969. Geographic variation of chromosome forms in *Spalax*, a subterranean rodent of restricted mobility. La Kromosomo 75: 2442-2443.

Gorman G.C., Jovanovic V., **Nevo E.** and McCollum F.C. 1970. Conservative karyotypes among lizards of the genus *Lacerta* from the Adriatic islands. Genetika 2: 149-154.

Wertheim G. and **Nevo E.** 1971. Helminths of birds and mammals from Israel: III. Helminths from chromosomal forms of the mole rat, *Spalax ehrenbergi*. J. Helminth 45: 161-169.

Nevo E. 1972. Climatic adaptation in size of the green toad (*Bufo viridis*). Isr. J. Med. Sci. 8: 1010.

Nevo E. and Blondheim S.A. 1972. Acoustic isolation in the speciation of mole crickets. Annals Entomol. Soc. America 65: 980-981.

Nevo E. and Shaw C.R. 1972. Genetic variation in a subterranean mammal, *Spalax ehrenbergi*. Biochem. Genet. 7: 235-241.

Nevo E., Gorman G., Soule M., Yang S.Y., Clover R. and Jovanovic V. 1972. Competitive exclusion between insular *Lacerta* species (Sauria, Lacertidae): Notes on experimental introductions. Oecologia 10: 183-190.

Schneider H. and **Nevo E.** 1972. Bio-acoustic study of the yellow-lemon tree frog, *Hyla arborea savignyi* Audouin. Zool. Jb. Physiol. Bol. 76(S): 497-506.

Nevo E. 1973. Test of selection and neutrality in natural populations. Nature 244: 573-575.

Nevo E. 1973. Adaptive variation in size of cricket frogs. Ecology 54: 1271-1281.

Nevo E. 1973. Adaptive color polymorphism in cricket frogs. Evolution 27: 353-367.

Capranica R.R., Frishkopf L.S. and **Nevo E.** 1973. Encoding of geographic dialects in the auditory system of the cricket frog. Science 182: 1272-1275.

Nevo E. 1973. Variation and evolution in the subterranean mole rat, *Spalax*. Isr. J. Zool. 22: 207-208.

Nevo E., Kim Y.J., Shaw C.R. and Thaeler C.S.Jr. 1974. Genetic variation, selection and speciation in *Thomomys talpoides* pocket gophers. Evolution 28: 1-23.

Nevo E. and Shkolnik A. 1974. Adaptive metabolic variation of chromosome forms in mole rats, *Spalax*. Experientia 30: 724-726.

Nevo E. and Sarich V. 1974. Immunology and evolution in the mole rat, *Spalax*. Isr. J. Zool. 23: 210-211.

Nevo E., Dessauer H.C. and Chuang K.C. 1975. Genetic variation as a test of natural selection. Proc. Natl. Acad. Sci. USA 72: 2145-2149.

Nevo E., Naftali G. and Guttman R. 1975. Aggression patterns and speciation. Proc. Natl. Acad. Sci. USA 72: 3250-3254.

Guttman R., Naftali G. and Nevo E. 1975. Aggression patterns in three chromosome forms of the mole rat, *Spalax ehrenbergi*. Anim. Behav. 23: 485-493.

Pevet P., Kappers J.A. and Nevo E. 1975. Etude ultrastructurale des pinealocytes de deux mammiferes souterrains aveugles, la taupe (*Talpa europaea*, Insectivore) et le (*Spalax ehrenbergi*, Rongeur): Importance des syntheses proteiques. J. Physiologie 76: 6B.

Gorman G.C., Soule M.S., Yang Y. and Nevo E. 1975. Evolutionary genetics of insular adriatic lizards. Evolution 29: 52-71.

Dessauer H.C., Nevo E. and Chuang K.C. 1975. High genetic variability in an ecologically variable vertebrate, *Bufo viridis*. Biochem. Genet. 13: 651-661.

Karlin S. and Nevo E. (eds.). 1976. *Population Genetics and Ecology*. Academic Press, N.Y., 832 pp.

Nevo E. and Schneider H. 1976. Mating call pattern of Green toads in Israel and its ecological correlate. J. Zool., Lond. 178: 133-145.

Nevo E. 1976. Adaptive strategies of genetic systems in constant and varying environments. In: *Population Genetics and Ecology*. Karlin S. and Nevo E. (eds.). Academic Press, N.Y., pp. 141-158.

Nevo E. and Bar Z. 1976. Natural selection of genetic polymorphisms along climatic gradients. In: *Population Genetics and Ecology*. Karlin S. and Nevo E. (eds.). Academic Press, N.Y., pp. 159-184.

Kornfield I.L. and Nevo E. 1976. Likely pre-Suez occurrence of a Red Sea fish *Aphanius dispar* in the Mediterranean. Nature 264: 289-291.

Pevet P., Kappers J.A. and Nevo E. 1976. The pineal gland of the mole rat (*Spalax ehrenbergi*, Nehring). Cell Tiss. Res. 174: 1-24.

Nevo E. 1976. Genetic variation in constant environments. Experientia 32: 858-859.

Nevo E. and Bar-El H. 1976. Hybridization and speciation in fossorial mole rats. Evolution 30: 831-840.

Nevo E., Bodmer M. and Heth G. 1976. Olfactory discrimination as an isolating mechanism in speciating mole rats. Experientia 32: 1511-1512.

Nevo E. and Heth G. 1976. Assortative mating between chromosome forms of the mole rat, *Spalax ehrenbergi*. Experientia 32: 1509-1510.

Bar Z. and **Nevo E.** 1976. Natural selection of shell banding polymorphism in *Theba pisana* (Mollusca) along climatic gradients. Isr. J. Zool. 25: 214-215.

Nevo E., Shimony T. and Libni M. 1977. Thermal selection of allozyme polymorphisms in barnacles. Nature 267: 699-701.

Brown A., **Nevo E.** and Zohary D. 1977. Association of alleles at esterase loci in wild barley *Hordeum spontaneum* L. Nature 268: 430-431.

Nevo E. 1978. Genetic variation in natural populations: Patterns and theory. Theor. Pop. Biol. 13: 121-177.

Nevo E. and Cleve H. 1978. Genetic differentiation during speciation. Nature 275: 125-126.

Brown A.H.D., Zohary D. and **Nevo E.** 1978. Outcrossing rates and heterozygosity in natural populations of *Hordeum spontaneum* Koch in Israel. Heredity 41: 49-62.

Brown A.H.D., Zohary D. and **Nevo E.** 1978. Genetic structure of *Hordeum spontaneum* in Israel. Barley Genetics Newsletter.

Brown, A.H.D., **Nevo E.**, Zohary D. and Dagan O. 1978. Genetic variation in natural populations of wild barley (*Hordeum spontaneum*). Genetica 49: 97-108.

Estes R., Spinar Z.V. and **Nevo E.** 1978. Early Cretaceous pipid tadpoles from Israel (Amphibia: Anura). Herpetologica 34: 374-393.

Nevo E. 1978. The diet of fossorial mole rats, *Spalax*, in nature and laboratory. In: *Handbook of Nutrition and Food*. M. Recheigl (ed.). CSR Press, Cleveland, Ohio, USA.

Nevo E., Shimony T. and Libni M. 1978. Pollution selection of allozyme polymorphisms in barnacles. Experientia 34: 1562-1564.

Nevo E. 1979. Adaptive convergence and divergence of subterranean mammals. Ann. Rev. Ecol. Syst. 10: 269-308.

Nevo E., Brown A.H.D. and Zohary D. 1979. Genetic diversity in the wild progenitor of barley in Israel. Experientia 35: 1027-1029.

Nevo E., Zohary D., Brown A.H.D. and Haber M. 1979. Genetic diversity and environmental associations of wild barley, *Hordeum spontaneum*, in Israel. Evolution 33: 815-833.

Malogolowkin-Cohen Ch., Levene H. and **Nevo E.** 1979. Climatic determinants of the distribution and abundance of *Drosophila subobscura* and other species in Israel. Brazil J. Genet. 2: 109-123.

Nevo E. and Yang S.Y. 1979. Genetic diversity and climatic determinants of tree frogs in Israel. Oceologia 41: 47-63.

Nevo E. and Yang S.Y. 1979. Genetic structure and variation of marsh frog populations in Israel. Isr. J. Zool. 28: 55.

Nevo E., Guttman R., Haber M. and Erez E. 1979. Habitat selection in evolving mole rats. Oecologia 43: 125-138.

Nevo E. 1979. Nature as an evolutionary test field in space and time. Biology Instruction in the Field, Biology Teachers Conference, December, 1979, pp. 3-4.

Kahler A.L., Allard R.W., Krzakowa M.R., Wehrhahn C. and **Nevo E.** 1980. Associations between isozyme phenotypes and environment in the slender wild oat (*Avena barbata*) in Israel. Theor. Appl. Genet. 56: 31-47.

Balemans M.G.M., Pevet P., Legerstee W.C. and **Nevo E.** 1980. Preliminary investigations on melatonin and 5-methoxytryptophol synthesis in pineal, retina, and harderian gland of the mole rat and in the pineal of the mouse "eyeless". J. Neural Transmission 49: 247-255.

Hertz P.E., Huey R.B. and **Nevo E.** 1980. Thermal dependence of defensive behavior in a lizard. Amer. Zool. 20: 791.

Brown A.H.D., Feldman M.W. and **Nevo E.** 1980. Multilocus structure of natural populations of *Hordeum spontaneum*. Genetics 96: 523-536.

Nevo E., Perl T., Beiles A., Wool D. and Zoller U. 1981. Genetic structure as a potential monitor of marine pollution. In: Intl. Comm. for the Scientific Exploration of the Mediterranean Sea, Monaco (C.I.E.S.M.) and the United Nations Environment Programme, Nirobi (U.N.E.P.). *Workshop on Pollution of the Mediterranean (V Journees Etud. Pollutions)*, Cagliari, Sardinia, Italy. pp. 61-68.

Nevo E., Perl T., Beiles A. and Wool D. 1981. Mercury selection of allozyme genotypes in shrimps. Experientia 37: 1152-1154.

Nevo E., Brown A.H.D., Zohary D., Storch N. and Beiles A. 1981. Microgeographic edaphic differentiation in allozyme polymorphisms of wild barley (*Hordeum spontaneum*, Poaceae) Pl. Syst. Evol. 138: 287-292.

Nevo E., Bar-El Ch., Bar Z. and Beiles A. 1981. Genetic structure and climatic correlates of desert landsnails. Oecologia 48: 199-208.

Heth G. and **Nevo E.** 1981. Origin and evolution of ethological isolation in subterranean mole rats. Evolution 35: 259-274.

Nevo E. 1981. Genetic variation and climatic selection in the lizard *Agama stellio*, in Israel and Sinai. Theor. Appl. Genet. 60: 369-380.

Hertz P.E. and **Nevo E.** 1981. Thermal biology of four Israeli agamid lizards in early summer. Isr. J. Zool. 30: 190-210.

Nevo E. 1981. E. Interdisciplinarity: Theory and practice. UNESCO Symposium on "Interdisciplinarity in Higher Education", organized by the European Center for Higher Education, 24-26 November, 1981, Bucharest.

Lavie B. and **Nevo E.** 1981. Genetic diversity in marine molluscs: A test of the niche-width variation hypothesis. Mar. Ecol. 2: 335-342.

Nevo E. 1981. Evolution of subterranean mammals. Isr. J. Zool. 29: 196-197.

Nevo E., Golenberg E., Beiles A., Brown A.H.D. and Zohary D. 1982. Genetic diversity and environmental associations of wild wheat, *Triticum dicoccoides*, in Israel. Theor. Appl. Genet. 62: 241-254.

Nevo E., Bar-El C., Beiles A. and Yom-Tov Y. 1982. Adaptive microgeographic differentiation of allozyme polymorphism in landsnails. Geneica 59: 61-67.

Nevo E., Guttman R., Haber M. and Erez E. 1982. Activity patterns of evolving mole rats. J. Mammal. 63: 453-463.

Hertz P.E., Huey R.B. and **Nevo E.** 1982. Fight versus flight: Body temperature influences defensive responses of lizards. Anim. Behav. 30: 676-679.

Brzoska J., Schneider H. and **Nevo E.** 1982. Territorial behavior and vocal response in male *Hyla arborea savignyi* (Amphibia: Anura). Isr. J. Zool. 31: 27-37.

Nevo E., Heth G. and Beiles A. 1982. Differential survivorship of evolving chromosomal species of mole rats, *Spalax*: An unplanned laboratory experiment. Evolution 36: 1315-1317.

Nevo E., Heth G. and Beiles A. 1982. Population structure and evolution in subterranean mole rats. Evolution 36: 1283-1289.

Nevo E. 1982. Speciation in subterranean mammals. In: *Mechanisms of Speciation*. Barigozzi C. (ed.). Alan R. Liss, Inc., New York, pp. 191-218.

Nevo E. 1982. Genetic structure and differentiation during speciation in fossorial gerbil rodents. Mammalia 46: 523-530.

Nevo E. and Yang S.Y. 1982. Genetic diversity and ecological relationships of marsh frog populations in Israel. Theor. Appl. Genet. 63: 317-330.

Lavie B. and **Nevo E.** 1982. Heavy metal selection of phosphoglucose isomerase allozymes in marine gastropods. Mar. Biol. 71: 17-22.

Nevo E., Lavie B. and Ben-Shlomo R. 1982. Can allozyme polymorphisms monitor marine environmental quality? In: *The 6th Pollution Study Journal (VI Journees Etud. Pollutions)*. The Intl. Comm. for the Scientific Exploration of the Mediterranean Sea (C.I.E.S.M.), Cannes, pp. 801-805.

Nevo E. 1983. Population genetics and ecology: The interface. In: *Evolution from Molecules to Men*. Bendall D.S. (ed.). Cambridge Univ. Press, Cambridge, England, pp. 287-321.

Nevo E., Beiles A., Storch N., Doll H. and Andersen B. 1983. Microgeographic edaphic differentiation in hordein polymorphisms of wild barley. Theor. Appl. Genet. 64: 123-132.

Nevo E., Lavie B. and Ben-Shlomo R. 1983. Selection of allelic isozyme polymorphisms in marine organisms: Pattern, theory and application. In: *Isozymes: Current Topics in Biological and Medical Research*, Vol. 10: *Genetics and Evolution*. Ratazzi M. C., Scandalios J. G. and Whitt G. S. (eds.). Alan R. Liss, Inc, N.Y., pp. 69-92.

Nevo E. 1983. Adaptive significance of protein variation. In: *Systematic Association*; Special Vol. 24, *Protein Polymorphism: Adaptive and Taxonomic Significance*. Oxford G. S. and Rollinson D. (eds.). Academic Press, N.Y., pp. 239-282.

Pontecorvo G., **Nevo E.**, Gaudio L. and Carfagna M. 1983. Biochemical study of phosphoglucomutase polymorphisms adaptive value in the shrimp *Palaemon elegans*. Atti Assoz. Genetica Italiana 29: 197-198.

Haim A., Heth G., Pratt H. and **Nevo E.** 1983. Photoperiodic effects on thermoregulation in a "blind" subterranean mammal. J. Exp. Biol. 107: 59-64.

Nevo E., Bar-El C. and Bar Z. 1983. Genetic diversity, climatic selection and speciation of *Sphincterochila* landsnails in Israel. Biol. J. Linn. Soc. 19: 339-373.

Hertz P.E., Huey R.B. and **Nevo E.** 1983. Homage to Santa Anita: Thermal sensitivity of sprint speed in agamid lizards. Evolution 37: 1075-1084.

Hickman G.C., **Nevo E.** and Heth G. 1983. Geographic variation in the swimming ability of *Spalax ehrenbergi* (Rodentia: Spalacidae) in Israel. J. Biogeogr. 10: 29-36.

Nevo E. and Schneider H. 1983. Structure and variation of *Rana ridibunda* mating call in Israel (Amphibia: Anura). Isr. J. Zool. 32: 45-60.

Moseman J.G., **Nevo E.** and Zohary D. 1983. Resistance of *Hordeum spontaneum* collected in Israel to infection with *Erysiphe graminis hordei*. Crop Science 23: 1115-1119.

Nevo E. 1983. Genetic resources of wild emmer wheat: Structure, evolution and application in breeding. In: *Proceedings of the 6th Intl. Wheat Genetics Symposium* (December, 1983). Sakamoto S. (ed.). Kyoto University, Kyoto, Japan, pp. 421-431.

Nevo E. 1983. Genetic diversity and the evolution of life and man. In: *Racism, Science and Pseudo-Science*, Proc. of Symposium to examine pseudo-scientific theories invoked to justify racism and racial discrimination, Athens, 30 March-3 April, 1981. United Nations Educational, Scientific and Cultural Organization (U.N.E.S.C.O.). Imprimerie des Presses Universitaires de France, Vendome, pp. 77-92.

Nevo E., Beiles A. and Ben-Shlomo R. 1984. The evolutionary significance of genetic diversity: Ecological, demographic and life history correlates. In: *Lecture Notes in Biomathematics*, Levin S. (mang. ed.), Vol. 53: "Evolutionary Dynamics of Genetic Diversity," Mani G. S. (ed.), Springer- Verlag, Berlin, Heidelberg, pp. 13-213.

Haim A., Heth G., Avnon Z. and **Nevo E.** 1984. Adaptive physiological variation in nonshivering thermogenesis and its significnce in speciation. J. Comp. Physiol. 154: 145-147.

Nevo E., Ben-Shlomo R. and Lavie B. 1984. Mercury selection of allozymes in marine organisms: Prediction and verification in nature. Proc. Natl. Acad. Sci. USA 81: 1258-1259.

Arieli R., Arieli M., Heth G. and **Nevo E.** 1984. Adaptive respiratory variation in four chromosomal species of mole rats. Experientia 40: 512-514.

Nevo E., Moseman J.G., Beiles A. and Zohary D. 1984. Correlation of ecological factors and allozymic variations with resistance to *Erysiphe graminis hordei* in *Hordeum spontaneum* in Israel: Patterns and application. Plant Syst. Evol. 145: 79-96.

Bonhomme F., Catalan J., Britton-Davidian J., Chapman V.M., Moriwaki K., Nevo E. and Thaler L. 1984. Biochemical diversity and evolution in the genus *Mus*. Biochem. Genet. 22: 275-303.

Kleinschmidt T., Nevo E. and Braunitzer G. 1984. The primary structure of the hemoglobin of the mole rat (*Spalax ehrenbergi*, Rodentia, Chromosome species 60). Hoppe-Seyler's Z. Physiol. Chem. 365: 531-537.

Redi C.A., Garagna S., Nevo E. and Heth G. 1984. Variabilita cariotipica e spermatogenesi in *Spalax ehrenbergi* (Nehring). Boll. Zool. 51 (suppl. 1984): 93.

Nizetic D., Figueroa F., Muller H.J., Arden B., Nevo E. and Klein J. 1984. Major histocompatibility complex of the mole rat. I. Serological and biochemical analysis. Immungenetics 20: 443-451.

Beiles A., Heth G. and Nevo E. 1984. Origin and evolution of assortative mating in actively speciating mole rats. Theor. Pop. Biol. 26: 265-270.

Lavie B., Nevo E. and Zoller U. 1984. Differential viability of phosphoglucose isomerase allozyme genotypes of marine snails in nonionic detergent and crude oil-surfactant mixtures. Environ. Res. 35: 270-276.

Moseman J.G., Nevo E., Gerechter-Amitai Z.K., El-Morshidy M.A. and Zohary D. 1984. Resistance of *Triticum dicoccoides* to infection with *Erysiphe graminis tritici*. Euphytica 33: 41-47.

Schneider H., Nevo E., Heth G., Simson S. and Brzoska J. 1984. Auditory discrimination tests of female Near Eastern tree frogs and reevaluation of the systematic position (Amphibia, Hylidae). Zool. Anz. 213: 306-312.

Nevo E., Beiles A., Gutterman Y., Storch N. and Kaplan D. 1984. Genetic resources of wild cereals in Israel and vicinity: I. Phenotypic variation within and between populations of wild wheat, *Triticum dicoccoides*. Euphytica 33: 717-735.

Nevo E., Beiles A., Gutterman Y., Storch N. and Kaplan D. 1984. Genetic resources of wild cereals in Israel and vicinity: II. Phenotypic variation within and between populations of wild barley, *Hordeum spontaneum*. Euphytica 33: 737-756.

Pevet P., Heth G., Haim A. and Nevo E. 1984. Photoperiod perception in the blind mole rat (*Spalax ehrenbergi*, Nehring): Involvement of the harderian gland, atrophied eyes and melatonin. J. Exp. Zool. 232: 41-50.

Gurnett A.M., O'Connell J.P., Harris D.E., Lehmann H., Joysey K.A. and Nevo E. 1984. The myoglobin of rodents: *Lagostomus maximus* (viscacha) and *Spalax ehrenbergi* (mole rat). J. Prot. Chem. 3: 445-454.

Moseman J.G., **Nevo E.**, Gerechter-Amitai Z.K., El-Morshidy M.A. and Zohary D. 1985. Resistance of *Triticum dicoccoides* collected in Israel to infection with *Puccinia recondita tritici*. Crop Sci. 25: 262-265.

Haim A., Heth G., **Nevo E.**, Gruener N. and Goldstein T. 1985. Urine analysis of three rodent species with emphasis on calcium and magnesium bicarbonate. Comp. Biochem. Physiol. 80A: 503-506.

Haim A., Heth G. and **Nevo E.** 1985. Adaptive thermoregulatory patterns in speciating mole rats. Acta Zool. Fennica 170: 137-140.

Nevo E. 1985. Ecological and populational correlates of allozyme polymorphisms in mammals. Acta Zool. Fennica 170: 25-29.

Nevo E. 1985. Genetic differentiation and speciation in spiny mouse, *Acomys*. Acta Zool. Fennica 170: 131-136.

Nevo E. 1985. Speciation in action and adaptation in subterranean mole rats: Patterns and theory. Boll. Zool. 52: 65-95.

Baker R., Lavie B. and **Nevo E.** 1985. Natural selection for resistance to mercury pollution. Experientia 41: 697-699.

Wahrman J., Richler C., Gamperl R. and **Nevo E.** 1985. Revisiting *Spalax*: Mitotic and meiotic chromosome variability. Isr. J. Zool. 33: 15-38.

Nizetic D., Figueroa F., **Nevo E.** and Klein J. 1985. Major histocompatibility complex of the mole rat. II. Restriction fragment polymorphism. Immunogenetics 22: 55-67.

Quax-Jeuken Y., Bruisten S., Bloemendal H., de Jong W. W. and **Nevo E.** 1985. Evolution of crystallins: Expression of lens-specific proteins in the blind mammals mole (*Talpa europaea*) and mole rat (*Spalax ehrenbergi*). Mol. Biol. Evol. 2: 279-285.

Nevo E., Moseman J.G., Beiles A. and Zohary D. 1985. Patterns of resistance of Israeli wild emmer wheat to pathogens. I. Predictive method by ecology and allozyme genotypes for powdery mildew and leaf rust. Genetica 67: 209-222.

Nevo E., Atsmon D. and Beiles A. 1985. Protein resources in wild barley, *Hordeum spontaneum*, in Israel: Predictive method by ecology and allozyme markers. Pl. Syst. Evol. 150: 205-222.

Rittenhouse J., **Nevo E.** and Marcus F. 1985. Distribution of a COOH-terminal amino acid extension of liver fructose-1,6-biphosphatase among rodent species. Comp. Biochem. Physiol. 82B: 507-509.

Nevo E. 1985. Diversity is the key to life and human culture. Lecture delivered at the inauguration of the "Ancell-Teicher Research Foundation for Genetics and Molecular Evolution," Board of Governors' Meeting, Univerisity of Haifa, 28 May, 1985.

Nevo E. and Capranica R.R. 1985. Evolutionary origin of ethological reproductive isolation in cricket frogs, *Acris*. In: *Evolutionary Biology*, Vol. 19. Hecht M. K., Wallace B. and Prance G.T. (eds.), Plenum Press, N.Y., pp. 147-214.

Kleinschmidt T., **Nevo E.**, Goodman M. and Braunitzer G. 1985. Mole rat hemoglobin: Primary structure and evolutionary aspects in a second karyotype of *Spalax ehrenbergi*, Rodentia, (2n=52). Biol. Chem. Hoppe-Seyler 366: 679-685.

Lavie B. and **Nevo E.** 1986. Genetic diversity of marine gastropods: Contrasting strategies of *Cerithium rupestre* and *C. scabridum* in the Mediterranean Sea. Mar. Ecol. Prog. Ser. 28: 99-103.

Nevo E., Grama A., Beiles A. and Golenberg E.M. 1986. Resources of high-protein genotypes in wild wheat, *Triticum dicoccoides* in Israel: Predictive method by ecology and allozyme markers. Genetica 68: 215-227.

Nevo E., Beiles A., Kaplan D., Storch N. and Zohary D. 1986. Genetic diversity and environmental associations of wild barley, *Hordeum spontaneum* (Poaceae), in Iran. Pl. Syst. Evol. 153: 141- 164.

Nevo E., Zohary D., Beiles A., Kaplan D. and Storch N. 1986. Genetic diversity and environmental associations of wild barley, *Hordeum spontaneum*, in Turkey. Genetica 68: 203-213.

Nevo E., Beiles A. and Zohary D. 1986. Genetic resources of wild barley in the Near East: Structure, evolution and application in breeding. Biol. J. Linn. Soc. Lond. 27: 355-380.

Flavell R.B., O'Dell M., Sharp P., **Nevo E.** and Beiles A. 1986. Variation in the intergenic spacer of ribosomal DNA of wild wheat, *Triticum dicoccoides* in Israel. Mol. Biol. Evol. 3: 547-558.

Nevo E. 1986. Evolutionary behavior genetics in active speciation and adaptation of fossorial mole rats. In: *Accademia Nazionale dei Lincei*, Vol. 259, "Proceedings of the International Meetings on Variability and Behavioral Evolution", Rome, 23-26.11.1983. Accademia Nazionale dei Lincei, Rome, pp. 39-109.

Lavie B. and **Nevo E.** 1986. The interactive effects of cadmium and mercury pollution on allozyme polymorphisms in the marine gastropod, *Cerithium scabridum*. Mar. Poll. Bull. 17: 21-23.

Nevo E., Beiles A., Kaplan D., Golenberg E.M., Olsvig-Whittaker L. and Naveh Z. 1986. Natural selection of allozyme polymorphisms: A microsite test revealing ecological genetic differentiation in wild barley. Evolution 40: 13-20.

Nevo E., Capanna E., Corti M., Jarvis J.U.M. and Hickman G.C. 1986. Karyotype differentiation in the endemic subterranean mole rats of South Africa (Rodentia, Bathyergidae). Z. Saugetierkunde 51: 36-49.

Arieli R., Heth G., **Nevo E.**, Zamir Y. and Neutra O. 1986. Adaptive heart and breathing frequencies in four ecologically differentiating chromosomal species of mole rats in Israel. Experientia 42: 131-133.

Nevo E. 1986. Mechanisms of adaptive speciation at the molecular and organismal levels. In: *Evolutionary Processes and Theory*, Karlin S. and **Nevo E.** (eds.), Academic Press, N.Y., pp. 439-474.

Arieli R., Heth G., **Nevo E.** and Hoch D. 1986. Hematocrit and hemoglobin concentration in four chromosomal species and some isolated populations of actively speciating subterranean mole rats in Israel. Experientia 42: 441-443.

Heth G., Pevet P., **Nevo E.** and Beiles A. 1986. The effect of melatonin administration and short exposures to cold on body temperature of the blind subterranean mole rat (Rodentia, *Spalax ehrenbergi*, Nehring). J. Exp. Zool. 238: 1-9.

Lavie B. and **Nevo E.** 1986. Genetic selection of homozygote allozyme genotypes in marine gastropods exposed to cadmium pollution. Sci. Tot. Environ. 57: 91-98.

Redi C. A., Garagna S., Heth G. and **Nevo E.** 1986. Descriptive kinetics of spermatogenesis in four chromosomal species of the *Spalax ehrenbergi* superspecies in Israel. J. Exp. Zool. 238: 81-88.

Nevo E. 1986. Pollution and genetic evolution in marine organisms: Theory and practice. In: *Environmental Quality and Ecosystem Stability*, Vol. III A/B. Dubinsky Z. and Steinberger Y. (eds.), Bar-Ilan Univ. Press, Ramat-Gan, Israel, pp. 841-848.

Lavie B. and **Nevo E.** 1986. Selection against rare alleles in mercury pollution: Multilocus structure across five marine gastropod species. In: *Environmental Quality and Ecosystem Stability*, Vol. III A/B. Dubinsky Z. and Steinberger Y. (eds.), Bar-Ilan Univ. Press, Ramat-Gan, Israel, pp. 859-864.

Nevo E., Beiles A., Heth G. and Simson S. 1986. Adaptive differentiation of body size in speciating mole rats. Oecologia 69: 327-333.

Heth G., Frankenberg E. and **Nevo E.** 1986. Adaptive optimal sound for vocal communication in tunnels of a subterranean mammal (*Spalax ehrenbergi*). Experientia 42: 1287-1289.

Nevo E., Heth G. and Beiles A. 1986. Aggression patterns in adaptation and speciation of subterranean mole rats. J. Genet. 65: 65-78.

Nevo E., Gerechter-Amitai Z., Beiles A. and Golenberg E.M. 1986. Resistance of wild wheat to stripe rust: Predictive method by ecology and allozyme genotypes. Pl. Syst. Evol. 153: 13-30.

Nevo E., Noy R., Lavie B., Beiles A. and Muchtar S. 1986. Genetic diversity and resistance to marine pollution. Biol. J. Linn. Soc. 29: 139-144.

Karlin S. and **Nevo E.** (eds.). 1986. *Evolutionary Processes and Theory*. Academic Press, N.Y., 832 pp.

Nevo E., Corti M., Heth G., Beiles A. and Simson S. 1986. Chromosomal polymorphisms in the *Spalax ehrenbergi* complex and their evolutionary significance. Boll. Zool. 53: 67.

Nevo E. 1986. Genetic resources of wild cereals and crop improvement: Israel, a natural laboratory. Isr. J. Bot. 35: 255-278.

Heth G., **Nevo E.** and Beiles A. 1987. Adaptive exploratory behaviour: Differential patterns in species and sexes of subterranean mole rats. Mammalia 51: 27-37.

Heth G., Frankenberg E., Raz A. and **Nevo E.** 1987. Vibrational communication in subterranean mole rats (*Spalax ehrenbergi*). Behav. Ecol. Sociobiol. 21: 31-33.

Nof N., Lavie B., Heth G. and **Nevo E.** 1987. Adaptive temperature dependent activity patterns in marine gastropods. Isr. J. Zool. 34: 61-66.

Lavie B., Noy R. and **Nevo E.** 1987. Genetic variability in the marine gastropods *Patella coerulea* and *Patella aspera*: Patterns and problems. Mar. Biol. 96: 367-370.

Nevo E., Heth G., Beiles A. and Frankenberg E. 1987. Geographic dialects in blind mole rats: Role of vocal communication in active speciation. Proc. Natl. Acad. Sci. USA 84: 3312-3315.

Noy R., Lavie B. and **Nevo E.** 1987. The niche-width variation hypothesis revisited: Genetic diversity in the marine gastropods, *Littorina punctata* (Gmelin) and *L. neritoides* (L.). J. Exp. Mar. Biol. Ecol. 109: 109-116.

Nevo E., Ben-Shlomo R., Beiles A., Jarvis J.U.M. and Hickman G.C. 1987. Allozyme differentiation and systematics of the endemic subterranean mole rats of South Africa (*Rodentia, Bathyergidae*). Bioch. Syst. Ecol. 15: 489-502.

Lavie B. and **Nevo E.** 1987. Differential fitness of allelic isozymes in the marine gastropods *Littorina punctata* and *Littorina neritoides* exposed to the environmental stress of the combined effects of cadmium and mercury pollution. Environ. Manag. 11: 345-349.

Nevo E. 1987. Plant genetic resources: Prediction by isozyme markers and ecology. In: *Isozymes: Current Topics in Biological Research*, Rattazi M. C., Scandalios J. G. and Whitt G. S. (eds). Volume 16: Agriculture, Physiology and Medicine, pp. 247-267.

Golenberg E.M. and **Nevo E.** 1987. Multilocus differentiation and population structure in a selfer, wild emmer wheat, *Triticum dicoccoides*. Heredity 58: 451-456.

Schoepfer R., Figueroa F., Nizetic D., **Nevo E.** and Klein J. 1987. Evolutionary diversification of class II P loci in the *Mhc* of the mole rat, *Spalax ehrenbergi*. Mol. Biol. Evol. 4: 287-299.

Flynn L.J., **Nevo E.** and Heth G. 1987. Incisor enamel microstructure in subterranean mole rats (*Spalax ehrenbergi* superspecies): Adaptive and phylogenetic significance. J. Mamm. 68: 500-507.

Nevo E., Noy R., Lavie B. and Muchtar S. 1987. Levels of genetic diversity and resistance to pollution in marine organisms. FAO Fisheries Report No. 352 Supplement 1985: 175-182.

Nizetic D., Figueroa F., Dembic Z., **Nevo E.** and Klein J. 1987. Major histocompatibility complex gene organization in the mole rat *Spalax* *ehrenbergi*: Evidence for transfer of function between class II genes. Proc. Natl. Acad. Sci. USA 84: 5828-5832.

Hendriks W., Leunissen J., **Nevo E.**, Bloemendal H. and de Jong W.W. 1987. The lens protein alpha-A crystallin in the blind mole rat, *Spalax ehrenbergi*: Evolutionary change and functional constraints. Proc. Natl. Acad. Sci. USA 84: 5320-5324.

Vincek V., Nizetic D., Golubic M., Figueroa F., **Nevo E.** and Klein J. 1987. Evolutionary expansion of Mhc class I genes in the mole rat, *Spalax ehrenbergi*. Mol. Biol. Evol. 4: 483-491.

Nevo E., Lavie B. and Noy R. 1987. Mercury selection of allozymes in two species of marine gastropods: Prediction and verification in nature iron. Monitor. Assess. 9: 233-238.

Nevo E. and Payne P.I. 1987. Wheat storage proteins: Diversity of HMW glutenin subunits in wild emmer from Israel. I. Geographical patterns and ecological predictability. Theor. Appl. Genet. 74: 827-836.

Suzuki H., Moriwaki K. and **Nevo E.** 1987. Ribosomal DNA (rDNA) spacer polymorphism in mole rats. Mol. Biol. Evol. 4: 602-610.

Honeycutt L.R., Edwards V.S., Nelson K. and **Nevo E.** 1987. Mitochondrial DNA variation and the phylogeny of African mole rats (Rodentia: Bathyergidae). Syst. Zool. 36: 280-292.

Heth G., Beiles A. and **Nevo E.** 1988. Adaptive variation of pelage color within and between species of the subterranean mole rat *(Spalax ehrenbergi)* in Israel. Oecologia 74: 617-622.

Nevo E., Beiles A. and Krugman T. 1988. Natural selection of allozyme polymorphisms: a microgeographic climatic differentiation in wild emmer wheat, *Triticum dicoccoides*. Theor. Appl. Genet. 75: 529-538.

Nevo E., Corti M., Heth G., Beiles A. and Simson S. 1988. Chromosomal polymorphisms in subterranean mole rats: origins and evolutionary significance. Biol. J. Linn. Soc. Lond. 33: 309-322.

Bruns U., Muller M., Hofer W., Heth G. and **Nevo E.** 1988. Inner ear structure and electrophysiological audiograms of the subterranean mole rat, *Spalax ehrenbergi*, Hearing Research 33: 1-10.

Heth G., Frankenberg E. and **Nevo E.** 1988. "Courtship" call of subterranean mole rats *(Spalax ehrenbergi)*: Physical analysis. J. Mamm. 69: 121-125.

Nevo E. 1988. Natural selection in action: The interface of ecology and genetics in adaptation and speciation at the molecular and organismal levels. In: *The Zoogeography of Israel*. Yom-Tov Y. and Tchernov E. (eds.). Dr. Junk Publ., Holland, pp. 411-438.

Ben-Shlomo R., Figueroa F., Klein J. and **Nevo E.** 1988. Mhc class II DNA polymorphisms within and between chromosomal species of the *Spalax ehrenbergi* superspecies in Israel. Genetics 119: 141-149.

Nevo E. 1988. Genetic diversity in nature: Patterns and theory. Evol. Biol. 23: 217-247.

Nevo E. and Beiles A. 1988. Ribosomal DNA nontranscribed spacer polymorphism in subterranean mole rats: Genetic differentiation, environmental correlates, and phylogenetic relationships. Evol. Ecol. 2: 139-156.

Nevo E. 1988. Genetic differentiation in evolution. *ISI Atlas of Science,* Anim. Pl. Sci. 1988: 195-202.

Nevo E., Beiles A. and Kaplan D. 1988. Genetic diversity and environmental associations of wild emmer wheat, in Turkey, Heredity 61: 31-45.

Ben-Shlomo R. and Nevo E. 1988. Isozyme polymorphism as monitoring of marine environments: The interactive effect of cadmium and mercury pollution in the shrimp, *Palaemon elegans*. Mar. Poll. Bull. 19: 314-317.

Nevo E., Lavie B. and Ben-Shlomo R. 1988. Population genetic structure of marine organisms as detectors and monitors of marine pollution. In: *UNEP/FAO: Toxicity, persistence and bioaccumulation of selected substances to marine organisms (Activity G)*. MAP Technical Reports Series No. 24, UNEP, Athens, 1988, pp. 37-49.

Auffray J.C., Tchernov E. and Nevo E. 1988. Origine du commensalisme de la souris domestique (*Mus musculus domesticus*) vis-a-vis de l'homme. C.R. Acad. Sc. Paris. III: 517-522.

Nevo E., Tchernov E. and Beiles A. 1988. Morphometrics of speciating mole rats: Adaptive differentiation in ecological speciation. Z. Zool. Syst. Evol. Forsch. 26: 286-314.

Nevo E. and Beiles A. 1988. Genetic parallelism of protein polymorphism in nature: Ecological test of the neutral theory of molecular evolution. Biol. J. Linn. Soc. Lond. 35: 229-245.

Nevo E., Beiles A. and Krugman T. 1988. Natural selection of allozyme polymorphisms: a microgeographic differentiation by edaphic, topographical and temporal factors in wild emmer wheat *Triticum dicoccoides*. Theor. Appl. Genet. 76: 737-752.

Yahav S., Simson S. and Nevo E. 1988. Adaptive energy metabolism in four chromosomal species of subterranean mole rats. Oecologia 77: 533-536.

Nevo E., Pirlot P. and Beiles A. 1988. Brain size diversity in adaptation and speciation of subterranean mole rats, Z. Zool. Syst. Evol. Forsch. 26: 467-479.

Edoute Y., Arieli R. and Nevo E. 1988. Evidence for improved myocardial oxygen delivery and function during hypoxia in the mole rat. J. Comp. Physiol. 158: 575-582.

Lavie B. and Nevo E. 1988. Multilocus genetic resistance and susceptibility to mercury and cadmium pollution in the marine gastropod, *Cerithium scabridum*, Aqua. Tox. 13: 291-296.

Filippucci M.G., Rodino E., Nevo E. and Capanna E. 1988. Evolutionary genetics and systematics of the garden dormouse, *Eliomys* Wagner, 1840. 2 - Allozymic diversity and differentiation of chromosomal forms. Boll. Zool. 55: 47-54.

Filippucci M.G., Simson S., **Nevo E.** and Capanna E. 1988. The Chromosomes of Israeli garden dormouse, *Eliomys melanurus* Wagner, 1840 (Rodentia, Gliridae). Boll. Zool. 53: 31-33.

Nevo E., Ben-Shlomo R. and Maeda N. 1989. Haptoglobin DNA polymorphism in subterranean mole rats of the *Spalax ehrenbergi* in Israel. Heredity 62: 85-90.

Nevo E. and Beiles A. 1989. Genetic diversity of wild emmer wheat in Israel and Turkey: Structure, evolution and application in breeding. Theor. Appl. Genet. 77: 421-455.

Pirlot P. and **Nevo E.** 1989. Brain organization and evolution in subterranean mole rats. Z. Zool. Syst. Evol. Forsch. 27: 58-64.

Nevo E. 1989. Modes of speciation: The nature and role of peripheral isolates in the origin of species. In: *Genetics, Speciation and the Founder Principle*, Giddings L.V., Kaneshiro K.Y. and Anderson W.W. (eds.). Oxford Univ. Press, Oxford, pp. 205-236.

Nevo E. 1989. Genetic resources of wild emmer wheat revisited: genetic evolution, conservation and utilization. In: *Proceedings Seventh Int. Wheat Genetics Symp*, Miller T.E. and Koebner R.M.D. (eds.), Inst. of Pl. Sci. Res., Cambridge, pp. 121-126.

Burda H., Bruns V. and **Nevo E.** 1989. Middle ear and cochlear receptors in the subterranean mole rat, *Spalax ehrenbergi*. Hearing Res. 39: 225-230.

Nevo E. 1989. Natural selection of body size differentiation in spiny mice, *Acomys*. Zeit. Saugetierkunde 52: 81-99.

Schuller C., Neuteboom B., Wubbels G.H., Beintema J.J. and **Nevo E.** 1989. The amino acid sequence of pancreatic ribonuclease from the mole rat *Spalax ehrenbergi* chromosomal species ($2n=60$). Biol. Chem. Hoppe-Seyler 370: 533-589.

Nevo E., Simson S., Beiles A. and Yahav S. 1989. Adaptive variation in structure and function of kidneys of actively speciating mole rats. Oecologia 79: 366-371.

Nevo E. and Filippucci M.G. 1989. Genetic differentiation between Israeli and Greek population of the marsh frog, *Rana ridibunda*. Zool. Anz. 221: 418-424.

Heth G., Golenberg E.M. and **Nevo E.** 1989. Foraging strategy of a subterranean rodent *Spalax ehrenbergi*. Oecologia 79: 496-505.

Yahav S., Simson S. and **Nevo E.** 1989. Total body water and adaptive water turnover rate in four chromosomal mole rat species of the *Spalax ehrenbergi* superspecies in Israel. J. Zool. Lond. 218: 461-469.

Nevo E. and Piechaczyk M. 1989. Glyceraldehyde-3-phosphate dehydrogenase multigene family of mole rats: Evolutionary and phylogenetic patterns. Mammalia 53: 295-300.

Capanna E., Civitelli M.V., Hickman B.C. and **Nevo E.** 1989. The chromosomes of *Amblysomus hottentotus* (Smith 1829) and *A. iris* Thomas &

Schwann 1905; a first report for the golden moles of Africa (Insectivora Chrysochloridae). Trop. Zool. 2: 1-12

Nevo E. and Beiles A. 1989. Sexual selection and natural selection in body size differentiation of subterranean mole rats. Z. Zool. Syst. Evol. Forsch. 27: 263-269.

Nevo E. and Beiles A. 1989. Genetic diversity in the desert: Patterns and testable hypotheses. J. Arid Envir. 17: 241-244.

Nevo E. and Lavie B. 1989. Differential viability of allelic isozymes in the marine gastropods *Cerithium scabridum* exposed to the environmental stress of nonionic detergent and crude oil-surfactant mixtures. Genetica 78: 205-213.

Catzeflis F.M., Nevo. E., Ahlquist J.E. and Sibley C.G. 1989. Relationships of the chromosomal species in the Eurasian mole rats of the *Spalax ehrenbergi* Group as determined by DNA-DNA hybridization, and an estimate of the spalacid-murid divergence time. J. Mol. Evol. 29: 223-232.

Nevo E. and Lavie B. 1989. Selection of allozyme genotypes of two species of marine gastropods (*genus Littorina*) in experiments of environmental stress by nonionic detergent and crude oil-surfactant mixtures. Genet. Select. Evol. 21: 295-302.

Corke H., Nevo E. and Atsmon D. 1989. Variation in vegetative parameters related to the nitrogen economy of wild barley, *Hordeum spontaneum* in Israel. Euphytica 39: 227-232.

Savic I. and Nevo E. 1990. The spalacidae: Evolutionary history, speciation, and population biology. In: *Evolution of Subterranean Mammals at the Organismal and Molecular Levels*. Nevo E. and Reig O.A. (eds.). Alan R. Liss, Inc., N.Y., pp. 129-153.

Nevo E., Filippucci M.G. and Beiles A. 1990. Genetic diversity and its ecological correlates in nature: Comparison between subterranean, fossorial and aboveground small mammals. In: *Evolution of Subterranean Mammals at the Organismal and Molecular Levels*. Nevo E. and Reig O.A. (eds.). Alan R. Liss, Inc., N.Y., pp. 347-366.

Jekel P.A., Ciabatti C., Schuller C., Beintema J.J. and Nevo E. 1990. Ribonuclease in different chromosomal species of the mole rat, superspecies *Spalax ehrenbergi*: Concentration in the pancreas and primary structure. In: *Evolution of Subterranean Mammals at the Organismal and Molecular Levels*. Nevo E. and Reig O.A. (eds.). Alan R. Liss, Inc., N.Y., pp. 367-381.

De Jong W.W., Hendriks W., Sanyal S. and Nevo E. 1990. The eye of the blind mole rat (*Spalax ehrenbergi*): Regressive evolution at the molecular level. In: *Evolution of Subterranean Mammals at the Organismal and Molecular Levels*. Nevo E. and Reig O.A. (eds.). Alan R. Liss, Inc., N.Y., pp. 383-395.

Nevo E. and Klein J. 1990. Structure and evolution of Mhc in subterranean mammals of the *Spalax ehrenbergi* superspecies in Israel. In: *Evolution of Subterranean Mammals at the Organismal and Molecular Levels*. Nevo E. and Reig O. A. (eds.). Alan R. Liss, Inc., N.Y., pp. 397-411.

Nevo E. and Reig O. A. (Eds.). 1990. *Evolution of Subterranean Mammals at the Organismal and Molecular Levels*. Alan R. Liss, Inc., N.Y.

Carver B. F. and Nevo E. 1990. Genetic diversity of photosynthetic characters in native populations of *Triticum dicoccoides*. Photosyn. Res. 25: 119-128.

Nevo E. 1990. Molecular Evolutionary Genetics of isozymes: Patterns, theory and application. In: *Isozymes: Structure, Function and Use in Biology and Medicine*. Ogita Z.I. and Markert C.L. (eds.) Wiley-Liss, Inc., N.Y., pp. 701-742.

Sanyal S., Jansen H.G., de Grip W.J., Nevo E. and de Jong W.W. 1990. The eye of the blind mole rat, *Spalax ehrenbergi*: Rudiment with hidden function? Invest. Ophtalmol. Vis. Sci. 31: 1398-1404.

Nevo E. 1990. Evolution of nonvisual communication and photoperiodic perception in speciation and adaptation of blind subterranean mole rats. Behaviour 114: 249-276.

Burda H., Nevo E. and Bruns V. 1990. Adaptive differentiation of ear structures in subterranean mole rats of the *Spalax ehrenbergi* superspecies in Israel. Zool. Jb. Syst. 117: 369-382.

Moseman J.G., Nevo E. and El-Morshidy M.A. 1990. Reactions of *Hordeum spontaneum* to infection with 2 cultures of *Puccinia hordei* from Israel and the United States. Euphytica 49: 169-175.

Filippucci M.G., Simson S. and Nevo E. 1990. Evolutionary biology of the genus *Apodemus* Kaup, 1829 in Israel. Allozymic and biometric analyses with description of a new species: *Apodemus hermonensis* (Rodentia, Muridae). Boll. Zool. 56: 361-376.

Nevo E., Joh K., Hori K. and Beiles A. 1990. Aldolase DNA polymorphism in subterranean mole rats: Genetic differentiation and environmental correlates. Heredity 65: 307-320.

Snipes R.L., Nevo E. and Sust H. 1990. Anatomy of the caecum of the Israeli mole rat, *Spalax ehrenbergi* (Mammalia). Zool. Anz. 224: 307-320.

Yahav S., Simson S. and Nevo E. 1990. The effect of protein and salt loading on urinary concentrating ability in four chromosomal species of *Spalax ehrenbergi*. J. Zool. Lond. 222: 341-347.

Nevo E., Kishi K. and Beiles A. 1990. Genetic polymorphism of urine deoxyribonuclease-I isomerases of subterranean mole rats, *Spalax ehrenbergi* superspecies in Israel: Ecogeographical patterns and correlates. Bioch. Genet. 28: 561-570.

Nevo E. 1991. Evolutionary theory and processes of active speciation and adaptive radiation in subterranean mole rats, *Spalax ehrenbergi* superspecies in Israel. Evol. Biol. 25: 1-125.

Nevo E. 1991. Evolution of vocal and vibrational communications in blind, photoperiod-perceptive, subterranean mole rats: Structure and function. In: *Le Rongeur et L'Espace (The Rodent and its Environment)*, Le Berret M. and le Guelte L. (eds.). Proceedings of the International Meeting, Lyon, 1989. pp. 15-34.

Nevo E., Heth G. and Pratt H. 1991. Seismic communication in a blind subterranean mammal: A major somatosensory mechanism in adaptive evolution underground. Proc. Natl. Acad. Sci. USA 88: 1256-1260.

Snape J.W., Nevo E., Parker B.B., Leckie D. and Morgunov A. 1991. Herbicide response polymorphism in wild populations of emmer wheat. Heredity 66: 251-257.

Nevo E., Carver B.F. and Beiles A. 1991. Photosynthetic performance in wild emmer wheat, *Triticum dicoccoides*: ecological and genetic predictability. Theor. Appl. Genet. 81: 445-460.

Nevo E., Gerechter-Amitai Z. and Beiles A. 1991. Resistance of wild emmer wheat to stem rust: Ecological, pathological and allozyme associations. Euphytica 53: 121-130.

Greenbaum I.F., Hale D.W., Sudman P.D. and Nevo E. 1991. Synaptonemal complex analysis of mole rats (*Spalax ehrenbergi*): unusual polymorphisms of chromosome 1. Genome 33: 898-902.

Nevo E. 1991. Complex pollution effects of two heavy metals (Mercury and Cadmium). The genetic structure of populations. In: *UNEP/FAO, MAP Technical Reports No. 48*, Athens, 1991. pp. 45-63.

Nevo E. and A. Beiles. 1991. Genetic diversity and ecological heterogeneity in amphibian evolution. Copeia 3: 565-592.

Heth G., Frankenberg E., Pratt H. and Nevo E. 1991. Seismic communication in the blind subterranean mole rats: Patterns of head thumping and of their detection in the *Spalax ehrenbergi* superspecies in Israel. J. Zool. Lond. 224: 633-638.

Ellis R.P., Forster B.P., Thomas W.T.B. and Nevo E. 1991. The use of *Hordeum spontaneum* Koch in barley improvement. 6th Intern. Barley Genetics Symp. July, 22-27, 1991, Helsingborg, Sweden. Barley Genetics 6: 65-67.

Snape J.W., Leckie D., Parker B.B. and Nevo E. 1991. The genetical analysis and exploitation of differential responses to herbicides in crop species. In: *Herbicide Resistance in Weeds and Crops*, Casley J. C., Cussans G.W. and Atkin R.K. (eds.). Butterworth-Heinemann, Oxford. pp. 305- 317.

Filippucci M.G., Hickman G.C., Capanna E. and Nevo E. 1991. Genetic diversity and differentiation of the endemic subterranean golden moles of

South Africa (Chrysochloridae, Mammalia). Bioch. Syst. Ecol. 19: 461-466.

Jana S. and **Nevo E.** 1991. Variation in response to infection with *Erysiphe graminis* hordei and *Puccinia hordei* in some wild barley populations in a centre of diversity. Euphytica 57: 133-140.

Arieli R. and **Nevo E.** 1991. Hypoxic survival differs between two mole rat species (*Spalax ehrenbergi*) of humid and arid habitats. Comp. Biochem. Physiol. 3: 543-545.

Nevo E. 1991. Genetic diversity and the conservation of endangered species. In: *Mammals in the Palaearctic Desert: Status and trends in the Sahara-Gobian region.* McNeedly J. A. and Neronov V.M. (eds.). Symposium on "Conservation of rare and endangered species". UNESCO Programme on Man and the Biosphere. pp. 79-92.

Hunger R.M., Sherwood J.L., Pennington R.E., Carver B.F. and **Nevo E.** 1992. Reaction of native populations of *Triticum dicoccoides* to wheat soilborne mosaic. In: *Biological & Cultural Tests for Control of Plant Diseases.* Vol. 7. APS Press.

Nevo E. 1992. Origin, evolution, population genetics and resources for breeding of wild barley, *Hordeum spontaneum*, in the Fertile Crescent. In: *Barley: Genetics, Molecular Biology and Biotechnology*, Shewry P. (ed.). C.A.B. International. pp. 19-43.

Burda H., Filippucci G.M., Macholan M., **Nevo E.** and Zima J. 1992. Biological, allozyme, and karyotype differentiation of African mole-rats (*Cryptomys*, Bathyergidae) from Zambia. Z. Saugetierk. Suppl. 57.

Nevo E., Gorham J. and Beiles A. 1992. Variation for [22]Na uptake in wild emmer wheat, *Triticum dicoccoides* in Israel: Salt tolerance resources for wheat improvement. J. Exp. Bot. 43: 511-518.

Schneider H., Sinsch U. and **Nevo E.** 1992. The lake frogs in Israel represent a new species. Zool. Anz. 228: 97-106.

Nevo E. 1992. Evolutionary processes and theory: the ecological-genetics interface. In: *Environmental Quality and Ecosystem Stability*; Vol. V/B - Ecosystem Stability. Gasith A., Adin A., Steinberger Y. and Garty J. (eds.) ISEEQS Publication. pp. 724-731.

Nevo E. and Beiles A. 1992. Amino acid resources in the wild progenitor of wheats, *Triticum dicoccoides* in Israel: Polymorphisms and predictability by ecology and isozymes. Pl. Breeding 108: 190-201.

Nevo E. and Beiles A. 1992. Selection for class II Mhc heterozygosity by parasites in subterranean mole rats. Experientia 48: 512-515.

Chalmers K.J., Waugh R., Waters J., Forster B.P., **Nevo E.** and Powell W. 1992. Grain isozyme and ribosomal DNA variability in *Hordeum spontaneum* populations from Israel. Theor. Appl. Genet. 84: 313-322.

Nevo E., Snape J.W., Lavie B. and Beiles A. 1992. Herbicide response polymorphisms in wild emmer wheat: Ecological and isozyme correlations. Theor. Appl. Genet. 84: 209-216.

Olsvig-Whittaker L.S., Naveh Z., Giskin M. and Nevo E. 1992. Microsite differentiation in a Mediterranean oak savannah. J. Veget. Sci. 3: 209-216.

Necker R., Rehkamper G. and Nevo E. 1992. Electrophysiological mapping of body representation in the cortex of the blind mole rat. Neuroreport. 3: 505-508.

Nevo E., Noy-Meir I., Beiles A., Krugman T. and Agami M. 1992. Natural selection of allozyme polymorphisms: A micro-geographical spatial and temporal ecological differentiations in wild emmer wheat. Isr. J. Bot. 40: 419-449.

Nevo E., Ben-Shlomo R., Beiles A., Hart C.P. and Ruddle F.H. 1992. Homeobox DNA polymorphisms (RFLPs) in subterranean mammals of the *Spalax ehrenbergi* superspecies in Israel: Patterns, correlates and evolutionary significance. J. Exp. Zool. 263: 430-441.

Nevo E., Ordentlich A., Beiles A. and Raskin I. 1992. Genetic divergence of heat production within and between the wild progenitors of wheat and barley: evolutionary and agronomical implications. Theor. Appl. Genet. 84: 958-962.

Heth G., Nevo E., Ikan R., Weinstein V., Ravid U. and Duncan H. 1992. Differential olfactory perception of enantiomeric compounds by blind subterranean mole rats (*Spalax ehrenbergi*). Experientia 48: 897-902.

De Grip W.J., Janssen J.J.M., Foster R.G., Korf H.W., Rothschild K.J., Nevo E. and de Caluwe G.L.J. 1992. Molecular analysis of photoreceptor protein function. In: *Signal Transduction in Photoreceptor Cells*. Hargrave P.A., Hofmann K.P. and Kaupp U.B., (eds.). Springer-Verlag, Berlin. pp. 43-59.

Menzies R.A., Heth G., Ikan R., Weinstein V. and Nevo E. 1992. Sexual pheromones in lipids and other fractions from urine of the male mole rat, *Spalax ehrenbergi*. Physiol. and Behav. 52: 741-747.

Menzies R., Cohen Y., Lavie B. and Nevo E. 1992. Niche adaptation in two marine gastropods, *Monodonta turbiformis* and *M. turbinata*. Boll. Zool. 59: 297-302.

Nevo E. and Beiles A. 1992. mtDNA polymorphisms: Evolutionary significance in adaptation and speciation of subterranean mole rats. Biol. J. Linn. Soc. Lond. 47: 385-405.

Nevo E., Simson S., Heth G. and Beiles A. 1992. Adaptive pacifistic behaviour in subterranean mole rats in the Sahara Desert, contrasting to and originating from polymorphic aggression in Israeli species. Behaviour 123: 70-76.

Cooper H.M., Herbin M. and **Nevo E.** 1993. Mosaic evolution: ocular regression conceals adaptive progression of the visual system in a blind subterranean mammal. Nature 361: 156-159.

Ben-Shlomo R., Shin H.S. and **Nevo E.** 1993. Period - homologous sequence polymorphisms in subterranean mammals of the *Spalax ehrenbergi* superspecies in Israel. Heredity 70: 111-121.

Nevo E. 1993. Genetic diversity revisited. In: *Current Contents; This Week's Citation Classic.* 4: 8.

Butler P.M., **Nevo E.**, Beiles A. and Simson S. 1993. Variations of molar morphology in *Spalax ehrenbergi* superspecies: adaptive and phylogenetic significance. J. Zool. Lond. 229: 191-216.

Nevo E., Nishikawa K., Furuta Y., Gonokami Y. and Beiles A. 1993. Genetic polymorphisms of alpha- and beta-amylase isozymes in wild emmer wheat, *Triticum dicoccoides*, in Israel. Theor. Appl. Genet. 85: 1029-1042.

Couch L., Duszynski D.W. and **Nevo E.** 1993. Coccidia (Apicomplexa), genetic diversity, and environmental unpredictability of four chromosomal species of the subterranean superspecies *Spalax ehrenbergi* (molerat) in Israel. Parasitology 79: 181-189.

Simson S., Lavie B. and **Nevo E.** 1993. Penial differentiation in speciation of subterranean mole rats *Spalax ehrenbergi* in Israel. J. Zool. Lond. 229: 493-503.

Cooper H.M., Herbin M. and **Nevo E.** 1993. Visual system of a naturally microphthalmic mammal: The blind mole rat *Spalax ehrenbergi*. J. Comp. Neurology 328: 313-350.

Nevo E., Honeycutt R.L., Yonekawa H., Nelson K. and Hanzawa N. 1993. Mitochondrial DNA polymorphisms in subterranean mole rats of the *Spalax ehrenbergi* superspecies in Israel, and its peripheral isolates. Mol. Biol. Evol. 10: 590-604.

Nevo E. 1993. Evolutionary processes and theory: the ecological-genetics interface. Wat. Sci. Tech. 27: 489-496.

Nevo E., Krugman T. and Beiles A. 1993. Genetic resources for salt tolerance in the wild progenitors of wheat (*Triticum dicoccoides*) and barley (*Hordeum spontaneum*) in Israel. Pl. Breeding. 110: 338-341.

Lavie B., Achituv Y. and **Nevo E.** 1993. The niche-width variation hypothesis reconfirmed: Validation by genetic diversity in the sessile intertidal cirripeds *Chthamalus stellatus* and *Euraphia depressa* (Crustacea, Chthamalidae). Z. Zool. Syst. Evol.-Forsch. 31: 110-118.

Ben-Shlomo R. and **Nevo E.** 1993. Myosin heavy chain DNA polymorphisms of subterranean mole rats of the *Spalax ehrenbergi* superspecies in Israel. Genet. Select. Evol. 25: 211-227.

Nevo E., Meyer H. and Piechulla B. 1993. Diurnal rhythms of the chlorophyll a/b binding protein mRNAs in wild emmer wheat and wild barley (Poaceae) in the Fertile Crescent. Pl. Syst. Evol. 185: 181-188.

Shpun S., Hoffman J., Nevo E. and Katz U. 1993. Is the distribution of *Pelobates syriacus* related to its limited osmoregulatory capacity? Comp. Biochem. Physiol. 105A: 135-139.

Nevo E. 1993. Mode, tempo and pattern of evolution in subterranean mole rats of the *Spalax ehrenbergi* superspecies in the Quaternary of Israel. Quaternary International 19: 13-19.

Ellis R.P., Nevo E. and Beiles A. 1993. Milling energy polymorphism in *Hordeum spontaneum* Koch in Israel, and its potential utilization in breeding for malting quality. Pl. Breeding 111: 78-81.

Ellis R.P., Thomas W.T.B., Forster B.P., Macaulay M. and Nevo E. 1993. The use of *Hordeum spontaneum* Koch in barley improvement. Scottish Crop Res. Inst. Ann. Report 1992: 20-23.

Zhao W., Beintema J.J., Hofsteenge J. and Nevo E. 1993. The primary structure of pancreatic ribonuclease from mole rat superspecies *Spalax leucodon*. Mol. Phylogen. and Evol. 2: 270-273.

Nevo E. 1993. Evolutionary novelties. In: J.B.S. Haldane commemoration volume. Zoological Society, Calcutta. pp. 27-40.

Nevo E. 1993. Genetic resources of wild emmer, *Triticum dicoccoides* for wheat improvement: News and Views. 8th Intern. Wheat Genet. Symp., 20-25 July, 1993, Beijing. Proc. Symp. pp. 79-87.

The T.T., Nevo E. and McIntosh R.A. 1993. Responses of Israeli wild emmers to selected Australian pathotypes of *Puccinia* species. Euphytica 71: 75-81.

Nevo E., Krugman T. and Beiles A. 1994. Edaphic natural selection of allozyme polymorphisms in *Aegilops peregrina* at a Galilee microsite in Israel. Heredity 72: 109-112.

Filippucci G.M., Burda H., Nevo E. and Kocka J. 1994. Allozyme divergence and systematics of common mole rats (*Cryptomys*, Bathye-rgidae, Rodentia) from Zambia. Z. Saugetierkunde 59: 42-51.

Nevo E., Filippucci M.G. and Beiles A. 1994. Genetic polymorphisms in subterranean mammals (*Spalax ehrenbergi* superspecies) in the Near East revisited: Patterns and theory. Heredity 72:465-487.

Forster B.P., Handley L.L., Pakniyat H., Scrimgeour C.B., Nevo E. and Raven J. 1994. Genetic and ecological factors controlling carbon isotope discrimination in barley. Aspects of Applied Biology 38: 139-143.

Broza M. and Nevo E. 1994. Selective land snail predation by the spiny mouse, *Acomys cahirinus*, in Nahal Oren, Mt. Carmel Israel. Isr. J. Zool. 40: 173-176.

Nevo E., Filippucci M.G., Redi C., Korol A. and Beiles A. 1994. Chromosomal speciation and adaptive radiation of mole rats in Asia Minor correlated with increased ecological stress. Proc. Natl. Acad. Sci. USA 91: 8160-8164.

Nevo E. 1994. Adaptive speciation at the molecular and organismal levels and its bearing on Amazonian biodiversity. Evol. Biologica 7: 207-249.

Vuillez P., Herbin M., Cooper H.M., **Nevo E.** and Pevet P. 1994. Photic induction of Fos-immunoreactivity in the suprachiasmatic nuclei of the blind mole rat (*Spalax ehrenbergi*). Brain Res. 654: 81-84.

Nevo E. 1994. Evolutionary significance of genetic diversity in nature: Environmental stress, pattern and theory. In: *Isozymes: Organization and Roles in Evolution, Genetics and Physiology.* Markert C.L., Scandalios J.G., Lim H.A. and Serov O.L. (eds.). World Scientific, New Jersey. pp. 267-296. (7th Intern. Cong. on Isozymes, 6-13.9.1992, Novosibirsk, Russia).

Rehkamper G., Necker R. and **Nevo E.** 1994. Functional anatomy of the thalamus in the blind mole rat *Spalax ehrenbergi:* An architectonic and electrophysiologically controlled tracing study. J. Compara. Neurol. 347: 570-584.

Gutterman Y. and **Nevo E.** 1994. Germination comparison study of *Hordeum spontaneum* regionally and locally in Israel: A population in the Negev Desert Highlands and from two opposing slopes on Mediterranean Mount Carmel. Barley Genet. Newslett. 22: 18-19.

Lavie B., Stow V., Krugman T., Beiles A. and **Nevo E.** 1994. Fitness in wild barley from two opposing slopes of a Mediterranean microsite at Mount Carmel, Israel. Barley Genet. Newslett. 23: 12-14.

Handley L.L., **Nevo E.**, Raven J.A., Martinez-Carrasco R., Scrimgeour C. M., Pakniyat H. and Forster. B.P. 1994. Chromosome 4 controls potential water use efficiency (delta 13 C) in barley. J. Exper. Bot. 45: 1661-1663.

Gutterman Y. and **Nevo E.** 1994. Temperatures and ecological-genetic differentiation affecting the germination of *Hordeum spontaneum* caryopses harvested from three populations: the Negev Desert and opposing slopes on Mediterranean Mount Carmel. Isr. J. Pl. Sci. 42: 183-195.

Kirzhner, V.M., Korol A.B., Ronin Y.I. and **Nevo E.** 1994. Cyclical behaveior of genotype frequencies in a two-locus population under fluctuating haploid selection. Proc. Natl. Acad. Sci. USA 91: 11432-11436.

Lindenlaub T., Burda H. and **Nevo E.** 1995. Convergent evolution of the vestibular organ in the subterranean mole rats, *Cryptomys* and *Spalax*, as compared with aboveground rat, *Rattus*. J. Morph. 224: 303-311.

Vinogradova O.N., Kovalenko O.V., Wasser S.P., **Nevo E.**, Kislova O.A. and Belikova O.A. 1995. Biodiversity of Cyanophyta in Israel: Preliminary studies at "Evolution Canyon", Lower Nahal Oren, Mt. Carmel Natural Preserve. Algologia 5: 46-55; Hydrobiol. J. 33: 109-123, 1997.

Nevo E., Filippucci M.G., Redi C., Simson S., Heth G. and Beiles A. 1995. Karyotype and genetic evolution in speciation of subterranean mole rats of the genus *Spalax* in Turkey. Biol. J. Linn. Soc. 54: 203-229.

Pavliček T. and Nevo E. 1995. Genetic diversity and the width of the food niche of *Phytophagous* insects. Biologia, Bratislava, 50: 143-149.

Vinogradova O.N., Kovalenko O.V., Wasser S.P. Nevo E., Tsarenko P.M., Stupina V.V. and Kondratyuk E.S. 1995. Algae of the Mount Carmel National Park (Israel). Algologia 5: 178-192.

Wasser S.P., Nevo E., Vinogradova O.N., Navrotskaya I.L., Ellanskaya I.A., Volz P.A., Kovalenko O.V., Virchenko V.M., Tsarenko P.M., Kondratyuk S.Ya. and Stupina V.V. 1995. Contribution to study of algae, fungi and mosses in "Evolution Canyon" at Nahal Oren, Mount Carmel Natural Preserve, Israel. Ukr. Bot. J. 52: 354-371.

Ben-Shlomo R., Ritte U. and Nevo E. 1995. Activity pattern and rhythm in the subterranean mole rat superspecies *S. ehrenbergi*. Behav. Genet. 25: 239-245.

Cooper H.M., Herbin M., Nevo E. and Negroni J. 1995. Neuroanatomical consequences of microphthalmia in mammals. Les Seminaires ophtalmologiques d'IPSEN, Vision et adaptation, Christen Y., Doly M. and Droy-Lefaix M.-T. (eds). Elsevier. 6: 127-139.

Nevo E. 1995. Evolution and Extinction. Encyclopedia of Environmental Biology. Academic Press, Inc., N.Y. 1: pp. 717-745.

Pavliček T. and Nevo E. 1995. Gene diversity and divergence of a beetle, *Carabus hemprichi* from a microclimatically contrasting Mediterranean microsite, "Evolution Canyon", Nahal Oren, Mt. Carmel, Israel. Isozyme Bull. 28: 49.

Nevo E. 1995. Aridity stress and climatic unpredictability as selective agents in molecular and organismal adaptations in nature. In: *Time Scales of Biological Responses to Water Constraints*. Roy J., Aronson J. and di Castri F. (eds.). SPB Academic Publishing, Amsterdam. pp. 141-166.

Kirzhner V.M., Korol A.B., Ronin Y.I. and Nevo E. 1995. Genetic supercycles caused by cyclical selection. Proc. Natl. Acad. Sci. USA 92: 7130-7133.

Pavliček T. and Nevo E. 1995. Genetic diversity of a beetle *Oxythyrea neomi* in a microsite: A test of correlation in nature between diversity and environmental unpredictability. Zool. Jahrb. Syst.121: 505-513.

Pagnotta M.A., Nevo E., Beiles A. and Porceddu E. 1995. Wheat storage proteins: glutenin diversity in wild emmer, *Triticum dicoccoides*, in Israel and Turkey. 2. DNA diversity detected by PCR. Theor. Appl. Genet. 91: 409-414.

Nevo E., Pagnotta M.A., Beiles A. and Porceddu E. 1995. Wheat storage proteins: glutenin DNA diversity in wild emmer wheat, *Triticum dicoccoides*, in Israel and Turkey. 3. Environmental correlates and allozymic associations. Theor. Appl. Genet. 91: 415-420.

Zuccotti M., Garagna S., Redi C.A., Simson S., **Nevo E.** and Capanna E. 1995. Spermatogenesis of *Spalax ehrenbergi* natural hybrids (Rodentia, Spalacidae). Mammalia 59: 119-125.

Argamaso S.M., Froehlich A.C., McCall M.A., **Nevo E.**, Provencio I. and Foster R.G. 1995. Photopigments and circadian systems of vertebrates. Biophys. Chem. 56: 3-11.

Palamar-Mordvintseva G.M., Wasser S.P. and **Nevo E.** 1995. Conjugatophyceae of some water bodies of Israel. Algologia 5: 378-385.

Nevo E. 1995. Mammalian evolution underground. The ecological-genetic-phenetic interfaces. Acta Theriol. 3: 9-31.

Nevo E. 1995. Asian, African and European biota meet at "Evolution Canyon", Israel: Local tests of global biodiversity and genetic diversity patterns. Proc. Roy. Soc. Lond. B 262: 149-155.

Shanas U., Heth G., **Nevo E.**, Shalgi R. and Terkel J. 1995. Reproductive behaviour in the female blind mole rat (*Spalax ehrenbergi*). J. Zool. Lond. 237: 195-210.

Joppa L.R., **Nevo E.** and Beiles A. 1995. Chromosome translocations in wild populations of tetraploid emmer wheat in Israel and Turkey. Theor. Appl. Genet. 91: 713-719.

Butler P.M., **Nevo E.**, Beiles A. and Simson S. 1995. Molar variation in mole rats (*Spalax*) in relation to speciation and selection. In: *Anthropology and Evolution. Aspects of Dental Biology: Palaeontology*, Moggi-Cecchi J. (ed.). Florence: Intern. Inst. for the Study of Man. pp. 355-356.

Wasser S.P., **Nevo E.**, Vinogradova O.N., Navrotskaya I.L., Ellanskaya I.A., Volz P.A., Virchenko V.M., Tsarenko P.M. and Kondratyuk, S.Ya. 1995. Biodiversity of cryptogamic plants and fungi in "Evolution Canyon", Nahal Oren, Mt. Carmel. Israel. Isr. J. Pl. Sci. 43: 367-383.

Rottenberg A., Zohary D. and **Nevo E.** 1995. Isozymic relationships between cultivated artichoke and the wild relatives. Genet. Resourc. and Crop Evol. 43: 59-62.

Herbin M., Rio J.P., Reperant J., Cooper H.M., **Nevo E.** and Lemire M. 1995. Ultrastructural study of the optic nerve in blind mole rats (Spalacidae, *Spalax*). Soumis a Visual Neurosci. 12: 253-261.

Suzuki H., Wakana S., Yonekawa H., Moriwaki K., Sakurai S. and **Nevo E.** 1996. Variations in ribosomal DNA and mitochondrial DNA among chromosomal species of subterranean mole rats. Mol. Biol. Evol. 13: 85-92.

Hisoriev H., Wasser S.P., **Nevo E.** and Stupina V.V. 1996. In addition to the flora of *Euglenophyta* of Israel. Algologia 6: 49-55.

Krahmalnyj A.F., Wasser S.P. and **Nevo E.** 1996. New *Dinophyta* species for Israel. Algologia 6: 82-85.

Vinogradova O. N., Kovalenko O. V., Wasser S. P. and **Nevo E.** 1996. New for Israel representatives of *Chroococcophyceae* (Cyanophyta) from the national park Mount Carmel and Dead Sea area. Algologia 6: 97-101.

Pavliček T. and **Nevo E.** 1996. Population-genetic divergence of a diplopod in a Mediterranean microsite, Mount Carmel, Israel. Pedobiologia 40: 12-20.

Ben-Shlomo R., Ritte U. and **Nevo E.** 1996. Circadian rhythm and the per ACNGGN repeat in the mole rat, *Spalax ehrenbergi*. Behav. Genet. 26: 177-184.

Derzhavets E.M., Korol A.B. and **Nevo E.** 1996. Increased male recombination rate in *D. melanogaster* correlated with population adaptation to stressful conditions. Drosophila Information Service 77: 92-94.

Derzhavets E.M., **Nevo E.** and Korol A.B. 1996. Potential for the P-M hybrid dysgenesis in differentiated *D. melanogaster* population at "Evolution Canyon", Lower Nahal Oren, Mount Carmel. Drosophila Information Service 77: 124-126.

Korol A.B., Ronin Y.I., Tadmor Y., Bar-Zur A., Kirzhner V.M. and **Nevo E.** 1996. Estimating variance effect of QTL: an important prosepct to increase the resolution power of interval mapping. Genet. Res. Cambridge. 67: 187-194.

Kondratyuk, S.Ya., Navrotskaya I.L., Zelenko S.D., Wasser S.P. and **Nevo E.** 1996. *The First Checklist of Lichen-Forming and Lichenicolous Fungi of Israel*. **Nevo E.** and Wasser S. (eds.). Peledfus Pub. House, Haifa, 140 pp.

Ivanitskaya E., Shenbrot G. and **Nevo E.** 1996. *Crocidura ramona* sp. nov. (Insectivora, Soricidae): a new species of shrew from the central Negev desert, Israel. Z. Saugetierkunde 61: 93-103.

Ganem G. and **Nevo E.** 1996. Ecophysiological constraints associated with aggression, and evolution toward pacifism in *Spalax ehrenbergi*. Behav. Ecol. Sociobiol. 38: 245-252.

Volz P.A., Wasser S.P. and **Nevo E.** 1996. New and rare soil micromycetes for the biota of Israel. Ukr. Bot. J. 53: 51-58.

Tsarenko P.M., Stupina V.V., Wasser S.P., **Nevo E.**, Kovalenko O.V., Kondratiuk E.S., Hisoriev H., Krahmalny A.F. and Krinitz L. 1996. Species diversity of algae in water bodies of Hula Valley (northern Israel). Algologia 6: 182-194.

Korol A.B., Kirzhner V.M., Ronin Y.I. and **Nevo E.** 1996. Cyclical environmental changes as a factor maintaining genetic polymorphism. 2. Diploid selection for an additive trait. Evolution 50: 1432-1441.

Kawahara T., **Nevo E.**, Yamada T. and Zohary D. 1996. III. Genetic stock: Collection of wild *Aegilops* species in Israel. Wheat Inform. Service 82: 36-45.

Kirzhner V.M., Korol A.B. and **Nevo E.** 1996. Complex dynamics of multilocus systems subjected to cyclical selection. Proc. Natl. Acad. Sci. USA 93: 6532-6535.

Ronin Y., Korol A.B., Fahima T., Kirzhner V.M. and **Nevo E.** 1996. Sequential estimation of linkage between PCR-generated markers and a target gene employing stepwise bulked analysis. Biometrics 52: 1428-1439.

Nevo E., Ben-Shlomo R., Beiles A., Ronin Y., Blum S. and Hillel J. 1996. Genomic adaptive strategies: DNA fingerprinting and RAPDs reveal ecological correlates and genetic parallelism to allozyme and mitochondrial DNA diversities in the actively speciating mole rats in Israel. In: *Gene Families: Structure, Function, Genetics Evolution*, Holmes R.S. and Lim H.A. (eds.). World Scientific, Singapure, pp. 55-70.

Ivanitskaya E., Gorlov I., Gorlova O. and **Nevo E.** 1996. Chromosome markers for *Mus macedonicus* (Rodentia, Muridae) from Israel. Hereditas 124: 145-150.

Dudka I.O., Wasser S.P. and **Nevo E.** 1996. A Preliminary report on aquatic *Oomycetes* and *Hyphomycetes* from Israel. Ukr. Bot. J. 53: 517-527.

Corti M., Fadda C., Simson S. and **Nevo E.** 1996. Size and shape variation in the mandible of the fossorial rodents *Spalax ehrenbergi*. A procrustes analysis of three dimensions. In: *Advances in Morphometrics*, Marcus L. F., Corti M., Loy A., Slice D. and Naylor G. (eds). Plenum Press, N.Y. pp. 303-320.

Offer Y., Harris Y., Beiles A. and **Nevo E.** 1996. Biodiversity of ants at "Evolution Canyon", Nahal Oren, Mt. Carmel, Israel. Isr. J. Entomol. 30: 115-119.

Navrotskaya I.L., Kondratyuk S.Ya., Wasser S.P., **Nevo E.** and Zelenko S.D. 1996. Lichens and lichenicolous fungi new for Israel and other countries. Isr. J. Pl. Sci. 44: 181-193.

Nevo E., Vinogradova O.N. and Wasser S.P. 1996. Stress and Evolution: Diversity and divergence of Cyanophyta at "Evolution Canyon" (Mt Carmel National Park, Israel). Algologia 6: 225-234.

Tsarenko P.M., Wasser S.P., **Nevo E.** and Krinitz L. 1996. New for the flora of Israel species of *Chlorococcales* (Chlorophyta). Algologia 6: 295-302 (in Russian); In English: Hydrobiol. J. 1997. 33: 80-94.

Blaustein L., Engert N., Steiner E., **Nevo E.** and Warburg M.R. 1996. Israel's endangered urodele species: preliminary studies on their distribution across Mount Carmel and their influences on community structure of temporary pools. Proc. Isr. Soc. Ecol. Environ. Qual. 6: 514-516.

Vinogradova O.N., Kovalenko O.V., Wasser S.P. and **Nevo E.** 1996. Cyanophyta: Checklist of Continental Species from Israel. **Nevo E.** and Wasser S.P. (eds.). Peledfus Publ. House, Haifa, 105 pp.

L

Kawahara T. and **Nevo E.** 1996. Screening of spontaneous major translocations in Israeli populations of *Triticum dicoccoides* Koern. Wheat Inform. Service 83: 28-30.

Crosatti C., **Nevo E.**, Stanca A.M. and Cattivelli L. 1996. Genetic analysis of the COR14 proteins accumulation in wild (*Hordeum vulgare* L. subsp. *spontaneum*) and cultivated (*Hordeum vulgare*) barley. Theor. Appl. Genet. 93: 957-981.

Ben-Shlomo R., Fahima T. and **Nevo E.** 1996. Random amplified polymorphic DNA of the *Spalax ehrenbergi* superspecies in Israel. Isr. J. Zool. 42: 317-326.

Blaustein L., Kotler B.P. and **Nevo E.** 1996. Rodent species diversity and microhabitat use along opposing slopes of Lower Nahal Oren, Mount Carmel, Israel. Isr. J. Zool. 42: 327-334.

Nevo E., Raz S. and Beiles A. 1996. Biodiversity of reptiles at "Evolution Canyon", Lower Nahal Oren, Mount Carmel, Israel. Isr. J. Zool. 42: 395-402.

Pavliček T. and **Nevo E.** 1996. Genetic divergence in populations of the beetle, *Carabus hemprichi* from microclimatically opposing slopes of "Evolution Canyon": a Mediterranean microsite, Mount Carmel, Israel. Isr. J. Zool. 42: 403-410.

Broza M. and **Nevo E.** 1996. Differentiation of the snail community on the north- and south-facing slopes of Lower Nahal Oren (Mount Carmel, Israel). Isr. J. Zool. 42: 411-424.

Rankevich D., Lavie B., **Nevo E.**, Beiles A. and Arad Z. 1996. Genetic and physiological adaptations of the prosobranch landsnail *Pomatias olivieri* to microclimatic stresses on Mt. Carmel, Israel. Isr. J. Zool. 42: 425-442.

Pavliček T., Csuzdi C., Smooha G., Beiles A. and **Nevo E.** 1996. Biodiversity and microhabitat distribution of earthworms at "Evolution Canyon", a Mediterranean microsite, Mount Carmel, Israel. Isr. J. Zool. 42: 449-454.

Heth G., **Nevo E.** and Todrank J. 1996. Seasonal changes in urinary odors and in responses to them by blind subterranean mole rats. Physiol. Behav. 60: 963-968.

Heth G., Beauchamp G., **Nevo E.** and Yamazaki K. 1996. Species, population and individual specific odors in urine of mole rats (*Spalax ehrenbergi*) detected by laboratory rats. Chemoecology 7: 107-111.

Ben-Shlomo R., **Nevo E.**, Ritte U., Steinlechner S. and Klante G. 1996. 6-sulphatoxymelatonin secretion in different locomotor activity types of the blind mole rat *Spalax ehrenbergi* J. Pineal Res. 21: 243-250.

Palamar-Mordvintseva G.M., Wasser S.P. and **Nevo E.** 1996. On the flora of Zygnematales (Conjugatophyceae) of Israel. Algologia 6: 401-406 (in Russian): In English: Int. J. Algae 1999, 1: 52-58.

Mann M.D., Rehkamper G., Reinke H., Frahm H.D., Necker R. and **Nevo E.** 1997. Size of somatosensory cortex and of somatosensory thalamic nuclei of the naturally blind mole rat, *Spalax ehrenbergi.* J. Brain Res. 38: 47-59.

Handley L.L., Robinson D.R., Forster B.P., Ellis R.P., Scrimgeour C.M., Gordon D.C., **Nevo E.** and Raven J.A. 1997. Shoot delta ^{15}N correlates with genotype and salt stress in barley. Planta 201: 100-102.

Krugman T., Levy O., Snape J.W., Rubin B., Korol A.B. and **Nevo E.** 1997. Comparative RFLP mapping of the chlorotoluron resistance gene (Su1) in cultivated wheat (*Triticum aestivum*) and wild wheat (*Triticum dicoccoides*). Teor. Appl. Genet. 94: 46-51.

Widmer H.P., Hoppeler H., **Nevo E.**, Taylor C.R. and Weibel E.W. 1997. Working underground: Respiratory adaptations in the blind mole rat. Proc. Natl. Acad. Sci. USA 94: 2062-2067.

Baum B.R., **Nevo E.**, Johnson D.A. and Beiles A. 1997. Genetic diversity in wild barley (*Hordeum spontaneum* Koch) in the Near East: a molecular analysis using Random Amplified Polymorphic DNA (RAPD). Genet. Resourc. Crop Evol. 44: 147-157.

Pavlíček T., Csuzdi C. and **Nevo E.** 1997. The first recorded earthworms from the Negev and Sinai Deserts. Isr. J. Zool. 43: 1-3.

Nevo E., Kirzhner V.M., Beiles A. and Korol A.B. 1997. Selection versus random drift: Long-term polymorphism persistence in small populations (evidence and modelling). Phil. Tran. R. Soc. Lond. B 352: 381-389.

Couch L., Blaustein L., Duszynski D.W., Shenbrodt G. and **Nevo E.** 1997. A new coccidian from *Acomys cahirinus* Desmarest, 1819, from Evolution Canyon, Lower Nahal Oren, Mount Carmel, Israel. J. Parasitol. 83: 276-279.

Nevo E., Apelbaum-Elkaher I., Garty J. and Beiles A. 1997. Natural selection causes microscale allozyme diversity in wild barley and a lichen at "Evolution Canyon" Mt. Carmel, Israel. Heredity 78: 373-382.

Wasser S.P., Duckman I. and Lewinsohn D. 1997. Catalogue of Cultures (Higher Basidiomycota). **Nevo E.** (ed.). Peledfus Publishing House, Haifa, 50 pp.

Klauer G., Burda H. and **Nevo E.** 1997. Adaptive differentiations of the skin of the head in a subterranean rodent, *Spalax ehrenbergi.* J. of Morph. 233: 53-66.

Wasser S.P. and Weis A.L. 1997. Medicinal mushrooms (*Ganoderma lucidum* (Leyss.: Fr.) Karst., Reishi Mushroom. **Nevo E.** (ed.). Peledfus Pub. House, Haifa, 39 pp.

Tsarenko P.M., Stupina V.V., Mordvintseva G.M., Wasser S.P. and **Nevo E.** 1997. Chlorophyta – Checklist of Continental Species from Israel. **Nevo E.** and Wasser S.P. (eds.). Peledfus Publ. House, Haifa, 150 pp.

Pakniyat H., Powell W., Baird E., Handley L.L., Robinson D., Sorimgeour C.M., **Nevo E.**, Hackett C.A., Caligari P.D.S. and Forster B.P. 1997.

AFLP variation in wild barley (*Hordeum spontaneum* C. Koch) with reference to salt tolerance and associated ecogeography. Genome 40: 332-341.

Pavliček T., Chikatunov V., Lopatin I. and **Nevo E.** 1997. New records of leaf beetles from Israel. Phytoparasit. 25: 337-338.

Kato K., Mori Y., Beiles A. and **Nevo E.** 1997. Geographical variation in heading traits in wild emmer wheat, *Triticum dicoccoides*. I. Variation in vernalization response and ecological differentiation. Theor. Appl. Genet. 95: 546-552.

Handley L.L., Robinson D.R., Scrimgeour C.M., Gordon D., Forster B.P., Ellis R.P. and **Nevo E.** 1997. Correlating molecular markers with physiological expression in *Hordeum*, a developing approach using stable isotopes. New Phythol. 137: 159-163.

Forster B.P., Russell J.R., Ellis R.P., Handley L.L., Robinson D., Hackett C.A., **Nevo E.**, Waugh R., Gordon D.C., Keith R. and Powell W. 1997. Locating genotypes and genes for abiotic stress tolerance in barley: a strategy using maps, markers and the wild species. New Phytol. 137: 141-147.

Owuor E.D., Fahima T., Beiles A., Korol A.B. and **Nevo E.** 1997. Population genetics response to microsite ecological stress in wild barley *Hordeum spontaneum*. Mol. Ecol. 6: 1177-1187.

Chikatunov V., Lillig M., Pavliček T., Blaustein L. and **Nevo E.** 1997. Biodiversity of insects at a microsite "Evolution Canyon", Nahal Oren, Mt. Carmel, Israel. Coleoptera: Tenebrionidae. J. Arid Envir. 37: 367-377.

Wasser S.P. and Weis A.L. 1997. Medicinal Mushrooms (*Lentinus edodes* (Berk.) Sing.). Shiitake Mushroom. **Nevo E.** (ed.). Peledfus Publ. House, Haifa, 95 pp.

Ivanitskaya E., Coskun Y. and **Nevo E.** 1997. Banded karyotypes of mole rats (*Spalax*, Spalacidae, Rodentia) from Turkey: a comparative analysis. J. Zool. Syst. Evol. Res. 35: 171-177.

Nevo E. 1997. Evolution in action across phylogeny caused by microclimatic stresses at "Evolution Canyon". Theor. Pop. Biol. 52: 231-243.

Vinogradova O.N., Kovalenko O.V., Wasser S.P, **Nevo E.**, Kislova O.A. and Belikova O.A. 1997. Biodiversity of Cyanophya in Israel. Preliminary studies at "Evolution Canyon", Lower Nahal Oren, Mt. Carmel Natural Preserve. Hydrobiol. J. 33: 109-123.

Weinberg H.Sh., **Nevo E.**, Korol A.B., Fahima T., Rennert G. and Shapiro S. 1997. Molecular changes in the offspring of liquidators who emigrated to Israel from the Chernobyl disaster area. Envir. Health Perspectives 105: 1479-1481.

Ellanskaya I.A., Volz P.A., **Nevo E.**, Wasser S.P. and Sokolova E. V. 1997. Soil micromycetes of "Evolution Canyon", Lower Nahal Oren, Mt. Carmel National Park, Israel. Microbios 92: 19-33.

Frahm H.D., Rehkamper G. and **Nevo E.** 1997. Brain structure volumes in the mole rat, *Spalax ehrenbergi* (Spalacidae, Rodentia) in comparison to the rat and subterrestrial insectivores. J. Brain Res. 38:209-222.

Nevo E., Rashkovetsky E., Pavlicek T. and Korol A.B. 1998. A complex adaptive syndrome in *Drosophila* caused by microclimatic contrasts. Heredity 80:9-16.

Fahima T., Roder M.S., Grama A. and **Nevo E.** 1998. Microsatellite DNA polymorphism divergence in *Triticum dicoccoides* accessions highly resistant to yellow rust. Theor. Appl. Genet. 96: 187-195.

Somersalo S., Makela P., Rajala A., **Nevo E.** and Peltonen-Sainio P. 1998. Morpho-physiological traits characterizing environmental adaptation of *Avena barbata*. Euphytica 99: 213-220.

Nevo E., Baum B., Beiles A. and Johnson D.A. 1998. Ecological correlates of RAPD DNA diversity of wild barley, *Hordeum spontaneum*, in the Fertile Crescent. Genet. Resour. and Crop Evol. 45: 151-159.

Korol A.B., Ronin Y.I., **Nevo E.** and Hayes P. 1998. Multi-interval mapping of correlated trait complexes. Heredity 80: 273-284.

Kirzhner V.M., Lembrikov B.I., Korol A.B. and **Nevo E.** 1998. Supercycles, strange attractors and chaos in a standard model of population genetics. Physica A 249: 565-570.

Nevo E. 1998. Evolution of a visual system for life without light: Optimization via tinkering in blind mole rats. In: *Principles of Animal Design. The Optimization and Symmorphosis Debate*. Weibel E. R., Taylor C.R. and Bolis C. (eds.). Cambridge University Press. pp. 288-298.

Korol A.B., Ronin Y.I. and **Nevo E.** 1998. Approximated analysis of QTL-environment interaction with no limits on the number of environments. Genetics 148: 2015-2028.

Volz P.A., Rosenzweig N., Blackburn R.B., Wasser S.P. and **Nevo E.** 1998. Cobalt 60 radiation and growth of eleven species of micro-fungi from "Evolution Canyon", Lower Nahal Oren, Israel. Microbios 91: 191-201.

Lamb B.C., Saleem M., Scott W., Thapa N. and **Nevo E.** 1998. Inherited and environmentally-induced differences in mutation frequencies between wild strains of *Sordaria fimicola* from "Evolution Canyon". Genetics 149: 87-99.

Kirzhner V. M., Korol A. B. and **Nevo E.** 1998. Complex limiting behaviour of multilocus genetic systems in cyclical environments. J. Theor. Biol. 190: 215-225.

Kato K., Tanizoe C., Beiles A. and **Nevo E.** 1998. Geographical variation in heading traits in wild emmer wheat, *Triticum dicoccoides*. II. Variation in heading date and adaptation to diverse eco-geographical conditions. Hereditas 128: 33-39.

Broza M., Blondheim S. and **Nevo E.** 1998. New species of mole crickets of the *Gryllotalpa gryllotalpa* group (Orthoptera: Gryllotalpidae) from

Israel, based on morphology, song recordings, chromosomes and cuticular hydrocarbons, with comments on the distribution of the group; in Europe and the Mediterranean region. Syst. Entomol. 23: 125-135.

Yang H., **Nevo E.** and Tashian E. 1998. Unexpected expression of carbonic anhydrase I and selenium-binding protein as the only major non-heme proteins in erythrocytes of the subterranean mole rats (*Spalax ehrenbergi*). Fed. of European Bioch. Soc. 430: 343-347.

Nevo E. 1998. Molecular evolution and ecological stress at global, regional and local scales: The Israeli perspective. J. Exp. Zool. 282: 95-119.

Buchalo A.S., **Nevo E.** and Wasser S.P., Oren A. and Molitoris H.P. 1998. Fungal life in the extremely hypersaline water of the Dead Sea: First records. Proc. R. Soc. Lond. B 265: 1461-1465.

Nevo E. 1998. Genetic diversity in wild cereals: Regional and local studies and their bearing on conservation *ex-situ* and *in-situ*. Genet. Resour. and Crop Evol. 45: 355-370.

Pavliček T., Chikatunov V., Kravchenko V., Dorchin J. and **Nevo E.** 1998. *Xylotrechus stebbingi* Gahan: a new species for Israeli beetle fauna (Coleoptera: Cerambycidae). Mitt. Internat. Entomol. Ver. 23: 73-74.

Li Y.C., Krugman T., Fahima T., Beiles A. and **Nevo E.** 1998. Genetic diversity of alcohol dehydrogenase 3 in wild barley population at the "Evolution Canyon" microsite, Nahal Oren, Mt. Carmel, Israel. Barley Genet. Newsl. Barley Genet. Newslett. 28: 58-60.

Tobler I., Herrmann M., **Nevo E.**, Cooper H.M. and Achermann P. 1998. Rest-activity rhythm of the blind mole rats *Spalax ehrenbergi* under different lighting conditions. Behav. Brain Res. 96: 173-183.

Vinogradova O.N., Kovalenko O.V., Wasser S.P., **Nevo E.** and Weinstein-Evron M. 1998. Species diversity gradient to darkness stress in blue-green algae (Cyanobacteria): A microscale test in a prehistoric cave, Mount Carmel, Israel. Isr. J. Pl. Sci. 46: 229-238.

Korol A.B., Kirzhner V.M. and **Nevo E.** 1998. Dynamics of recombination modifiers caused by cyclical selection: Interaction of forced and auto-oscillations. Genet. Res., 72: 135-147.

Nevo E., Travleev A., Belova N.A., Tsatskin A., Pavliček T., Kulik A.F., Tsvetkova N.N. and Yemshanov D.C. 1998. Edaphic interslope and valley bottom differences at "Evolution Canyon", Lower Nahal Oren, Mount Carmel, Israel. Catena 33: 241-254.

Nevo E. 1998. Book review of: *Environmental Stress, Adaptation and Evolution.* Bijlsma R. and Loeschcke V. (eds.). Birkhauser Verlag, Basel. Heredity 81: 591-597.

Ivanitskaya E. and **Nevo E.** 1998. Cytogenetics of mole rats of the *Spalax ehrenberg* superspecies from Jordan (Spalacidae, Rodentia). Z. Saugetierk. 63: 336-346.

David-Gray Z.K., Janssen J.W.H., **Nevo E.** and Foster R.G. 1998. Light detection in a "blind" mammal. Nature Neurosci. 12-98: 655-656.

Bolshakova M.A., Musatenko L.I., **Nevo E.**, Sytnik K.M., Pavliček T. and Beharav A. 1998. Environment abiotic factors effect the phytohormonal level in leaves of some woody plants. Ukr. Bot. J. 55: 578-584.

Nevo E., Filippucci G.M., Pavliček T., Gorlova O., Shenbrot G., Ivanit-skaya E. and Beiles A. 1998. Genotypic and phenotypic divergence of rodents (*Acomys cahirinus* and *Apodemus mystacinus*) at "Evolution Canyon": Micro- and macroscale parallelism. Acta Theriol. 5: 9-34.

Hisoriev H., Wasser S. P., **Nevo E.** and Stupina V.V. 1999. Additions to the flora of Euglenophyta of Israel. Int. J. Algae 1: 63-75.

Ronin Y.I., Korol A.B. and **Nevo E.** 1999. Single- and multiple-trait mapping analysis of linked QTL: some asymptotic analytical approxi-mations. Genetics 151: 387-396.

Pavliček T., Chikatunov V., Kravchenko V., Zahradnik P. and **Nevo E.** 1999. New records of deathwatch beetles (Anobiidae) from Israel. Zool. in the Middle East 17: 77-78.

Csuzdi Cs., Pavliček T. and **E. Nevo.** 1999. A new earthworm species *Dendrobaena rothschildae* sp. nov. from Israel, and comments on the distribution of *Dendrobaena* species in the Levant (Oligochaeta: Lumbri-cidae). Opusc. Zool. Budapest 31: 1-5.

Lioupis A., **Nevo E.** and Wallis M. 1999. Cloning and characterisation of the gene encoding mole rat (*Spalax ehrenbergi*) growth hormone. J. Mol. Endocrinol. 22: 29-36.

Lu Z., Krugman T., Neumann P.M. and **Nevo E.** 1999. Physiological charac-terization of drought tolerance in wild barley (*Hordeum spontaneum*) from the Judean Desert. Barley Genet. Newslett. 29: 36-40.

Fahima T., Sun G.L., Beharav A., Krugman T., Beiles A. and **Nevo E.** 1999. RAPD polymorphism of wild emmer wheat population, *Triticum dicoccoides*, in Israel. Theor. Appl. Genet. 98: 434-447.

Auffray J.C., Renaud S., Alibert P. and **Nevo E.** 1999. Developmental stability and adaptive radiation in the *Spalax ehrenbergi* superspecies in the Near East. J. Evol. Biol. 12: 207-221.

Palamar-Mordvintseva G.M., Wasser S.P. and **Nevo E.** 1999. On the flora of Zygnematales (Conjugatophyceae) of Israel. Int. J. Algae 1: 52-58.

Nevo E., Fragman O., Dafni A. and Beiles A. 1999. Biodiversity and interslope divergence of vascular plants caused by microclimatic differences at "Evolution Canyon", Lower Nahal Oren, Mount Carmel, Israel. Isr. J. Pl. Sci. 47: 49-59.

Peng J.H., Fahima T., Roder M.S., Li Y.C., Dahan A., Grama A., Ronin Y.I., Korol A.B. and **Nevo E.** 1999. Microsatellite tagging of stripe-rust resistance gene *YrH52* derived from wild emmer wheat, *Triticum* *dicoccoides,* and suggestive negative crossover interference on chromosome 1B. Theor. Appl. Genet. 98: 862-872.

Li Y.C., Fahima T., Beiles A., Korol A.B. and **Nevo E.** 1999. Microclimatic stress and adaptive DNA differentiation in wild emmer wheat, *Triticum dicoccoides.* Theor. Appl. Genet. 99: 873-883.

Pavliček T. and **Nevo E.** 1999. Book review of: The Yponomeutinae (Lepidoptera) of the world exclusive of the Americas. In: Vestnik Zoologii (Journal of Schmalhausen Institute of Zoology) 33(1-2): 112.

Kirzhner V.M., Korol A.B. and **Nevo E.** 1999. Abundant multilocus polymorphisms caused by genetic interaction between species on *trait-for-trait* basis. J. Theor. Biol. 198: 61-70.

Avivi A., Resnik M.B., Joel A., **Nevo E.** and Levi A.P. 1999. Adaptive hypoxic tolerance in the subterranean mole rat *Spalax ehrenbergi*: the role of vascular endothelial growth factor. FEBS Let. 452: 133-140.

Burda H., Begall S., Grutjen O., Scharff A., **Nevo E.**, Beiles A., Cerveny J. and Prucha K. 1999. How to eat a carrot? Convergence in the feeding behavior of subterranean rodents. Naturwissenschaften 86: 325-327.

Andreyuk E.I., Antipchuk A.F., Iutinskaja G.A., Fahima T. and **Nevo E.** 1999. Species characteristics of the soil bacterial communities at "Evolution Canyon", Nahal Oren, Mount Carmel Natural Preserve, Israel. J. Microbiology, Kiev, 61: 3-9.

Chikatunov V., Pavliček T. and **Nevo E.** 1999. Coleoptera of "Evolution Canyon", Lower Nahal Oren, Mount Carmel, Israel. I. Families: Buprestidae, Carabidae, Cerambycidae, Glaphyridae, Hybosoridae, Hydrophilidae, Lucanidae, Scarabaeidae, Tenebrionidae and Trogidae. Pensoft Series Faunistica No. 14; Sofia-Moscow, 174 pp.

Blaustein L., Garb J.E., Shebitz D. and **E. Nevo**. 1999. Microclimate, developmental plasticity and community structure in artificial temporary pools. Hydrobiologia 392: 187-196.

Sicard D., Woo S.S., Arroyo-Garcia R., Ochoa O., Nguyen D., Korol A.B., **Nevo E.** and Michelmore R. 1999. Molecular diversity at the major cluster of disease resistance genes in cultivated and wild *Lactuca* spp. TAG 99: 405-418.

Rottenberg A., Zohary D. and **Nevo E.** 1999. Patterns of isozyme diversity and vegetative reproduction of willows in Israel. Int. J. Pl. Sci. 160: 561-566.

Vicient C. M., Suoniemi A., Anamthawat-Jonsson K., Tanskanen J., Beharav A., **Nevo E.** and Schulman A.H. 1999. Retrotransposon *BARE*-1 and its role in genome evolution in the genus *Hordeum*. The Plant Cell 11: 1769-1784.

Nevo E., Beiles A. and Spradling Th. 1999. Molecular evolution of cytochrome *b* of mole rats, *Spalax ehrenbergi* superspecies in Israel. J. Mol. Evol. 49: 215-226.

Owuor E.D., Fahima T., Beharav A., Korol A.B. and **Nevo E.** 1999. RAPD divergence caused by microsite edaphic selection in wild barley. Genetica 105: 177-192.

Nevo E. 1999. *Mosaic Evolution of Subterranean Mammals: Regression, Progression and Global Convergence.* Oxford University Press, Oxford, 413 pp.

David-Gray, Zoe K., Cooper H.M., Jannsen J.W.H., **Nevo E.** and Foster R.G. 1999. Spectral tuning of a circadian photopigment in a subterranean "blind" mammals (*Spalax ehrenbergi*). FEBS Letters 461: 343-347.

Harry M., Rashkovetsky E., Pavliček T., Baker S., Derzhavets E.M., Capy P., Cariou M.L., Lachaise D., Asada N. and **Nevo E.** 1999. Fine-scale biodiversity of Drosophilidae in "Evolution Canyon" at the Lower Nahal Oren Microsite, Israel. Biologica 54: 685-705.

Csuzdi Cs., Pavliček T. and **Nevo E.** 1999. A new earthworm species, *Dendrobaena rothschildae* sp. n. from Israel, and comments on the distribution of *Dendrobaena* species in the Levant (Oligochaeta: Lumbricidae). Opusc. Zool. Budapest 31: 25-32.

Turpeinen T., Kulmala J. and **Nevo E.** 1999. Genome size variation in *Hordeum spontaneum* populations. Genome 42: 1094-1099.

Chague V., Fahima T., Dahan A., Sun G.L., Korol A.B., Ronin Y., Grama A., Roder M.S. and **Nevo E.** 1999. Isolation of microsatellite and RAPD markers flanking the *Yr15* gene of wheat using NILs and bulked segregant analysis. Genome 42: 1050-1056.

Heth G., Todrank J. and **Nevo E.** 2000. Do *Spalax ehrenbergi* blind mole rats use food odours in searching for and selecting food? Ethol. Ecol. and Evol. 12: 75-82.

Peng J.H., Fahima T., Roder M.S., Li Y.C., Grama A. and **Nevo E.** 2000. Microsatellite high-density mapping of the stripe rust resistance gene *YrH52* region on chromosome 1B and evaluation of its marker-assisted selection in the F2 generation in wild emmer wheat. New Phytol. 146: 141-154.

Vinogradova O.N., Kovalenko O.V., Wasser S.P., **Nevo E.**, Weinstein-Evron M. 2000. Cyanoprocaryotes/Cyanobacteria of Jamal Cave, Nahal Me'arot Nature Reserve, Mount Carmel, Israel. Int. J. Algae 2: 41-50.

Marhold S., Beiles A., Burda H. and **Nevo E.** 2000. Spontaneous directional preference in a subterranean rodent, the blind mole rat, *Spalax ehrenbergi.* Folia Zool. 49: 7-18.

Rottenberg A., **Nevo E.** and Zohary D. 2000. Genetic variability in sexually dimorphic and monomorphic populations of *Populus euphratica* (Salicaceae). Can J. For. Res. 30: 482-486.

Nevo E., Ivanitskaya E., Filippucci M.G. and Beiles A. 2000. Speciation and adaptive radiation of subterranean mole rats, *Spalax ehrenbergi* super-species, in Jordan. Biol. J. Linn. Soc. 69: 263-281.

Li Y.C., Roder M.S., Fahima T., Kirzhner V.M., Beiles A., Korol A.B. and **Nevo E.** 2000. Natural selection causing microsatellite divergence in wild emmer wheat at the ecologically variable microsite at Ammiad, Israel. Theor. Appl. Genet. 100: 985-999.

Nevo E., Beiles A., Korol A.B., Ronin Y.I., Pavliček T. and Hamilton W.D. 2000. Extraordinary multilocus genetic organization in mole crickets, gryllotalpidae. Evolution 54: 586-605.

Wasser S.P., **Nevo E.**, Sokolov D., Timor-Tismenetsky M. and Reshetnikov S. 2000. The regulation of dietary supplements from medicinal mushrooms. In: Van Griensven (ed.), Science and Cultivation of Edible Fungi, Balkema, Rotterdam, pp. 789-801.

Li Y.C., Fahima T., Korol A.B., Peng J., Roder M.S., Kirzhner V.M., Beiles A. and **Nevo E.** 2000. Microsatellite diversity correlated with ecological-edaphic and genetic factors in three microsites of wild emmer wheat in North Israel. Mol. Biol. Evol. 17: 851-862.

Kalendar R., Tanskanen J., Immonen S., **Nevo E.** and Alan Schulman H. 2000. Genome evolution of wild barley (*Hordeum spontaneum*) by BARE-1 retrotransposon dynamics in response to sharp microclimatic divergence. Proc. Natl. Acad. Sci. USA 97: 6603-6607.

Chikatunov V., Pavliček T., Lopatin I. and **Nevo E.** 2000. Biodiversity and microclimatic divergence of chrysomelid beetles at "Evolution Canyon", Lower Nahal Oren, Mt. Carmel, Israel. Biol. J. Linn. Soc. 69: 139-152.

Nevo E., Bolshakova M.A., Martyn G.I., Musatenko L.I., Sytnik K.M., Pavliček T. and Beharav A. 2000. Drought and light anatomical adaptive leaf strategies in three woody species caused by microclimatic selection at "Evolution Canyon", Israel. Isr. J. Pl. Sci. 48: 33-46.

Van Rijn C.P.E., Heersche I., Van Berkel Y.E.M., **Nevo E.**, Lambers H. and Poorter H. 2000. Growth characteristics in *Hordeum spontaneum* populations from different habitats. New Phytol. 146: 471-481.

Bolshoy A. and **Nevo E.** 2000. Ecological genomics of DNA: upstream bending in prokaryotic promoters. Genome Res. 10: 1185-1193.

Buchalo A.S., **Nevo E.**, Wasser S.P., Molitoris H.P., Oren A. and Volz P.A. 2000. Fungi discovered in the Dead Sea. Mycol. Res. 104: 129-133.

Pavliček T., Smooha G. and **Nevo E.** 2000. The niche-width variation hypothesis revisited: Microscale testing of the earthworm *Bimastos syriacus* (Rosa) and multispecies comparison. Zool. Anzeiger 239: 21-26.

Nevo E. 2000. Speciation: Chromosomal Mechanisms. In: *Encyclopedia of Life Sciences*. Macmillan Reference Ltd., London, pp.1-11.

Volkovitsh M., Pavliček T., Chikatunov V. and **Nevo E.** 2000. Species diversity and microsite divergence of insects at "Evolution Canyon", Lower Nahal Oren, Mt. Carmel, Israel (Coleoptera: Buprestidae). Zool. in the Middle East 20: 125-136.

Vinogradova O.N., Poem-Finkel M., **Nevo E.** and Wasser S.P. 2000. Diversity of *Cyanoprocaryota* in Israel. First data on blue-green algae of dry limestones of western Upper Galilee. Intern. J. Algae 2: 27-45.

Vinogradova O.N., Kovalenko O.V., Wasser S.P. and **Nevo E.** 2000. Blue-green algae (Cyanoprocaryota) of Jamal Cave, National Park Mount Carmel, Israel. Algologia 10 (1): 82-90.

Peng J.H., Korol A.B., Fahima T., Roder M.S., Ronin Y.I., Li Y.C. and **Nevo E.** 2000. Molecular genetic maps in wild emmer wheat, *Triticum dicoccoides*: genome-wide coverage, massive negative interference, and putative quasi-linkage. Genome Res. 10: 1509-1531.

Buchalo A.S., **Nevo E.**, Wasser S.P. and Volz P.A. 2000. Newly discovered halophilic fungi in the Dead Sea (Israel). In: Seckbach J. (ed.). *Journey to Diverse Microbial Worlds*. pp. 239-252. Kluwer Academic Publ., Dortrecht, The Netherlands.

Korol A.B., Rashkovetsky E., Iliadi K., Michalak P., Ronin Y.I. and **Nevo E.** 2000. Nonrandom mating in *Drosophila melanogaster* laboratory populations derived from closely adjacent ecologically contrasting slopes at "Evolution Canyon". Proc. Natl. Acad. Sci. USA 97: 12637-12642.

Belyayev A., Raskina O., Korol A.B. and **Nevo E.** 2000. Coevolution of A and B-genomes in allotetraploid *Triticum dicoccoides*. Genome 43: 1021-1026.

Li Y.C., Fahima T., Peng J.H., Roder M.S., Kirzhner V.M., Beiles A., Korol A.B. and **Nevo E.** 2000. Edaphic microsatellite DNA divergence in wild emmer wheat, *Triticum dicoccoides* at a microsite: Tabigha, Israel. Theor. Appl. Genet. 101: 1029-1038.

Lewinsohn D., **Nevo E.**, Hadar Y., Wasser S.P. and Beharav A. 2000. Ecogeographical variation in the *Pleurotus eryngii* complex in Israel. Mycol. Res. 104: 1184-1190.

Wasser S.P., **Nevo E.**, Sokolov D., Reshetnikov S.V. and Timor-Tismenetsky M. 2000. Dietary supplements from medicinal mushrooms: diversity of types and variety of regulations. Int. J. Med. Mushr. 2: 1-19.

Tsarenko P.M., Vinogradova O.N., Stupina V.V., Wasser S.P. and **Nevo E.** 2000. Diversity of algae in the continental part of Israel. Int. J. Algae 2: 20-39.

Reshetnikov S.V., Wasser S.P., **Nevo E.** and Duckman I. 2000. Medicinal value of the genus *Tremella* Pers. (Heterobasidiomycetes) (Review). Int. J. Med. Mushr. 2: 169-193.

Nevo E. and Wasser S.P. (eds.) 2000. Cyanoprocaryotes and Algae of Continental Israel. (Biodiversity of Cyanoprocaryotes, Algae and Fungi of Israel). A.R.A. Gantner Verlag, Ruggell/Liechtenstein.

Rashkovetsky E., Iliadi K., **Nevo E.** and Korol A.B. 2000. Fitness related traits in *Drosophila melanogater* subpopulations from the opposite slopes of Nahal Oren canyon. DIS 83: 138-140.

Ivandic V., Hackett C.A., Zhang Z.J., Staub J.E., **Nevo E.**, Thomas W.T.B. and Forster B.P. 2000. Phenotypic responses of wild barley to experimentally imposed water stress. J. Exper. Bot. 51: 2021-2029.

Grishkan I., **Nevo E.**, Wasser S.P. and Pavliček T. 2000. Spatiotemporal distribution of soil microfungi in "Evolution Canyon", Lower Nahal Oren, Mount Carmel, Israel. Isr. J. Pl. Sci. 48: 297-308.

Ellanskaya I.A., **Nevo E.**, Wasser S.P., Volz P.A. and Sokolova E.V. 2000. Species diversity of soil micromycetes in two contrasting soils at the Tabigha microsite (Israel). Isr. J. Pl. Sci. 48: 309-315.

Tsarenko P.M., Wasser S.P., **Nevo E.** and Krienitz L. 2000. Species of chlorococcales (*Chlorophyta*) new for the flora of Israel. Int. J. Algae 2: 53-63.

Nevo E. 2001. Genetic diversity. In: *Encyclopedia of Biodiversity* 3: 195-213, Academic Press.

Nevo E. 2001. Konstantin Merkuryevich Sytnik: Scientist, Visionary, Architect and Humanist - for his 75th Brithday. Ukr. Bot. J. 58: 383-385.

Avivi A., Joel A. and **Nevo E.** 2001. The lens protein α-B-crystallin of the blind subterranean mole-rat: high homology with sighted mammals. Gene 264: 45-49.

Masyuk N.P., Lilitskaya G.G., Wasser S.P. and **Nevo E.** 2001. Green flagellate algae of Israel. New and rare species. Int. J. Algae 3: 48-61.

Korol A.B., Ronin Y.I., Itskovich A.M., Peng J. and **Nevo E.** 2001. Enhanced efficiency of quantitative trait loci mapping analysis based on multivariate complexes of quantitative traits. Genetics 157: 1789-1803.

Nevo E. 2001. Hamilton W.D. – Evolutionary theorist: Life and vision (1936-2000). Theor. Pop. Biol. 59: 21-25.

Belyayev A., Raskina O. and **Nevo E.** 2001. Chromosomal distribution of reverse transcriptase-containing retroelements in two *Triticeae* species. Chromosome Res. 9: 129-136.

Háva J., Pavliček T., Chikatunov V. and **Nevo E.** 2001. Dermestid beetles in "Evolution Canyon", Lower Nahal Oren, Mt. Carmel, including new records for Israel. Phytoparas. 29: 97-101.

Weinberg H.Sh., Korol A.B., Kitrzhner V.M., Avivi A., Fahima T., **Nevo E.**, Shapiro S., Rennert G., Piatak O., Stepanova E.I. and Skvarskaja E. 2001. Very high mutation rate of offspring of the Chernobyl accident liquidators. Proc. Royal. Soc. Lond. B 268: 1001-1005.

Shmueli O., Gdalyahu A., Sorokina K., **Nevo E.**, Avivi A. and Reiner O. 2001. DCX in PC12 cells: CREB-mediated transcription and neurite outgrowth. Hum. Mol. Genet. 10: 1061-1070.

Nevo E. 2001 Evolution of genome-phenome diversity under environmental stress. Proc. Natl. Acad. Sci. USA 98: 6233-6240.

Li Y.C., Fahima T., Krugman T., Beiles A., Roder M.S., Korol A.B. and **Nevo E.** 2001. Parallel microgeographic patterns of genetic diversity and divergence revealed by allozyme, RAPD, and microsatellites in *Triticum dicoccoides* at Ammiad, Israel. Conservation Genet. 1: 191-207.

Kravchenko V., Hacker H.H. and **Nevo E.** 2001. List of Noctuoidea (Lepidoptera) collected in Israel. In: *Esperiana Buchreihe zur Entomologie* Bd. 8: 459-474.

Wasser S.P., Reshetnikov S.V.G., Solomko E.F., Buchalo A.S. and **Nevo E.** 2001. For higher Basidiomycetes mushrooms grown (as Biomass) in submerged culture. International Publication; Patent Cooperation Treaty, PCT No. WO 01/32003 A1.

Volz P.A., Ellanskaya I.A., Grishkan I., Wasser S.P. and **Nevo E.** 2001. Biodiversity of Cyanoprocaryotes, Algae and Fungi of Israel: Soil Microfungi of Israel; Subramanian C.V. and Wasser S.P. (eds.). A.P.A. Gantner Verlag, Ruggel/Liechtenstein.

Peng J.H., Fahima T., Roder M.S., Huang Q.Y., Dahan A., Li Y.C., Grama A. and **Nevo E.** 2001. High-density molecular map of chromo-some region harboring stripe-rust resistance genes *YrH52* and *Yr15* derived from wild emmer wheat, *Triticum dicoccoides*. Genetica 109: 199-210.

Li Y.C., Krugman T., Fahima T., Beiles A., Korol A.B. and **Nevo E.** 2001. Spatiotemporal allozyme divergence caused by aridity stress in a natural population of wild wheat, *Triticum dicoccoides*, at the Ammiad microsite, Israel. Theor. Appl. Genet. 109: 853-864.

Molitoris H.P., Buchalo A.S., Kurchenko I., **Nevo E.**, Rawal B.S., Wasser S.P. and Oren A. 2001. Physiological diversity of the first filamentous fungi isolated from the hypersaline Dead Sea. In: Aquatic Mycology across the Millennium, Hyde K.D., Ho W.H. and Pointing S.B. (eds.). Fungal Diversity 5: 55-70.

Turpeinen T., Tenhola T., Manninen O., **Nevo E.** and Nissila E. 2001. Microsatellite diversity associated with ecological factors in *Hordeum spontaneum* populations in Israel. Mol. Ecol. 10: 1577-1591.

Nilima S. and **Nevo E.** 2001. Cryopreservation of wild isolates of *Nostoc linckia* (Roth) Born. et Flah. Int. J. Algae 3: 46-51.

Kis-Papo T., Grishkan I., Oren A., Wasser S.P. and **Nevo E.** 2001. Spatio-temporal diversity of filamentous fungi in the hypersaline Dead Sea. Mycol. Res. 105: 749-756.

Nevo E., Ivanitskaya E. and Beiles A. 2001. Adaptive radiation of blind subterranean mole rats: naming and revisiting the four sibling species of the *Spalax ehrenbergi* superspecies in Israel: *Spalax galili* (*2n*=52), *S. golani* (*2n*=54), *S. carmeli* (*2n*=58) and *S. judaei* (*2n*=60). Bachkhuys Publishers, Leiden, The Netherlands.

Nevo E., Korol A.B., Beiles A. and Fahima T. 2001. Evolution of Wild Emmer and Wheat Improvement. Population Genetics, Genetic Resources, and Genome Organization of Wheat's Progenitor, *Triticum dicoccoides*. Springer-Verlag, Berlin. 364 pp.

Satish N., Krugman T., Vinogradova O.N., **Nevo E.** and Kashi Y. 2001. Genome evolution of the cyanobacterium *Nostoc linckia* under sharp microclimatic divergence at "Evolution Canyon", Israel. Microb. Ecol. 42: 306-316.

Janssen J.W.H., Bovee-Geurts P.H.M., Peeters Z.P.A., Bowmaker J.K., Cooper H.M., David-Gray Z.K., **Nevo E.** and DeGrip W.J. 2001. A fully functional Rod visual pigment in a *Blind* mammal: A case for adaptive functional reorganization? J. Biol. Chem. 275: 38674-38679.

Nevo E. 2001. Genetic resources of wild emmer, *Triticum dicoccoides*, for wheat improvement in the third millennium. Isr. J. Pl. Sci. 49: 77-91.

Lewinsohn D., **Nevo E.**, Wasser S.P., Hadar Y. and Beharav A. 2001. Genetic diversity in populations of *Pleurotus eryngii* complex in Israel. Mycol. Res. 105: 941-951.

Belyayev A., Raskina O. and **Nevo E.** 2001. Evolutionary dynamics and chromosomal distribution of repetitive sequences chromosomes of *Aegilops speltoides* revealed by genomic in situ hybridization. Heredity 86: 738-742.

Belyayev A., Raskina O. and **Nevo E.** 2001. Detection of alien chromosomes from S-genome species in the addition/substitution lines of bread wheat and visualization of A-, B- and D-genomes by GISH. Hereditas 135: 119-122.

Michalak P., Minkov I., Helin A., Lerman D.N., Bettencourt B.R., Feder M.E., Korol A.B. and **Nevo E.** 2001. Genetic evidence for adaptation-driven incipient speciation of *Drosophila melanogaster* along a microclimatic contrast in "Evolution Canyon", Israel. Proc. Natl. Acad. Sci. USA 98: 13195-13200.

Ryndin A., Kirzhner V., **Nevo E.** and Korol A.B. 2001. Polymorphism maintenance in populations with mixed random mating and apomixis subjected to stabilizing and cyclical selection. J. Theor. Biol. 212: 169-181.

Iliadi K., Iliadi N., Rshkovetsky E., **Nevo E.** and Korol A.B. 2001. Sexual and reproductive behaviour of *Drosophila melanogaster* from a microclimatically interslope differentiated population of "Evolution Canyon" (Mount Carmel, Israel). Proc. R. Soc. Lond. B 268: 2365-2374.

Krugman T., Satish N., Vinogradova O.N., Beharav A., Kashi Y. and **Nevo E.** 2001. Genome diversity in the cyanobacterium *Nostoc linckia* at "Evolution Canyon" Israel, revealed by inter-HIP1 size polymorphisms. Evol. Ecol. Res. 3: 899-915.

Avivi A., Albrecht U., Oster H., Joel A., Beiles A. and **Nevo E.** 2001. Biological clock in total darkness: The Clock/MOP3 circadian system of the blind subterranean mole rat. Proc. Natl. Acad. Sci. USA 98: 13751-13756.

Melamed-Frank M., Terzic A., Carrasco A.J., **Nevo E.**, Avivi A. and Levy A.P. 2001. Reciprocal regulation of expression of pore-forming KATP channel genes by hypoxia. Mol. Cell Biochem. 225: 145-150.

Nevo E. 2001. Molecular evolution and environmental-stress. In: Xue, G. et al. (eds.). Gene Families: Studies of DNA, RNA, Enzymes and Proteins. World Scientific Publ. Co. pp. 73-87.

Mikhailyuk T.I., Tsarenko P.M., **Nevo E.** and Wasser S.P. 2001. Additions to the study of aerophytic eukaryotic algae of Israel. Int. J. Algae 3: 19-39.

Vinogradova O.N., Kovalenko O.V., Wasser S.P. and **Nevo E.** 2001. New representatives of Chroococcophyceae (Cyanophyta) from the Mount Carmel National Park and Dead Sea area, Israel. Int. J. Algae 3: 95-99.

Meir F., Fragman O. and **Nevo E.** 2001. Biodiversity and interslope divergence of vascular plants caused by sharp microclimatic stresses at "Evolution Canyon" II, Lower Nahal Keziv, Upper Galilee, Israel. Isr. J. Pl. Sci. 49: 285-295.

Van Rijn C.P.E., Vanhala T.K., **Nevo E.**, Stam P., van Eeuwijk F.A. and Poorter H. 2001. Association of AFLP markers with growth-related traits in *Hordeum spontaneum*. In: Cynthia van Rijn Ph.D. Thesis: A physiological and genetic analysis of growth characteristics in *Hordeum spontaneum*. Universiteit Utrecht, Faculteit Biologie, Utrecht, pp. 75-93.

Saleem M., Lamb B.C. and **Nevo E.** 2001. Inherited differences in crossing-over and gene conversion frequencies between wild strains of *Sordaria fimicola* from "Evolution Canyon". Genetics 159: 1573-1593.

Gershenson Z., Kravchenko V., Pavliček T. and **Nevo E.** 2001. New records of Yponomeutoid moths in Israel (Lepidoptera: Yponomeutidae, Plutellidae, Argyresthiidae). Mitt. Internat. Entomol. Ver. 26: 147-153.

Fahima T., Roder M.S., Wendehake K., Kirzhner V.M. and **Nevo E.** 2002. Microsatellite polymorphism in natural populations of wild emmer wheat, *Triticum dicoccoides*, in Israel. Theor. Appl. Genet. 104: 17-29.

Dvornyk V., Vinogradova O. and **Nevo E.** 2002. Long-term microclimatic stress causes rapid adaptive radiation of *kaiABC* clock gene family in a cyanobacterium, *Nostoc linckia*, from "Evolution Canyons" I and II, Israel. Proc. Natl. Acad. Sci. USA 99: 2082-2087.

Gupta P.K., Sharma P.K., Balyan H.S., Roy J.K., Sharma S., Beharav A. and **Nevo E.** 2002. Polymorphism at rDNA loci in barley and its relation with climatic variables. Theor. Appl. Genet. 104: 473-481.

Raskina O., Belyayev A. and **Nevo E.** 2002. Repetitive DNAs of wild emmer wheat (*Triticum dicoccoides*) and their relation to S-genome species: molecular cytogenetic analysis. Genome 45: 391-401.

Liu Z., Sun Q., Ni Z., **Nevo E.** and Yang T. 2002. Molecular characterization of a novel powdery mildew resistance gene *Pm30* in wheat originating from wild emmer. Euphytica 123: 21-29.

Pavliček T., Vivanti S., Fishelson L. and **Nevo E.** 2002. Biodiversity and microsite divergence of insects at "Evolution Canyon", Nahal Oren, Mt. Carmel, Israel II. Orthoptera: Acrididae. J. Entomol. Res. Soc. 4: 25-39.

Ivandic V., Hackett C.A., **Nevo E.**, Keith R., Thomas W.T.B. and Forster B.P. 2002. Analysis of simple sequence repeats (SSRs) in wild barley from the Fertile Crescent: associations with ecology, geography and flowering time. Plant Mol. Biol. 48: 511-527.

Kanazin V., Talbert H., See D., DeCamp P., **Nevo E.** and Blake T. 2002. Discovery and assay of single-nucleotide polymorphisms in barley (*Hordeum vulgare*). Plant Mol. Biol. 48: 529-537.

Kovalenko O.V., **Nevo E.**, Wasser S.P., Tsarenko P.M. and Bleich S.A. 2002. New data on the diversity of *Cyanoprocaryota* in inland waters of Israel. Int. J. Algae 4: 41-50.

Lewinsohn D., Wasser S.P., Reshetnikov S.V., Hadar Y. and **Nevo E.** 2002. The *Pleurotus eryngii* species-complex in Israel: distribution and morphological description of a new taxon. Mycotaxon 81: 51-67.

Wang W., Brunet F.G., **Nevo E.** and Long M. 2002. Origin of *sphinx*, a young chimeric RNA gene in *Drosophila melanogaster*. Proc. Natl. Acad. Sci. USA 99: 4448-4453.

Wasser S.P. 2002. Biodiversity of cyanoprocaryotes, algae and fungi of Israel. Family Agaricaceae (Fr.) Cohen (Basidiomycetes) of Israel mycobiota. I. Tribe Agariceae Pat. **Nevo E.** and Volz P.A. (eds.). A.R.A. Gantner Verlag, Ruggell/Liechtenstein.

Hough R.B., Avivi A., Davis J., Joel A., **Nevo E.** and Piatigorsky J. 2002. Adaptive evolution of small heat shock protein/αB-crystallin promoter activity of the blind subterranean mole rat, *Spalax ehrenbergi*. Proc. Natl. Acad. Sci. USA 99: 8145-8150.

Heth G., Todrank J., Begall S., Koch R., Zilbiger Y., **Nevo E.**,. Braude S.H and Burda H. 2002. Odours underground: subterranean rodents may not forage "blindly". Behav. Ecol. Sociobiol. 52: 53-58.

Iliadi K.G., Iliadi N.N., Rashkovetsky E.L., Girin S.V., **Nevo E.** and Korol A.B. 2002. Sexual differences for emigration behavior in natural populations of *Drosophila melanogaster*. Behav. Genet. 32: 173-180.

Tsarenko P.M., Wasser S.P. and **Nevo E.** 2002. *Chlorophyta* of the continental part of Israel: a brief analysis and new data. Int. J. Algae 4: 75-92.

Zhang F., Gutterman Y., Krugman T., Fahima T. and **Nevo E.** 2002. Differences in primary dormancy and seedling revival ability for some *Hordeum spontaneum* genotypes of Israel. Isr. J. Pl. Sci. 50: 271-276.

Holovachov O., Bostrom S., Susulovsky A. and **Nevo E.** 2002. Description of *Cervidellus capricornis* sp. n. (Nematoda: Cephalobidae) from Israel. Nematol. Medit. 29: 223-230.

Przybos E., **Nevo E.** and Pavliček T. 2002. *Paramecium tredecaurelia* of the *Paramecium aurelia* Complex in Israel. Folia Biologica (Krakow) 50: 221-222.

Cernuda-Cernuda R., DeGrip W.J., Cooper H.M., **Nevo E.**, and Garcia-Fernandez J.M. 2002.The retina of *Spalax ehrenbergi*. Novel histological features supportive of a modified photosensory role. Investigative Ophthalmology & Visual Sci., 43: 2374-2383.

Finkel M., Chikatunov V. and **Nevo E.** 2002. Biodiversity and interslope divergence of Tenebrionidae species caused by sharp microclimatic stresses at "Evolution Canyon" II, Lower Nahal Keziv, western Upper Galilee, Israel. J. Mediter. Ecol. 3: 29-36.

Shnyukova E.I., **Nevo E.**, Wasser S.P. and Zolotareva E.K. 2002. Effect of different sources of nitrogen on production of exopolysaccharides from *Nostoc linckia* (Roth) Born. et Flah. (Cyanophyta). Int. J. Algae 4: 86-98.

Finkel M., Chikatunov V. and **Nevo E.** 2002. Coleoptera of "Evolution Canyon" II: Lower Nahal Keziv, Western Upper Galilee, Israel. Pensoft, Sofia-Moscow.

Hannibal J., Hindersson P., **Nevo E.** and Fahrenkrug J. 2002. The circadian photopigment melanopsin is expressed in the blind subterranean mole rat, *Spalax*. NeuroReport 13: 1411-1414.

Li Y.C.,. Roder M.S, Fahima T., Kirzhner V., Beiles A., Korol A.B. and **Nevo E.** 2002. Climatic effects on microsatellite diversity in wild emmer wheat *(Triticum dicoccoides)* at the Yehudiyya microsite, Israel. Heredity 89: 127-132.

Eitam A., Pavliček T., Blaustein L. and **Nevo E.** 2002. Effects of inter- and intraslope contrasts at "Evolution Canyon", Israel, on a gall aphid and its natural enemies. J. Mediter. Ecol. 3: 47-52.

Avivi A., Oster H., Joel A., Beiles A., Albrecht U. and **Nevo E.** 2002. Circadian genes in a blind subterranean mammal II: Conservation and uniqueness of the three *Period* homologs in the blind subterranean mole rat, *Spalax ehrenbergi* superspecies. Proc. Natl. Acad. Sci. USA 99: 11718-11723.

Kirzhner V.M, Korol A.B., Bolshoy A. and **Nevo E.** 2002. Compositional spectrum revealing patterns for genomic sequence characterization and comparison. Physica A. 312: 447-457.

Przybos E., Pavliček T. and **Nevo E.** 2002. Distribution of species of the *Paramecium aurelia* complex in Israel. Acta Protozool. 41: 293-295.

Kravchenko V., Pavliček T., Chikatunov V. and **Nevo E.** 2002. Seasonal and spatial distribution of butterflies (Lepidoptera-Rhopalocera) in "Evolution Canyon", Lower Nahal Oren, Mt. Carmel, Israel. Ecol. Mediter. 28: 98-112.

Huang Q., Beharav A., Li Y., Kirzhner V. and **Nevo E.** 2002. Mosaic micro-ecological differential stress causes adaptive microsatellite divergence in wild barley, *Hordeum spontaneum*, at Neve Yaar, Israel. Genome 45: 1216-1229.

David-Gray Z.K., Bellingham J., Munoz M., Avivi A., **Nevo E.** and Foster R.G. 2002. Adaptive loss of ultraviolet-sensitive/violet-sensitive (UVS/VS) cone opsin in the blind mole rat (*Spalax ehrenbergi*). Europ. J. Neurosci. 16: 1186-1194.

Oster H., Avivi A., Joel A., Albrecht U. and **Nevo E.** 2002. A switch from diurnal to noctural activity in *Spalax ehrenbergi* is accompanied by an uncoupling of light input and the circadian clock. Curr. Biol. 12: 1919-1922.

Liviero L., Maestri E., Gulli M., **Nevo E.** and Marmirioli N. 2002. Ecogeographic adaptation and genetic variation in wild barley, application of molecular markers targeted to environmentally regulated genes. Genet. Resour. and Crop Evol. 49: 133-144.

Wasser S.P., Didukh M.Ya., Amazonas M.A.L. de A., **Nevo E.**, Stamets P. and da Eira A.F. 2002. Is a widely cultivated culinary-medicinal mushroom Royal Sun *Agaricus* (the Heimematsutake mushroom) indeed *Agaricus blazei* Murrill? Int. J. Med. Mushr. 4: 267-290.

Guoxiong C., Krugman T., Fahima T., Korol A.B. and **Nevo E.** 2002. Comparative study of morphological and physiological traits related to drought resistance between xeric and mesic *Hordeum spontaneum* lines in Israel. Barley Genet. Newslet. 32: 22-33.

Li Y.C., Korol A.B., Fahima T., Beiles A. and **Nevo E.** 2002. Microsatellites: genomic distribution, putative functions and mutational mechanisms: a review. Mol. Ecol. 11: 2453-2465.

Gershenson Z., Pavliček T., Chikatunov V. and **Nevo E.** 2002. New records of Yponomeutoid moths (Lepidoptera, Yponomeutidae, Plutellidae) from Israel. Vestnik Zoologii, Kiev 36: 77-80.

Kovalenko O.V., **Nevo E.** and Wasser S.P. 2002. New for Israel taxa of blue-green algae (*Cyanoprocaryota*). Int. J. Algae 4: 98-110.

Peleg O., Brunak S., Trifonov E.N., **Nevo E.** and Bolshoy A. 2002. RNA secondary structure and sequence conservation in C1 region of human immunodeficiency virus type 1 *env* gene. AIDS Research and Human Retroviruses 18: 867-878.

Baek H.J., Beharav A. and **Nevo E.** 2003. Ecological-Genomic diversity of microsatellites in wild barley, *Hordeum spontaneum*, populations in Jordan. Theor. Appl. Genet. 106: 397-410.

Grishkan I.B., Korol A.B., **Nevo E.** and Wasser S.P. 2003. Ecological stress and sex evolution in soil microfungi. Proc. J. Roy. Soc. Lond., B. 270: 13-18.

Kis-Papo T., Oren A., Wasser S.P. and **Nevo E.** 2003. Survival of filamentous fungi in hypersaline Dead Sea water. Microb. Ecol. 45: 183-190.

Didukh M., Wasser S.P., **Nevo E.** and Ur Y. 2003. New species of *Leucocoprineae* and *Lepioteae* (Basidiomycetes, Agaricales, s. l.) in Israel. Docum. Mycol. 32: 39-58.

Peng J.H., Ronin Y.I., Fahima T., Roder M., Li Y., **Nevo E.** and Korol A.B. 2003. Domestication quantitative trait loci in *Triticum dicoccoides*, the progenitor of wheat. Proc. Natl. Acad. Sci. USA 100: 2489-2494.

Dvornyk V., Vinogradova O. and **Nevo E.** 2003. Origin and evolution of circadian clock genes in prokaryotes. Proc. Natl. Acad. Sci. USA 100: 2495-2500.

Temina M., Wasser S.P. and **Nevo E.** 2003. A contribution to the species diversity of lichens in Israel. Flora Mediter. 12: 285-298.

Buerstmayr H., Stierschneider M., Steiner B., Lemmens M., Griesser M., **Nevo E.** and Fahima T. 2003. Variation for resistance to head blight caused by *Fusarium graminearum* in wild emmer (*Triticum dicoccoides*) originating from Israel. Euphytica 130: 17-23.

Li Y.C., Fahima T., Roder M.S., Kirzhner V.M., Beiles A., Korol A.B. and **Nevo E.** 2003. Genetic effects on microsatellite diversity in wild emmer wheat (*Triticum dicoccoides*) at the Yehudiyya microsite, Israel. Heredity 90: 150-156.

Dvornyk V. and **Nevo E.** 2003. Genetic polymorphism of cyanobacteria under permanent natural stress: A lesson from the "Evolution Canyons". Res. Microbiol. 154: 79-84.

Owuor E.D., Beharav A., Fahima T., Kirzhner V.M., Korol A.B. and **Nevo E.** 2003. Microscale ecological stress causes RAPD molecular selection in wild barley, Neve Yaar microsite, Israel. Genet. Resour. and Crop Evol. 50: 213-224.

Roguin A., Avivi A., Nitechi S., Rubinstein I., Levy N.S., Abassi Z.A., Resnick M.B., Lache O., Melamed-Frank M., Joel A., Hoffman A., **Nevo E.** and Levy A.P. 2003. Restoration of blood flow by using continuous perimuscular infiltration of plasmid DNA encoding subterranean mole rat *Spalax ehrenbergi* VEGF. Proc. Natl. Acad. Sci. USA 100: 4644-4648.

Nagornaya S.S., Babich T.V., Podgorsky V.S., Beharav A., **Nevo E.** and Wasser S.P. 2003. Yeast interslope divergence in soils and plants of "Evolution Canyon", Lower Nahal Oren, Mount Carmel, Israel. Isr. J. Pl. Sci. 51: 55-57.

Pavliček T., Sharon D., Kravchenko V., Saaroni H. and **Nevo E.** 2003. Microclimatic interslope differences underlying biodiversity contrasts in "Evolution Canyon", Mt. Carmel, Israel. Isr. J. Earth Sci. 52: 1-9.

Grishkan I., **Nevo E.**, Wasser S.P. and Beharav A. 2003. Adaptive spatio-temporal distribution of soil microfungi in "Evolution Canyon" II, Lower Nahal Keziv, western Upper Galillee, Israel. Biol. J. Linn. Soc. Lond. 78: 527-539.

Belyayev A., Raskina O. and **Nevo E.** 2003. Evolutionary dynamics of repetitive DNA fraction in two wild Triticeae species. In: Plant Genome: Biodiversity and Evolution; Vol. 1, part A – Phanerogams. Science Publishers, Inc., USA, pp. 37-56.

Turpeinen T., Vanhala V., **Nevo E.** and Nissila E. 2003. AFLP genetic polymorphism in wild barley (*Hordeum spontaneum*) populations in Israel. Theor. Appl. Genet. 106: 1333-1339.

Xie C., Sun Q., Ni Z., Yang T., **Nevo E.** and Fahima T. 2003. Chromosomal location of a *Triticum dicoccoides*-derived powdery mildew resistance gene in common wheat by using microsatellite markers. Theor. Appl. Genet. 106: 341-345.

Zerova M.D., Seryogina L.Y., Melika G., Pavliček T. and **Nevo E.** 2003. New genus and new species of cynipid gall inducing wasp (Hymenoptera: Cynipidae) and new species of Chalcid wasps (Hymenoptera: Chalcidoidea) from Israel. J. Ent. Res. Soc. 5: 35-40.

List of Contributors

Francisco J. Ayala
Department of Ecology and Evolutionary Biology
University of California
Irvine, CA 92697-2525, USA
Tel: (949) 8248 293
Fax: (949) 8242 474
E-mail: fjayala@uci.edu

Evgeniy S. Balakirev
Department of Ecology and Evolutionary Biology
University of California
Irvine, CA 92697-2525, USA
Academy of Ecology, Marine Biology, and Biotechnology
Far Eastern State University
Vladivostok 690600, Russia
Tel: (949) 8248 293
Fax: (949) 8242 474

Walter Bodmer
Cancer and Immunogenetics Laboratory, Cancer Research UK
Weatherall Institute of Molecular Medicine, John Radcliffe Hospital
Headington, Oxford OX3 9DS, England
Tel: (44) 186 5279 405
Fax: (44) 186 5279 437
E-mail: walter.bodmer@hertford.oxford.ac.uk

Yohay Carmel
Faculty of Agricultural Engineering
Technion, Israel Institute of Technology,
Haifa 31905, Israel
Tel: (972) 4 8292 620
Fax: (972) 4 8221 529
E-mail: yohay@tx.technion.ac.il

James F. Crow
Department of Genetics
University of Wisconsin
Madison, 445 Henry Mall, WI 53706-1574, USA
Tel: (608) 2634 438
Fax: (608) 2622 976
E-mail: jfcrow@facstaff.wisc.edu

Yoav Eitan
Department of Genetics
The Hebrew University of Jerusalem
Givat Ram, Jerusalem 91904, Israel
Tel: (972) 2 6585 104
Fax: (972) 2 6586 975

Vladimir M. Frenkel
Institute of Evolution
University of Haifa, Mount Carmel, Haifa 31905, Israel
Tel: (972) 4 8288 040
Fax: (972) 4 8246 554

Andrew J. Gentles
Stanford University
Department of Mathematics
Stanford, CA 94305-2125, USA
Tel: (650) 7256 284
Fax: (650) 7254 066
E-mail: andrew@grinch.stanford.edu

Morris Goodman
Department of Anatomy and Cell Biology
Wayne State University, School of Medicine
540 E. Canfield Ave.
Detroit, MI 48201, USA
Tel: (313) 5771 338
Fax: (313) 5771 080, 5773 125
E-mail: mgoodwayne@aol.com

Samuel Karlin
Stanford University
Department of Mathematics
Stanford, CA 94305-2125, USA
Tel: (650) 7256 284
Fax: (650) 7254 066
E-mail: sam@grendel.stanford.edu

Valery M. Kirzhner
Institute of Evolution
University of Haifa, Mount Carmel, Haifa 31905, Israel
Tel: (972) 4 8288 040
Fax: (972) 4 8246 554
E-mail: valery@esti.haifa.ac.il

Abraham B. Korol
Institute of Evolution
University of Haifa, Mount Carmel, Haifa 31905, Israel
Tel: (972) 4 8240 449
Fax: (972) 4 8246 554
E-mail: korol@esti.haifa.ac.il

Valentin A. Krassilov
Institute of Evolution
University of Haifa, Mount Carmel, Haifa 31905, Israel
Tel: (972) 4 8249 799
Fax: (972) 4 8246 554
E-mail: krassilo@research.haifa.ac.il

Charu G. Kumar
The Department of Animal Sciences
The W. M. Keck Center for Comparative and Functional Genomics
The University of Illinois
Urbana-Champaign, Urbana, IL 61801, USA
Tel: (217) 3335 998
Fax: (217) 2445 617

Noam Lahav
Institute of Evolution
University of Haifa, Mount Carmel, Haifa 31905, Israel
Faculty of Agricultural, Food and Environmental Quality Sciences
The Hebrew University of Jerusalem
P.O. Box 12, Rehovot 76100, Israel
Tel: (972) 2 5343 047
Fax: (972) 2 9475 181
E-mail: lahav@agri.huji.ac.il

Joshua H. Larson
The Department of Animal Sciences
The W. M. Keck Center for Comparative and Functional Genomics
The University of Illinois at Urbana-Champaign
Urbana, IL 61801, USA
Tel: (217) 3335 998
Fax: (217) 2445 617

Harris A. Lewin
The Department of Animal Sciences
The W. M. Keck Center for Comparative and Functional Genomics
The University of Illinois at Urbana-Champaign

210 Edward R. Madigan Laboratory
1201 West Gregory Drive
Urbana, IL 61801, USA
Tel: (217) 3335 998
Fax: (217) 2445 617
E-mail: h-lewin@uiuc.edu

Robert M. May
Department of Zoology
University of Oxford
South Park Road
Oxford OXI 3PS, England
Tel: (018) 6 5271 170
Fax: (018) 6 5310 447
E-mail: rmay@vax.oxford.ac.uk

Ernst Mayr
Museum of Comparative Zoology
Harvard University
26 Oxford St.
Cambridge, MA 02138, USA

Zev Naveh
Faculty of Agricultural Engineering
Technion, Israel Institute of Technology
Haifa 31905, Israel
Tel: (972) 4 8292 620
Fax: (972) 4 8221 529
E-mail: znave@techunix.technion.ac.il

Yuval Ne'eman
School of Physics and Astronomy
Raymond and Beverly Sackler Faculty of Exact Sciences
Tel Aviv University
Tel Aviv 69978, Israel
Tel. (972) 3 640 9579, 640 9580
Fax (972) 3 642 4264
E-mail: matildae@tauex.tau.ac.il

Aharon Oren
Division of Microbial and Molecular Ecology
The Institute of Life Science
The Hebrew University of Jerusalem
Jerusalem 91904, Israel

Tel: (972) 2 658 4951
Fax: (972) 2 652 8008
E-mail: orena@cc.huji.ac.il

Mark Pagel
School of Animal and Microbial Sciences
University of Reading
Reading RG6 6AJ, England
Tel: (44) 186 5279 405
Fax: (44) 186 5279 437

Peter A. Parsons
Department of Genetics, Faculty of Science, Technology and Engineering
La Trobe University
Bundoora, Victoria, Australia Vic. 3083
PO Box 906, UNLEY, SA 5061, Australia
Fax: (618) 8 2725 557
E-mail: pparsons@senet.com.au

Karl Skorecki
Rappaport Family Institute for the Medical Sciences
P.O.Box 9697,
Haifa 31096, Israel
Tel: (972) 4 829 5365
Fax: (972) 4 855 2296
E-mail: skorecki@techunix.technion.ac.il

Moshe Soller
Department of Genetics
The Hebrew University of Jerusalem
Givat Ram, Jerusalem 91904, Israel
Tel: (972) 2 6585 104
Fax: (972) 2 6586 975
E-mail: soller@vms.huji.ac.il

Alan R. Templeton
Department of Biology
Washington University
St. Louis, MO 63130-4899, USA
Tel: (314) 935 6868
Fax: (314) 935 4432
E-mail: temple_a@biology.wustl.edu

Edward N. Trifonov
Genome Diversity Center, Institute of Evolution
University of Haifa, Mount Carmel, Haifa 31905, Israel
Tel: (972) 4 8288 096
Fax: (972) 4 8246 554
E-mail: trifonov@research.haifa.ac.il

Maty Tzukerman
Rappaport Family Institute for the Medical Sciences
P.O. Box 9697
Haifa 31096, Israel
Tel: (972) 4 829 5365
Fax: (972) 4 855 2296

Solomon P. Wasser
Institute of Evolution
University of Haifa, Mount Carmel, 31905 Haifa, Israel
Tel: (972) 4 8249 218
Fax: (972) 4 8288 197
E-mail: spwasser@research.haifa.ac.il
M.G. Kholodny Institute of Botany
National Academy of Sciences of Ukraine,
Tereshchenkivksa St. 2, 01601 Kiev-MSP-1, Ukraine

Derek E. Wildman
Department of Anatomy and Cell Biology
Center for Molecular Medicine and Genetics
Wayne State University
School of Medicine
540 E. Canfield Ave.
Detroit, MI 48201, USA
Tel: (313) 5773 418
Fax: (313) 5775 218
E-mail: dwildman@genetics.wayne.edu

Ivan V. Zmitrovich
V.L. Komarov Botanical Institute
Russian Academy of Sciences
2, Popova St.
St. Petersburg 197376, Russia
Tel: 7 812 5246 714
E-mail: iva@ IZ6284.spb.edu

Acknowledgments

This volume presents the contributions of 32 scholars who have collaborated in writing eighteen articles to celebrate the 75th birthday of Professor Eviatar (Eibi) Nevo. My warmest thanks to each and every one of my colleagues for their contribution and commitment to the success of this interdisciplinary volume.

It is also my pleasant duty to extend grateful thanks to the President of Haifa University, Professor Yehuda Hayut, Professor Aharon Ben-Zeev, the Rector, and to the Dean of Research, Professor Moshe Zeidner for their generous financial support. This support has greatly contributed to the celebration of the occasion by facilitating the publication of the *Festschrift* and its distribution to contributors and colleagues.

We appreciate the excellent editing of the book by Mrs. Robin Permut.

A special thanks to Michael Margulis for his prodigious efforts in preparing the indices, taxonomic lists, and the overall camera-ready manuscript of this book.

Solomon P. Wasser

Part One

Evolution of Life and Evolutionary Theory

EVOLUTION AND LIFE'S EMERGENCE UNDER PREBIOTIC CONDITIONS AND IN A TEST-TUBE

Noam Lahav

Institute of Evolution, University of Haifa, Mt. Carmel, Haifa 31905, and Faculty of Agricultural, Food and Environmental Quality Sciences, The Hebrew University of Jerusalem, P.O. Box 12, Rehovot 76100, Israel

Abstract: Any attempt to study the transition from inanimate to animate matter must include a definition of a living entity and its central attributes. The definition of life used in the present work is based on the idea that the organizational principle, characterizing both past and present living entities, is a feedback system in an adequate geochemical environment and in the context of an appropriate chemical composition. By extending this principle to the molecular level of the transition from inanimate to animate matter, the origin of life is viewed as the establishment of the first "living" entity (or population of "living" entities) characterized by feedback loops and compartmentation and amenable to Darwinian evolution. Thus, the emergence of compartmentalized, catalyzed, and template-directed feedback loops in an inanimate, fluctuating, and rhythmical environment is identical with the emergence of living entities. It is proposed that, as such, it is also amenable to artificial synthesis in a test-tube.

1. INTRODUCTION

The transition from inanimate to animate matter is the outcome of the spontaneous emergence of coupled reaction cycles and their evolution processes in the framework of an adequate biogeochemical system. Because

S.P. Wasser (ed.), Evolutionary Theory and Processes: Modern Horizons,
Papers in Honour of Eviatar Nevo
N. Lahav. Evolution and Life's Emergence Under Prebiotic Conditions and in a Test-Tube, 3-15.
© 2004 *Kluwer Academic Publishers.*

of the complexity of living entities and their multitude of features and expressions, any attempt to study the emergence of living entities from inorganic matter must include a definition of the living entity and its central attributes. Moreover, focusing on just one attribute of life might be misleading, because such a single feature needs to be studied in the context of its environment and the complex relationships between different attributes of living entities. For instance, the generation of an RNA strand such as obtained by Ferris (2002) or Reader and Joyce (2002), cannot be considered a process of living entity emergence without taking into account additional central attributes of life and their interactions in a given biogeochemical system. The complexity of the transition from inanimate to animate matter and the lack of a paradigm for this process are to be blamed for the lack of a scientific breakthrough in our understanding of the origin of life. Indeed, the era of laboratory research focusing on simulations of the emergence and evolution of living entities in a test-tube has hardly been discussed.

Implied in a study of the origin of life is the principle of biological continuity. Central to the latter guideline is the assumption that the fundamental features of living entities, i.e., chemical composition, replication, translation, and compartmentation can be studied both theoretically and experimentally at the molecular level and in simplified systems. Obviously, scenarios for the emergence of living entities are model-dependent. One such theory, probably the most detailed model published to date, is the theory of Emergence of Template-Informational and Functionality (ETIF) (Lahav et al., 2001; see below). The current working hypothesis is that this theory provides the basis for an attempt to synthesize living entities in a test-tube.

Our definition of life is based on the idea (Lahav et al., 2001; see also Korzeniewsky, 2001) that the organizational principle behind all living entities is a feedback-loop system in an adequate geochemical environment, irrespective of its location on the phylogenetic tree and in the context of an adequate chemical composition. Implied in this approach is the plausible assumption that it can also be applied to primordial living entities. By extending this approach to the molecular level of the transition from inanimate to animate matter, the origin of life is viewed as the establishment of the first "living" entity (or populations of living entities) characterized by a feedback loop and compartmentation (Lahav et al., 2001) and amenable to Darwinian evolution. In the absence of experimental data, it seems adequate to assume, at first, that the chemical constituents of those feedback systems are not necessarily unique and limited to specific chemical groups such as those known from biology. Relevant examples for this assumption are the RNA-world theory (see Joyce, 2002), the ETIF theory (Lahav et al., 2001), and the Clay World theory (Cairn-Smith, 1982).

1.1   Goals

The goals of the present contribution are: (i) to discuss the mechanisms of life's emergence according to the ETIF and RNA-world theories; (ii) to explore the molecular mechanisms of Darwinian evolution during the transition from inanimate to animate matter, and (iii) to pave the road for the next generation of experimental work designed to synthesize "living" entities in a test-tube.

2.   DISCUSSION

2.1   Prebiotic biogeochemical feedback systems, their formation and evolution

It is generally accepted that certain rhythmical environments, i.e., small puddles undergoing hydration-dehydration cycles, characterized many primordial geochemical environments, just as they characterize certain areas on our planet today. Such an environment has the potential to serve as the arena for life's emergence. The solid phase of this arena includes one or more minerals capable of catalyzing organic reactions under certain conditions (see Lahav et al., 1978; Ferris, 2002). Such environments can support the formation of populations of chemical entities undergoing reaction cycles. These are featured, for instance, by processes of energy input and dissipation, during which reaction cycles are established and destroyed, controlled by the dynamics of the environment and the rate constants of the main reactions. For instance, in a fluctuating system containing water, glycine, and clay minerals (Lahav et al., 1978; Plankenstern et al., 2002) each cycle is characterized by the following (schematic) stages and central reactions:

Wetting and cooling stage: (i) dissolution of glycine and oligo-glycine, and (ii) their redistribution between solid and aqueous phases.

Heating stage: (i) dehydration and (ii) solute redistribution between the solid and aqueous phases.

Hot-dry stage: (i) catalyzed condensation of glycine on the dehydrated clay surfaces to form oligo-glycine (ii) oligo-glycine decomposition.

By the same token, wetting-drying processes where the organic molecule is a building block of proto-RNA can take place in a similar neighboring puddle (see Ferris, 2002). Obviously, such a single cycle is still far from the beginning of life.

2.2   Emergence of primordial living entities according to the ETIF theory

The two central functions of this system are based on templates (proto-RNAs) and catalysts (proto-RNAs and/or peptides). Furthermore, catalysis and template reactions, as well as the tendency of the fluctuating environment to form sequences of coupled reaction cycles, eventually bring about the next stage of ordered sequences of reactions, discussed below.

Short proto-RNAs and short peptides accumulate when the production rates of these chemical entities are higher than their decomposition rates. Gradually, a small arsenal of short proto-RNAs and peptides capable of carrying out catalyzed template-directed synthesis is established. Feedback systems are initiated, thus forming a population of simple living entities. This is the beginning of the transition from inanimate to animate matter and at the same time – the beginning of Darwinian evolution.

The tendency to form coupled reactions with ordered cycles reflects the built-in cyclic nature of the organic reactions in the fluctuating environment, and it is characterized by molecular populations of similar properties. Gradually, the molecules of these systems are shaped by their interaction with their micro-environment, where some of these feedback systems have the potential to affect the survivability of the feedback entities into which they were incorporated. Those feedback systems, which increase the survivability of existing molecular systems become integral parts of the Darwinian wherewithal and selection process at the molecular level.

The latter feature, by which Darwinian selection takes place, is conceived as an outcome of the dynamics of the populations of the chemical entities under study. These chemical entities undergo cyclic processes, where some of the new feedback systems form "libraries" of similar entities resonating to the rhythm of the geochemical system and competing for "food". The selection is the outcome of this competition, where "survival of the fittest" applies to a population of entities (Schwabe, 1985; Woese, 1998).

Thus, according to the present biogeochemical model, organic molecules and their condensation products, i.e., proto-nucleotides and their proto-RNAs, or amino acids and their peptides are formed spontaneously in the fluctuating environment, but these syntheses are not template and sequence-directed at first. Gradually, some of the above oligomers reach a size which enables them to participate in template-directed reactions. This critical size is suggested to be of the order of six or seven building blocks for peptides as inferred from various considerations (Trifonov et al., 2001; see also Lahav et al., 2001). The critical size of proto-RNAs is assumed to be somewhat bigger than that of peptides (Lahav and Nir, 1997; Nir and Lahav, 1997). When this stage is reached, a new era begins where feedback loops are formed and either accumulated or destroyed. The stable systems are capable of undergoing template-and-sequence-directed (TSD) reactions and the first living entities proceed with their incessant historical march according to the Darwinian survival-of-the-fittest model.

The ETIF theory assumes a primordial, fluctuating, aqueous (wet/cool-dry/hot) environment – the arena of the first manifestation of living entities and emergence of Darwinian evolution. During the wet period, this fluctuating system provides the rhythmic features of hydrogen-bond formation and dissociation reactions (between complementary proto-RNA strands as well as their building blocks) in the low and high temperature ranges, respectively. During the hot-dry period, this system provides the surface area including adsorbed organic templates (Lahav and White, 1980; Lahav, 1999) to which the soluble organic molecules are attached. This process is followed by catalyzed thermal condensation and oligomer formation on the mineral surfaces. During the next stage of the wetting-drying cycle, a certain portion of the bonded oligomers remain attached to the mineral surfaces and, as such, are hardly involved in the subsequent stages. However, some of the bonded oligomers dissolve and return to the liquid phase, thus continuing the sequence of reactions (including condensation) involved in the formation of feedback loop systems. The latter organizational features also define the boundary of a "living" entity, namely: a living entity includes the compartmentalized molecules that are essential for the successful functions of a feedback loop system. The formation of the first heritable TSD feedback-loop systems thus heralds the appearance of the first living entities. Assuming that mistakes and variations in the information transfer processes were rather common in this era of primordial life (unavoidable in fact), the emergence of life processes coincides with the emergence of the Darwinian era and is characterized by features such as functions, selection, information transfer, memory, and death. Moreover, the sophistication of primordial feedback loops gradually increases and a process of "takeover" starts where the direct involvement and effects of the environment are gradually replaced by organic processes related to life in the framework of Darwinian evolution.

2.3   Death

The primordial feedback loop of a TSD system is the first manifestation of information transfer (including "memory" and other related parameters) in the system under study; its decomposition signifies another aspect of life as well, namely, death at the molecular-organizational level. In contrast to the emergence of life, which is the formation of feedback loop organizations, death may be described as the disruption of the molecular organization of a feedback loop system (which is defined as a living entity). Thus, both life and death are defined in molecular-organizational terms. Such a definition helps approaching some apparent paradoxes, i.e., life's definition of a mule, or a male rabbit. An additional fundamental aspect of life is the irreversibility of the transition between "life" and "death". These issues will not be further discussed here.

2.4   Small RNAs

The size of proto-RNAs and peptides is determined essentially by two opposing tendencies, namely, molecular stability on the one hand, and efficiency of their corresponding reactions on the other. Oligomers that are too small are less effective in the reactions under study, whereas the synthesis of oligomers that are too big takes time, thus exposing them to decomposition processes. Hence, in an experiment aimed at studying the role of the size ranges of the central templates and catalysts, the efficiency of the two kinds of oligomers needs to be determined experimentally.

In view of the recently discovered importance of small RNAs (RNAi) in a variety of reactions (see Couzin, 2002), it would be interesting to study their relevance in the origin-of-life processes. The members of this family of RNAs range in length from 21 to 29 nucleotides. This size range would seem to be too big, in prebiotic terms for the synthesis of RNA oligomers, according to the procedure used, for instance, by Ferris (2002). However, it is conceivable that small proto-RNAs, i.e., below this range, were synthesized during the prebiotic era and gradually became essential in various control reactions. The relevance of the small RNAs in the transition between inanimate and animate matter is not yet known.

2.5   Emergence of "living" entities according to the RNA-world theory

According to the RNA-world theory, proto-RNAs can serve as templates and catalysts. Assuming feedback systems based on RNAs and also including co-factors, metals and phosphorus, allow for the possibility that ancient living forms based on RNA preceded DNA-protein life somewhere in the early history of life's emergence (for a recent review see Joyce, 2002). Indeed, it was recently shown (Moore and Steitz, 2002) that the region on the ribosome where peptide bond formation takes place by peptidyl transferase activity contains only RNA in the vicinity of the active site. This seems to suggest that RNA catalysis preceded that by protein-based enzymes.

It should be noted that implied in this suggestion is the presence of amino acids in the vicinity of the primordial catalytic process under consideration. If the geochemical system for this process is a fluctuating environment of the kind proposed by the ETIF theory, then it is conceivable to expect condensation of amino acids to form non-directed peptides before the emergence of template-directed syntheses and feedback systems. Furthermore, a template-directed catalyst, i.e., a peptide, is not in and of itself considered a living system according to the definition of the ETIF theory (Lahav et al., 2001). One possible pathway to the stage of living entities in this system is to couple the translation reaction with replication of proto-RNAs. Central aspects of this system have been explored by computer modeling (Nir and Lahav, 1997; Lahav and Nir, 1997). Still, there are several open questions regarding the relationships between the RNA-world theory on the one hand and the feedback systems under study on the other (see below).

2.6   Reaction cycles

Chemical reaction cycles are the result of interactions between specific compounds under specific environmental conditions. The ETIF theory focuses on two kinds of central molecules, which are assumed to have been central during the origin-of-life processes as they are in modern biology. These are proto-RNAs serving as templates on the one hand and proto-RNAs and/or peptides serving as catalysts on the other. The fluctuating environment is rather common now as it is assumed to have been at the time of life's emergence.

It was recently argued (Orgel, 2000) that the spontaneous establishment of the reverse citric acid cycle with its 11 steps, under prebiotic condition, is highly unlikely. The reverse citric acid cycle is considered a feature of an early primordial metabolic cycle and primordial life (Wachtershauser, 1992; see also Lahav, 1999). Since the spontaneous primordial establishment of this cycle during life's emergence is unlikely, then the very early stages of the origin of life should be further explored in an attempt to find still simpler candidates for the first feedback system. These candidates would encompass the molecular wherewithal of information transfer and metabolic cycle emergence. One such suggested candidate is the coupled feedback loop system of the TSD moiety according to the ETIF hypothesis (Lahav et al., 2001; see below). More experimental data are needed to evaluate the simplicity of this proposed candidate.

Proto-RNAs are presumed to have been central catalysts at this early stage of life's emergence. However, the easy prebiotic synthesis of non-directed short peptides makes their presence in the prebiotic environment under consideration rather likely during the transition from inanimate to animate matter. Could these short peptides have been involved in primordial catalytic functions of the early feedback systems?

It is noted that according to the ETIF theory, a system capable of peptide-bond formation is not considered a living system. At least one additional function is needed for such a system to form a feedback system. For instance, the primordial system suggested as a candidate for the most ancient feedback loop is characterized by translation and replication. The catalysts of this system are presumably proto-RNAs, as well as peptides, and other feedback loops can be added to these sub-cycles (Figures 1 and 2 in Lahav et al., 2001). Thus, it is speculated that the first feedback system consisted of two coupled sub-cycles, namely, replication and translation. Until more data are available, both proto-RNAs and peptides are assumed to have been catalysts (see below), and proto-RNAs are assumed to have functioned as templates. Note that proto-RNA moieties could have also served as primordial cofactors, thus enriching the catalytic arsenal of the emerging entities.

Is it conceivable that short catalytic peptides were involved in the transition from inanimate to animate matter functioning, for instance - as either scaffolds or catalysts, in the primordial feedback loop systems, and in the formation of the first living entities?

2.7   Transition from RNA-world to RNA-peptide world

In speculating about the transition from the RNA-world to the RNA-peptide world, one has to employ the assumptions of simplicity and biological continuity. As shown above, the priority of proto-RNAs in the primordial peptidyl transferase activity suggests that peptides were also available in the same environment. Simplicity suggests that during the transition from inanimate to animate matter, the size of the central oligomers, i.e., proto-RNAs and peptides, involved in this evolutionary stage, was very small.

Moreover, the above suggestion regarding the primordial peptidyl transferase implies the presence of amino acids in the prebiotic environment under consideration. The synthesis of amino acids under prebiotic conditions is likely under the geochemical conditions of many prebiotic systems. Peptide formation is also possible, for instance, by thermal condensation catalyzed by minerals (Lahav et al., 1978) in a fluctuating system, which enhances the formation of reaction cycles. Some of the proto-RNAs and peptides thus formed are expected to act as catalysts in some of the reactions involved in the above cycles. At first, these reactions were not template-directed. From time to time, however, oligomers capable of functioning as templates (proto-mRNAs, proto-tRNAs), as well as proto-ribozymes and primordial peptide catalysts, were synthesized in the fluctuating system under study. In some of the reaction cycles, one or more products were capable of enhancing the synthesis of one or more of the reactions taking place in this environment. Gradually, environments relatively rich in template-directed reactions, with their cycles, became more and more dominant in the compartmentalized micro-environments under study. This process culminated in the formation of one or more feedback loop systems and the simultaneous emergence of auto-catalysis and Darwinian evolution, i.e., the emergence of living entities.

2.8   Biogeochemical simplicity of the ETIF model

The simplicity of the ETIF theory is exemplified by the following details regarding the main system constituents (Lahav, 1999; Lahav et al., 2001):

Organic molecules. Amino acids (most predominantly glycine and alanine; see Trifonov, 1999) and the building blocks of proto-RNAs (Ferris, 2002) were formed spontaneously in the prebiotic environment. Condensation of such small building blocks to form oligomers, i.e., proto-RNAs and peptides, took place in this prebiotic environment. In view of the low reactivity of glycine and alanine, it is suggested that other amino acids were also involved in the formation of the primordial peptides, albeit in lower concentrations.

Minimal types of TSD molecules. The minimal number of types of TSD molecular species is 2, i.e., proto-RNAs and primordial peptides.

Minimal number of functions. The minimal number of functions performed by the above TSD oligomers is 3, namely, loading of the amino acid on proto-tRNA, replication of proto-RNAs, and TSD synthesis of catalysts.

2.9   Proto-peptidyl transferase and feedback systems

The extant peptidyl transferase center is the most ancient functional center of the ribosome and is characterized by the absence of proteins (Moore and Steitz, 2002). The detailed chemical mechanism of the emergence of this ancient catalyst in a prebiotic environment is not known. A general characterization includes the guidelines of simplicity and biochemical continuity, as well as the presumed geochemical environment. Thus, based on experimental data collected by many research groups during the last five decades of study, it is assumed that the transition from inanimate to animate matter is characterized by a small number of amino acids, predominantly (though not exclusively) glycine and alanine (see Trifonov, 1999; Trifonov et al., 2001) and their short peptides as well as proto-nucleic acids (Ferris, 2002). The presence of means of compartmentation such as lipid vesicles is assumed to be an old feature of these environments (Deamer et al., 1994), and thus the fluctuating environment imposes its rhythm on the main reactions of this geochemical system.

Some of these chemical entities, i.e., some of the short proto-RNAs and peptides, are capable of forming reaction cycles dictated by the relevant environmental rhythm (see above). These cycles are formed spontaneously and include template-directed processes. However, as long as they do not merge into unified systems encompassing replication and translation processes, they are not considered "alive" according to our earlier definition. If some of these entities are capable of forming feedback systems, a new era is initiated, one in which living entities emerge, and is in line with our definition of life. According to the ETIF theory, the central information-transfer reactions of the first living entities were replication and translation (by means of short proto-RNA templates) on the one hand and short catalysts (proto-ribozymes and catalytic peptides) on the other.

The emergence of the ability of proto-peptidyl transferase to form TSD peptide bonds, and thus also short peptides, is perceived as a gradual process since it is illogical to expect the formation of the appropriate structures and functions in a very small number of steps. Back-extrapolating from the modern peptidyl transferase to the earliest proto-peptidyl transferase, assumes that the latter entity was a proto-RNA molecule capable of forming a hairpin (Moller and Janssen, 1990; Lahav, 1997).

The sources of the amino acids used in this process can be either prebiotic synthesis such as in the Miller-Urey reaction (both terrestrial and extra-terrestrial) or primordial metabolic processes. The latter source is unlikely at the first stages because it means the emergence of primordial metabolism before the emergence of TSD syntheses, i.e., translation and replication. At the same time, the translation and replication processes can be understood in the context of reaction cycles, where the products of one or more molecules of the cycle affect the entire system, either by involvement in reactions such as the emergence and operation of auto-catalysis, or by means of a new product of the cycle under consideration, or by means of a feedback system. This implies that from the stage of emergence of proto-peptidyl transferase, this catalyst functioned in an environment containing amino acids and short peptides.

2.10   Synthesis of primordial living entities in a test-tube

The two experimental approaches discussed in the literature with regard to life's synthesis are: (i) the *whole system* methodology and (ii) the *constructionist* methodology (see Lahav, 1999). When the first strategy was suggested and discussed (see Lahav, 1985), it reflected our ignorance at that time. More specifically – it reflected the lack of a general paradigm of the origin of life. Gradually, theories and observations encompassing more and more aspects of life's origin were published, pioneered by White (1980), Lahav and White (1980), and White and Raab (1983), and later continued by

Lahav and Nir (1997), Nir and Lahav (1997), Lahav et al. (2001; 2003, in preparation). The impression one gets from these theoretical considerations is that the ETIF model, including the clues provided by Trifonov's school of thought (see above), is detailed enough to serve as the basis for new experimental work aimed at the synthesis of primordial "living" entities in a test-tube (Lahav et al., 2001, 2003). Methodologically, this proposed strategy is a combination of the *whole system* strategy on the one hand and the *constructionist* approach on the other.

Apparently, the feasibility of this experimental approach draws on the definition of life as a feedback system combined with TSD reactions in a fluctuating (wet-dry) environment. For instance, a more specific experimental system is made-up of a mixture of a clay mineral such as kaolinite, RNA-like building blocks and their oligomers rich in G and C bases, and a few amino acids and their short peptides.

The presence of a feedback reaction in the physico-chemical process characterizing life's emergence can be determined experimentally, i.e., by chemical analyses.

The two central reactions of the primordial feedback are replication and translation, where the main proto-nucleotides and amino acids, as well as the size of the primordial catalytic peptides have been discussed by Trifonov (1999), Trifonov et al. (2001), Lahav et al. (2001, 2003). With these clues regarding the constituents of the primordial feedback systems and their environment, it is tempting to suggest that the appropriate experimental system, as well as an adequate computer model (Nir and Lahav, 1997), can be used to establish an experimental system designed to study the emergence of primordial living entities.

The establishment of such an experimental system, designed to simulate the transition from inanimate to animate matter, would be the first attempt to synthesize primordial "living" entities in a test-tube. Obviously, it includes an attempt to shorten the long pathway needed to synthesize "living" entities according to the *constructionist* strategy.

3.   CONCLUSIONS AND QUESTIONS

The study of the origin of life has reached a stage where there are enough clues for an attempt to synthesize simple "living" entities in a test-tube. The adoption of the ETIF theory as a model for the emergence of the first "living" entities in the prebiotic world immediately suggests a long list of novel questions by which the plausibility of this model should be evaluated, analyzed, and tested. Some of these questions are as follows:

1. Is the feedback loop system under consideration a general organizational attribute, which encompasses living forms in the entire evolutionary range

starting with the emergence of the first living entities and ending with extant living entities?

2. Is the feedback system of life independent of its constituents?
3. What are the differences between the feedback systems of the RNA-world theory and the ETIF theory? Is there continuity between these two kinds of feedback systems which represent consecutive evolutionary stages?
4. Can a feedback system be formed in an RNA world and in the absence of amino acids and catalytic peptides?
5. What are the evolutionary advantages of peptides (and proteins) that brought about the takeover process of the "proto-RNA" world by the "proto-RNA-peptide" world?
6. Assuming, as is widely accepted, that the RNA-world proceeded the RNA-peptide era, is it possible to estimate the length of the time interval between these two eras?
7. According to the ETIF school, glycine and alanine are central to the emergence of biohomochirality during the transition from inanimate to animate systems. Can the RNA-world theory explain the initiation and establishment of biohomochirality?
8. Based on the relative facility of synthesizing prebiotic peptides in the laboratory compared to synthesizing proto-RNAs, it seems that the first peptides were relatively abundant in the prebiotic environment. Does this imply that peptides could serve not only as scaffolds for the proto-RNA catalytic sites, but also as catalysts?
9. What are the main intermediate stages between, for example, the presumed proto-peptidyl transferase and the extant peptidyl transferase?

4.   ACKNOWLEDGMENT

I thank Prof. E. N. Trifonov, Dr. A. C. Elitzur, and Dr. S. Nir for many helpful discussions.

5.   REFERENCES

Cairns-Smith A.G. 1982. Genetic takeover and mineral origins of life. Cambridge University Press, Cambridge.
Couzin J. 2002. Small RNAs make big splash. Science 298, 2296-2297.
Deamer D.W., Harang-Mahon E. and Bosco G. 1984. Self-assembly and function of primitive membrane structures. In: Bengtson S. (ed.) Early life on Earth. Nobel Symposium 84. Columbia University Press. New York. Pp. 107 – 123.
Deamer D.W. 1997. The first living systems: A bioenergetic perspective. Microbiol Mol Biol Rev. 61, 239-261.

Ferris J.P. 2002. Montmorillonite catalysis of 30-50 mer oligonucleotides: Laboratory demonstration of potential steps in the origin of the RNA world. Orig Life Evol Biosphere 32, 311-332.

Joyce G.F. 2002. The antiquity of RNA-based evolution. Nature 418, 214-221.

Korzeniewsky B. 2001. Cybernetics formulation of the definition of life. J Theor Biol. 209, 275-286.

Lahav N. 1985. The synthesis of primitive "living" forms: Definitions, goals, strategies, and evolution synthesizers. Orig Life Evol Biosphere 16, 129-149.

Lahav N. 1999. Biogenesis: Theories of life's origin. Oxford University Press. New York; Oxford.

Lahav N., Nir S. and Elitzur A. C. 2001. The emergence of life on Earth. Progr Biophys Mol Biol. 75, 75-120.

Lahav N., Trifonov E.N., Nir S. and Elitzur A.C. 2003. Synthesis of simple living entities in a test-tube (In preparation).

Lahav N. and White H.D. 1978. A possible role of fluctuating clay-water systems in the production of ordered prebiotic oligomers. J Mol Evol. 16, 11-21.

Lahav N., White H.D. and Chang S. 1978. Peptide formation in the prebiotic era: Thermal condensation of glycine in fluctuating clay environments. Science 201, 67-69.

Moller W. and Janssen G.M.C. 1990. Transfer RNAs for primordial amino acids contain remnants of primitive code at positions 3 to 5. Biochimie. 72, 361-368.

Moore P.B. and Steitz T.A. 2002. The involvement of RNA in ribosome function. Nature 418, 229-235.

Nir S. and Lahav N. 1997. Emergence of template-and-sequence-directed (TSD) syntheses: II. A computer simulation model. Orig Life Evol. Biospher. 27, 567-584.

Orgel L.E. 2000. Self-organizing biochemical cycles. Proc Natl Acad Sci USA 97, 12503-12507.

Plankenstem K., Righi A. and Rode B. 2002. Glycine and diglycine as possible catalytic factors in the prebiotic evolution of peptides. Orig Life Evol Biosphr. 32, 225-230.

Reader J.S. and Joyce G.F. 2002. A ribozyme composed of only two different nucleotides. Nature 420, 841-844.

Schwabe C. 1985. On the basis of the studies of the origin of life. Orig Life 15, 213-216.

Trifonov E.N. 1999. Glycine clock: Eubacteria first, Archaea next, Protocista, Fungi, Planta and Animalia at last. Genetherapy Mol Biol. 4, 313-322.

Trifonov E.N., Kirzhner A., Kirzhner V.M. and Berezovski I. 2000. Distinct stages of protein evolution as suggested by protein sequence analysis. J Mol Evol. 53, 394-401.

Wachtershauser G. 1992. Groundworks for evolutionary biochemistry: The iron-sulfur world. Progr. Biophys Mol Biol. 58, 85-201.

White D.H. 1980. A theory for the origin of self-replicating system: I. Natural selection of the Autogen from short, random oligomers. J Mol Evol. 16, 121-147.

Woese C. 1998. The universal ancestor. Proc Natl Acad Sci USA 95, 6854-6859.

AGING AND THE ENVIRONMENT: THE STRESS THEORIES

Peter A. Parsons
La Trobe University, Bundoora, Vic. 3083, Australia

Abstract: Unstable and often extreme environments have always been a feature of life on Earth. Droughts, floods, and temperature extremes lead to nutritional inadequacy especially of the young. Taking into account the energy consequences of stressful environments, the free-radical theory of aging becomes a general stress theory of aging, recently expressed as a deprivation-syndrome theory, which highlights resource shortages. This ecological scenario contrasts with the more benign protected circumstances of the laboratory, domesticated and island populations, and of the well-nourished humans of the modern era. Some other evolutionary theories of aging, especially the mutation accumulation and antagonistic pleiotropy theories, appear to be implicitly assuming, predominantly protected environments. Empirical work on *Drosophila* populations from Lower Nahal Oren, Israel, by Nevo and collaborators, are in accord with the stress theory of aging. More information incorporating stress levels in the wild is a high priority.

1. INTRODUCTION

It is not well appreciated that natural environments have always been hostile to organisms. White (1993) amassed an enormous body of evidence, especially in herbivorous animals indicating that the abundance of organisms is principally determined by resource limitations for the young. He writes, "Most die just because they fail, right at the start, ever to gain a foothold - to gain access to enough of the limiting resource to survive and grow". That is,

S.P. Wasser (ed.), Evolutionary Theory and Processes: Modern Horizons,
Papers in Honour of Eviatar Nevo
P.A. Parsons. Aging and the Environment: the Stress Theories, 17-33.
© 2004 *Kluwer Academic Publishers.*

organisms have the capacity to produce many offspring but environmental stresses, principally climatic, intervene to cause nutritional inadequacy so that few offspring ultimately survive and reproduce. A permanent ecological theatre occurs consisting of unstable and often extreme environments especially droughts, floods, and temperature extremes severely restricting nutrition. Furthermore, recent evidence suggests that the Mesoproterozoic from 1800 million years ago to 800 million years ago were nutritionally hard times (Anbar and Knoll, 2002; Kerr, 2002). Therefore nutritional inadequacy is a feature of the living and fossil biota.

Stress is a quantitative trait that can be defined operationally as an environmental probe causing a potentially injurious change to a biological system with major impacts on evolutionary processes (Hoffmann and Parsons, 1991). Variation in the intensity of stress can be expressed on a fitness-stress continuum, where fitness is inversely related to stress level, and stress resistance is an adaptive process resisting mortality (Parsons 1992, 2002a, b). Therefore, since the environment is fundamental to understanding evolutionary change, ecological conditions must be accurately assessed. The emphasis in this paper concerns aging under the stressful scenario in natural populations.

In investigations on stress and evolution initiated over 40 years ago, the variability and genetics of morphological and fitness traits in *Drosophila melanogaster* from natural populations were considered, incorporating stress-resistance traits under extreme environments. Of particular importance in the aging context is the early demonstration of polymorphism for stress-resistance genes in natural populations including the ability to withstand desiccation and temperature shocks (Hosgood and Parsons, 1968; Parsons, 1974). Recent work on heat shock proteins, hsps, in natural populations follows from these early observations.

From the early 1970s, field studies were carried out simultaneously on the cosmopolitan Australian *Drosophila* fauna of disturbed habitats and on the endemic *Drosophila* fauna principally from the rainforests of the humid tropics of Australia. The abiotic variables of temperature extremes and desiccation are pivotal in determining species distributions and diversities. For example, stress-resistant generalist *Drosophila* species dominate at rainforest edges and in disturbed forests, suggesting that extinctions and replacements are rarer, but more numerous stress-sensitive species can occur after relatively small increases in temperature/desiccation stress. In parallel under laboratory conditions, widespread species tolerate wide ranges of environmental stress assayed by resistance to desiccation and high and low temperatures. Hence, an efficient predictor of *Drosophila* species distributions can be obtained from laboratory experiments on resistance to extreme stresses (Parsons, 1982).

The two components of drought stress and low temperature stress are also important determinants of plant distributions (Osmond et al., 1987; Parsons, 1997). The striking parallels between the major determinants of *Drosophila* distributions and plants follow from the dependence of *Drosophila* species on plant resources (Parsons, 1982). The short life span of *Drosophila* has the advantage that field estimates of survival are possible. Capture-recapture experiments on domesticated species in the region of Leeds, England, indicate the normality of stressful conditions since life expectancy was reduced to an order of magnitude less than that achievable under laboratory conditions (Rosewell and Shorrocks, 1987). Furthermore, *D. melanogaster* under field conditions in France are commonly starved (Boulétreau, 1978). Stress therefore appears to be a universal feature of natural populations. Clearly, evolutionary theories of aging need to emphasize adaptation under this ecological scenario, in addition to and in contrast with the more benign circumstances of laboratory, domesticated and island populations, and of well-nourished humans of the modern era.

The paradox to be faced is the predicament of using experiments carried out under laboratory conditions as models for evolutionary change in natural populations. However, studies under environmental extremes are not simple. Survival in the face of substantial mortalities, when fecundity is severely restricted but fertility is maintained, is an important but difficult experimental regime to replicate. For example, following exposure of six members of the *melanogaster* subgroup of species of the genus *Drosophila* to high temperature/desiccation stress, some flies remained fertile even when mortality from stress was exceedingly high, which indicates that survival of extreme stress is a good guide to population continuity (Stanley et al., 1980; Parsons, 1987, 1992). A field example comes from French populations of *D. melanogaster* under low-temperature stress and resource shortages, where individual survival can occur during periods of low fecundity in unfavorable seasons (Boulétreau-Merle et al., 1992). Regrettably, environmental manipulations incorporating challenging extremes have received little attention, even in the light of increasing interest in the physiology of life-history tradeoffs. Indeed, most such *Drosophila* studies are carried out under the unrestricted nutrition of the laboratory (Zera and Harshman, 2001). Comparative data across environments incorporating the abiotic extremes of the wild are needed to cross integration levels from the molecular to life-history traits, especially survival and longevity. I now turn to aging theories under the stressful environments of the wild.

2. THE FREE RADICAL THEORY OF AGING

Harman (2001) wrote that "Aging is the accumulation of diverse deleterious changes in the cells and tissues with advancing age that increase the risks of disease and death". In developed countries the aging process is slow early in life, but then rapidly increases with age. It depends upon chemical reactions in normal metabolism that increase the chance of death exponentially with increasing age even under optimal living conditions. Harman (1956) put forward the free radical theory of aging, which is based upon the premise that a single common process, modifiable by genetic and environmental factors, is responsible for aging and death of all living beings. Free radicals, however incited, are proposed to underlie the progressive deterioration of biological systems over time due to their innate ability to produce change.

Free radical reactions are ubiquitous in living systems. Life apparently evolved spontaneously from basic chemicals formed by free radical reactions largely initiated by ionizing radiation from the sun, so that life span evolved in parallel with the ability of organisms to deal with damage from free radicals. Under this theory, the aging process is determined by the sum of the deleterious free radical reactions occurring continuously throughout the cells and tissues of organisms.

Oxygen is combined with enzymatically degraded food products to produce energy in the mitochondria. In this process a number of potentially reactive oxygen intermediates, or oxygen radicals, are generated. Oxygen radicals produced in the mitochondria include the superoxide anion, hydrogen peroxide, and hydroxyl free radicals, which can attack and seriously damage a wide range of cellular molecules including membrane lipids, cellular proteins, and DNA (Clark, 1999; Lane, 2002).

Free radicals therefore generate oxidative stress. Arking (2001) found that extended longevity phenotypes formed by selection in *D. melanogaster* show reduced oxidative damage. For the ecologically important trait, desiccation resistance, large increases in resistance rapidly occurred in *D. melanogaster* in selection experiments associated with increased longevity. Furthermore, the selected strains were resistant to other stresses including starvation, high temperatures, intense ^{60}Co-γ irradiation, and anoxia and toxic concentrations of ethanol (Hoffmann and Parsons, 1993). Among other empirical examples, Rose et al. (1992) found increased longevity following selection for both resistances to desiccation and starvation. More generally, resistance to a variety of intrinsic and extrinsic stressors is strongly correlated with life span in many species. For instance, in *Caenorhabditis elegans*, genes whose alterations cause life extension, or gerontogenes, show increased resistance to oxidative stress, thermal stress, and UV stress (Cypser and Johnson, 2002).

While a range of stresses and organisms has been studied, the primary or operative factor is oxidative stress (Beckman and Ames, 1998; Finkel and Holbrook, 2000). Individuals with the potential for a long life should therefore carry genes for oxidative stress resistance, which in itself, correlated with resistance to a wide array of stresses.

3. METABOLIC EFFICIENCY UNDER STRESS AND AGING

A metabolic consequence of stress is the induction of heat shock proteins, hsps, which are present in every organism from bacteria to man. Furthermore, cross-resistances among stresses (Smith-Sonneborn, 1993; Boriss and Loeschcke, 2003) indicate underlying common metabolic pathways.

In the wild, the ability of organisms to live in stressful habitats can be related to their ability to produce hsps in response to the stress (Coleman et al., 1995; Gehring and Wehner, 1995). Assuming that the production and expression of hsps is metabolically costly (Krebs and Loeschcke, 1994), hsps production level should depend upon the level of stress in natural populations. In fact, field and laboratory experiments, especially in *D. melanogaster*, indicate that hsp 70 levels in natural populations reflect evolutionary adaptation to periods of extreme stress (Krebs and Feder, 1997; Feder, 1999). Furthermore, the expression of hsp 70 and induced heat-shock resistance declines with age, but least in stress-resistant individuals (Arking, 1998; Tatar, 1999; Sorensen and Loeschcke, 2002).

In mole rats, *Spalax ehrenbergi*, in Israel, Nevo et al. (1994) demonstrated parallel patterns across habitats for data ranging from the nuclear and DNA levels of integration to the whole organism level. These patterns are underlain by ecological factors with an emphasis on aridity stress, which therefore plays a major role in adaptation by mole rats. The convergence of the organismic and molecular levels to give parallel patterns implies that the 36 gene loci surveyed cover most of the metabolic potential of mole rats, which is reflected at the organismic level. Furthermore, the inference that stress resistance is associated with metabolic potential implies that in extreme environments, the preservation of organisms having maximum metabolic potential to withstand the energy costs of stress should be at a premium. Therefore, stress applied at the organismic level has evolutionary consequences at the physiological and molecular levels.

One direct effect of stress is to increase the expenditure of metabolic energy, which implies a cost since energy is diverted from maintenance and production to repair and recovery (Odum et al., 1979). In fact, organisms struggle to exist in harsh environments that are nutritionally, and hence, energetically inadequate. Nature has rarely provided optimal conditions for growth, and when it did, the process of growth itself would deplete resources, making the environment rapidly unfavorable. Constant cycling between "feast and famine" favored the selection of organisms that survived from nutritional duress and other stressful adversities. Heininger (2001) combined this scenario into a general deprivation response. He argues that responses to environmental challenges became established as the driving force of adaptation and speciation in the period of prokaryotic evolution, and are reflected in the various extinction events of metazoan evolution. The expression of stress responses is interdependent and involves hsps, oxidative defenses and DNA repair enzymes, which combine to confer metabolic efficiency in response to and for the accommodation of stress. For Heininger, the deprivation syndrome is the fundamental force driving evolutionary charge.

In many stress selection experiments in insects, the metabolic rate fell when associated with increased stress resistance and with increased longevity and survival (Donahaye, 1993; Hoffmann and Parsons, 1993). In larval fitness studies in *D. melanogaster*, Kohane (1988) suggested that the finding of reduced metabolic rate following stress selection implies a target of selection of stress directed towards energy carriers. In any case, Kauffman (1993) argues for a connected metabolism underlying life in descriptions of biological systems in energy terms, which ultimately arise from the role of ATP in providing energy for work under aerobic conditions. The free radical theory of aging can be accommodated into this scenario because the rate of oxygen consumption in mitochondria is the major determinant of ATP production. Arkings (2001) finding of multiple genetic strategies to cope with similar levels of oxidative stress, not necessarily involving reduced metabolic rate, can be reduced to these terms. For caloric restriction, Yu and Chung (2001) conclude that an energy utilization mechanism geared for higher efficiency under resource limitation is primary to metabolic rate variations. Similarly, metabolic alterations and shifts in energy allocation in the direction of metabolic efficiency are co-requisites for the expression of longevity genes in *Drosophila* (Arking, 1998; Arking et al., 2002).

Stress is an environmental change or probe that initially brings about a loss in energy efficiency. Organisms respond to changes in their environments by matching nutrient and energy input and biosynthesis to needed levels for survival and reproduction in their habitats (Emlen et al., 1998), a process underlain by a connected metabolism. Similarly Toussaint et al. (2002) conclude that corollaries of the second law of thermodynamics allow the proposal that all living systems have common energy traits. This law requires that any process underway in a system irreversibly degrades the quality of the energy in that system. Stress and/or aging results in lower energy flow meaning lower global metabolic activity until ultimately a critical threshold is reached where cells no longer survive. Akashi and Gojobori (2002) provide evidence that the amino acid composition of *Escherischia coli* and *Bacillus subtilis* reflects the action of natural selection to enhance metabolic efficiency based upon accurate calculations of the energy costs of amino acid biosynthesis. Evolutionary adaptation should, therefore, favor cells that are selected for efficiency in the use of free energy under given ecological conditions. The promotion of survival and, hence, longevity is dependent upon high metabolic and especially energy efficiency, which is an expression of fitness defined predominantly in energy terms (Van Valen, 1991; Brown et al., 1993).

4.   EVOLUTIONARY APPROACHES TO AGING: THE STRESS THEORIES

Under the multiplicity of environmental hazards of natural populations, fitness to survive to old age presumably depends upon genes conferring high stress resistance, which promote high-vitality, homeostasis, and energy and metabolic efficiency.   This is the stress theory of aging (Parsons, 1995, 2002a, 2003). Under this theory, stress targets energy carriers so that evolutionary changes in longevity depend mainly upon selection for stress resistance. For example, in humans, vitality and survival are inherited more directly than life span (Vaupel, 1988; Yashin and Iachine, 1995). Therefore, stress resistance has a primary role in adaptation and in the determination of life span from microorganisms to man. In fact, the stress theory of aging can be expressed as the evolutionary selection of energy efficiency in the face of the cost of free radical stress in the environments of natural populations (Parsons, 1996a, b, 2000). Alternatively, Novoseltsev et al. (2001) developed a homeostatic model of oxidative damage where the aging process derives from age-related accumulation of damage produced by oxidative stress, which progressively reduces the homeostatic ability (i.e., energy and metabolic efficiency) of the organism.

Heininger (2002) envisages that evolution is driven by a perpetual conflict of individuals competing for limited resources. The resulting deprivation syndrome is a fundamental determinant of evolutionary change including the aging process. In a deprived world of limited resources, he argues that aging evolved as a somatic survival pathway via selection for stress resistance and driven by germ-soma conflict. Increased stress resistance favors survival in a variety of adverse circumstances, extends longevity, slows aging and delays death, as in the stress theory of aging. In summary, aging becomes a deprivation syndrome driven by conflicting germ-soma requirements shaped by compromises determined by environmental conditions.

The disposable soma theory of aging holds that the achieved life span reflects an optimal fitness balance between investment in reproduction and soma maintenance (Kirkwood, 1997; Kirkwood and Austad, 2000). This balance is subject to environmental modulation, which appears in its extreme form in Heininger's (2002) conclusion that aging is a deprivation syndrome driven by germ-soma conflict. Aging then is not the consequence of a loss of selective forces, but is a naturally selected stress-resistance dependent phenomenon, which allows individuals to resist the death sentence impact on germ cells. For example, for the expression of extended longevity genes in *D. melanogaster*, Arking et al. (2002) argue that metabolic alteration is necessary to enhance the ability of organisms to shift energy from reproduction to somatic maintenance.

5.   DISCUSSION

5.1   Free-living vs. protected populations

Interactions between organisms and environments are central for understanding evolution. In free-living populations, environmental perturbations mainly of climatic origin emphasize physical factors as the determinants of evolutionary and ecological processes, both in the living and fossil biota. Furthermore, inadequate nutrition is usual especially following drought, so that animals struggle to survive to adulthood and reproduce in hostile environments. Consequently, many organisms are born but few should survive due to a combination of climatic stress interacting with and causing nutritional stress (White, 1993).

In contrast, Darwin (1859) noted that domesticated plants and animals generally differ much more from each other, than do individuals of any one species or variety in a state of nature, and concluded that this greater variability is simply due to our domestic productions having been raised under conditions of life not as uniform as, and somewhat different from, those to which the parent species have been exposed under nature. He also notes that there is some probability ... that this variability may be partly connected with an excess of food. These conditions indicate that under domestication, conventional ecological limitations from climatic extremes and nutritional limitations should be ameliorated.

Various categories of populations occur under such benign or protected circumstances:

(1) Domestic populations, in which there is adaptation to man and the environments so provided, so that climatic extremes are avoided and nutrition is adequate to excessive.

(2) Laboratory populations, which live in abiotically controlled environments with adequate to unlimited resources, and form an extreme version of (1).

(3) Modern human populations, which to varying degrees are protected from the hostile environments of hunter-gatherers and pre-industrial populations (Williams and Nesse, 1991).

(4) Island populations, which are exposed to lesser climatic extremes than in neighboring mainland habitats.

In the shift from free-living to more protected populations substantial adaptive shifts have been documented. For example, when the olive fruit fly, *Dacus oleae*, was introduced to artificial laboratory environments consisting of a diet of yeast hydrolyzate, sucrose, dry egg yolk, and an antibiotic, one allele at the alcohol dehydrogenase, ADH, locus increased from around 1% to 40%, while the allele most common in natural populations decreased from around 65% to 30% in six generations. Additional changes occurring within a few generations of artificial rearing included reduced population size, high acceptance of the artificial oviposition site by females, an increase in larval viability and growth in the artificial medium, shorter time to sexual maturation of laboratory-reared males, earlier occurrence of matings during the photophase, and high oviposition rates at early ages (Zouros et al., 1982). The effect of contrasting nutritional regimes, therefore, warrants more attention since there must be substantial to stressful shifts in metabolic demands in many protected populations compared with free-living populations.

Following introduction of *D. melanogaster* to the laboratory, Frankham and Loebel (1992) found that a competitive index of reproductive fitness in a large population after 12 months was twice that of a recently collected wild population from the same locality, and Briscoe et al. (1992) found that levels of variation measured phenotypically and genetically declined in large populations over periods up to 23 years. Laboratory populations of *D. melanogaster* can therefore diverge from the wild populations from which they originated.

Adaptation to captivity has been documented in fish, rodents, plants, bacteria, and several *Drosophila* species (Price, 1984; Kohane and Parsons, 1988; Frankham, 1995). Some increases in sensitivity to environmental extremes have occurred presumably in response to a lack of exposure to the stressful extremes of free-living populations. For example, Hoffmann et al. (2001) found that desiccation and starvation resistance in *D. melanogaster* declined during three years of laboratory culture. In contrast, Krebs et al. (2001) found little effect of domestication on thermotolerance in over 50 +

generations of laboratory culture. However, Shabalina et al. (1997) found a rapid decline in fitness measured by survival in 30 generations, which was much smaller when competitive conditions were tough.

Examples from bacteria to man, therefore, demonstrate the contrast between the environments of protected populations and those of free-living populations upon which the stress theories of aging are based. Unfortunately, there are few examples of aging under the field conditions of free-living populations.

5.2   Stress and longevity in free-living populations

An important field exception to this situation comes from Lower Nahal Oren Canyon or "Evolution Canyon", Israel. In *D. melanogaster*, a highly significant association between stress resistance assessed by heat treatment with viability and longevity, occurred in a comparison of flies from a heat-stressed south-facing habitat with an adjacent, more benign, north-facing habitat (Nevo et al., 1998). The slopes of Lower Nahal Oren are separated by 100m at the bottom and 400m at the top. The viability and longevity data, therefore, indicate that strong microclimatic natural selection overrides migration in *Drosophila* at this microsite. Furthermore, the multivariate adaptive complex includes divergences for oviposition temperature preference and sexual and reproductive behavior, as well as the genetic differentiation of microsatellites and the regulatory regions of hsp 70Ba, which encodes the major inducible heat-shock protein of *Drosophila*. Adaptation to the contrasting microclimates is therefore responsible for major genetic and phenotypic divergence between the populations (Nevo et al., 1998; Korol et al., 2000; Michalak et al., 2001). This dramatic result very much depends on a realistic assessment of stress levels in a natural population and is an approach worth employing extensively elsewhere.

D. melanogaster is a widespread species characteristically exposed to abiotic extremes. In contrast, island habitats tend to be more benign since the surrounding water ameliorates seasonal temperature extremes so that risks from environmental hazards should be reduced. In a sense, the ecological comparison of widespread species with island populations appears analogous to the contrast of wild with domesticated laboratory populations. Austad (1993, 1997) compared mainland and island opossum populations from Georgia and found that the island opossums averaged about 25% greater longevity than those on the mainland, and their maximum longevity was around 50% greater. In addition, the rate of tendon aging, a direct measure of senescence, was slower in the island population, as was the rate of reproductive decline. Therefore, under an environment of reduced stress, the island population has evolved a slower aging life-history process than on the mainland. In contrast, aging in species such as *D. melanogaster* may be better understood by assuming that organisms in natural habitats are normally close to their limits of survival where energy efficiency is at a high premium. However, island populations can provide reference points for comparisons with more rigorous environments (Parsons, 2002a).

In the next section, aging theories under protected environments, or where such environments are assumed, will therefore be considered briefly.

5.3   Aging theories in protected environments

For most of the time humans have occupied the Earth, there has been a long-term pattern of high and unstable death rates. High mortality occurred at younger ages so that only a small subgroup survived to older ages. This pattern changed abruptly early in the twentieth century in many populations when substantial reductions in infant, child, and maternal death rates occurred, consequent upon medical technology, combined with improved living conditions and nutrition (Olshansky et al., 1990). Compared with the typical environments of earlier phases of human history, many present-day populations live under relatively benign conditions, as do most protected populations. Diets have become adequate and relatively uniform on a daily and seasonal basis and predation and diseases minimized.

Muller (1950) drew attention to the accumulating mutation load that is likely from improvements in living conditions in humans of the modern era due to the survival and reproduction of individuals who would not have contributed to the gene pool in earlier generations. Similarly, Hamilton (2001) was concerned about the burden imposed by mutationally challenged people on the less challenged and able-bodied. Parsons (1978) expressed concern about the acquisition of deleterious genes due to modern medicine creating vulnerability under any future environmental deterioration.

Crow (1997) regards mutation accumulation to be a problem with an extremely long-time scale, especially for numerous small mutations of relatively small fitness effect.   In addition, he cautions that mutation accumulation should be viewed in the light of more immediate stresses such as habitat loss, water depletion, and food shortage. Similarly, Lynch et al. (1999) considered that deleterious mutation accumulation is likely to be minimal in societies living under relatively harsh conditions and that populations exposed to modern medical practices could be asymptotically approaching a benign-environment maximum.

The survival of a novel variant should be inversely related to the magnitude of the phenotypic change, or in the present context, the energy cost of the change in a stressful environment. Consequently, the more extreme the environment, the less the change can be accommodated by variant organisms for their survival. Therefore, the accumulation of deleterious mutants predicted under the mutation accumulation theory of aging may be unimportant in most natural populations because the energy cost for their survival would be restrictive especially with increasing age. However, in modern human populations, mutants of relatively small effect could accumulate and impact negatively on longevity as aging proceeds.

Hughes et al. (2002) conducted experiments in *D. melanogaster* on aging and found strong support for mutation accumulation theory. They concluded that an observed increase in genetic variation with age occurs because genes with deleterious effects with a late age of onset are unopposed by natural selection. In other words, under relatively benign laboratory conditions mutations can accumulate as is postulated in modern human populations, but this is a less likely scenario in societies living under relatively harsh conditions (Lynch et al., 1999). Hughes et al., (2002) also considered the antagonistic pleiotropy hypothesis, whereby senescence occurs because of the selection of traits with positive fitness early in life changing to negative later in life, which implies the selection of alleles trading early-life fitness with late-life costs (Williams 1957). Some evidence for antagonistic pleiotropy has been obtained in the laboratory for *Drosophila* under controlled and generally benign conditions (Abrams and Ludwig 1995). The best evidence for antagonistic pleiotropy should occur in well-nourished humans of the modern era because of benign environments, which permit the survival of the old by medical and nutritional interventions. Under their laboratory regime, Hughes et al. (2002) argued that the mutation accumulation and antagonistic pleiotropy theories are not mutually exclusive, but favor the former since their data are compatible with the accumulation of many deleterious alleles.

In natural populations, only the fitter organisms can cope with environmental hazards and can accumulate the resources necessary for repetitive reproduction and care for offspring, so that death from the expectation of high mortalities under the normally inadequate environment is delayed (White, 1993). In this light, the antagonistic pleiotropy and the mutation accumulation theories of aging are difficult to apply to wild populations. Indeed, in contrast with human populations of the modern era, animal populations subjected to a world of predators, parasites, food shortages, and various abiotic stresses would be less able to manifest antagonistic pleiotropy because of severely restricted life spans (Toupance et al., 1998). Under these conditions, the genetics of survival under environmental hazards become the predominant influence, and this is the underlying premise of the stress theories of aging.

6. CONCLUSIONS

Assuming a world dominated by abiotic stress, the distribution and abundance of organisms can be reduced to an energy balance, or trade-off, between the energy costs of stressful environments and energy gains from resources. This largely reductionist model is based on a primary target of selection of stress at the level of energy carriers. In this sense, stress is an environmental probe with genetic consequences affecting fitness, that is, for life history traits including longevity and especially survival. This ecological approach to adaptation, especially towards its limits, forms the background for the stress theories of aging. Some other theories of aging appear to assume more protected environments where exposures to abiotic stresses are ameliorated and nutrition is abundant to unlimited. Therefore, I combine the basic tenets advanced by Darwin and his successors with energy analyses assuming a far more stressful world than envisaged by many of the architects of evolution. The outstanding contribution of Eviatar Nevo and his collaborators on the populations of "Evolution Canyon" form a most important case study of the evolution of aging in a stressed world.

7. REFERENCES

Abrams P.A. and Ludwig D. 1995. Optimality theory, Gompertz law, and the disposable soma theory of senescence. Evolution 49, 1055-1066.

Akashi H. and Gojobori T. 2002. Metabolic efficiency and amino acid composition in the proteomes of *Escherichia coli* and *Bacillus subtilis*. Proc Natl Acad Aci USA 99, 3695-3700.

Anbar A.D. and Knoll A.K. 2002. Proterozoic ocean chemistry and evolution: a bioinorganic bridge. Science 297, 1137-1142.

Arking, R. 1998. Biology of Aging. Sinauer Associates, Sunderland, Massachusetts.

Arking, R .2001.Gene expression and regulation in the extended longevity phenotypes of *Drosophila*. Ann NY Acad Sci. 928, 157-167.

Arking R., Buck S., Novoseltev V.N., Hwangbo D.S. and Lane M. 2002. Genomic plasticity, energy allocations, and the extended longevity pheotypes of *Drosophila*. Age Res Rev. 1, 202-228.

Austad S.N. 1993. Retarded senescence in an insular population of Virginia opossums (*Didelphis virginiana*). J Zool Lond. 229, 695-708.

Austad S.N. 1997. Why we age. John Wiley, New York.

Beckman K.B. and Ames B.N. 1998. The free radical theory of aging matures. Physiol Rev. 78, 547-581.

Boriss H. and Loeschcke V. 2003. Complexity theory provides metaphors for a better under-standing of biological processes. In: The Significance of Complexity, Buhl H. and van Kooten Neikerk eds. Ashgate, London.

Boulétreau J. 1978. Ovarian activity and reproductive potential in a natural population of *Drosophila melanogaster*. Oecologia 33, 319-342.

Boulétreau-Merle J., Fouillet P. and Terrier O. 1992. Clinal and seasonal variations in initial retention of virgin *Drosophila* melanaster females as a strategy for fitness. Evol Ecol. 6, 223-242.

Briscoe D.A., Malpica J.M., Robertson A., Smith G.J., Frankham R., Banks R.G. and Barker J.S.F. 1992. Rapid loss of genetic variation in large captive populations of *Drosophila* flies: implications for the management of captive populations. Conserv Biol. 6, 416-425.

Brown J.H., Marquet P.A. and Taper M.L. 1993. Evolution of body size. Consequences of an energetic definition of fitness. Amer Natl. 142, 373-384.

Clark W.R. 1999. The means to an end: The Biological Basis of Aging and Death. Oxford, University Press, Oxford.

Coleman J.S., Heckathorn S.A. and Hallberg R.L. 1995. Heat-shock proteins and thermotolerance: linking ecological and molecular perspectives. Trends Ecol Evol. 10, 305-306.

Crow J.F. 1997. The high spontaneous mutation rate: Is it a health risk? Proc Natl Acad Sci. USA. 94, 8380-8386.

Cypser J.R. and Johnson T.E. 2002. Multiple stressors in *Caenorhabditis* elegans induce stress hormesis extended longevity. J Gerontol Biol Sci. 57A, B109-B114.

Darwin C. 1859. On The Origin of Species by Natural Selection. J Murray, London.

Donahaye E. 1993. Biological differences between strains of *Tribolium* castaneum selected for resistance to hypoxia and hypercarbia, and the unselected strain. Physiol Entomol. 18, 247-250.

Emlen J.M., Freeman D.C., Mills A. and Graham J.H. 1998. How organisms do the right thing: the attractor hypothesis. Chaos. 8, 717-726.

Feder M.E. 1999. Engineering candidate genes in studies of adaptation: the heat-shock protein Hsp70 in *Drosophila melanogaster*. Amer Natl. 154, 555-566.

Finkel T. and Holbrook N.J. 2000. Oxidants, oxidative stress and the biology of ageing. Nature 408, 239-247.

Frankham R. 1995. Conservation genetics. Ann Rev Genet. 29, 305-327.

Frankham R. and Loebel D.A. 1992. Modelling problems in conservation genetics using captive *Drosophila* populations: rapid genetic adaptation to captivity. Zoo Biology 11, 333-342.

Gehring W.J. and Wehner R. 1995. Heat shock protein synthesis and thermotolerance in *Cataglyphis*, an ant from the Sahara Desert. Proc Natl Acad Sci. USA 92, 2994-2998.

Hamilton W.D. 2001. Narrow roads of gene land. Vol. 2. Evolution of sex. Oxford University Press, Oxford.

Harman D. 1956. Aging - a theory based on free radical and radiation chemistry. J Gerontol. 11, 298-300.

Harman D. 2001 Aging: overview. Ann NY Acad Sci. 928, 1-21.

Heininger K. 2001. The deprivation syndrome is the driving force of phylogeny, ontogeny and oncogeny. Rev Neurosciences 12, 217-287.

Heininger K. 2002. Aging is a deprivation syndrome driven by a germ-soma conflict. Ageing Res Rev. 33, 481-536.

Hoffmann A.A., Hallas R., Sinclair C. and Partridge L. 2001. Rapid loss of stress resistance in *Drosophila melanogaster* under adaptation to laboratory culture. Evolution 55, 436-438 .

Hoffmann A.A. and Parsons P.A. 1991. Evolutionary genetics and environmental stress. Oxford University Press, Oxford.

Hoffmann A.A. and Parsons P.A. 1993. Selection for adult desiccation resistance in *Drosophila melanogaster*: fitness components, larval resistance and stress correlations. Biol J Linn Soc. 48, 43-54.

Hosgood S.M.W. and Parsons P.A. 1968. Polymorphism in natural populations of *Drosophila* for the ability to withstand temperature shocks. Experientia 24, 727-729.

Hughes K.A., Alipaz J.A., Drnevich S.M. and Reynolds R.M. 2002. A test of evolutionary theories of aging. Proc Natl Acad Sci. USA 99, 14286-14291.

Kauffman, S.A. 1993. The Origins of Order: Self-organization and Selection in Evolution. Oxford University Press, New York.

Kerr R.A. 2002. Could poor nutrition have held life back? Science 297, 1104-1105.

Kirkwood T.B.L. 1997. The origins of human ageing. Phil Trans R Soc Lond B. 352, 1765-1772.

Kirkwood T.B.L. and Austad S.N. 2000.Why do we age? Nature 408, 233-238.

Kohane M.J. 1988. Stress, altered energy availability and larval fitness in *Drosophila melanogaster*. Heredity 60, 273-281.

Kohane M.J. and Parsons P.A. 1988. Domestication: evolutionary change under stress. Evol Biol. 23, 31-48.

Korol A., Rashkovetsky E., Iliadi K., Michalak P., Ronin Y. and Nevo E. 2000. Nonrandom mating in *Drosophila melanogaster* laboratory populations derived from closely adjacent ecologically contrasting slopes at "Evolution Canyon". Proc Natl Acad Sci USA 97, 12637-12642.

Krebs R.A. and Feder M.E. 1997. Natural variation in the expression of the heat-shock protein Hsp 70 in a population of *Drosophila melanogaster* and its correlation with tolerance of ecologically relevant thermal stress. Evolution. 51, 173-179.

Krebs R.A. and Loeschcke V. 1994. Costs and benefits of activation of the heat-shock response in *Drosophila melanogaster*. Funct Ecol . 8, 730-737.

Krebs R.A., Roberts S.P., Bettencourt B.R. and Feder M.E. 2001. Changes in thermotolerance and Hsp 70 expression with domestication in *Drosophila melanogaster*. J Evol Biol. 14, 75-82.

Lane N. 2002. Oxygen: The molecule that made the World. Oxford University Press, Oxford.

Lynch M., Blanchard J., Houle D., Kibota T., Schultz S., Vassilieva L. and Willis J. 1999. Perspective: spontaneous deleterious mutation. Evolution 53, 645-663.

Michalak P., Minkov I., Helin A., Lerman D.N., Bettencourt B.R., Feder M.E., Korol A.B. and Nevo E. 2001. Genetic evidence for adaptation-driven incipient speciation of *Drosophila melanogaster* along a microclimatic contrast in "Evolution Canyon", Israel. Proc Natl Acad Sci USA 23, 13195-13200.

Muller H.J. 1950. Our load of mutations. Am J Hum Genet. 2, 111-176.

Nevo E., Filippucci M.G. and Beiles A. 1994. Genetic polymorphisms in subterranean mammals (*Spalax ehrenbergi* superspecies) in the Near East revisited: Patterns and theory. Heredity 72, 465-497.

Nevo E., Rashkovetsky E., Pavlicek T. and Korol A. 1998. A complex adaptive syndrome in *Drosophila* caused by microclimatic contrasts. Heredity 80, 9-16.

Novoseltsev V.N., Novoseltseva J. and Yashin A.I. 2001. A homeostatic model of oxidative damage explains paradoxes observed in earlier aging experiments: a fusion and extension of older theories of aging. Biogerontology 2, 127-138.

Odum E.P., Finn J.T. and Franz E.H. 1979. Perturbation theory and the subsidy stress gradient. BioScience 29, 349-352.

Olshansky S.J., Carnes B.A. and Cassel C. 1990. In search of Methuselah: estimating the upper limits to human longevity. Science 2508, 634-640.

Osmond C. B., Austin M.P., Berry J.A., Billings W.D., Boyer J.S., Dacey J.W.H. et al. 1987. Stress physiology and the distribution of plants. BioScience 37, 38-48.

Parsons P.A. 1974. Genetics of resistance to environmental stresses in *Drosophila* populations. Am Rev Genet. 7, 239-265.

Parsons P.A. 1978. The genetics of aging in optimal and stressful environments. Exp Gerontol. 13, 357-363.

Parsons P.A. 1982. Evolutionary ecology of Australian *Drosophila*: a species analysis. Evol Biol. 14, 297-350.

Parsons P.A. 1987. Evolutionary rates under environmental stress. Evol Biol. 21, 311-347.

Parsons P.A. 1992. Evolutionary adaptation and stress: the fitness gradient. Evol Biol. 26, 191-223.

Parsons P.A. 1995. Inherited stress resistance and longevity: a stress theory of ageing. Heredity 75, 216-221.

Parsons P.A. 1996a. The limit to human longevity: an approach through a stress theory of ageing. Mech Age Dev. 87, 211-218.

Parsons P.A. 1996b. Stress, resources, energy balances, and evolutionary change. Evol Biol. 29, 39-72.

Parsons P.A. 1997. Evolutionary change: a phenomenon of stressful environments. In: The Web of Life Vol. I., Padmanaban G., Biswas M., Shaila M.S. and Vishveshwara S. eds., Harwood Academic Publishers, Amsterdam.

Parsons P.A. 2000. Caloric restriction, metabolic efficiency and hormesis. Human Exp Toxicol 19, 345-347.

Parsons P.A. 2002a. Life span: does the limit to survival depend upon metabolic efficiency under stress. Biogerontol. 3, 233-241.

Parsons P.A. 2002b. Aging: the fitness-stress continuum and genetic variability. Exp Aging Res. 28, 1-13.

Parsons P.A. 2003. From the stress theory of aging to energetic and evolutionary expectations for longevity. Biogerontol. 4 (in press).

Price E.O. 1984. Behavioral aspects of animal domestication. Quart Rev Biol. 39, 1-32.

Rose M.R., Vu L.N., Pank S.U. and Groves J.L.Jr. 1992. Selection on stress resistance increases longevity in *Drosophila melanogaster*. Exp Gerontol. 27, 241-250.

Rosewell J. and Shorrocks B. 1987. The implication of survival rates in natural populations of *Drosophila*: capture-recapture experiments on domestic species. Biol J Linn Soc. Lond 32, 373-384.

Shabalina S.A, Yampolsky L.Y. and Kondrashov A.S. 1997. Rapid decline of fitness in panmitic populations of *Drosophila melanogaster* maintained under relaxed natural selection. Proc Natl Acad Sci USA 94, 13034-13039.

Smith-Sonneborn J. 1993. The role of the "stress protein response" in hormesis. BELLE Newsletter 1, 4-9.

Sorensen J.G. and Loeschcke V. 2002. Decreased heat-shock resistance and down-regulation of Hsp70 expression with increasing age in adult *Drosophila melanogaster*. Funct Ecol. 16, 379-384.

Stanley S.M., Parsons P.A., Spence G.E. and Weber L. 1980.Resistance of species of the *Drosophila melanogaster* subgroup to environmental extremes. Aust J Zool. 28, 413-421.

Tatar M. 1999. Evolution of senescence: longevity and the expression of heat shock proteins. Amer Zool. 39, 920-927.

Toupance B., Godelle B., Gouyon P.H. and Schächter F. 1998. A model for antagonistic pleiotropic gene action for mortality and advanced age. Am J Hum Genet. 62, 1525-1534.

Toussaint O., Remacle J., Dierick J.F., Pascal T., Frippiat C., Royer V. and Chainiaux F. 2002. Approach of evolutionary theories of ageing, stress, senescence-like phenotypes, calorie restriction and hormesis from the point of view of far-from-equilibrium theormodynamics. Mech Ageing Dev. 138, 937-946.

Van Valen L.M. 1991. Biotal evolution: A manifesto. Evol Theory 10, 1-13.

Vaupel J.W. 1988. Inherited frailty and longevity. Demography 25, 277-287.

White T.C.R. 1993. The Inadequate Environment: Nitrogen and the Abundance of Animals. Springer-Verlag, Berlin.

Williams G.C. 1957. Pleiotropy, natural selection and the evolution of senescence. Evolution 11, 398-411.

Williams G.C. and Nesse R.M. 1991. The dawn of Darwinian medicine. Quart Rev Biol. 66, 1-22.

Yashin A.I. and Iachine I. 1995. How long can humans live? Lower bound for biological limit of human longevity calculated from Danish twin data using correlated frailty method. Mech Ageing Dev. 80, 147-169.

Yu B.P. and Chung H.Y. 2001. Stress resistance by caloric restriction for longevity. Ann NY Acad Sci. 928, 39-47.

Zera A.J. and Harshman L.G. 2001. The physiology of life history trade-offs in animals. Ann Rev Ecol Syst. 32, 95-126.

Zouros E., Loukas M., Economopolous A. and Mazomenos B. 1982. Selection at the alcohol dehydrogenase locus of the olive fruit fly *Dacus oleae* under artificial rearing. Heredity 48, 169-185.

ASSESSING POPULATION SUBDIVISION

James F. Crow

Genetics Department, University of Wisconsin, Madison, WI 53706, USA

Abstract: Nei's G_{ST} is widely used as a measure of population subdivision, especially now that molecular data are abundant. The standard assumption is that G_{ST} measures the degree of subdivision independent of allele frequencies. In this article, I show that this is indeed true, provided there is no selection or mutation and that migration, splitting, and fusion are independent of allele frequencies. G_{ST} has other desirable properties in addition: It is determined mainly by the absolute number of migrants per generation; it approaches equilibrium rapidly; and it can be used, with molecular markers, to assess the degree of altruism that would be expected with the current population structure.

1. INTRODUCTION

Among the myriad of articles that Eviatar Nevo has written, there is one that I found especially impressive and useful (Nevo, 1984). This was written at a time when the population genetics world was deluged with reports of isozyme variability based on electrophoretic studies. Nevo's paper includes measures of diversity (= heterozygosity, h) and polymorphism for no less than 1111 species representing a wide variety of plants and animals. There were 14 or more loci (mean 23) in each case and a minimum of 10 individuals per species (mean 199). It was by far the largest collection ever assembled. Overall, invertebrates ($h = 0.100$) were more variable than vertebrates (0.054), with plants in between (0.075). For mammals, *Dro-*

S.P. Wasser (ed.), Evolutionary Theory and Processes: Modern Horizons,
    Papers in Honour of Eviatar Nevo
J. F. Crow. Assessing Population Subdivision, 35-42.
© 2004 *Kluwer Academic Publishers.*

sophila, and coelenterates the values were 0.041, 0.123, and 0.140. Nevo also correlated these values with various ecological, demographical, and life history measurements. The study was remarkable for its extent and thoroughness. Nevo did not, in this paper, consider population structure.

Of course, he didn't stop there. He did many more studies, several of which considered variability within and between populations. A recent example among many that I might have chosen is Huang et al. (2002). It is this kind of population structure analysis that I wish to discuss, and to honor Eibi in this way.

2. MEASURING INBREEDING AND RANDOM DRIFT

The traditional measurement of reduction of homozygosity because of inbreeding or random drift is Wright's F. In this measure, the heterozygosity is given by $H = H_0(1-F)$, where H_0 is the heterozygosity of a random mating or founding population. Thus $1-F$ is a measure of relative, rather than absolute heterozygosity. F can also be defined as the probability that two alleles in an individual are *identical by descent* that is derived from a common ancestral gene or one from the other, as opposed to *identity in state*, having the same molecular structure or phenotypic manifestation (Malécot, 1948; Crow, 1954). With the coming of abundant molecular markers it is often convenient to adopt the "infinite allele" or "infinite site" model, in which case identity by descent and identity in state are equivalent. I shall assume that the model is appropriate and therefore make no distinction between the two kinds of identity.

Natural populations are not usually the randomly-mating units that simple theory assumes, but are divided into more or less isolated subgroups. Historically, such population structure was measured by Wright's F statistics, F_{IS}, F_{IT}, and F_{ST} (Wright, 1951). These were defined in terms of correlations and relative heterozygosities, which are appropriate for phenotypic traits, especially quantitative ones. Now that molecular data provide information at levels down to the nucleotide, it is natural to define relationships at the gene level, in terms of gene identity.

3. MEASURING POPULATION SUBDIVISION

I shall now consider the effect of having the population divided into subpopulations (demes) of various sizes. From Maruyama (1970, 1977, p. 130), let f_T be the probability that two alleles chosen randomly from the entire population are identical (by descent or in state). Let f_S be the probability that two alleles drawn randomly from the same population are identical; f_S is the average over all subpopulations weighted by size. Let $h_T = 1 - f_T$ and $h_S = 1 - f_S$. Then, following Nei (1973, 1977), define

$$G_{ST} = \frac{f_S - f_T}{1 - f_T} = \frac{h_T - h_S}{h_T} \qquad (1)$$

I have used G_{ST} rather than the near-equivalent F_{ST} (Wright, 1951) to emphasize that this method is intended for molecular data.

It is generally assumed that G_{ST} measures the extent of population substructure without regard to allele frequencies. But I am not aware of an explicit statement of the conditions under which this is true. I first tested this assumption with some numerical examples. I started out with various populations having specified values of h_S and h_T and followed them through successive generations. In one numerical example the population first split into two subpopulations of size 3,000 and 5,000. Each of these mated at random for 10 generations with random drift because of their finite size. Then they split again and again drifted for several generations. Two of the subpopulations fused and another split again with intervening periods of random drift. Between the latter two there was continued migration. Throughout the process the population sizes changed irregularly; some increased, others decreased. At the end, I computed h_T and h_S. In this example, the values were:

Initial $h_T = h_S$	Final h_T	Final h_S	G_{ST}
0.1	0.09963	0.09017	0.094935
0.9	0.89663	0.81150	0.094935

Other examples gave similar results. Always G_{ST} did not depend on the initial allele frequencies, as reflected in heterozygosity.

Having satisfied myself of the correctness of the principle, I decided to demonstate it analytically. The most direct way is to start with an equation from Malécot (1951). This model assumes a population of N_i monoecious individuals in subpopulation i. Mating within each subpopulation is random; hence, a proportion $1/N_i$ of the individuals in deme i are self-fertilized. I am also assuming a gamete migration model.

Let $m_{ij}(t)$ be the probability that a gamete in deme i was produced by an individual in deme j. Then m_{ii} is the probability that a gamete in i is not a migrant. Let $f_{ij}(t)$ be the probability that two alleles drawn randomly, one from deme i and the other from deme j, are identical. If $i = j$, they are from the same deme. Although the m's and N's vary with time, t, the dependence on t will be suppressed in the formulae and the ensuing generation is indicated by a prime. Letting $v_k = 1/(2N_k)$ and using the Malécot equation (Eq. 15 from Nagylaki,1989), we obtain

$$f_{ij}' = \sum_{k,l} m_{ik} m_{jl} f_{kl} + \sum_k m_{ik} m_{jk} v_k (1 - f_{kk}) \qquad (2)$$

Nagylaki (1989) has given the conditions under which this equation is exact. It is a very good approximation to a diploid-migration model with self-fertilization prohibited, unless the demes are very small (Nagylaki, 1983). I am assuming no mutation or selection, and that migration, fusion, and splitting are independent of genotype.

Since $h_{ij} = 1 - f_{ij}$, (2) becomes

$$h_{ij}' = 1 - \sum_{k,l} m_{ik} m_{jl} + \sum_{k,l} m_{ik} m_{jl} h_{kl} - \sum_k m_{ik} m_{jk} v_k h_{kk} \,.$$

But the second term on the right is the product of two terms, each of which sums to one, so

$$h_{ij}' = \sum_{k,l} m_{ik} m_{jl} h_{kl} - \sum_k m_{ik} m_{jk} v_k h_{kk} \qquad (3)$$

The m's and v's can change arbitrarily from generation to generation. By choosing appropriate values of the m's, various patterns of migration, fusion, and splitting can be accommodated. Changes in deme size are reflected in the values of v.

We start with a randomly mating population. In the next two generations

$$h_{ij}' = \sum_{k,l} m_{ik} m_{jl} h_0 - \sum_k m_{ik} m_{jk} v_k h_0$$

and

$$h_{ij}'' = \sum_{k,l} m_{ik} m_{jl} h_{kl}' - \sum_k m_{ik} m_{jk} v_k h_{kk}'$$

But $h_{kl}' = \Phi h_0$ and $h_{kk}' = \varphi h_0$, where Φ and φ are functions of the m's and v's, but not of the allele frequencies, which are contained in h_0. Thus in the second and all succeeding generations

$$h_{ij}(t) = \Gamma h_0 \qquad (4)$$

in which Γ is a function of the population size, structure, and migration pattern, but not of allele frequencies. Γ can be quite complicated, but fortunately we do not need to understand it if our purpose is to assess the degree of population subdivision.

Thus, h_T and h_S both have h_0 as a factor but otherwise are not determined by allele frequencies. So, from (1), G_{ST} is the ratio of two quantities, each containing h_0 as a factor, which cancels. Thus G_{ST} is a property of the structure, history, and demographics of the population, but not of allele frequencies.

Therefore, we should expect that if the model holds, G_{ST} would have the same value for any locus, provided that no mutation or selection occurred since the founding population. We should be able to estimate G_{ST} from data on any locus and the validity of the assumptions can be tested by agreement of values for different loci.

The conditions of strict neutrality and zero mutation are not likely to be met in any real population. Nevertheless, G_{ST} should provide an approximate measure of population structure that is useful for nearly neutral loci and for time periods that are less than the reciprocal of the mutation rate. I should emphasize that this treatment is deterministic. In the real world, stochastic fluctuations intrude.

Of course G_{ST} can be used simply as an empirical measure of population subdivision, as it often is, without a theoretical justification. One example of such use is to assess the importance of population structure for forensic DNA analysis and the application of formulae to correct for such structure (NAS, 1996). In that book θ is used for G_{ST}.

4. OTHER USEFUL PROPERTIES OF G$_{ST}$

The quantities f_S and f_T depend on at least five parameters: mutation rates, migration rates, number of groups, size of groups, and number of alleles. These quantities are rarely known, even for simple models with equal group sizes and symmetrical mutation rates. In contrast, if, as is usually the case, the mutation rate is small relative to the migration rate, the number of alleles is large, and the number of groups is large, we have the simple formula

$$G_{ST} \approx \frac{1}{4Nm + 1} \tag{5}$$

in which N is the group size and m the migration rate, assumed to be uniform in all directions. This and the following properties are discussed in Crow and Aoki (1984).

Note that Nm is the absolute number of migrants per generation, demonstrating the well-known property that the effectiveness of migration depends on the absolute number of migrants, rather than the group sizes or proportion of migrants. If the absolute number of migrants is much greater than one per generation, the population is essentially panmictic.

A second useful property is that G_{ST} approaches equilibrium much faster than the component quantities f_T and f_S , so an equilibrium assumption is much more likely to be realistic. Even as the two component quantities change slowly, G_{ST} quickly comes close to its equilibrium value and stays close.

Nusha Keyghobadi, Jens Roland, Stephen F. Matter, and Curtis Strobeck (personal communication) studied populations of an alpine meadow-dwelling butterfly, *Parnassius smintheus,* over a long time period. Their data are consistent with G_{ST} changing more rapidly than h_S, in agreement with predictions of the theory. As far as I know, this is the first empirical demonstration of this theoretical prediction in an evolving natural population.

Crow and Aoki (1984) explored numerically two contrasting models: (1) the island model of Wright in which migrants move randomly among the groups, and (2) Kimura's stepping-stone model in which migrants move to a neighboring group. The second model is more realistic, especially since it can include the first by introducing the additional possibility of long-range migration. In the island model, G_{ST} is essentially independent of mutation rate and population size. In the stepping-stone model, G_{ST} is nearly independent of the mutation rate and is only weakly dependent on the total population size.

Finally, G_{ST} can be used to obtain the coefficient of relationship, $r = 2G_{ST}/(1 + G_{ST})$. Hamilton's (1964) condition for altruism to spread is $c/b < r$, where c and b are the cost to the altruist and the benefit to the population, both small quantities and measured in units of fitness. Aoki and I assumed that individuals behave altruistically toward members of their own group, rather than specifically recognizing kin, which we thought might be more realistic. From rather fragmentary protein and DNA polymorphism data on Japanese macaques, we estimated that if c/b is less than about 1/6 the trait would increase in this population. Of course this is a very crude estimate, but it shows a way in which molecular data can be used to infer the degree of population subdivision and therefore the opportunity for group selection to override the selection on individuals. If the requisite behavioral genes exist in the population, this relationship could be used to assess the likelihood for altruistic traits to evolve.

Although r gives the maximum c/b for which altruistic traits would increase, of deeper interest is the distribution of c/b values of incorporated traits. If a continuum of c/b values is possible, the average value at equilibrium is $r/2$ (Engels, 1983).

5.   ACKNOWLEDGMENTS

My greatest debt is to Tom Nagylaki for suggesting the Malécot equation as a starting point for the derivation. I thank Nusha Keyghobadi for permission to refer to his unpublished butterfly data. Bill Engels and Carter Denniston contributed to my understanding of this problem, and Carter Denniston kindly read through the manuscript and caught several errors.

6.   REFERENCES

Crow J.F. 1954. Breeding structure of populations. II. Effective population number. Pp. 543-556 in Statistics and Mathematics in Biology, edited by Kempthorne, Bancroft, Gowen and Lush. Iowa State College Press, Ames, Iowa.

Crow J.F. and Aoki K. 1984. Group selection for a polygenic behavioral trait: estimating the degree of population subdivision. Proc Natl Acad Sci USA 81, 6073-6077.

Engels W.R. 1983. Evolution altruistic behavior by kin selection: an alternative approach. Proc Natl Acad Sci USA 80, 515-518.

Hamilton W.D. 1964. The genetical evolution of social behavior. J Theor Biol. 7, 1-52.

Huang Q., Beharav Q., Li Y., Kirzhner V. and Nevo E. 2002. Mosaic microecological differential stress causes adaptive microsatellite divergence in wild barley, *Hordeum spontaneum* at Neve Yaar, Israel. Genome. 45, 1216-1229.

Malécot G. 1948. Les Mathématiques d l'Hérédté. Masson, Paris.

Malécot G. 1975. A traitement stochastique des problémes linéaries (mutation, linkage, migration) en Génetique de Populations. Ann Univ Lyon Sci Sect A 14, 79-117.

Maruyama T. 1970. Effective number of alleles in a subdivided population. Theor Popul Biol. 1, 273-306.

Maruyama T. 1977. Stochastic problems in population genetics. Lect Notes Biomath. 17, 1-245. Springer-Verlag, Berlin.

Nagylaki T. 1983. The robustness of neutral models of geographic variation. Theor Popul Biol. 24, 268-294.

Nagylaki T. 1989. Gustav Malécot and the transition from classical to modern population genetics. Genetics 122, 253-268.

N.A.S. 1996. The evaluation of forensic DNA evidence. National Academy Press, Washington, D.C.

Nei M. 1973. Analysis of gene diversity in subdivided populations. Proc Natl Acad Sci USA 70, 3321-3323.

Nei M. 1977. F-statistics and analysis of gene diversity in subdivided populations. Ann Hum Genet. 41, 225-233.

Nevo E., Beiles A. and Ben-Shlomo R. 1984. The evolutionary significance of genetic diversity: ecological, demographic and life history correlates. Lect Notes Biomath. Springer-Verlag, Berlin.

Wright S. 1951. The genetical structure of populations. Annals Eugen. 15, 323-354.

CONVERGENT EVOLUTION IN EXTREMELY HALOPHILIC PROKARYOTES: A COMPARISON BETWEEN *SALINIBACTER RUBER* (BACTERIA) AND THE HALOBACTERIACEAE (ARCHAEA)

Aharon Oren

Division of Microbial and Molecular Ecology, The Institute of Life Sciences, and The Moshe Shilo Minerva Center for Marine Biogeochemistry, The Hebrew University of Jerusalem, 91904 Jerusalem, Israel

Abstract: *Salinibacter* is an aerobic, red, extremely halophilic bacterium that was recently isolated from saltern crystallizer ponds in Spain. Phylogenetically, *Salinibacter* belongs to the *Flavobacterium - Cytophaga - Bacteroides* group. It is one of the most halophilic organisms belonging to the domain Bacteria, and it is unable to grow at salt concentrations below 150 g/l. A comparison of *Salinibacter* with the extremely halophilic representatives of the Halobacteriaceae (Archaea) shows many striking similarities. Both groups maintain high intracellular K^+ and Cl^- concentrations. No organic osmotic solutes were detected in significant concentrations in the cytoplasm of either group, and the intracellular enzymatic machinery was found to be functional in the presence of molar concentrations of salt, an adaptation based on salt-requiring proteins with a high excess of acidic amino acids. Comparison of additional properties shows many additional similarities such as the absorption spectrum of the pigments, the use of (variations of) the Entner-Doudoroff pathway for sugar degradation, and even a similar G+C percentage of their DNA. These comparative studies suggest that we may be dealing here with an excellent example of convergent evolution, in which similar features have evolved in phylogenetically unrelated organisms as an adaptation to life at the highest salt concentrations.

S.P. Wasser (ed.), Evolutionary Theory and Processes: Modern Horizons,
       Papers in Honour of Eviatar Nevo
A. Oren. Convergent Evolution in Extremely Halophilic Prokaryotes: a Comparison between Salinibacter ruber (Bacteria) and the Halobacteriaceae (Archaea), 43-64.
© 2004 *Kluwer Academic Publishers.*

1.   INTRODUCTION

Microorganisms living at high-salt concentrations need to adjust the osmotic pressure of their cytoplasm to that of their saline habitat. Because biological membranes are permeable to water, no microorganism living in a low-water activity brine can maintain a dilute, high-water activity cytoplasm. Moreover, the presence of a slightly hypertonic cytoplasm is a prerequisite for the maintenance of turgor pressure.

The intracellular enzymatic machinery of non-salt-adapted micro-organisms (and of many salt-adapted ones as well, as discussed below) is salt-sensitive. Presence of molar concentrations of salts has a devastating influence on the activity of many enzymes and on their solubility. To overcome this problem, two fundamentally different strategies have evolved enabling microorganisms to live in high-salt environments (Oren, 1999a). The first is based on the intracellular accumulation of salts at concentrations at least as high as those present in the medium. Potassium chloride rather than sodium chloride is then the main osmotically active compound inside the cells. Maintenance of molar concentrations of KCl inside the cell implies that all intracellular enzymes must be functional in the presence of high-salt. Far-reaching adaptations of all cellular components are therefore necessary. As a result, organisms using this "high-salt-in" strategy have lost the ability to live in low-salt environments. An alternative strategy of coping with life in hypersaline environments is to exclude salt from the cytoplasm, while providing the necessary osmotic balance by synthesizing low-molecular-weight organic molecules or accumulating such molecules from the surrounding medium. No drastic modification of the intracellular machinery is required in this case, as these organic "compatible solutes" do not greatly inhibit the activity of conventional, non-salt-adapted enzymes. Moreover, this "low-salt-in" strategy often bestows considerable flexibility upon the organisms, as the intracellular concentrations of osmotic solutes can be regulated to adjust for changes in the extracellular salt concentration.

Use of organic osmotic solutes is widespread in the microbial world. A wide range of such solutes is found including polyols (glycerol, arabitol), amino acids (glutamate, proline), amino acid derivatives (glycine betaine and many others), tetrahydropyrimidines (ectoine, hydroxyectoine), and sugars and derivatives (sucrose, trehalose, glucosylglycerol). We find osmo-adaptation based on such compatible solutes in representatives of all three domains of life. Salt-adapted eukaryotic microalgae and fungi invariably use this option, as do most halophilic and halotolerant aerobic heterotrophic representatives of the domain Bacteria (Ventosa et al., 1998), the oxygenic and anoxygenic phototrophic Bacteria, and the halophilic species among the methanogenic Archaea (Oren, 1999a).

The "high-salt-in" strategy is relatively rarely found in nature. This mode of osmotic adaptation is best known from the aerobic halophilic Archaea of the family Halobacteriaceae. Molar concentrations of KCl are found inside the cells of *Halobacterium* and related genera (Christian and Waltho, 1962; Matheson et al., 1976). Intracellular Na$^+$ concentrations are kept at relatively low levels. To be soluble and functional at such high KCl concentrations, the proteins of these Archaea have a high content of acidic amino acids (glutamate, aspartate), a low content of basic amino acids (lysine, arginine), and they contain relatively low levels of hydrophobic amino acids (Lanyi, 1974; Dennis and Shimmin, 1997). A similar mode of osmotic adaptation, based on accumulation of inorganic ions rather than organic osmotic solutes, has been found in the anaerobic fermentative halophiles of the order Halanaerobiales (domain Bacteria; low G+C subgroup of the Firmicutes). They share with the aerobic Archaea of the family Halobacteriaceae a large excess of acidic amino acids in their proteins (Oren, 1986), and their enzymes require high-salt concentrations for optimal activity (Rengpipat et al., 1988; Oren and Gurevich, 1993).

In the search for microorganisms adapted to life at the highest salt concentrations, the NaCl-saturated crystallizer ponds of solar salterns used for the production of salt from seawater are a rich hunting ground. Such salterns are found worldwide in subtropical and tropical coastal areas (Rodriguez-Valera et al., 1981; Javor, 1989; Oren, 1993; Litchfield et al., 1999). Many types of halophilic and halotolerant microorganisms have been isolated from such salterns. The crystallizer brines are generally pink-red as a result of the presence of dense communities of pigmented halophilic microorganisms. The red coloration of the brines has thus far been attributed to the presence of two types of pigmented microorganisms: Archaea of the family Halobacteriaceae, which contain 50-carbon carotenoids of the bacterioruberin series; and the unicellular green alga *Dunaliella salina*, which under suitable conditions accumulates large amounts of β-carotene.

Until recently there was little evidence that heterotrophic representatives of the domain Bacteria may play a significant role in the microbial community inhabiting salt-saturated environments. Halophilic or highly halotolerant representatives of the bacterial domain were known (Ventosa et al., 1998), but these compete poorly with the Archaea at the highest salt concentrations (Rodriguez-Valera et al., 1980). Moreover, inhibitor studies suggested that all or nearly all heterotrophic activity in saltern crystallizers can be attributed to Archaea (Rodriguez-Valera et al., 1985; Oren, 1990a,b; Pedrós-Alió et al., 2000). However, the prevailing view that Bacteria do not contribute to the saltern crystallizer ecosystem now needs revision following the isolation of *Salinibacter ruber*, a red, aerobic, heterotrophic representative of the domain Bacteria. The organism was first isolated from saltern crystallizer ponds in Spain (Antón et al., 2002). Phylogenetically this species is most closely related to the genus *Rhodothermus* (order Cyto-phagales; *Cytophaga/Flavobacterium* branch of the Bacteria). *S. ruber* is an obligate halophile that grows optimally between 200 and 300 g/l salts, and does not grow below 150 g/l NaCl. The organism may be quite abundant in saltern crystallizer ponds, even to the extent that its red pigment may contribute to the coloration of the brines (Oren and Rodríguez-Valera, 2001). Related red halophilic Bacteria have now been isolated from a salt crust that develops around saline pools in Death Valley, CA (Hollen et al., 2003), and these may probably represent a new species of the genus *Salinibacter*.

Salinibacter not only inhabits similar environments as the halophilic Archaea of the family Halobacteriaceae; examination of its properties has shown that it shares many characteristics with the Archaea rather than with the halophilic or halotolerant aerobic heterotrophic representatives of the domain Bacteria, to which it phylogenetically belongs. Many of these properties point to the possibility of convergent evolution of two phylogenetically disparate groups of organisms toward adaptation to life at the uppermost salt concentrations. This chapter discusses both the similarities and the differences between these two types of red, extremely halophilic, microorganisms.

2.   THE DISCOVERY OF *SALINIBACTER*

The first indications for the existence of extremely halophilic representatives of the domain Bacteria were obtained around 2000, when bacterial 16S rRNA genes of a previously unknown type were recovered from biomass collected from saltern ponds in Spain. In NaCl-saturated ponds, a novel phylotype was found, phylogenetically clustering within the *Cytophaga-Flavobacterium-Bacteroides* group (Antón et al., 2000). The closest known relative was *Rhodothermus*, a genus of slightly halophilic, thermophilic bacteria isolated from marine hot springs (Alfredsson et al., 1988; Sako et al., 1996; Silva et al., 2000). Related 16S rDNA sequences have recently been amplified from a microbial mat within a salt crust developing around hypersaline pools at the Badwater site, Death Valley National Park, CA, USA (Hollen et al., 2003).

When fluorescent probes, designed specifically to detect the bacteria harboring this novel phylotype, were applied to samples of saltern crystallizer biomass, it became clear that these organisms are abundant in this ecosystem: between 5 and 25% of the prokaryotes found in crystallizer ponds at several locations in Spain belonged to this type (Antón et al., 2000; Oren, 2002b). The organisms reacting with the probe are rod-shaped (Antón et al., 1999), and are of a type that had been known to occur in these ponds for some time already, coexisting with the flat square-shaped Archaea that dominate the community (Guixa-Boixareu et al., 1996).

When saltern crystallizer brine samples were enriched with low concentrations of yeast extract, the number of cells harboring the novel bacterial phylotype increased. By this experimental approach, the optimum salinity for their growth was determined to be between 20 and 25%, i.e., in the same range as most halophilic Archaea of the family Halobacteriaceae (Antón et al., 2000). At the time this new type of rod-shaped extreme halophile was designated "*Candidatus Salinibacter*", awaiting the isolation of the organism.

Isolation of this novel type of halophile from saltern crystallizer ponds in Spain and on the Balearic Islands soon followed. The organism produces red-orange colonies on agar plates containing low nutrient, high salt media. Two techniques used independently led to the recognition of colonies of this new bacterium. One approach was based on the use of a fluorescent 16S rRNA-based probe, designed specifically to react with the new bacterial phylotype previously amplified from the Spanish saltern ponds, to detect colonies harboring this or related phylotypes. Analysis of the polar lipids of randomly selected colonies developing on agar plates also led to the isolation of a number of similar strains, recognizable by their bacterial lipid signature. The new species was described as *Salinibacter ruber*. The type strain is strain M31, isolated from a saltern on Mallorca (DSM 13855^T; CECT 5946^T) (Antón et al., 2002). The 16s rRNA gene sequences of the isolates were found to be nearly identical to those sequences recovered earlier from the saltern biomass without prior cultivation (Antón et al., 2000).

The isolation of *Salinibacter* has opened the way to a characterization of its properties. The species consists of brightly red-colored, motile, straight, or slightly curved rods measuring 2-6 x 0.4 μm. Cells stain Gram-negative. With a minimal salt concentration of 150 g/l required for growth, they are among the most halophilic types of the domain Bacteria known. Salt concentrations between 200 and 300 g/l support optimal growth. All strains could grow in solutions saturated with NaCl. Maintenance of cell shape does not depend on the presence of high-salt concentrations, and suspension in distilled water does not lead to cell lysis, in contrast to the lysis of non-coccoid species of the Halobacteriaceae under such conditions. Optimal growth with doubling times of 14-18 h is obtained at temperatures of 37-47 °C and pH 6.5-8.0.

Salinibacter is an obligate aerobe, shows positive oxidase and catalase reactions, and does not reduce nitrate. Growth requirements are probably complex. Amino acids appear to be the preferred nutrients for growth. Certain sugars stimulate growth (see below). Sugar metabolism is not accompanied by the formation of acidic products. Most strains hydrolyze starch and gelatin. High-nutrient levels are inhibitory for growth.

S. ruber is sensitive to many antibiotics that commonly inhibit Bacteria including penicillin G, ampicillin, chloramphenicol, streptomycin, novobiocin, rifampicin, and ciprofloxacin. Kanamycin, bacitracin, tetracycline, and colistin do not inhibit growth, nor do anisomycin and aphidicolin, two potent inhibitors of the growth of halophilic Archaea of the family Halobacteriaceae.

The DNA of *S. ruber* isolates has a G+C content of 66.3 to 67.7 mol%, this being in the same range as found in the Halobacteriaceae.

Two *Salinibacter* isolates were recently obtained from the Badwater site, Death Valley National Park, CA (Hollen et al., 2003). Their 16S rRNA sequence is identical and shows 95-96% similarity to that of the sequences directly amplified from that environment (F.A. Rainey, Louisiana State University, Baton Rouge, personal communication). The similarity of the 16S sequences of these isolates and the type strain of *Salinibacter ruber* is only 93-94% suggesting that they should be classified in a separate species within the genus.

The isolation of *Salinibacter* represents a rare case in which an organism, numerically important in this habitat but previously uncultured, having first been recognized on the basis of 16S rDNA sequences amplified from the environment, has been obtained in pure culture and can be studied. A combination of molecular biological approaches and culturing techniques has thus led to the identification of an important member of the microbial community in salt-saturated ecosystems. The red colonies formed by *S. ruber* on high-salt agar plates closely resemble those of representatives of the Halobacteriaceae. It is therefore very possible that colonies of *Salinibacter* have been observed many times in the past, but were mistakenly considered to consist of red halophilic Archaea.

3.   A COMPARISON BETWEEN *SALINIBACTER RUBER* AND THE HALOBACTERIACEAE

3.1   Ion metabolism

On the basis of its phylogenetic position it was originally expected that *Salinibacter* would use organic osmotic solutes rather than inorganic ions to provide the cells with osmotic balance. With the exception of the anaerobic fermentative bacteria classified within the order Halanaerobiales (low G+C branch of the Firmicutes), the halophilic and halotolerant representatives of the domain Bacteria maintain intracellular ion concentrations generally much below the levels present in their medium (Ventosa et al., 1998; Oren,

2002a). In contrast, the halophilic Archaea of the order Halobacteriales that live in the same habitat as *Salinibacter* use KCl as osmotic solute (Christian and Waltho, 1962; Matheson et al., 1976; Oren, 1999b, 2000) and do not produce or accumulate organic solutes.

It was therefore surprising to find extremely high concentrations of both K^+ and Cl^- within the cells of *Salinibacter ruber*. Flame photometric measurements of K^+ in perchloric acid extracts of cells grown at a total salt concentration of 250 g/l yielded values between 11.4 and 15.2 μmol K^+/mg protein (Antón et al., 2002). Values in the same range (12.0 ± 0.7 μmol K^+/mg protein) were measured in the halophilic archaeon *Halobacterium salinarum*. K^+/protein ratios (in g per g) between 0.6-1.9 and 0.5-1.3 were reported earlier in the Archaea *Halobacterium salinarum* and *Haloferax mediterranei* (Pérez-Fillol and Rodríguez-Valera, 1986), values equivalent to 15-49 and to 13-33 μmol/mg protein. For comparison, depending on growth conditions, between 1.1 and 2.2 μmol K^+ per mg protein were found in the aerobic *Halomonas elongata* (γ-Proteobacteria), an organism that synthesizes ectoine and/or accumulates glycine betaine as osmotic solutes. Intermediate values (6.3 μmol K^+/mg protein) were found in *Halanaerobium praevalens*, a representative of the Halanaerobiales shown earlier to use KCl to provide osmotic balance (Oren, 1986; Oren et al., 1997).

To confirm the presence of high intracellular K^+ concentrations and to assess the level of intracellular Cl^- in *Salinibacter* cells, X-ray microanalysis of single cells was performed in the electron microscope. This technique has extensively been used in the elemental analyses of marine and other bacteria (Heldal et al., 1985; Norland et al., 1995) and has confirmed the presence of high KCl concentrations in *Halanaerobium praevalens* (Oren et al., 1997). The measurements confirmed the presence of high intracellular K^+ concentrations in *S. ruber*, and showed that intracellular Cl^- approximately balances the sum of K^+ and Na^+ (Oren et al., 2002). Apparent concentrations of K^+, Na^+, Mg^{2+}, and Cl^- in *Salinibacter* grown in medium containing 3.3 M NaCl were 0.59 M, 0.38 M, 0.29 M, and 0.9 M, respectively. The corresponding values for *Halobacterium salinarum* were 0.93 M, 0.22 M, 0.03 M, and 1.14 M. That the apparent KCl concentration in *H. salinarum* cells was lower than the 4-5 M expected (Christian and Waltho, 1962; Matheson et al., 1976), may be due to certain artifacts inherent to the procedure such as the possibility that ions leaked out from the cells during centrifugation. Far lower intracellular K^+, Na^+, and Cl^- concentrations were measured in the halophilic Bacteria *Salinivibrio costicola* and *Halomonas elongata*, as expected in view of the fact that these organisms produce and accumulate organic osmotic solutes.

3.2   Compatible solutes

All halophilic and halotolerant Bacteria characterized thus far (with the exception of the anaerobic fermentative Halanaerobiales) use organic compatible solutes rather than inorganic salts for osmotic stabilization (Galinski, 1993, 1995; Ventosa et al., 1998). Therefore, it was predicted that *Salinibacter* may use a similar strategy. Its phylogenetically closest known relative, the marine thermophilic *Rhodothermus marinus*, which has been isolated from undersea hot springs, uses α-mannosylglycerate and α-mannosylglyceramide to provide osmotic balance (Silva et al., 1999).

A search for intracellular organic osmotic solutes in *Salinibacter* was therefore made. Cell extracts were analyzed by ^{13}C-NMR and HPLC techniques. Moreover, amino acids and other amino-reactive compounds were identified and quantified by HPLC following derivatization with 9-fluorenyl-methoxycarbonyl chloride (FMOC). Direct HPLC analysis of extracts of *Salinibacter*, using a refractive index detector to identify organic solutes, showed no osmotic compounds in significant concentrations. ^{13}C-NMR analyses showed signals attributable to low levels of glutamate and of glycine betaine. $N\alpha$-acetyllysine was detected in low concentrations. HPLC analysis of FMOC-reactive compounds showed glutamate to be present at relatively low concentrations of about 5.8 and 4.5 mg/g dry weight in cells grown in 150 and in 250 g/l NaCl, respectively, equivalent to about 19 and 24 mg/g cell protein. $N\alpha$-acetyllysine was the only compound whose concentration increased with the medium salinity. However, its cytoplasmic levels were low - from undetectable in cells grown at 150 g/l NaCl to 1.2 mg/g dry weight or 6.5 mg/g protein in cells grown at 250 g/l NaCl. Other, unidentified amino-reactive compounds were present, especially in cells grown in the presence of yeast extract, but their concentration was very low (Oren et al., 2002). Calculations of the total amounts of organic solutes detected in the cell extracts show that all these solutes are present in low concentrations only and do not contribute much to the overall osmotic balance.

3.3   Acidic proteins

Microorganisms that maintain high intracellular KCl concentrations, such as *Halobacterium* and related archaeal genera, possess salt-adapted enzymes and other proteins. These proteins are characterized by a high content of acidic amino acids (aspartate, glutamate), a low content of basic amino acids (lysine, arginine), a low content of hydrophobic amino acids, and a relatively high content of "borderline hydrophobic" amino acids such as serine and threonine (Lanyi, 1974; Oren, 2002a). Serine appears to be an important determinant of hydrophobicity because of its compact size and borderline hydrophobic-hydrophilic character (Dennis and Shimmin, 1997). Such proteins are fully functional at high-salt concentrations. They generally require high salt for stability and activity, and they often denature at low-salt concentrations.

Different mechanisms have been proposed to explain how the halophilic behavior of these proteins depends on their peculiar amino acid composition. Acidic residues are highly hydrated, and they can coordinate the organization of a hydrated salt ion network at the surface of the protein (Lanyi, 1974; Eisenberg and Wachtel, 1987; Eisenberg et al., 1992; Dym et al., 1995; Eisenberg, 1995; Dennis and Shimmin, 1997; Madigan and Oren, 1999). Salt bridges between acidic residues and strategically positioned basic residues further provide structural rigidity to the proteins, and they are thus important determinants in the stabilization of their tertiary structure (Dym et al., 1995; Dennis and Shimmin, 1997). An ordered water molecule network and presence of inter-subunit salt bridges are of importance as well (Richard et al., 2000). The low number of surface-exposed lysine residues further reduces the hydrophobic character of the protein's surface (Britton et al., 1998).

Intracellular proteins of halophilic and halotolerant microorganisms that do not maintain high intracellular KCl concentrations, but use organic osmotic solutes instead, do not require such a far-reaching adaptation. Their amino acid composition is not very different from that of the proteins of non-halophilic microorganisms. A slightly increased content of acidic amino acids is, however, sometimes found. The bulk protein of *Halomonas elongata*, for example, showed a certain excess of acidic amino acids and a lowered frequency of basic amino acids, values being intermediate between those of the non-halophilic *Escherichia coli* and the halophilic archaeon *Haloferax mediterranei*. It was suggested that the increased acidity of the *Halomonas* proteins was a result of convergent evolution (Gandbhir et al., 1995; see also Oren, 1995).

Comparison of the amino acid composition of the bulk protein of *Salinibacter ruber* with that of archaeal aerobic halophiles (*Halobacterium salinarum*, *Haloarcula marismortui*), the bacterial anaerobic halophile *Halanaerobium praevalens*, the aerobic *Halomonas elongata*, and the non-halophilic *Escherichia coli* showed that *Salinibacter* proteins have properties very similar to those of the Halobacteriales. The acid hydrolysis procedure used to hydrolyze the proteins did not allow separate determination of glutamate and glutamine and of aspartate and asparagine. Therefore the comparison was based on the relative content of "Glx" + "Asx". The Glx + Asx content of *Salinibacter*, *Halobacterium*, and *Haloarcula* bulk protein was 29.2, 31.8, and 32.3 mol%, respectively. For comparison, values for *Halomonas* and *Escherichia* were both around 26.0 mol%. The apparent excess of acidic amino acids ([Asx + Glx] – [Lys + Arg], again including the contribution of Asn and Gln), is also high in *Salinibacter*: 21.2 mol%, as compared to 25.4; 26.2 mol% for the two Archaea tested; and 17.0 and 15.9 mol% for *Halomonas* and *Escherichia*. *Halanaerobium praevalens* showed intermediate values for the above indices (Oren and Mana, 2002). The content of hydrophobic amino acids in the proteins of *Salinibacter* is low: [Leu + Val + Ile + Phe] = 22.5 mol%, being only little higher than the values of 21.6 and 22.0 mol% determined for *Halobacterium* and *Haloarcula* but substantially lower than the 24.1 and 23.8 mol% for *Halomonas* and *Escherichia*. The *Salinibacter* bulk protein has a high serine content, namely 5.5 mol%, as high as the values measured in *Halobacterium* and *Haloarcula* (5.3 and 5.6 mol%) and higher than those of the non-halophilic *Escherichia* (4.6 mol%) or the moderately halophilic *Halomonas* (5.2 mol%) (Oren and Mana, 2002).

3.4   Salt-dependent and salt-independent enzymes

Most intracellular enzymes of the halophilic Archaea function optimally in the presence of molar concentrations of salts (KCl, NaCl), such as are present inside the cells' cytoplasm, and they lose their activity when suspended in low-salt solutions. This is in sharp contrast to the enzymes of the aerobic, moderately halophilic Bacteria, which maintain low-salt concentrations intracellularly while using organic solutes for osmotic balance. Salt-requiring enzymes are found as well in the anaerobic Bacteria of the order Halanaerobiales, a group that also uses the "high-salt-in" strategy of osmotic adaptation (Oren, 1986; Rengpipat et al., 1988; Oren and Gurevich, 1993).

So far, eight enzymatic activities of *Salinibacter ruber* have been tested for their relationship to salt: two enzymes of the central dissimilatory metabolism (isocitrate dehydrogenase, malate dehydrogenase), a central enzyme in amino acid metabolism (glutamate dehydrogenase), enzymes involved in sugar and glycerol dissimilation (hexokinase, glucose-6-phosphate dehydrogenase, glycerol dehydrogenase, glycerol kinase), and the fatty acid synthetase complex, essential for the biosynthesis of lipids. The findings were as follows:

• Isocitrate dehydrogenase: Two activities were detected, one being NAD-dependent and one NADP-dependent (Oren and Mana, 2002). The NAD-dependent enzyme was salt-dependent. Optimal activity was measured at KCl concentrations between 0.5 and 2 M. At 3.3 M KCl activity was about two-thirds of that measured under optimum conditions. NaCl was slightly less effective as an activator. The NADP-dependent isocitrate dehydrogenase activity was little influenced by NaCl concentration in the range 0-3.2 M while increasing KCl concentrations effected a slight stimulation. It is unknown whether the two activities are catalyzed by different enzymes or whether there is a single enzyme with dual coenzyme specificity. For comparison, the NADP-dependent isocitrate dehydrogenase from *Halobacterium salinarum* was reported to be halophilic. Optimal activity was obtained at molar concentrations of KCl or NaCl (Baxter and Gibbons, 1956; Aitken and Brown, 1969; Aitken et al., 1970). No data are available on the presence and properties of NAD-dependent isocitrate dehydrogenase in halophilic Archaea.

- Malate dehydrogenase: NAD-dependent malate dehydrogenase activity of *Salinibacter* was optimal in the absence of salt, and increasing KCl and NaCl concentrations lowered the activity. However, the enzyme was still functional at the high-salt concentrations detected in the cytoplasm: approximately 25% of the maximum activity remained at NaCl and KCl concentrations above 3 M (Oren and Mana, 2002). For comparison, the NAD-dependent malate dehydrogenase of *Halobacterium salinarum* is optimally active in 2-3 M KCl, and little activity is found below 1 M salt (Baxter and Gibbons, 1956; Holmes and Halvorson, 1965; Aitken and Brown, 1969). The enzyme from *Haloarcula marismortui* is inactivated below 2 M NaCl. Crystallographic and physico-chemical studies have shown this enzyme to be a tetramer whose structure is stabilized by an ordered water molecule network kept together by the numerous acidic residues on the surface and by intersubunit salt bridges (Mevarech and Neumann, 1977; Dym et al., 1995; Eisenberg, 1995; Richard et al., 2000).

- Glutamate dehydrogenase: Little activity of NAD-dependent glutamate dehydrogenase (as assayed by the reductive amination of α-ketoglutarate) was obtained in the absence of salt. Increasing KCl concentrations lowered the activity still more, and no activity could be detected above 2.5 M KCl. However, high NaCl concentrations were highly stimulatory, and optimum activity was found at 3-3.5 M (Oren and Mana, 2002). Indications have been obtained for the presence of two different enzymes, one salt-sensitive and one salt-dependent (Maria-José Bonete, University of Alicante, personal communication). No NADP-dependent glutamate dehydrogenase activity was detected in *S. ruber* extracts. For comparison, *Halobacterium salinarum* has both NAD-dependent and NADP-dependent glutamate dehydrogenase activity. The NAD-dependent enzyme probably has a catabolic function in the oxidative deamination of glutamate. It is optimally active in 3.2 M NaCl or in 0.8 M KCl. High KCl concentrations inhibit the deamination reaction, while 4 M NaCl supported excellent activity. The amination reaction was also inhibited by KCl concentrations above 0.8 M while activity increased with NaCl concentration up to 4 M (Bonete et al., 1986, 1987). Other reports state that the enzyme was still active in 4 M KCl (Britton et al., 1998). The NADP-specific enzyme, assayed by the reductive amination of α-ketoglutarate, was optimally active in 1.6 M KCl and required at least 0.5 M KCl or NaCl for activity

(Bonete et al., 1987). The NADP-dependent glutamate dehydrogenase of *Haloferax mediterranei* functioned optimally at 1-2 M KCl or NaCl. Activity in the absence of salt was about one-third of the optimal value (Ferrer et al., 1996).

- Hexokinase: activity of hexokinase was detected in cell extracts of *S. ruber*, in cells grown in both the presence and the absence of glucose. Its activity was markedly inhibited by salt, and no activity was recorded in the presence of 2.1 M KCl or 2.8 M NaCl. A similar salt sensitivity was reported for the hexokinases of *Haloferax mediterranei* and *Haloarcula vallismortis* (Rawal et al., 1988).

- Glucose-6-phosphate dehydrogenase: *S. ruber* has a constitutive NADP-linked glucose-6-phosphate dehydrogenase activity. The enzyme is markedly salt-dependent: activity was optimal above 1.5-2 M NaCl or KCl, and no activity was found below 0.8 M salt (Oren and Mana, 2003). This enzyme probably functions in the Entner-Doudoroff pathway for glucose dissimilation. It has no parallel in the archaeal domain, as those halophilic Archaea that degrade glucose use a modification of the Entner-Doudoroff pathway in which the glucose is first oxidized to gluconate while the phosphorylation step is postponed (see below).

- Glycerol kinase: *S. ruber* cells grown in the presence of glycerol contained glycerol kinase activity. The activity was similar from 0.6-2.8 M KCl. No activity was detected when glycerol was absent from the growth medium (Jonathan Sher, Lili Mana and Aharon Oren, unpublished results). Constitutive presence of glycerol kinase has been documented in many halophilic Archaea (Oren and Gurevich, 1994). Those assays were performed in 3 M KCl; the effect of lowering the salt concentrations in the assay mixtures was not tested at the time.

- Fatty acid synthetase: The fatty acid synthetase complex of *Salinibacter ruber* is optimally active at 0.5-1.5 M salt, and its activity is much reduced in the absence of salt (Oren et al., 2003). It is thus the second truly halophilic fatty acid synthetase known. A fatty acid synthetase complex with a high-salt optimum (about 1 M KCl or NaCl) was earlier characterized from the halophilic anaerobic *Halanaerobium praevalens* (Oren and Gurevich, 1993). As halophilic Archaea have lipids with isoprenoid hydrophobic chains, the synthesis of their membrane lipids does not depend on fatty acid synthetase. Still, fatty acid synthetase activity was detected in *Halobacterium salinarum*, and surprisingly its activity was found to be strongly salt-inhibited (Pugh et al., 1971). It was recently shown that straight-chain fatty acids are used by these halophilic Archaea, not as building blocks for membrane lipids but to acylate certain membrane proteins, making them more hydrophobic (Pugh and Kates, 1994).

Summarizing, nearly all enzymes of *Salinibacter* that have been studied function at high-salt concentrations, and many are markedly stimulated by molar salt concentrations. In this respect they resemble the enzymes of the Halobacteriales. However, significant differences do occur in the relation to salt of the individual enzymes: some function best at low-salt, while still retaining considerable activity at high-salt concentrations such as occurs in the cytoplasm of *Salinibacter*.

3.5   Sugar metabolism

The Halobacteriaceae generally prefer amino acids to sugars as substrates. Some species appear altogether unable to grow on carbohydrates. *Halobacterium salinarum* is an example, although there are reports that addition of carbohydrates may result in some growth stimulation (Gochnauer and Kushner, 1969). Other species, notably those belonging to the genera *Haloferax* and *Haloarcula*, can grow on sugars as a single carbon and energy source.

The enzymology of carbohydrate metabolism in halophilic Archaea was first elucidated in *Halorubrum saccharovorum* (Tomlinson and Hochstein, 1972; Hochstein, 1978). Breakdown of glucose proceeds through a modification of the Entner-Doudoroff pathway, in which the phosphorylation step is postponed (glucose → gluconate → 2-keto-3-deoxygluconate, followed by phosphorylation to 2-keto-3-deoxy-6-phosphogluconate, which is subsequently split into pyruvate and glyceraldehyde-3-phosphate) (Tomlinson et al., 1974; Rawal et al., 1988; Sonawat et al., 1990; Johnsen et al., 2001). For comparison, glucose-6-phosphate, and 6-phosphogluconate are intermediates in the conventional Entner-Doudoroff pathway. The genes coding for glucose dehydrogenase and for 2-keto-3-deoxygluconate kinase have been identified in the genome of *Halobacterium* NRC-1, but the gene for 2-keto-3-deoxy-6-phosphogluconate aldolase remains to be assigned (Ng et al., 2000).

Those halophilic Archaea that degrade fructose use the Embden-Meyerhof pathway, in which fructose-1,6-bisphosphate aldolase is a key enzyme (D'Souza and Altekar, 1982; Dhar and Altekar, 1986a,b; Rawal et al., 1988; Altekar and Rangaswamy, 1990, 1991; Krishnan and Altekar, 1991). A ^{13}C-NMR study of sugar metabolism in *Halococcus saccharolyticus* confirmed that glucose is metabolized via the modified Entner-Doudoroff pathway while fructose is degraded through the Embden-Meyerhof pathway (Johnsen et al., 2001).

The special features of carbohydrate metabolism in the halophilic Archaea make a comparison with the enzymology of sugar dissimilation in *Salinibacter* of special interest. The original description of *S. ruber* states that the species may be unable to metabolize sugars, as addition of sugars and related compounds at a concentration of 5 g/l did not stimulate growth

(Antón et al., 2002). However, a few sugars (glucose, maltose, starch) significantly increase the growth yield when added at a concentration of 1 g/l (Oren and Mana, 2003). Fructose, galactose, sucrose, lactose, arabinose, and a number of other sugars do not stimulate growth. Glucose degradation starts only after other, more easily metabolizable substrates have been depleted from the medium. A diauxic growth curve is thus obtained in medium containing both yeast extract and glucose as carbon and energy sources.

Glucose dehydrogenase, an enzyme present in halophilic Archaea (Madan and Sonawat, 1996; Bonete et al., 1996; Ferrer et al., 2001) could not be demonstrated in *S. ruber*. However, the cells contain NADP-dependent glucose-6-phosphate hydrogenase activity when grown in both the presence and the absence of glucose. This enzyme functions optimally above 1.5-2 M NaCl or KCl, and is inactive below 0.8 M salt. Furthermore, a constitutive hexokinase activity was found, which was inhibited by high-salt concentrations (Oren and Mana, 2003). It is therefore suggested that *Salinibacter* metabolizes glucose by the classic Entner-Doudoroff pathway and not by the Embden-Meyerhof glycolytic pathway or by the modified Entner-Doudoroff pathway present in halophilic Archaea of the family Halobacteriaceae, in which the phosphorylation step is postponed. However, activity of 2-keto-3-deoxy-6-phosphogluconate aldolase could not be detected in extracts of *Salinibacter* cells, whether or not grown in the presence of glucose.

3.6   Pigments

Microorganisms inhabiting salt lakes and saltern crystallizer ponds are generally brightly pigmented. *Dunaliella* cells living in such environments are often orange-colored due to their high content of β-carotene granules located between the thylacoids of the chloroplast. Even more conspicuous is the pink-red color contributed to the brines by α-bacterioruberin and related C-50 carotenoids present in the cytoplasmic membrane of nearly all species of Halobacteriaceae. These pigments protect the cells against harmful sunlight radiation to which the organisms are exposed in their natural environment (Dundas and Larsen, 1962). Colorless species of Halobacteriaceae are rare, *Natrialba asiatica* being the only species described to date that lacks bacterioruberin carotenoids.

Salinibacter is colored red as well, and the color of its colonies on agar plates superficially resembles that of the halophilic Archaea. It is therefore quite possible that *Salinibacter* colonies have been observed many times in the past, but have erroneously been considered as belonging to species of Halobacteriaceae. However, the pigment of *S. ruber* differs from the archaeal bacterioruberins. Its absorption spectrum in methanol/acetone (1:1, v/v) shows a maximum at 478 nm and a shoulder at 506-510 nm (compare an absorption maximum at 496 nm, with a shoulder around 470 nm and minor peak at 530 nm, for extracts of *Halobacterium* and relatives). The principal (>96% of total) pigment of *S. ruber* is a novel acylated gluco-carotenoid, now named salinixanthin (Lutnæs et al., 2002). By methods of visible light spectroscopy, electron impact mass spectrometry, ^{1}H- nuclear magnetic resonance, circular dichroism, and gas chromatography-mass spectroscopy, and by chemical methods, the chemical structure of the pigment has been elucidated as (all-*E*, 2'*S*)-2'-hydroxy-1'-[6-*O*-(13-methyltetradecanoyl)- β - D – glycopyranosyloxy - 3',4' – didehydro-1',2'-dihydro-β,ψ-caroten-4-one (molecular formula ($C_{61}H_{92}O_9$). Note that the esterifying fatty acid, the branched 13-methyltetradecanoic acid, is also a major constituent of the membrane lipids of *S. ruber* (see below). This pigment greatly differs from the non-isoprenoid polyene pigments of the flexirubin type characteristic of many representatives of the *Cytophaga-Flavobacterium* group. However, it does resemble the pigment of the phylogenetically related *Rhodothermus marinus*, which has a similar absorption spectrum (Alfredsson et al., 1988). Moreover, it resembles a minor carotenoid (4-keto-phleixanthophyll) from the phylogenetically unrelated *Mycobacterium phlei*, which corresponds to the planar structure of the free glycoside of salinixanthin (Bjart-Frode Lutnæs, Synnøve Liaaen-Jensen and coworkers, Trondheim, unpublished results).

Because the *Salinibacter* can easily be resolved from halophilic archaeal pigments by HPLC, the presence of salinixanthin in natural hypersaline brines can easily be ascertained and quantified. In brine collected from saltern crystallizer ponds near Alicante, Spain, and on Mallorca, about 5% of the total prokaryote-derived carotenoid-like pigments could be attributed to *Salinibacter* (Oren and Rodríguez-Valera, 2001).

3.7   Lipids

As for membrane lipids, *Salinibacter* differs greatly from the aerobic halophilic Archaea. The polar lipids of the Archaea are isoprenoid ether derivatives of phosphatidylglycerol, the methyl ester of phosphatidyl-glycerophosphate, phosphatidylglycerosulfate (present only in part of the species), and different glycolipids, some of which carry sulfate groups on one of the sugar moieties. *Salinibacter* shows a polar lipid pattern characteristic of many halotolerant and non-halotolerant representatives of the bacterial domain, with phosphatidylcholine, phosphatidylethanolamine, phosphatidylglycerol and diphosphatidylglycerol. As explained above, some of the *S. ruber* isolates were initially recognized as belonging to the bacterial domain on the basis of their polar lipid patterns on thin layer chromatography plates (Antón et al., 2002).

The lipids of *Salinibacter* contain ester-linked fatty acyl chains that are typical of Bacteria. The major fatty acids detected are 16:1ω7, 18:1ω7, and methyl-branched acids (mainly *iso*-branched 15:0). The adaptive changes in both head-group and fatty acid composition in response to changes in salinity or temperature are complex and typical of other halophilic or halotolerant Bacteria (Oren et al., 2003).

4.   FINAL COMMENTS

The data that have accumulated since *Salinibacter* was discovered a few years ago show that its physiological properties closely resemble those of the Halobacteriaceae, the family of extremely halophilic aerobic Archaea. *Salinibacter*, being one of the most halophilic organisms belonging to the domain Bacteria, thereby behaves very different from the other halophilic and halotolerant representatives of the bacterial domain.

Both *Salinibacter* and *Halobacterium* and its relatives are aerobic heterotrophs that maintain high intracellular K^+ and Cl^- concentrations. We know only one other group within the domain Bacteria that accumulates KCl to provide osmotic balance and does not produce organic osmotic solutes. This is the order Halanaerobiales, families Halanaerobiaceae and Halobacteroidaceae (low G+C subgroup of the Firmicutes, thus phylogenetically not related with *Salinibacter*) (Oren, 1986; Rengpipat et al., 1988; Oren et al., 1997). The Halanaerobiales are fermentative bacteria that obtain only a little energy from the fermentation processes they perform. It has therefore been argued that the use of KCl as osmotic solute is dictated by bioenergetic considerations, as accumulation of KCl appears to be energetically cheaper than the production of organic compatible solutes (Oren, 1999a). Such bioenergetic constraints can hardly apply to the case of *Salinibacter*, which may be expected to obtain large amounts of energy from aerobic respiration processes.

No organic osmotic solutes were detected in significant concentrations in the cytoplasm of *Salinibacter*. In this respect it resembles most members of the Halobacteriaceae. The finding of 2-sulfotrehalose providing osmotic balance together with KCl in some haloalkaliphilic Archaea (Desmarais et al., 1997) appears to be an exception to the rule. Use of KCl as an osmotic solute requires far-reaching adaptations of the intracellular enzymatic machinery to be functional in the presence of molar concentrations of salt (salt-requiring proteins with a high excess of acidic amino acids) and does not allow much flexibility with respect to salt concentrations enabling growth. The advantage of the use of organic osmotic solutes is the often great ability of the cells to rapidly adapt to changes in the external salt concentration (Ventosa et al., 1998; Oren, 1999a). *Salinibacter*, similar to the halophilic Archaea, has abandoned this flexibility to be able to grow at the highest salt concentrations. Organisms that can grow at near-saturated salt concentrations while excluding salts from the cytoplasm and accumulating organic solutes are rare, probably because of the limited solubility of even the most suitable organic osmotic solutes. Glycerol, used by *Dunaliella* for this purpose, is not encountered in the prokaryotic world, probably because of the high permeability of bacterial membranes to glycerol. The most halophilic among the organic solute accumulating prokaryotes are probably the species of the anoxygenic photosynthetic bacterial genus *Halorhodospira*. *Halorhodospira* uses cocktails of osmotic compounds including glycine betaine, ectoine, and trehalose (Galinski and Herzog, 1990). Glycine betaine is a small molecule and very soluble. However, only a few non-photosynthetic prokaryotes synthesize this compound *de novo*.

Comparison of all the properties of *Salinibacter* with those of the Halobacteriales suggests that we may be dealing here with an excellent example of convergent evolution, in which similar features have evolved in phylogenetically unrelated organisms as an adaptation to life at the highest salt concentrations. The similarity extends to the absorption spectrum of their pigments (despite the different chemical nature of the pigments involved), to the use of (variations of) the Entner-Doudoroff pathway for sugar degradation, and even to the G+C percentage of their DNA.

The first genomes of halophilic Archaea have now been sequenced. The first genome released was that of *Halobacterium* strain NRC-1 (Ng et al., 2000). The genome of *Halobacterium salinarum* strain R1 is also complete (see http://www.halolex.mpg.de) and the sequencing of the genomes of *Natronomonas pharaonis* and *Haloferax volcanii* is nearing completion. In view of the similarity in the way of life of *Salinibacter* and the halophilic Archaea, a comparative genomic analysis of *Salinibacter* will undoubtedly shed much light on the process of convergent evolution that led to the ability to thrive in saturated salt, while giving up the ability to live at low salinities.

5. ACKNOWLEDGMENT

This study was supported by a grant from the Israel Science Foundation founded by the Israel Academy of Sciences and Humanities.

6. REFERENCES

Aitken D.M. and Brown A.D. 1969. Citrate and glyoxylate cycles in the halophil, *Halobacterium salinarum*. Biochimica et Biophysica Acta, 177, 351-354.

Aitken D.M., Wicken A.J. and Brown A.D. 1970. Properties of a halophil nicotinamide-adenine dinucleotide phosphate-specific isocitrate dehydrogenase. Preliminary studies of the salt relations and kinetics of the crude enzyme. Biochem J. 116, 125-134.

Alfredsson G.A., Kristjansson J.K., Hjorleifsdottir S. and Stetter K.O. 1988. *Rhodothermus marinus* gen. nov., a thermophilic, halophilic bacterium from submarine hot springs in Iceland. J Gen Microbiol. 134, 299-306.

Altekar W. and Rangaswamy V. 1990. Induction of a modified EMP pathway for fructose breakdown in a halophilic archaebacterium. FEMS Microbiol Letters 69, 139-144.

Altekar W. and Rangaswamy V. 1991. Ketohexokinase (ATP: D-fructose 1-phospho-transferase) initiates fructose breakdown via the modified EMP pathway in halophilic archaebacteria. FEMS Microbiol Letters 83, 241-246.

Antón J., Llobet-Brossa E., Rodríguez-Valera F. and Amann R. 1999. Fluorescence in situ hybridization analysis of the prokaryotic community inhabiting crystallizer ponds. Environm Microbiol. 1, 517-523.

Antón J., Rosselló-Mora R., Rodríguez-Valera F. and Amann R. 2000. Extremely halophilic Bacteria in crystallizer ponds from solar salterns. Appl Environm Microbiol. 66, 3052-3057.

Antón J., Oren A., Benlloch S., Rodríguez-Valera F., Amann R. and Rosselló-Mora R. 2002. *Salinibacter ruber* gen. nov., sp. nov., a novel extreme halophilic member of the Bacteria from saltern crystallizer ponds. Int J Syst Evol Microbiol. 52, 485-491.

Baxter R.M. and Gibbons N.E. 1954. The glyceroldehydrogenases of *Pseudomonas saliaria*, *Vibrio costicola*, and *Escherichia coli* in relation to bacterial halophilism. Can J Biochem Physiol. 32, 206-217.

Baxter R.M. and Gibbons N.E. 1956. Effects of sodium and potassium on certain enzymes of *Micrococcus halodenitrificans* and *Pseudomonas salinaria*. Can J Microbiol. 2, 599-606.

Bonete M.J., Camacho M.L. and Cadenas E. 1986. Purification and some properties of NAD$^+$-dependent glutamate dehydrogenase from *Halobacterium halobium*. Int J Biochem. 18, 785-789.

Bonete M.J., Camacho M.L. and Cadenas E. 1987. A new glutamate dehydrogenase from Halobacterium halobium with different coenzyme specificity. Int J Biochem. 19, 1149-1155.

Bonete M.J., Pire C., Llorca F.I. and Camacho M.L. 1996. Glucose dehydrogenase from the halophilic archaeon *Haloferax mediterranei*: enzyme purification, characterisation, and N-terminal sequence. FEBS Letters 383, 227-229.

Britton K.L., Stillman T.J., Yip K.S.P., Forterre P., Engel P.C. and Rice D.W. 1998. Insights into the molecular basis of salt tolerance from the study of glutamate dehydrogenase from *Halobacterium salinarum*. J Bio Chem. 293, 9023-9030.

Christian J.H.B. and Waltho J.A. 1962. Solute concentrations within cells of halophilic and non-halophilic bacteria. Biochimica et Biophysica Acta 65, 506-508.

Dennis P.P. and Shimmin L.C. 1997. Evolutionary divergence and salinity-mediated selection in halophilic Archaea. Microbiol Mol Biol Rev. 61, 90-104.

Desmarais D., Jablonski P.E., Fedarko N.S. and Roberts M.F. 1997. 2-*Sulfotrehalose*, a novel osmolyte in haloalkaliphilic archaea. J Bacteriol. 179, 3146-3153.

Dhar N.M. and Altekar W. 1986a. A class I (Schiff base) fructose-1,6-bisphosphate aldolase of halophilic archaebacterial origin. FEBS Letters 199, 151-154.

Dhar N.M. and Altekar W. 1986b. Distribution of class I and class II fructose bisphosphate aldolases in halophilic archaebacteria. FEMS Microbiol Letters 35, 177-181.

D'Souza S.E. and Altekar W. 1982. A halophilic fructose 1,6-bisphosphate aldolase from *Halobacterium halobium*. Ind J Biochem Biophys. 19, 135-138.

Dundas I.D. and Larsen H. 1962. The physiological role of the carotenoid pigments of *Halobacterium salinarium*. Archiv für Mikrobiologie 44, 233-239.

Dym O., Mevarech M. and Sussman J.L. 1995. Structural features that stabilize halophilic malate dehydrogenase from an archaebacterium. Science 267, 1344-1346.

Eisenberg H. 1995. Life in unusual environments: progress in understanding the structure and function of enzymes from extreme halophilic bacteria. Arch Biochem Biophys. 318, 1-5.

Eisenberg H. and Wachtel E.J. 1987. Structural studies of halophilic proteins, ribosomes, and organelles of bacteria adapted to extreme salt concentrations. Ann Rev Biophys Biophys Chem. 16, 69-92.

Eisenberg H., Mevarech M. and Zaccai G. 1992. Biochemical, structural, and molecular genetic aspects of halophilism. Advan Prot Chem. 43, 1-62.

Ferrer J., Pérez-Pomares F. and Bonete M.J. 1996. NADP-glutamate dehydrogenase from the halophilic archaeon *Haloferax mediterranei*: enzyme purification, N-terminal sequence and stability. FEMS Microbiol Letters 141, 59-63.

Ferrer J., Fisher M., Bürke J., Sadelnikova S.E., Baker B.J., Gilmour D.J., Bonete M.J., Pire C., Esclapez J. and Rice D.W. 2001. Crystallization and preliminary X-ray analysis of glucose dehydrogenase from *Haloferax mediterranei*. Acta Crystallographica D57, 1887-1889.

Galinski E.A. 1993. Compatible solutes of halophilic eubacteria: molecular principles, water-solute interaction, stress protection. Experientia 49, 487-496.

Galinski E.A. 1995. Osmoadaptation in bacteria. Advan Microbiol Physiol. 37, 273-328.

Galinski E.A. and Herzog R.M. 1990. The role of trehalose as a substitute for nitrogen-containing compatible solutes (*Ectothiorhodospira halochloris*). Arch Microbiol. 153, 607-613.

Gandbhir M., Rashed I., Marlière P. and Mutzel R. 1995. Convergent evolution of amino acid usage in archaebacterial and eubacterial lineages adapted to high salt. Res Microbiol. 146, 113-120.

Gochnauer M.B. and Kushner D.J. 1969. Growth and nutrition of extremely halophilic bacteria. Can J Microbiol. 15, 1157-1165.

Guixa-Boixareu N., Calderón-Paz J.I., Heldal M., Bratbak G. and Pedrós-Alió C. 1996. Viral lysis and bacterivory as prokaryotic loss factors along a salinity gradient. Aquatic Microbial Ecol. 11, 215-227.

Heldal M., Norland S. and Tumyr O. 1985. X-ray microanalytic method for measurement of dry matter and elemental content of individual bacteria. Appl Environm Microbiol. 50, 1251-1257.

Hochstein L.I. 1978. Carbohydrate metabolism in the extremely halophilic bacteria: the role of glucose in the regulation of citrate synthase activity. In Caplan S.R. and Ginzburg M. (Eds.), Energetics and structure of halophilic microorganisms (pp. 397-412). Elsevier/North Holland Biomedical Press, Amsterdam.

Hollen B.J., Bagaley D.R., Small A.M., Oren A., McKay C.P. and Rainey F.A. 2003. Investigation of the microbial community of the salt surface layer at Badwater, Death Valley National Park. Abstract, ASM annual meeting, Washington, D.C.

Holmes P.K. and Halvorson H.O. 1965. Properties of a purified halophilic malic dehydrogenase. J Bacteriol. 90, 316-326.

Javor B. 1989. Hypersaline environments. Microbiology and biogeochemistry. Springer-Verlag, Berlin.

Johnsen U., Selig M., Xavier K.B., Santos H. and Schönheit P. 2001. Different glycolytic pathways for glucose and fructose in the halophilic archaeon *Halococcus saccharolyticus*. Arch Microbiol. 175, 52-61.

Krishnan G. and Altekar W. 1991. An unusual class I (Schiff base) fructose-1,6-bisphosphate aldolase from the halophilic archaebacterium *Haloarcula vallismortis*. Europ J Biochem. 195, 343-350.

Lanyi J.K. 1974. Salt-dependent properties of proteins from extremely halophilic bacteria. Bacteriol Rev. 38, 272-290.

Litchfield C.D., Irby A. and Vreeland R.H. 1999. The microbial ecology of solar salt plants. In A. Oren (Ed.), Microbiology and biogeochemistry of hypersaline environments, pp. 39-52. CRC Press, Boca Raton.

Lutnæs B.F., Oren A. and Liaaen-Jensen S. 2002. New C_{40}-carotenoid acyl glycoside as principal carotenoid of *Salinibacter ruber*, an extremely halophilic eubacterium. J Nat Prod. 65, 1340-1343.

Madan A. and Sonawat H.P. 1996. Glucose dehydrogenase from *Halobacterium salinarium*: purification and salt dependent stability. Physiol Chem Physics Medic NMR 28, 15-28.

Madigan M.T. and Oren A. 1999. Thermophilic and halophilic extremophiles. Curr Opin Microbiol. 2, 265-269.

Matheson A.T., Sprott G.D., McDonald I.J. and Tessier H. 1976. Some properties of an unidentified halophile: growth characteristics, internal salt concentrations, and morphology. Can J Microbiol. 22, 780-786.

Mevarech M. and Neumann E. 1977. Malate dehydrogenase isolated from extremely halophilic bacteria of the Dead Sea. 2. Effect of salt on the catalytic activity and structure. Biochem. 16, 3786-3792.

Ng W.V., Kennedy S.P., Mahairas G.G., Berquist B., Pan M., Shukla H.D., Lasky S.R., Baliga N.S., Thorsson V., Sbrogna J., Swartzell S., Weir D., Hall J., Dahl T.A., Welti R., Goo Y.A., Leithauser B., Keller K., Cruz R., Danson M.J., Hough D.W., Maddocks D.G., Jablonski P.E., Krebs M.P., Angevine C.M., Dale H., Isenberger T.A., Peck R.F., Pohlschroder M., Spudich J.L., Jong K.-H., Alam M., Freitas T., Hou S., Daniels C.J., Dennis P.P., Omer A.D., Ebhardt H., Lowe T.M., Liang P., Riley M., Hood L. and DasSarma S. 2000. Genome sequence of *Halobacterium* species NRC-1. Proc Natl Acad Sci USA 97, 12176-12181.

Norland S., Fagerbakke K.M. and Heldal M. 1995. Light element analysis of individual bacteria by X-ray microanalysis. Appl Environm Microbiol. 61, 1357-1362.

Oren A. 1986. Intracellular salt concentrations of the anaerobic halophilic eubacteria *Haloanaerobium praevalens* and *Halobacteroides halobius*. Can J Microbiol. 32, 4-9.

Oren A. 1990a. Estimation of the contribution of halobacteria to the bacterial biomass and activity in a solar saltern by the use of bile salts. FEMS Microbiol Ecol. 73, 41-48.

Oren A. 1990b. The use of protein synthesis inhibitors in the estimation of the contribution of halophilic archaebacteria to bacterial activity in hypersaline environments. FEMS Microbiol Ecol. 73, 187-192.

Oren A. 1993. Ecology of extremely halophilic microorganisms. In R.H. Vreeland & L.I. Hochstein (Eds.), The biology of halophilic bacteria, pp. 25-53. CRC Press, Boca Raton.

Oren A. 1995. Comment on "Convergent evolution of amino acid usage in archaebacterial and eubacterial lineages adapted to high salt", by M. Gandbhir et al. (Res. Microbiol. 1995, 146, 113-120). Res Microbiol. 146, 805-806.

Oren A. 1999a. Bioenergetic aspects of halophilism. Microbiol Molec Biol Rev. 63, 334-348.

Oren A. 1999b. The halophilic Archaea - evolutionary relationships and adaptation to life at high salt concentrations. In S.P. Wasser (Ed.), Evolutionary theory and processes: modern perspectives. Papers in honour of Eviatar Nevo, pp. 345-361. Kluwer Academic Publishers, Dordrecht.

Oren A. 2000. The order Halobacteriales. In M. Dworkin, S. Falkow, E. Rosenberg, K.-H. Schleifer, & E. Stackebrandt (Eds.), The Prokaryotes. A handbook on the biology of bacteria: ecophysiology, isolation, identification, applications. 3rd. ed. (electronic publication; release 3.2). Springer-Verlag, New York.

Oren A. 2002a. Halophilic microorganisms and their environments. Kluwer Scientific Publishers, Dordrecht.

Oren A. 2002b. Molecular ecology of extremely halophilic Archaea and Bacteria. FEMS Microbiol Ecol. 39, 1-7.

Oren A. and Gurevich P. 1993. The fatty acid synthetase complex of *Haloanaerobium praevalens* is not inhibited by salt. FEMS Microbiol Letters 108, 287-290.

Oren A. and Gurevich P. 1994. Distribution of glycerol dehydrogenase and glycerol kinase in halophilic archaea. FEMS Microbiol Letters 118, 311-316.

Oren A. and Mana L. 2002. Amino acid composition of bulk protein and salt relationships of selected enzymes of *Salinibacter ruber*, an extremely halophilic bacterium. Extremophiles 6, 217-223.

Oren A. and Mana L. 2003. Sugar metabolism in the extremely halophilic bacterium *Salinibacter ruber*. FEMS Microbiol Letters 223, 83-87.

Oren A., Heldal M. and Norland S. 1997. X-ray microanalysis of intracellular ions in the anaerobic halophilic eubacterium *Haloanaerobium praevalens*. Can J Microbiol. 43, 588-592.

Oren A., Heldal M., Norland S. and Galinski E.A. 2002. Intracellular ion and organic solute concentrations of the extremely halophilic bacterium *Salinibacter ruber*. Extremophiles 6, 491-498.

Oren A. and Rodríguez-Valera F. 2001. The contribution of *Salinibacter* species to the red coloration of saltern crystallizer ponds. FEMS Microbiol Ecol. 36, 123-130.

Oren A., Rodríguez-Valera F., Antón J., Benlloch S., Rosselló-Mora R., Amann R., Coleman J. and Russell N.J. 2003. Red, extremely halophilic, but not archaeal: the physiology and ecology of *Salinibacter ruber*, a bacterium isolated from saltern crystallizer ponds. In A. Ventosa (Ed.), Halophilic microorganisms. Springer-Verlag, Berlin (in press).

Pedrós-Alió C., Calderón-Paz J.I., MacLean M.H., Medina G., Marassé C., Gasol J.M. and Guixa-Boixereu N. 2000. The microbial food web along salinity gradients. FEMS Microbiol. Ecol. 32, 143-155.

Pérez-Fillol M. and Rodríguez-Valera F. 1986. Potassium ion accumulation in cells of different halobacteria. Microbiología 2, 73-80.

Pugh E.L. and Kates M. 1994. Acylation of proteins of the archaebacteria *Halobacterium cutirubrum* and *Methanobacterium thermoautotrophicum*. Biochimica et Biophysica Acta 1196, 38-44.

Pugh E.L., Wassef M.K. and Kates M. 1971. Inhibition of fatty acid synthetase in *Halobacterium cutirubrum* and *Escherichia coli* by high salt concentrations. Can J Biochem. 49, 953-958.

Rawal N., Kelkar S.M. and Altekar W. 1988. Alternative routes of carbohydrate metabolism in halophilic archaebacteria. Ind J Biochem Biophys. 25, 674-686.

Rengpipat S., Lowe S.E. and Zeikus J.G. 1988. Effect of extreme salt concentrations on the physiology and biochemistry of *Halobacteroides acetoethylicus*. J Bacteriol. 170, 3065-3071.

Richard S.B., Madern D., Garcin E. and Zaccai G. 2000. Halophilic adaptation: novel solvent protein interactions observed in the 2.9 and 2.6 Å resolution structures of the wild type and a mutant of malate dehydrogenase from *Haloarcula marismortui*. Biochem. 39, 992-1000.

Rodriguez-Valera F., Ruiz-Berraquero F. Ramos-Cormenzana A. 1980. Behaviour of mixed populations of halophilic bacteria in continuous culture. Can J Microbiol. 26, 1259-1263.

Rodriguez-Valera F., Ruiz-Berraquero F. and Ramos-Cormenzana A. 1981. Characteristics of the heterotrophic bacterial populations in hypersaline environments of different salt concentrations. Microbi Ecol. 7, 235-243.

Rodriguez-Valera F., Ventosa A., Juez G. and Imhoff J.F. 1985. Variation of environmental features and microbial populations with salt concentrations in a multipond saltern. Microbial Ecol. 11, 107-115.

Sako Y., Takai K., Ishida Y., Uchida A. and Katayama Y. 1996. *Rhodothermus obamensis* sp. nov., a modern lineage of extremely thermophilic bacteria. Int J Syst Bacteriol. 46, 1099-1104.

Silva, Z., Borges, N., Martins, L.O., Wait, R., da Costa, M.S., & Santos, H. 1999. Combined effect of the growth temperature and salinity of the medium on the accumulation of compatible solutes by *Rhodothermus marinus* and *Rhodothermus obamensis*. Extremophiles, 3, 163-172.

Silva Z., Horta C., da Costa M.S., Chung A.P. and Rainey F.A. 2000. Polyphasic evidence for the reclassification of *Rhodothermus obamensis* Sako et al. 1966 as a member of the species *Rhodothermus marinus* Alfredsson et al. 1988. Int J Syst Evol Microbiol. 50, 1457-1461.

Sonawat H.M., Srivasta R., Swaminathan S. and Govil G. 1990. Glycolysis and Entner-Doudoroff pathways in *Halobacterium halobium*: Some new observations based on ^{13}C NMR spectroscopy. Biochem Biophys Res Comm. 173, 358-362.

Tomlinson G.A. Hochstein L.I. 1972. Studies on acid production during carbohydrate metabolism by extremely halophilic bacteria. Can J Microbiol. 18, 1973-1976.

Tomlinson G.A., Koch T.K. and Hochstein L.I. 1974. The metabolism of carbohydrates by extremely halophilic bacteria: glucose metabolism via a modified Entner-Doudoroff pathway. Can J Microbiol. 20, 1085-1091.

Ventosa A., Nieto J.J. and Oren A. 1998. Biology of moderately halophilic aerobic bacteria. Microbiol Mol Biol Rev. 62, 504-544.

COMPLEX DYNAMICS OF MULTILOCUS GENETIC SYSTEMS CAUSED BY CYCLICAL SELECTION

Valery M. Kirzhner, Vladimir M. Frenkel, and Abraham B. Korol
Institute of Evolution, University of Haifa, Mt. Carmel, Haifa 31905, Israel

Abstract: We demonstrate that simple cyclical selection, in particular, selection for a trait controlled by multiple additive, dominant, or semidominant loci can result in extremely complex limiting behavior (CLB) of population trajectories including supercycles and chaotic-like phenomena. These CLB may arise in a rather broad and natural class of multilocus systems, both haploid and diploid, with panmixia or more complex breeding systems (such as partial selfing or random mating mixed with asexual reproduction). The observed complex dynamics appear to manifest certain stability with respect to disturbances of parameters specifying the selection, recombination, and breeding system (fixed in population or genotype-dependent). This discovered diversity of multilocus dynamics by far exceeds the range of dynamic patterns described earlier for these ordinary selection models. It may represent a novel evolutionary mechanism increasing genetic diversity over long-term time periods. This novel mechanism could contribute to the observation that biological diversity has increased over geological time regardless of the well-known massive extinctions.

S.P. Wasser (ed.), Evolutionary Theory and Processes: Modern Horizons,
Papers in Honour of Eviatar Nevo
V. M. Kirzhner, V. M. Frenkel, and A. B. Korol. Complex Dynamics of Multilocus Genetic Systems Caused by Cyclical Selection, 65-108.
© 2004 *Kluwer Academic Publishers.*

1.   INTRODUCTION: POPULATION GENETIC MODELS, COMPLEX DYNAMICS, AND DETERMINISTIC CHAOS

Theoretical analysis of complex (cyclical) behavior of population trajectories begins with the classical ecological models of Lotka-Volterra dealing with variations of population size. This framework was first used to explain cyclical fluctuations of population size observed in nature and later more complex (chaotic) dynamic patterns (May, 1976). The population genetic models built on the Mendelian mechanism of inheritance arose at about the same time as the first ecological models (Fisher, 1930). It was shown that a purely genetic mechanism (i.e., with no selection, mutation, and migration) could never produce complex limiting behavior. This was proved for one- and two-locus models (see e.g., Nagylaki, 1992) and extended later to the case of multilocus models (Geiringer, 1949; Reiersol, 1962; Lyubich, 1971). It was found that under very general assumptions about ploidy, mutation, migration, and mating systems (including multiple sexes), each trajectory of population dynamics always converges to some limit point (Kirzhner and Lyubich, 1974). Hence, pure Mendelian inheritance without external forces produces no complex behavior of trajectories.

1.1   Complexity in simple constant selection models

The main "external" force acting on population is selection. Clearly, from the very beginning of population-genetic modeling, selection was among the major foci of theoretical analysis. One of the most famous results in this area was the discovery that trajectories of a single-locus panmictic population subjected to constant selection not only converge to a stable point, but the mean fitness function increases along the trajectory (Fisher's Fundamental Theorem: see Moran, 1962; Lyubich et al., 1976). Unfortunately, numerous subsequent attempts to find a "criterion function" of evolution growing along the trajectory in population-genetic models with few selective loci failed. Parallel efforts to build examples with complex limiting dynamics succeeded only when Akin (1979) found limiting cycles of trajectories in the continuous genetic model. Later, Hastings (1981) numerically showed the existence of a long-period cycle in the case of population with two diallelic selective loci with non-overlapping generations (see also Sacker and Bremen, 2003). Although Hastings's cycle is unstable to variations of the model parameters, the discovery of cycles in this model makes the existence of a criterion function growing during evolution very unlikely for any type of selection. It is worth mentioning that no other instances of complex behavior of multilocus population trajectories under constant selection have been found so far.

The convergence of trajectories to equilibrium points is proven only for very restrictive conditions. In particular, for populations with an arbitrary number of selected loci and alleles convergence is proven only in the case of additive selection (for this kind of selection Fisher's Theorem does hold: Kun and Lyubich 1979; see also Lyubich, 1992, and for very small deviations from this selection scheme: Nagylaki et al., 1999). Consequently, in general, the existence of stable complex behavior under constant selection and the exceptionality of this kind of behavior, remain open questions. In addition to these findings, some results about finiteness of the number of limiting points should be mentioned. It was shown (Lyubich et al., 2001) that the limiting set of population trajectories in the case of multilocus multiallelic model is finite almost for all selection regimes if there are no specific restrictions on selective coefficients. In particular, these conditions hold in the case of selection when fitnesses of all full homozygotes differ from each other and from fitnesses of heterozygotes. This result extends also to dominant and multiplicative selection (Lyubich et al., 2001; Kirzhner and Lyubich, 2003).

1.2 Frequency- and density-dependent selection

The foregoing discussion on selection and dynamic complexity of panmictic populations concerned constant selection regimes. If selection depends on parameters of population, in particular, on genotype frequencies in the current generation, behavior of the population trajectories can be complex even in single-locus cases (e.g., Altenberg, 1991). Density dependence was also found to generate dynamic complexity, and numerous examples have been provided in many ecological-genetic models (e.g., Charlesworth, 1971; May and Anderson, 1983; Doebly and deJong, 1999). Complex dynamics in these kinds of systems is more likely the rule than the exception. Various types of dynamic complexity have been described for these systems including cycles, T-cycles, quasi-periodical oscillations, and dynamic chaos. External manifestation of dynamic chaos cannot be distinguished from genuine stochastic behavior of the trajectory.

1.3 More about dynamic chaos

Non-regular dynamics is a rather common phenomenon in nature. For a long time such irregularity was explained by the assumption that a part of "significant parameters" needed for precise analysis are unavailable. This hypothesis of existence of "hidden parameters" was considered the main restriction on Laplacian determinism. As a result, one cannot predict system's behavior even if the lows of its dynamics are known; hence, such behavior is usually referred to as random. This also implies that with increasing information about the significant parameters, one would expect a better prediction. It appears that such "continuity" is not a general rule. Fundamental limitations to the foregoing expectations caused by nonlinear interactions are discussed in the natural sciences (primarily physics). Mathematical concepts underlying such an understanding were proposed more than 90 years ago by H. Poincaré.

One of the manifestations of these limitations is the impossibility to predict the system's dynamics despite any level of precision of information about the initial state. This phenomenon was first described by Lorenz in a numerical analysis of a mathematical model of convective movement in atmosphere and was referred to as "dynamic chaos". Later, dynamic chaos was detected in many other models of natural systems and processes. Each trajectory in such models is fully deterministic. Thus, chaotic behavior is not a property of each trajectory; rather, it is strong dependence of the trajectory on even small variations of the initial state. As in "stochastic situations", it is impossible to predict the trajectory of certain deterministic systems if the initial state is not determined exactly. In such systems, trajectories with arbitrarily close initial states will diverge exponentially with time.

1.4   Complex dynamics caused by cyclical selection

Periodic variation in parameter values of dynamic systems can also cause complex dynamics. Such a kind of effect may appear in some epidemiological models of infectious diseases (Olsen and Schaffer, 1990). Nonetheless, for purely genetic models this effect was unknown until our analysis of cyclical selection. In a series of studies we demonstrated that complex limiting behavior (supercycles and chaotic-like phenomena) may arise in a rather broad and natural class of multilocus genetic systems, both haploid and diploid, experiencing stabilizing selection with cyclically varying optimum (Kirzhner et al., 1994-1998; Korol et al., 1996, 1998; Ryndin et al., 2001). These include loci with purely additive, dominant, or semidominant effects with different types of distribution among and along chromosomes. The observed complex dynamics appeared to manifest certain stability with respect to disturbances of parameters specifying the structure of the selected system and environmental characteristics. Classic population genetic models resulting in complex behavior include, as a rule, some forms of frequency- or density-dependent selection (May and Anderson, 1983; Preygel and Korol, 1990; Altenberg, 1991; Kirzhner et al., 1999). These models directly or indirectly involve an ecological component. By contrast, in our standard model of cyclical selection the coefficients in the evolutionary operator do not depend on the system's phase variables. The selection model considered in our studies is a standard one in population genetics (Nagylaki, 1992). Nevertheless, it manifests *en masse* earlier undetected complex dynamic patterns.

This discovered mode of multilocus dynamics might represent a novel evolutionary mechanism increasing genetic diversity over long-term periods. This novel mechanism could contribute to the observation that biological diversity has increased over geological time regardless of the well-known massive extinctions.

2. COMPLEX LIMITING BEHAVIOR OF TRAJECTORIES CAUSED BY TEMPORALLY VARYING SELECTION

We review our results on complex limiting behavior (CLB) of populations with discrete (non-overlapping) generations subjected to temporally varying selection regimes. We consider population-genetic models in situations where the selection coefficients depend on a finite number of different environment states, where these states follow in a strict order and the selection coefficients are given for every environmental state. It appeared that complex behavior of trajectories is a fairly typical phenomenon in this kind of environment. This paper, reviewing our results on complex dynamics, mainly concerns a special (but very important) case of varying selection, i.e., stabilizing selection for a fitness-related trait with cyclically moving optimum. Both diploid and haploid models are considered for panmictic population models, although examples with deviation from panmixia (partial selfing or partial apomixis) are also presented. However, in the following two subsections, we first provide a few examples of CLB that arise even in the simplest genetic systems.

2.1 Two-locus diploid selection with two environmental states

Let us consider a standard model of a diploid two-locus diallelic (*A/a*, *B/b*) infinite population with panmixia and non-overlapping generations, which is subjected to periodically changing diploid selection. The selection determined by fitness matrix $W = W(t) = (w_{ij})_{i,j=1,2,3,4}$, with $w_{ji} = w_{ij}$, $w_{14} = w_{23}$. Here w_{ij} is a fitness of diploids with genotype consisting of haplotypes i and j (haplotype numbers $i, j = 1,2,3,4$ correspond to haplotypes *ab, Ab, aB, AB*).

The evolutionary operator for two consecutive generations in the haploid phase of the life cycle can be written in standard form, as:

$$x_1' = (W_1 x_1 - w_{23} rD)/\overline{W}$$
$$x_2' = (W_2 x_2 + w_{23} rD)/\overline{W}$$
$$x_3' = (W_3 x_3 + w_{23} rD)/\overline{W} \qquad (2.1)$$
$$x_4' = (W_4 x_4 - w_{23} rD)/\overline{W}$$

where x_i and x_i' ($i = 1,2,3,4$) are haplotype frequencies (ab, Ab, aB, AB, correspondingly) in consecutive generations; r is the rate of recombination between the selected loci A/a and B/b; $D = x_1 x_4 - x_2 x_3$ is a linkage disequilibrium between loci A/a and B/b; $W_i = \sum_j w_{ij} x_j$ is a marginal fitness of haplotype i ($i = 1,2,3,4$); $\overline{W} = \sum_i W_i x_i = \sum_{i,j} w_{ij} x_i x_j$ is a mean fitness of population (in the current generation).

For determining elements of the fitness matrix it is sufficient to determine selection coefficients for all diploid allele combinations:

	AA	Aa	aa
BB	w_{44}	w_{34}	w_{33}
Bb	w_{24}	w_{23}	w_{13}
bb	w_{22}	w_{12}	w_{11}

$$S = S(t) = \begin{pmatrix} w_{44} & w_{34} & w_{33} \\ w_{24} & w_{23} & w_{13} \\ w_{22} & w_{12} & w_{11} \end{pmatrix}. \qquad (2.2)$$

Following Kirzhner et al. (1995a), we considered a system with time structure of periodically changing selection consisted of $p = 2$ parts with lengths $\tau_1 = \tau_2 = 1$ (period length $T = \tau_1 + \tau_2 = 2$) generations. During part $\theta = 1, 2$ (τ_θ generations) action of selection is characterized by corresponding fitness coefficients, which can be described by corresponding matrices S_θ. Hence, we considered a system with two environmental states:

$$S_1 = \begin{pmatrix} 1 & 0.5218 & 0.3779 \\ 0.7718 & 0.4700 & 0.2110 \\ 0.7326 & 0.4170 & 0.1631 \end{pmatrix}, \quad S_2 = \begin{pmatrix} 0.1544 & 0.3105 & 0.4315 \\ 0.2063 & 0.3466 & 0.7532 \\ 0.2264 & 0.3886 & 1 \end{pmatrix}. \qquad (2.3)$$

This system displays two local stable polymorphic limit movements (depending on the starting point, i.e., displays bistability): a stable *cycle* of forced oscillation with the environmental period $T = 2$, and stable auto-oscillations (limiting *supercycle*) of about 1900 environmental periods (Fig. 1a).

Note that the effect of supercycling on the range of variations in allele frequencies at the selected loci is nontrivial. For the considered example, δP_A, the difference between the maximal and minimal values of allele frequencies along the supercycle for locus A/a was 0.55 and for B/b, δP_B was 0.37. This variation by far exceeds the difference between P_A and P_B values within any environmental period, with the range (0.00, 0.14) for δP_A and (0.00, 0.08) for δP_B. This pattern of behavior could be observed for recombination values $r \in (0.087, 0.11)$. At $r < 0.085$, the system displays only forced oscillations as stable polymorphic limiting movement. The ability to supercyclical auto-oscillations will result in an increase in range of population variation on a long-term scale (see Kirzhner et al., 1995a). This effect may complicate interpretations of gene frequency patterns in natural populations. The complexity of mean fitness ($\overline{W}$) behavior along the supercycle (Fig. 1b) may have ecological consequences, allowing for speculations on possible mechanisms of ecological cycles revealed in nature (Begon et al., 1996).

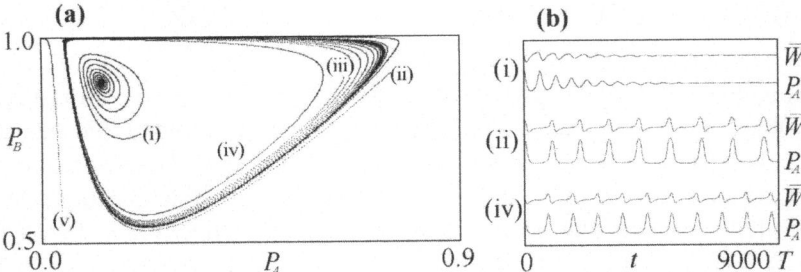

Figure 1. **Supercyclical behavior of a diploid two-locus population subjected to cyclical selection** with $\tau_1 = \tau_2 = 1$; S_1, S_2 as in Eq. 2.3; $r = 0.1$.

(a) Phase diagram of the system dynamics; the following trajectories are shown: (i) converging to the stable polymorphic point; (ii) converging to the stable supercycle from the "outside"; (iii) converging to the stable supercycle from the "inside"; (iv) an unstable supercycle separating the domains of attraction of the stable point and the stable supercycle; (v) trajectories converging to the trivial fixed point. The starting points $x^0 = (x_1^0, x_2^0, x_3^0, x_4^0)$ for trajectories in the figure are: $x_{(i)}^0 = (0.10, 0.10, 0.70, 0.10)$; $x_{(ii)}^0 = (0.02, 0.05, 0.28, 0.65)$; $x_{(iii)}^0 = (0.02, 0.05, 0.38, 0.55)$; $x_{(iv)}^0 \approx (0.2, 0.1, 0.4512, 0.2488)$; $x_{(v)}^0 = (0.36, 0.02, 0.6, 0.02)$.

(b) Temporal variation in P_A and population mean fitness ($\overline{W}$) changes (as geometric mean over the environmental period) for trajectories (i), (ii) and (iv).

Here and in all other figures, the points representing the system phase state (allele frequencies at the chosen two loci out of the modeled multilocus system) are sampled only at time multipliers of the environmental period length T. Thus, a full cycle of the environment is marked by some point of the period (e.g., the end-point). The starting point corresponds to the state of the environmental period #1. Frequency of alleles A, B, M, and A_n ($n=1,...,L$) we denote as P_A, P_B, P_M and P_n, respectively. Our model is based on diallelic loci A/a, B/b, M/m, and A_n/a_n, hence, frequencies of alleles a, b, m, and a_n are $1-P_A$, $1-P_B$, $1-P_M$, and $1-P_n$, respectively.

2.2   Single-locus multi-allelic diploid model

The case of multi-allelic loci can be transformed to a diallelic one, but with a higher number of loci. A single locus with four alleles $A_{(1)}$, $A_{(2)}$, $A_{(3)}$, and $A_{(4)}$ can be considered as two absolutely linked (i.e., with recombination rate between them $r = 0$) diallelic loci A/a and B/b. Haplotypes ab, Ab, aB, and AB could be considered as corresponding alleles of a single locus $A_{(1)}$, $A_{(2)}$, $A_{(3)}$, and $A_{(4)}$. The evolution of this system can be written in the form (2.1). Assuming for simplicity that $w(A_{(1)}A_{(4)}) = w(A_{(2)}A_{(3)})$, selection can be determined by matrix 3×3 of selection coefficients of diploid allele combinations in the diallelic case. We consider now a two-locus diallelic system with $r = 0$, subjected to cyclical selection with period $T = 2$ and selection determined by

$$S_1 = \begin{pmatrix} 1 & 0.5244 & 0.3781 \\ 0.7744 & 0.4500 & 0.2135 \\ 0.7316 & 0.4200 & 0.1600 \end{pmatrix}, \quad S_2 = \begin{pmatrix} 0.1500 & 0.3103 & 0.4314 \\ 0.2116 & 0.3300 & 0.7505 \\ 0.2260 & 0.3859 & 1 \end{pmatrix}. \quad (2.4)$$

The obtained dynamic pattern of the described system is shown in Fig. 2. The phase space contains a stable supercycle and a polymorphic unstable point. The last resides in the subspace $P_2 = 0$ (see Fig. 2), so that we can conclude that the supercyclical behavior is also possible in single-locus multi-allelic systems with as few as three alleles (Kirzhner et al., 1995a and Fig. 2). The existence of a limiting supercycle makes a rather improbable generalization of Fisher's Fundamental Theorem about the increase of population mean fitness along a trajectory in the case of single-locus selection in a fluctuating environment.

3.   COMPLEX DYNAMICS OF MULTILOCUS SYSTEMS SUBJECTED TO CYCLICAL STABILIZING SELECTION

Stabilizing selection is one of the major models of population genetics, and much effort has been applied to its theoretical analysis in the last decade (see Bürger, 1989; Turelli and Barton, 1990; Zhivotovsky and Feldman, 1992; Gavrilets and Hastings, 1994a,b; and references therein). Stabilizing selection with periodically changing optimum has also attracted some attention by theoreticians, especially in view of the interest in recombination evolution (Maynard Smith, 1988a; Korol and Preygel, 1989; Charlesworth, 1993; Korol et al., 1994; Kondrashov and Yampolsky, 1996; Bürger, 1999).

In particular, these authors have shown that cyclical selection can maintain recombination.

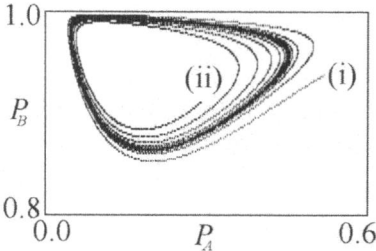

Figure 2. **Supercyclical behavior of a diploid single locus population with four alleles subjected to cyclical selection** with a period $T = 2$. In the case of selection determined by Eq. 2.4 a stable supercycle can be observed. The following trajectories are shown: **(i)** converging to the stable supercycle from the "outside"; **(ii)** converging to the stable supercycle from the "inside". The starting points are $x_{(i)}^0 = (0.05, 0.00, 0.45, 0.50)$; $x_{(ii)}^0 = (0.05, 0.00, 0.65, 0.30)$. Note that because of these starting points, the allele $A_{(2)}$ is not presented in the population. Hence, the described supercycle is situated in the subspace $A_{(2)}=0$.

3.1 Cyclical selection and polymorphism maintenance

Earlier, Lande (1976), assuming loose linkage and weak selection, reached the conclusion that stabilizing selection with temporally moving optimum does not help in the maintenance of genetic variation. Our results indicated that a change in the form of the fitness function, using additive genes with non-equal effects on the selected trait, or using genes with some dominance effect, relax the problem of polymorphism maintenance (Korol et al., 1994, 1996; Kirzhner et al., 1995c). As a rule, this is possible under sufficiently strong selection and/or relatively close linkage. Consequently, stable polymorphism is accompanied by linkage disequilibria. Kondrashov and Yampolsky (1996) considered cyclical selection for many loci of equal effect assisted by high mutation rate as a mechanism of polymorphism maintenance (see also Bürger, 1999). It is not known, however, whether or not polymorphism will be protected in an analogous mutation-free system. Our recent simulations showed that the proposed earlier mechanisms of polymorphism maintenance in infinite populations due to non-equal gene effects or semidominance may help in long-term polymorphism maintenance at many loci in finite populations as well (Nevo et al., 1997).

The results obtained in these studies demonstrated a remarkable property of cyclical selection concerning the conditions of polymorphism maintenance: the realistic patterns in nature, i.e., non-equal effects and deviation from purely additive within-locus scheme of gene action, appeared to promote polymorphism. Either of these two conditions is an important component of polymorphism protection in the considered model.

3.2   The model

We examined the behavior of an infinite population with panmixia, non-overlapping generations, and several linked diallelic loci, A_n/a_n $(n = 1,...,L)$ affecting the selected trait. Trait value u is defined for every genotype z as:

$u(z) = \sum_{n=1}^{L} u_n(z)$, where the effect of the n-th locus of the genotype z is specified as:

$$u_n(z) = \begin{cases} 2d_n, & A_n A_n \\ d_n(1+h_n), & A_n a_n \\ 0, & a_n a_n \end{cases} \tag{3.1}$$

Here d_n is an additive trait effect of allele A_n over allele a_n; h_n $(-1 \le h_n \le 1)$ is a dominance trait effect between alleles A_n and a_n. For genotype $z = z(i,j)$ consisting of haplotypes i and j, the fitness coefficients are defined as $w_{ij} = F(u(z(i,j)) - \hat{u})$, where $\hat{u}$ is the trait optimum selected for at current generation. For example, one can use

$$F(u(z) - \hat{u}) = \exp\left\{ -\frac{(u(z) - \hat{u})^2}{2\sigma^2} \right\}, \tag{3.2}$$

a fitness function that is widespread in population genetics (Gaussian fitness function).

The evolutionary operator for two consecutive generations in the haploid phase of the life cycle can be written in standard form as

$$x_i' = \left(\sum_{j,k} w_{jk} P_{jk \to i} x_j x_k \right) / \overline{W}, \tag{3.3}$$

where x_i and x_i' are gamete frequencies in adjacent generations; $\overline{W}$ is the mean fitness; $P_{jk \to i}$ $(P_{jk \to i} \ge 0,\ \sum_i P_{jk \to i} = 1)$ is the probability of producing gamete i by diploid resulting from the union of gametes with haplotypes i and j. The probability $P_{jk \to i}$ can easily be calculated as the sum of the probabilities of corresponding elementary recombination events.

The above system was studied numerically, under different types of cyclical selection regimes ($\hat{u} = \hat{u}(t) = \hat{u}(t + T)$), conditioned by an ordered set $((\hat{u}_\theta, \tau_\theta))_{\theta=1,2,\ldots,p}$, where $\hat{u}_\theta$ is the selected optimum at the θth environmental state, τ_θ is the longitude of the θth state, and $T = \tau_1 + \tau_2 + \ldots + \tau_p$ is the period length. A broad spectrum of CLB modes could be obtained with cyclical diploid selection for a trait controlled by a few (e.g., 3-4) linked loci (Kirzhner et al., 1996, 1998a,b).

3.3 CLB caused by diploid selection for additive genes with non-equal effects

Fig. 3 demonstrates a spectrum of CLBs for four types of positioning of selected loci: a block of four linked loci (Fig. 3a-c), a block of two tightly linked loci flanked by two loosely linked loci (Fig. 3d), two unlinked blocks, each consisting of two linked loci (Fig. 3e), and four unlinked loci (Fig. 3f).

A simple supercycle with a wide range of changes in the phase coordinates along the trajectory is shown in Fig. 3a. We found that this cycle can be obtained by varying the bifurcation parameter r (recombination rate). The system has a stable polymorphic point at $r = 0.14$. Numerical analysis of the Jacobian of the transformation in Eq. 3.3, iterated over the environmental period, reveals the existence of an eigenvalue $0.9973 \pm 0.0221i$, the remaining elements of the spectrum being real. This allows prediction of a cyclical movement with a period of about 280 environmental periods; in fact, a result close to this prediction was observed at $r = 0.3$: a supercycle with period length of 270 (Fig. 3a). A more complex supercycle in Fig. 3b consists of two 2-dimensional components that lie in different planes, with two alternative sets of 3-4 haplotypes predominating in the population. With the exception of a small domain close to the border set, any arbitrary initial point resulted in a trajectory converging to this supercycle. Moreover, this system also manifested a rather high parametric stability. Thus, within a certain range of variation in selection intensities and recombination rates, the majority of trajectories converged to CLB (e.g., this occurred in 50 cases from 50 random starts, for each of the following three sets: $\sigma^2 = 0.045$ and $0.1 < r < 0.5$; $\sigma^2 = 0.08$ and $0.1 < r < 0.4$; and $\sigma^2 = 0.125$ and $0.2 < r < 0.3$).

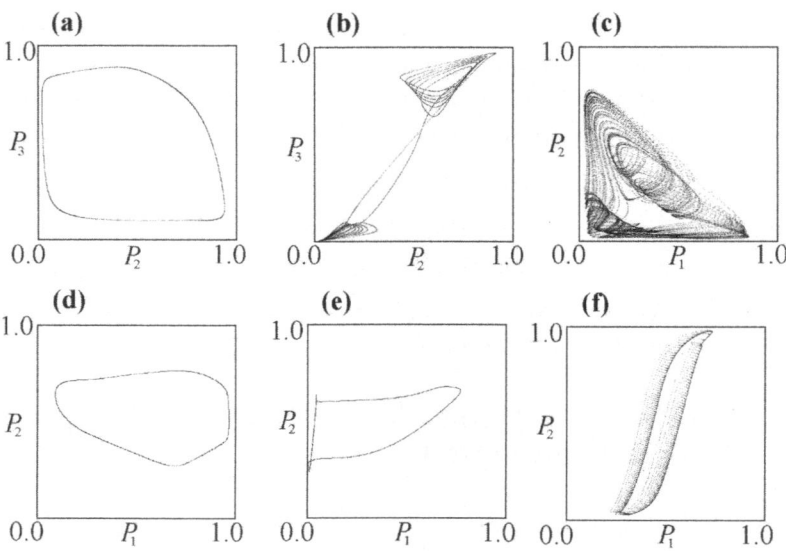

Figure 3. **Complex population trajectories of a population subjected to cyclical diploid selection for a trait controlled by four pure additive loci with unequal effects** ($h_n = 0$, $n = 1,...,4$): (a)-(d) a single linkage group, (e) two unlinked blocks each with two linked loci, (f) - four unlinked loci. In all cases, simple period structure with $p = 2$, $\tau_1 = \tau_2 = 1$, was employed. **(a)** A simple supercycle: $r_{11} = r_{23} = r_{34} = r = 0.3$; $d_1 = 3.4$, $d_2 = 1.6$, $d_3 = 0.4$, $d_4 = 0.1$; $\hat{u}_1 = 11$, $\hat{u}_2 = 0$; $\sigma^2 = 2$. Note the wide range of changes of allele frequencies along the supercycle; the scaled linkage disequilibria $D'_{n\ n+1}$ between the neighboring loci at the environmental state 2 varied along the supercycle in the range 0.06-0.43, 0.08-0.002, and 0.002-0.001 for D'_{12}, D'_{23}, and D'_{34}, respectively. **(b)** A complex two-component supercycle: $r_{11} = r_{23} = r_{34} = r = 0.05$; $d_1 = 1$, $d_2 = d_3 = d_4 = 0.3$; $\hat{u}_1 = 3.8$, $\hat{u}_2 = 0$; $\sigma^2 = 0.085$. With some changes in the model parameters, we can split this CLB into two separate cycles or obtain one central cycle. $D'_{n\ n+1}$ between the neighboring loci at the environmental state 2 varied along the supercycle in the range 0.01-0.40, 0.03-0.35, and 0.02-0.21 for D'_{12}, D'_{23}, and D'_{34}, respectively. **(c)** Non-cyclical complex trajectory. The parameter values used here were: $r_{11} = r_{23} = r_{34} = r = 0.004$; $d_1 = 2.3$, $d_2 = 1.8$, $d_3 = 1.6$, $d_4 = 1.2$; $\hat{u}_1 = 16.8$, $\hat{u}_2 = -3$; $\sigma^2 = 1.7$. As an initial point it can be taken as uniform distribution of haplotypes: $x_i^0 = \frac{1}{2^4} = 0.0625$, $i = 1,...,16$. **(d)** A supercycle in a system with a block of tightly linked loci flanked by two loosely linked loci: $r_{11} = 0.33$, $r_{23} = 0.03$, $r_{34} = 0.33$; $d_1 = 1.5$, $d_2 = 1.4$, $d_3 = 4.3$, $d_4 = 1.2$; $\hat{u}_1 = 16.9$, $\hat{u}_2 = -0.5$; $\sigma^2 = 2$. **(e)** A supercycle in a system with two unlinked blocks: (1 & 2) and (3 & 4): $r_{11} = 0.1$, $r_{23} = 0.5$, $r_{34} = 0.1$; $d_1 = 1.6$, $d_2 = 0.5$, $d_3 = 4.4$, $d_4 = 1.1$; $\hat{u}_1 = 0$, $\hat{u}_2 = 15.2$; $\sigma^2 = 2.645$. The scaled linkage disequilibria $D'_{n\ n+1}$ between the neighboring loci at the environmental state 2 varied along the supercycle in the range 0.001-0.10, 0.01-0.25, and 0.01-0.50 for D'_{12}, D'_{23}, and D'_{34}, respectively. **(f)** A fragment of a trajectory converging to a simple supercycle in a system with four unlinked loci: $r_{11} = r_{23} = r_{34} = r = 0.5$; $d_1 = 0.1$, $d_2 = 0.4$, $d_3 = 1.6$, $d_4 = 4.4$; $\hat{u}_1 = 13$, $\hat{u}_2 = 0$; $\sigma^2 = 1.4$.

A complex non-cyclical trajectory is presented in Fig. 3c. This limiting chaotic-like movement belongs to an 8-dimensional plane of haplotypes in the $2^L = 2^4 = 16$-dimensional space of haplotypes. A small perturbation of the coordinates of the initial point leads to increasing divergence over time of the resulting trajectory as compared to the initial (non-disturbed) trajectory. Some other 2-dimensional projections could be found where the trajectories look like chaotic attractors. Fig. 4a displays two domains of attraction; consequent switching of the trajectory between these domains appears to be non-regular, and brings to mind the classical Lorenz attractor (Fig. 4b).

(a)   (b)

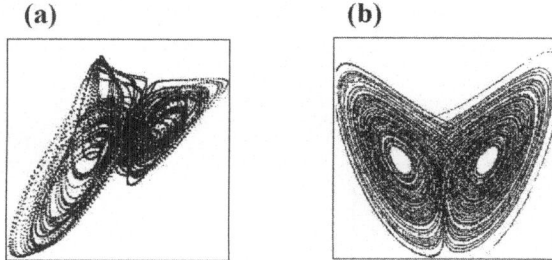

Figure 4. **Complex attractors from meteorology and population genetics.**
(a) The same system as in Fig. 3c, but projected to a special plane in the space of haplotypes;
(b) Lorenz attractor.

The next group of questions concerns the stability of CLB to random disturbances of parameters characterizing the environment. Two aspects were examined: changes in selected optimum trait values and changes in the period length. Kirzhner et al. (1998b) compared limiting trajectories in a system with $p = 2$ and $\tau_1 = \tau_2 = 1$ and in the disturbed ones. Deterministically selected value $\hat{u}_1$ was replaced by an evenly distributed random variable u_1 with mean value $\hat{u}_1$. The period structure was replaced by a stochastically changing to structure 2:1, 1:2, or 2:2 (i.e. $p = 2$, $\xi = (\tau_1, \tau_2, T) = (2,1,3)$, $(1,2,3)$, or $(2,2,4)$) with some probabilities. It was found that CLB could be quite resistant to such disturbances.

3.4   Haploid selection for additive genes with non-equal effects

Clearly, fluctuating selection (temporal or spatial) is of special interest for polymorphism maintenance models (Nevo, 1997, 2001a), especially haploid selection models, as the simplest mechanism of polymorphism maintenance in such systems (Kirzhner et al., 1994, 1995b). As applied to our subject (complex limiting behavior), haploid population models provide a unique opportunity to exclude any hidden causal effect of heterozygosity. In Fig. 5a we demonstrate a supercycle in a system with four equally-spaced closely-linked loci. This is an example of CLB caused by alternating variations in the selected trait value. Allowing for non-equal lengths of adjacent chromosomal intervals, results in a more complicated set of limiting trajectories. For example, we obtained two stable supercycles (one included in the other), as shown in Fig. 5b. Fig. 5c demonstrates a supercycle in a system with two unlinked 2-locus blocks. In contrast to diploid selection, no examples of CLB were found for four unlinked loci under haploid selection regimes. An important fact is that cyclical haploid selection may also produce more complex behavioral patterns. One such example is presented in Fig. 5d. In spite of its quite complex pattern, the presented attractor is close to a cycle; however, trajectories with close starting points tend to diverge with time.

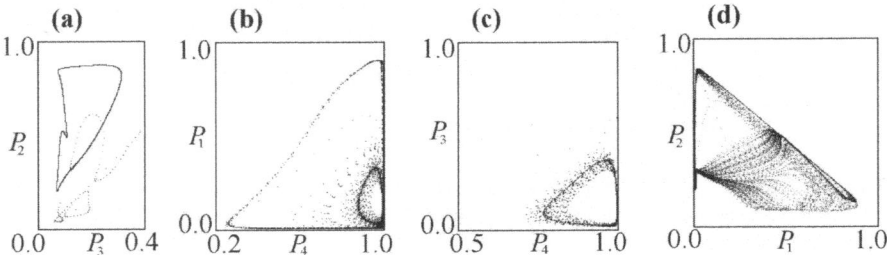

Figure 5. **Supercycles caused by cyclical haploid selection for a trait controlled by four additive loci with unequal effects.**

(a) A simple supercycle in a system with four linked loci: $r_{12} = r_{23} = r_{34} = r = 0.007$; $d_1=0.18$, $d_2=0.1$, $d_3=0.17$, $d_4=0.23$; $\tau_1 = \tau_2=3$, T=6; $\hat{u}_1=0.846$, $\hat{u}_2 = -0.15$; $\sigma^2 = 0.167$. **(b)** Two stable supercycles in a system with two loosely linked blocks ($r_{23}= 0.4$) each consisting of two tightly linked loci ($r_{12} =r_{34} = 0.0012$): $d_1=0.12$, $d_2=0.16$, $d_3=0.18$, $d_4=0.23$; $p = 2$; $\tau_1= \tau_2=3$, T=6; $\hat{u}_1=1.04$, $\hat{u}_2 = -0.20$; $\sigma^2 = 0.125$ **(c)** A supercycle in a system with two unlinked blocks each consisting of two tightly linked loci: $r_{12} = r_{23} = r_{34} = r = 0.0012$; $d_1=0.12$, $d_2=0.16$, $d_3=0.18$, $d_4=0.23$; $p = 2$; $\tau_1= \tau_2=3$, T=6; $\hat{u}_1=1.04$, $\hat{u}_2 = -0.18$; $\sigma^2 = 0.15$. **(d)** A chaotic limit behavior of system with four linked loci: $r_{12} = r_{23} = r_{34} = r = 0.002$; $d_1=0.18$, $d_2=0.1$, $d_3=0.17$, $d_4=0.23$; $p = 2$; $\tau_1= \tau_2=1$, T=2; $\hat{u}_1=0.7$, $\hat{u}_2 = -0.005$; $\sigma^2 = 0.0034$.

3.5   Diploid selection for semidominant or dominant genes

We consider here the same model, but assume that the selected trait is controlled by semidominant loci (i.e., $h_i \neq 0$) with equal effects. The main difference of the regimes produced by this assumption is that, in addition to the previously described behavioral patterns, T-supercycles (i.e., short supercycles, with <u>exact</u> repetition of the phase states after T environmental periods) and chaotic-like behavior are common modes of the manifested CLB (Fig. 6). Surprisingly, more than one mode of CLB can be manifested by one system, depending on the initial point.

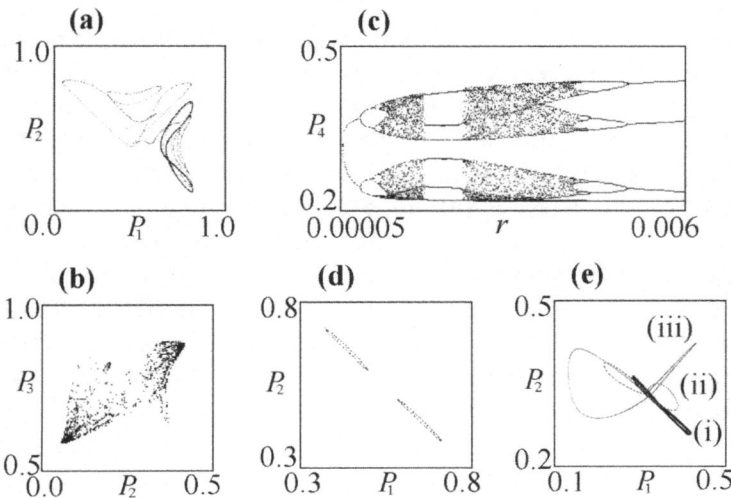

Figure 6. **Complex population trajectories of population subjected to cyclical selection for a trait controlled by semidominant loci. (a)** A trajectory converging to the supercycle: $r_{12} = r_{23} = r = 0.001$; $d_1=d_2=d_3=1$; $h_1=0.1$, $h_2=0.1$, $h_3=0.2$; $p = 2$; $\tau_1 = \tau_2 = 1$, T=2; $\hat{u}_1 = 6.7$, $\hat{u}_2 = 0.6$; $\sigma^2 = 0.12$. **(b)** Chaotic dynamics: $r_{12} = r_{23} = r_{34} = r = 0.03$; $d_1=d_2=d_3=d_4=1$; $h_1=h_2=h_3=h_4=0.5$; $p = 3$; $\tau_1= \tau_2= \tau_3=2$, $T=6$; $\hat{u}_1=4.0$, $\hat{u}_2 = 2.2$, $\hat{u}_3 = 0.1$; $\sigma^2 = 0.5$. The conclusion about a chaotic-like regime is derived from the known criterion of trajectory divergence caused by perturbation of the initial point (Lyapunov exponent $\lambda>0$). The range of r resulting in CLB was [0.01-0.05]. Within this range, T-supercycles were also observed. **(c)** Bifurcation diagram for the example (b) with r taken as bifurcation parameter. The initial point was $x_{a_1a_2a_3a_4} = 0.004$, $x_{a_1a_2A_3a_4} = 0.201$, $x_{a_1A_2A_3a_4} = 0.001$, $x_{a_1A_2a_3A_4} = 0.35$, $x_{a_1a_2A_3A_4} = 0.35$, $x_{A_1A_2A_3A_4} = 0.004$. **(d)** Example of CLB (super T-cycle) in the system subjected to selection with a long period (T=40): $r_{12} = r_{23} = r = 0.00003$; $d_1=d_2=d_3=1$; $h_1=h_2=h_3=0.25$; $p = 2$; $\tau_1= \tau_2=20$, $T=40$; $\hat{u}_1=0.3$, $\hat{u}_2 = 4.3$; $\sigma^2 = 0.22$. **(e)** Example of CLB multi stability: $r_{12} = r_{23} = r_{34} = r = 0.001$; $d_1=d_2=d_3=d_4=1$; $h_1=h_2=h_3=h_4= 0.2$; $p = 2$; $\tau_1= \tau_2=1$; $\hat{u}_1=6.34$, $\hat{u}_2 = 0.25$; $\sigma^2 = 4.54$. Limiting behavior shown for trajectories starting from $X^0=(x_j)_{j=1,...,16}$: $x_1+...+x_{16}=1$: **(i)** $x_{a_1a_2a_3a_4} = 0.1$, $x_{A_1a_2a_3a_4} = 0.31$, $x_{a_1A_2a_3a_4} = 0.02$, $x_{a_1a_2A_3a_4} = 0.03$, $x_{a_1a_2a_3A_4} = 0.04$, $x_{A_1A_2A_3A_4} = 0.5$; **(ii)** $x_{a_1a_2a_3a_4} = 0.1$, $x_{A_1a_2a_3a_4} = 0.01$, $x_{a_1A_2a_3a_4} = 0.32$, $x_{a_1a_2A_3a_4} = 0.03$, $x_{a_1a_2a_3A_4} = 0.04$, $x_{A_1A_2A_3A_4} = 0.5$; **(iii)** $x_{a_1a_2a_3a_4} = 0.1$, $x_{A_1a_2a_3a_4} = 0.01$, $x_{a_1A_2a_3a_4} = 0.02$, $x_{a_1a_2A_3a_4} = 0.33$, $x_{a_1a_2a_3A_4} = 0.04$, $x_{A_1A_2A_3A_4} = 0.5$.

Fig. 6a represents a simple supercycle, whereas a complex chaotic-like attractor is given in Fig. 6b. The bifurcation diagram of the latter (with recombination rate taken as the bifurcation parameter) consists of a series of transformations of chaotic-like CLB into T-supercycles and back (Fig. 6c). Fig. 6d demonstrates CLB in the system under selection with long period length ($T = 40$). The last example, Fig. 6e, represents multistability of CLB.

The considered class of semidominant (dominant) models displays a higher level of robustness of CLB with respect to disturbance of parameters defining the selection regime than pure additive models. We analyzed the effect of random fluctuations of the modal values of the selected trait. The deterministic selected value $\hat{u}_1$ was replaced by an evenly distributed random variable with mean value $\hat{u}_1$. One can easily see (Fig. 7) that with moderate deviations from the mean $\hat{u}_1$ (up to 25%), the resulting movement largely preserves the behavioral mode characteristic of the non-disturbed deterministic system (compare Figs. 7a and 7b). The same conclusion was reached with respect to disturbances caused by random fluctuations of the period length (compare Figs. 7a and 7c). Therefore, we can conclude that CLB manifests certain robustness with respect to moderate random disturbances of the selection regime.

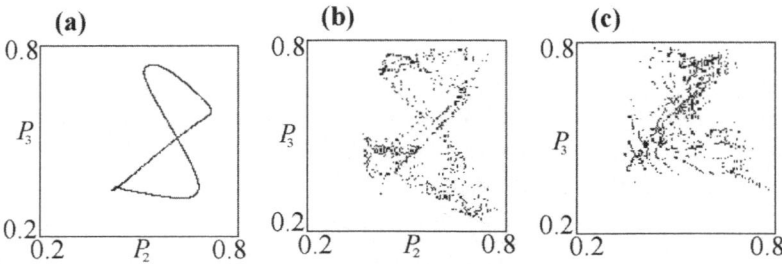

Figure 7. **Changes in phase diagrams of supercyclical movements in systems with dominant loci caused by random disturbances of environmental parameters:** The selected system consists of four semidominant loci $r_{12} = r_{23} = r_{34} = r = 0.001$; $d_1=d_2=d_3=d_4=1$; $h_1=h_2=h_3=0.18, h_4= -0.5; p = 2; \tau_1 = \tau_2=1; \hat{u}_1 =0.4, \hat{u}_2 = 6.08; \sigma^2 = 2$.
(a) Undisturbed process - a simple supercycle. **(b)** Random variations of $\hat{u}_1$ (range 25%). **(c)** Random variations of period length: $\Pr(\tau_1 = 2) = \Pr(\tau_2 = 2) = 0.1$.

Of special separate interest is the question of stability of CLB to random drift caused by finite population size. In contrast to the foregoing examples of stability against random disturbances of model parameters, introduction of fluctuations caused by finite population size actually means a change in the basic model. We consider this question using, as an example, a model with three semidominant loci. To analyze the robustness of CLB with respect to disturbances caused by finite population size, the following procedure was employed. At each generation the vector of haplotype frequencies was disturbed by a vector of "sampling errors" with the coordinates simulated as random normal variables with zero expectations and variances equal to $\sigma_N^2(x_i) = \frac{x_i(1-x_i)}{N}$, where N is the "simulated" population size. Another approach to simulate a finite population size was the introduction of a minimal haplotype frequency boundary x^0, so that at any generation any haplotype frequency that appeared to be less than x^0 was replaced by zero. Fig. 8 illustrates the results. As one can see, in the considered example, population sizes from $N = 5000$ and higher display the same pattern of CLB as the non-disturbed (infinite) population. An interesting phenomenon is that the finite population manifests simultaneously both CLBs to which the infinite population converges alternatively, i.e., when starting from different initial points. Clearly, the last phenomenon is possible due to random fluctuations of the trajectory, which allow it to appear alternately in the domain of attraction of each of the CLBs. With increasing N this effect disappears, and the trajectory becomes associated with only one of the possible CLBs.

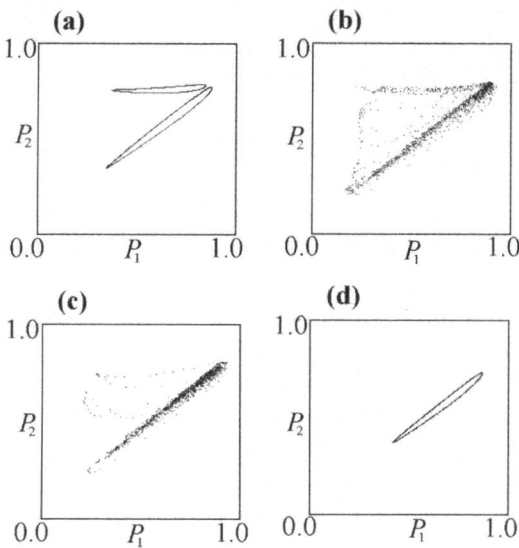

Figure 8. **Changes in phase diagrams of supercyclical movements caused by genetic drift.** **(a)** Two limiting supercyclical trajectories in the undisturbed (infinite) population depending on initial points ($L=3$; $r_{12} = r_{23} = r = 0.02$; $d_1 = d_2 = d_3 = 1$; $h_1 = h_2 = 0.15$, $h_3 = -0.195$; $p = 2$; $\tau_1 = \tau_2 = 1$; $\hat{u}_1 = 3.3$, $\hat{u}_2 = 0.3$; $\sigma^2 = 0.84$). Note that the size of the attraction domain of cycle 1 is several times larger than that of cycle 2. **(b)** The trajectory of the system with random disturbances of the haplotype frequencies simulating the effect of drift with population size $N=5000$. **(c)** The trajectory of the system with random disturbances of the haplotype frequencies simulating the effect of drift with population size $N=10000$. **(d)** The limiting set of the system's trajectories with a minimal boundary of haplotype frequencies: at each generation, all haplotype frequencies with a value less than $x_0 = 0.001$ were replaced by zero.

4.　CYCLICAL SELECTION, CLB, AND THE DYNAMICS OF RECOMBINATION MODIFIERS

The evolution of recombination remains an important unsolved problem in evolutionary genetics. The results of experiments on artificial selection for altered recombination rates (r) suggest that almost every population has enough stored genetic variability to ensure response to selection for changed r (reviews: Brooks, 1988; Korol et al., 1994). Theoretical analysis shows that under stable environmental conditions a panmictic population polymorphic for fitness-related loci should evolve toward the minimum possible level of recombination. Namely, the introduction of a new modifier allele affecting r into an equilibrium polymorphic population is accompanied by an increase in its frequency if it reduces r (Zhivotovsky et al., 1994; Barton, 1995). Hence, factors should exist opposing this trend.

4.1　Cyclical selection and evolution of recombination

A series of models have been proposed to explain the evolutionary mechanisms responsible for the persistence of recombination (and sex in general) in nature. These include selection in variable abiotic conditions, selection against harmful mutations, and frequency-dependent selection caused by interaction between antagonistic species (for reviews and classifications of the models see Maynard Smith, 1978, 1988a; Kondrashov, 1993; Korol et al., 1994; Otto and Michalakis, 1998; Otto and Lenormand, 2002). The basis of these models is the assumption that the gene pool is subjected to variable conditions (either external, due to changes in the selection regime, or internal, due to the mutation process). Previous theoretical studies have shown that temporal environmental variation can indeed promote increased recombination, although the revealed patterns appear to be very complex (Charlesworth, 1976, 1993; Maynard Smith, 1980, 1988b; Sasaki and Iwasa, 1987; Korol and Preygel, 1989; Bergman and Feldman, 1990; Korol et al., 1994; Kondrashov and Yampolsky, 1996; Feldman et al., 1997). A few experimental studies and observations have showed that selection for adaptively important traits may indeed result in changes in the recombination system, e.g., in increased rate of recombination or sex (Flexon and Rodell, 1982; Burt and Bell, 1987; Wolf et al., 1987; Gorodetsky et al., 1990; Gorlov et al., 1992; Korol and Iliadi, 1994; Korol, 1999, 2001; Saleem et al., 2001; Grishkan et al., 2003).

Our previous results on CLB (reviewed in sections 2 and 3) showed that recombination rate strongly affects the mode of CLB and the very existence of CLB. Therefore, it should be both interesting and instructive to analyze how this phenomenon will be expressed when, instead of a fixed parameter *r*, the rate of recombination is a genetically controlled trait, dependent on a polymorphic modifier ('*rec*-modifier'). Stronger motivation to analyze such models comes from the general interest in the evolution of sex and recombination. Indeed, the foregoing results on CLB were obtained with standard models of stabilizing selection with a cyclically moving optimum, exactly the same models that have been employed in the studies of recombination evolution in changing environments (Maynard Smith, 1980, 1988a,b; Korol and Preygel, 1989; Charlesworth, 1993; Korol et al., 1994; Barton, 1995).

4.2   Super-supercycling

Following Kirzhner et al. (1995a), we first provide an example of complex behavior associated with a polymorphic modifier locus affecting recombination rate between two diallelic selective loci. The model described in section 3.2 was extended to include the modifier (for the genetic scheme see Fig. 9a). We considered a two-state environment with period $T = \tau_1 + \tau_2 = 2 + 2 = 4$. The following matrices determining selection coefficients for all diploid allele combinations in both selective loci (see Eq. 2.2) were used:

$$S_1=\begin{pmatrix} 1 & 0.5218 & 0.3789 \\ 0.7718 & 0.4730 & 0.2110 \\ 0.7316 & 0.4170 & 0.1621 \end{pmatrix}, \quad S_2=\begin{pmatrix} 0.1544 & 0.3117 & 0.4315 \\ 0.2074 & 0.3423 & 0.7543 \\ 0.2264 & 0.3897 & 1 \end{pmatrix}.(4.1)$$

The evolutionary operator for two consecutive generations can be written in the form of Eq. 3.3 with $P_{jk \to i}$ dependent on alleles in *rec*-modifier locus *M/m* (haplotype numbers i and j determine alleles in *rec*-modifier locus, so we can save notation $P_{jk \to i}$). Assuming mutual independence of recombination events (no interference) we can determine all of $P_{jk \to i}$ using only recombination rates between selective loci r_{MM}, r_{Mm}, and r_{mm}, corresponding to allele combinations in *rec*-modifier locus and recombination rate r_c between *rec*-modifier and selective loci (see Fig. 9a).

Considering this model without *rec*-modifier (i.e. with r = const), we obtain for the selected system the same type of behavior as in Fig. 1 for $r \in [0.24, 0.33]$ (a stable supercycle). A decrease in r below 0.24 leads to tightening of the supercycle to the polymorphic fixed point $\hat{X} = (x_{ab}, x_{Ab}, x_{aB}, x_{AB}) = (0.19, 0.41, 0.00, 0.40)$, i.e., forced oscillations with $T = 4$. An increase in r above 0.33 results in a loss of polymorphism, whereas within the interval [0.24, 0.33] an increase in r leads to increasing of the superperiod. Thus, the system with *rec*-modifier displays two opposite tendencies: In the vicinity of $\hat{X}$, selection tends to increase recombination, provided linkage of *rec*-modifier with selected system is not very tight (see also: Feldman et al., 1980; Preygel and Korol, 1990; Korol et al., 1994). Alternatively, superoscillations with very long periods favor low recombination. A balance of these forces leads to the appearance of regular stable fluctuations in the 2-locus supercycle characteristics: repetitive changes in the period and amplitude of the supercyclical auto-oscillations. Thus, we could refer to such a regime as a stable *super-supercycle* (Fig. 9b). Corresponding oscillations of *rec*-modifier allele frequency are presented in Fig. 9c. We note a peculiarity of the dynamics of this system compared with those with fixed recombination rate: polymorphism for the *rec*-locus leads to preservation of a supercyclical mode of behavior over a much broader range of phase states (Kirzhner et al., 1995a).

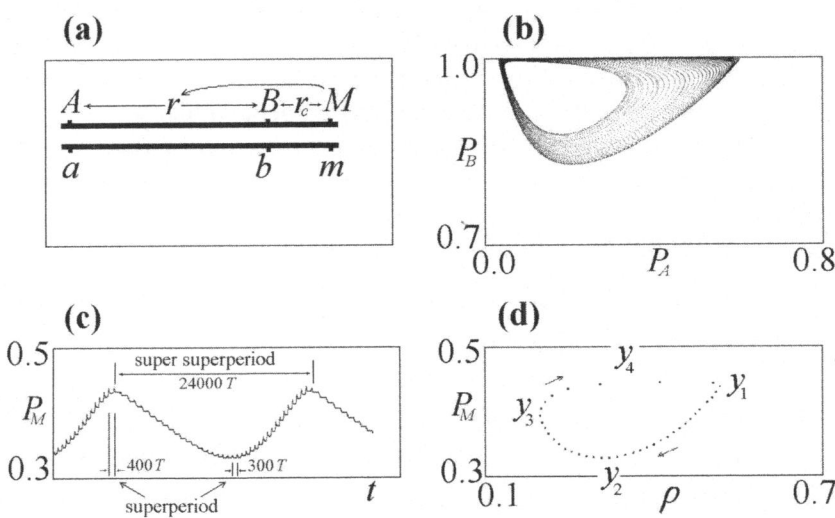

Figure 9. **Stable super-superoscillations in a model incorporating a two-locus selected system and a recombination modifier.** Selection defined by Eq. 4.1 with $\tau_1 = \tau_2 = 2$; r_c=0.0035; r_{mm}=0, r_{Mm}=0.25, r_{MM}=0.5; $X^0 = (x^0_{abm}, x^0_{Abm}, x^0_{aBm}, x^0_{ABm}, x^0_{abM}, x^0_{AbM}, x^0_{aBM}, x^0_{ABM}) =$ (0.10, 0.05, 0.20, 0.05, 0.05, 0.05, 0.20, 0.30). **(a)** The structure of the genetic system. **(b)** A fragment of the phase plane (P_A, P_B) with super-supercycle. **(c)** Time dependence of the frequency of allele M in *rec*-modifier. **(d)** Hysteresis-like behavior of the three-locus system: here ρ is a difference between the maximum of allele A frequency for every superperiod and allele A frequency in the point $\hat{X}$; P_M is mean frequency of allele M in *rec*-modifier during superperiod.

In the considered example, the super-supercycles move in the same domain of the phase space when attracting to or repelling from the fixed point ($\hat{X}$). However, some of the characteristics of this movement depend not only on the position of the system phase coordinates in the phase space, but also on the direction of the movement. By analogy with physics, we called this mode of the three-locus system behavior the "hysteresis" effect in population dynamics (Fig. 9d): Let us start from a remote, with respect to $\hat{X}$, supercycle (point y_1). Because of long periods, selection will promote evolution for decreasing recombination and, thereby, for ever-increasing attraction of the trajectory to the fixed point $\hat{X}$ (phase y_1-y_2). With decreasing of amplitude of the supercycle, the oppositely acting factor (forced oscillation with period $T = 2$) overcomes and recombination starts to increase (phase y_2-y_3) and (phase y_3-y_4). However, in the vicinity of y_3, recombination crosses a level that corresponds to repelling of the trajectories from $\hat{X}$. Then the supercycles with increasing amplitude and periods will again prevail over the forced oscillations, resulting in reduced recombination (phase y_4-y_1). Linkage of the modifier locus to the selected system, a kind of an "inertia factor", is the key component determining the very possibility of such a pattern. Moreover, without sufficiently close linkage, polymorphism for the modifier locus would hardly be possible, m tending to fixation during the slow supercycles (around phase y_1 in Fig. 9d).

4.3   *Rec*-modifier and stabilizing selection with cyclically moving optimum

We analyzed the effect of cyclical selection for a trait controlled by multiple additive, dominant, or semidominant loci on the dynamics of multilocus systems with polymorphic *rec*-modifier (Korol et al., 1998). The parameter sets were chosen in such a way that the selected system mani-fested relatively simple modes of CLB (supercycles) or a stable poly-morphism (in the sense explained above). Different types of *rec*-modifiers were considered: (a) modifiers of recombination in all intervals of the chromosome; and (b) modifiers with a non-even distribution of the effects, including: (*i*) modification of linkage between blocks of tightly linked loci, with no effect on the within-block recombination; and (*ii*) modification of linkage within blocks, with no effect on recombination between the blocks. The evolutionary equations of our standard multilocus selection model (Eq. 3.3) were extended to include the coordinates related to polymorphic selectively neutral *rec*-modifier.

For the sake of simplicity, let us assume equidistant distribution of the selected loci in the chromosome and equal effects of the modifier locus on recombination in each interval. We consider first a special mode of CLB of the selected system, supercycles, as a basis for analyzing the complications caused by the introduction of a polymorphic modifier into the system. The usual way to study the fate of the modifier locus, especially when using analytical tools, is to introduce at a low frequency a new modifier allele after reaching a (polymorphic) steady state for the selected system (Feldman et al., 1997). We are concerned with tracing the full dynamics; so starting the trajectories at arbitrary polymorphic initial points makes it possible to evaluate the volume of the CLB attraction domain. Three quite different types of *rec*-modifier effects on CLB will be demonstrated below (see Korol et al., 1998):

(a) modifiers as the source of CLB;

(b) modifiers as a source of the "next level" of CLB (super-superoscillations);

(c) modifiers as a factor of chaotization of population dynamics.

Likewise, we consider

(d) the effect of interaction between the "genetic architecture" of selected trait and regional distribution of modifier effects on CLB patterns.

For the case of uneven distribution of *rec*-modifier effects along the chromosome, we consider two situations: **(i)** modification of linkage between blocks of linked loci, with no effect on recombination within blocks; and **(ii)** modification of linkage within blocks of tightly linked loci with no effect on recombination between blocks.

Rec-modifier as source of CLB: the major motivation for studying the proposed model concerns the mode of modifier evolution as dependent on the system dynamics referred to as CLB. The existence of CLB in the system with fixed modifier is the precondition of such a consideration. However, an important question is whether or not the presence of a polymorphic modifier may by itself be the factor producing CLB: would it be possible to obtain CLB by injection of a new modifier allele into a system that is incapable by itself of manifesting this mode of dynamics at any fixed value of recombination? Korol et al. (1998) presented an example of such a system with three slightly dominant loci with non-equal effects on the selected trait: no CLB was found in computer simulations at any level of recombination ($0 \leq r \leq 0.5$), but a stable supercycle was obtained in dynamics of the same system supplied by a polymorphic *rec*-modifier ($r_{12}=r_{23}=r$, $r_{MM}=0.5$, $r_{Mm}=r_{mm}=0$, $r_{34}=r_c=0.1$). Moreover, with fixed recombination, the system trajectories go to fixation at the selected loci at any $r > 0.05$. Nevertheless, the attraction domain of this supercycle in the whole phase space of this system is very large, the frequency of allele M varies in the range of 0.914-0.971 along the limiting trajectory, and the corresponding range of the mean rate of recombination will be $0.417 < r < 0.471$ (assuming Hardy-Weinberg proportions at the modifier locus). Therefore, the polymorphic modifier is not only the source of complex limiting behavior; the very existence of (protected) polymorphism at loose linkage between the selected loci is possible here only in the form of CLB, caused by the presence of the polymorphic *rec*-modifier.

Super-supercycles caused by interaction of supercyclical dynamics and modifier movement: this mode of behavior arises when a polymorphic *rec*-modifier is introduced into a system that it is able to manifest supercyclical dynamics under some range(s) of the recombination rate. It is a rather trivial fact that the allele frequencies at the modifier locus may oscillate with a period equal to that of the supercycle. However, in addition to this movement the modifier locus may manifest auto-oscillations with a much longer period. In turn, the dynamics of the modifier causes long-periodical changes of the supercycle itself: see the example in section 4.2. In that example the model included a modifier of recombination and a 2-locus selected system, subjected to cyclical selection with a very special fitness matrix. Here we consider super-supercyclical dynamics based on a more natural class of models: multilocus systems subjected to stabilizing selection with cyclically moving optimum. Two examples are provided in Figs. 10 and 11. The supercyclical dynamics presented in Fig. 10a are robust with respect to changes of recombination rate approximately in the range between 0.002 and 0.04. Let the recombination rate in the system now be dependent on the *rec*-modifier, with $r_{mm}=r_{mM}=0$ and $r_{MM}=0.02$. Then, given a certain recombination level r_c between the modifier and the selected system ($r_c = 0.01$, in the present example), the super-supercycle shown in Fig. 10b,c will be obtained. Each large period in Fig. 10c consists of about 20 oscillations corresponding to the initial supercycles (i.e., the total period of the modifier is about 1000 environmental periods). The robustness of the revealed pattern with respect to random disturbances of the environmental period is also presented (Fig. 10d).

The second example concerns a system with complete positive interference (i.e., only single exchanges were allowed). The initial supercycle here is very simple (Fig. 11a). With the polymorphic modifier the behavior is much more complex. In the example of Fig. 11b,c the frequency of recombination within the selected system was $r_{mm}=0.01$, $r_{mM}=0.25$ and $r_{mM} = 0.5$. The modifier manifests superoscillations here corresponding to those of the selected system. However, this behavior of the modifier occurs in two fairly separate subsets, with transitions between them occurring in the form of short-time jumps (Fig. 11c). Note that the consequent visits of the two subsets take <u>dozens of thousands of generations</u>.

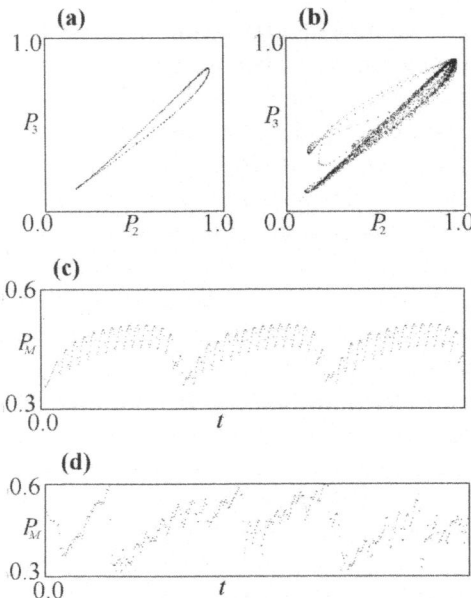

Figure 10. **Super-supercycles based on interaction of supercyclical dynamics of three semidominant selected loci and polymorphic *rec*-modifier.** The system consists of three selective and one *rec*-modifier locus ($A_4/a_4{=}M/m$). $d_1{=}d_2{=}d_3{=}1$, $d_4{=}0$; $h_1{=}0.3$, $h_2{=}0.3$, $h_3{=}0.4$, $h_4{=}0$; $\tau_1{=}\tau_2{=}1$; $\hat{u}_1{=}3.3$, $\hat{u}_2{=}0.3$; $\sigma^2 = 0.84$; $r_{12}{=}r_{23}{=}r$, $r_{MM}{=}0.02$, $r_{Mm}{=}r_{mm}{=}0$, $r_{34}{=}r_c{=}0.01$. **(a)** Phase diagram of the initial supercycle (the modifier is fixed for MM); the period of supercycle is $\approx$ 60 environmental periods. **(b)** Phase diagram of the system with polymorphic modifier. **(c)** Super-supercyclical dynamics of the modifier (represented across environmental periods). The brackets mark the points of a failure of stability of the basic supercycle caused by the current dynamics at the modifier. **(d)** Resistance of the super-supercycle to moderate disturbances of the environmental period with $\Pr(\tau_1 = 2) = \Pr(\tau_2 = 2) = 0.1$.

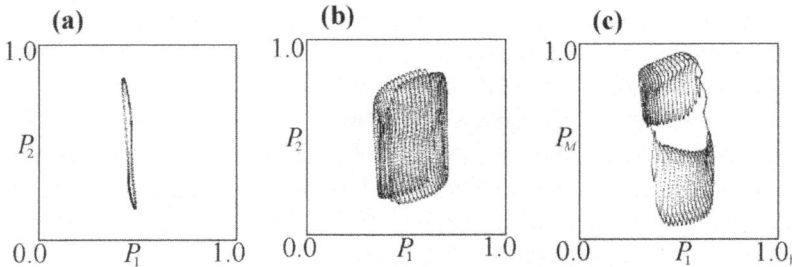

Figure 11. **Super-supercycles based on interaction of supercyclical dynamics of selected loci and polymorphic modifier.** The system consists of four selective and one *rec*-modifier locus ($A_5/a_5{=}M/m$). $r_{12} = r_{23} = r_{34} = r$; $d_1{=}0.01$, $d_2{=}0.04$, $d_3{=}0.16$, $d_4{=}3.4$, $d_5{=}0$; $h_1{=}h_2{=}h_3{=}h_4{=}h_5{=}0$; $\tau_1{=}\tau_2{=}1$; $\hat{u}_1{=}1.1$, $\hat{u}_2{=}0$; $\sigma^2 = 0.17$. Here, only single recombination events were allowed for within the chromosome segment containing the selected loci; $r_{MM}{=}0.5$, $r_{Mm}{=}0.25$, $r_{mm}{=}0.01$. The modifier was closely linked to the fourth selected locus, $r_c{=}0.0005$; **(a)** The supercycle obtained with fixed recombination rate $r{=}0.25$. **(b)** Phase diagram of the super-supercycle with a polymorphic modifier. **(c)** Superoscillations of the modifier consisting of two separate subsets.

The modifier as a cause of chaotization of population dynamics:
examples presented in Korol *et al.* (1998) demonstrate that limiting behavior
of selective system with *rec*-modifier can be much more complex than with
a fixed recombination rate. For characterization of complexity of trajectory
one can use Lyapunov exponent, λ (Wolf et al., 1987). If $\lambda>0$, then the
corresponding CLB can be classified as chaotic, where initially close starting
points produce diverging trajectories (Hastings et al., 1993). The presented
phenomenon does not mean that any kind of CLB should be characteristic of
the system with fixed *r* corresponding to either *mm* or *MM*. However, we can
assume that the existence of a range of such *r*-values is "exploited" by the
system with polymorphic recombination to produce more complex limiting
patterns.

Effect of interaction between the "genetic architecture" of selected trait
and regional distribution of modifier effects on CLB patterns

<u>*Modification of linkage between blocks of linked loci, with no effect on*</u>
<u>*recombination within blocks*</u>: this type of *rec*-modifier can be a source of
chaotic limiting behavior of the selective system even if it consists of purely
additive loci with unequal effects (Korol et al., 1998). Dynamic characteristics of the system can dramatically depend on the position of *rec*-modifier
relative to selective loci. This may derive from the fact that the rate of
changes of the strongest locus is higher than that of weaker loci, which in
turn may seriously affect the system dynamics because of "inertia-like"
effects caused by linkage between the modifier and the selected loci
(Kirzhner et al., 1995a; Korol et al., 1998).

<u>*Modification of linkage within blocks of tightly linked loci with no effect*</u>
<u>*on recombination between blocks*</u>: this situation of recombination
modification is complementary to that considered in the previous section,
where the modifier controlled the recombination rates between blocks
having no effect on within-block recombination. In this system, polymorphism for the described *rec*-modifier can be a source of CLB in the
dynamics of selected system (Korol et al., 1998). The proposed structures
provide an interesting opportunity to analyze the behavior of the modifier
and the whole system, when two contrasting versions of modifier linkage to
the selective blocks exist simultaneously in the system, i.e., *linked* and *freely*
recombined. Note, that from the viewpoint of recombination evolution
modeling, this type of system is much more realistic than multilocus systems
with one linkage group: it is natural to assume that loci affecting fitness-
related quantitative traits are spread over more than one locality of a
multichromosomal genome (Lewontin, 1974; Korol et al., 1994; Peng et al.,
2003). It also fits the concept and corresponding evidence for "fine" control
of recombination (Simchen and Stamberg, 1969; Chinnici, 1971).

5.   CLB AND THE MODE OF REPRODUCTION

Temporally varying selection is considered one of the potential mechanisms of evolution of sex and recombination. As shown above, cyclical selection for a trait controlled by multiple additive, dominant, or semidominant loci with a short period may induce auto-oscillations with a long period ("supercycles") and chaos-like phenomena. Such behavior was observed for a broad range of system parameters. The foregoing results on CLB were obtained for purely panmictic models. It appeared that recombination, either fixed or controlled by polymorphic modifier loci, plays a key role in determining the dynamic patterns. Therefore, we considered it interesting and instructive to extend the analysis with respect to the reproduction mode by including reproduction systems deviating from the pure panmixia. In particular, models with mixed panmixia and apomixis or mixed panmixia and selfing were considered.

5.1   Panmixia mixed with vegetative reproduction apomixis or selfing: the model

We consider a population genetics model with a mixed breeding system that includes panmixia, selfing, and asexual reproduction (e.g., apomixis or vegetative propagation). Let each individual with genotype composed of haplotyes i and j produce part $p_{\mathrm{panmixia},ij}$ of the progeny by random mating resulting in sex-derived zygotes, part $p_{\mathrm{selfing},ij}$ by pure self-fertilization, and the remainder $p_{\mathrm{asex},ij}$ being a result of precise copying of the maternal zygote; thus $p_{\mathrm{panmixia},ij} + p_{\mathrm{selfing},ij} + p_{\mathrm{asex},ij} = 1$ (Marshall and Weir, 1979; Hamilton, 1980; Hastings, 1984; Overath and Asmussen, 1998). If these parameters are equal for all genotypes, then indexes ij can be omitted. Denote by z_{ij} ($z_{ji} = z_{ij}$; $z_{ij} \geq 0$; $\sum_{i,j} z_{ij} = 1$) the frequency of zygotes of the (i, j) type; correspondingly, $z_{\mathrm{panmixia},ij}$, $z_{\mathrm{selfing},ij}$ and $z_{\mathrm{asex},ij}$ denote offspring frequency (for individuals of the (i, j)-type) which is produced by random mating, by selfing, and asexually, respectively. Therefore,

$$z_{ij} = z_{\mathrm{panmixia},ij} + z_{\mathrm{selfing},ij} + z_{\mathrm{asex},ij} = z_{ij}\, p_{\mathrm{panmixia},ij} + z_{ij}\, p_{\mathrm{selfing},ij} + z_{ij}\, p_{\mathrm{asex},ij} . \quad (5.1)$$

Transition to the next generation can be described as follows: x_i^{t+1} is a frequency of haplotype i among haploids produced by individuals of generation t.

$$x_i^{t+1} = \frac{\sum_{j,k} P_{jk \to l} z_{panmixia, jk}^t}{\sum_l \sum_{j,k} P_{jk \to l} z_{panmixia, jk}^t},$$ (5.2)

$$z_{ij}^{t+1} = \frac{w_{ij}}{\overline{W}^t} \left[z_{apomixis, ij}^t + \left(\sum_{k,l} z_{panmixia, kl}^t \right) x_i^{t+1} x_j^{t+1} + \sum_{k,l} P_{kl \to i} P_{kl \to j} z_{selfing, kl}^t \right],$$ (5.3)

where $P_{ij \to k}$ is the probability of obtaining a gamete of type k from zygote (i,j), w_{ij} is the fitness of zygote z_{ij}^t, and $\overline{W}^t$ is the mean fitness of population in generation t.

In our study, we consider behavior of the system under stabilizing Gaussian selection with periodically changing optimum (see section 3.2) (Charlesworth, 1989, 1993; Hedrick, 1998).

5.2   CLB in mixed breeding systems

Here, we provide a few examples of CLB in the described breeding systems with either fixed or modifier-dependent proportions of panmixia, selfing, and apomixis. These examples demonstrate stability of CLB with deviation from panmixia in breeding systems and with deviation in recombination rates. In particular, we wished to address the following questions: **(i)** Are CLBs robust with respect to deviations from panmixia, and how large can these deviations be? **(ii)** In case of robustness, how will the system's behavior depend on recombination? **(iii)** Do systems exist that display CLB for a mixed breeding system but not for pure panmixia? **(iv)** Can polymorphism of a modifier controlling the level of apomixis or selfing be itself a source of CLB in systems, which display no CLB at any fixed level of selfing or vegetative reproduction?

Robustness of CLB with respect to variations in the breeding system: to demonstrate robustness of CLB to variations in the breeding system, we consider an example of CLB in a system with pure panmixia and mixtures of panmixia with selfing and/or apomixis (Fig. 12). In the case of full panmixia (i.e., $p_{panmixia} = 1$) we obtain a stable supercycle (see Fig. 12a). Changing the breeding system from pure panmixia to a mixture of panmixia and vegetative reproduction (decreasing $p_{panmixia}$ and increasing p_{asex}), or panmixia and selfing (decreasing $p_{panmixia}$ and increasing $p_{selfing}$), had no effect on the type of CLB for a certain range of parameters. Namely, this system displays CLB for every level of $0 \leq p_{asex} < 0.20$ (Fig. 12b) and $0 \leq p_{selfing} < 0.26$ (Fig. 12c). Moreover, CLB was preserved even in the case of simultaneous deviation from full panmixia by a combination of asexual reproduction and selfing (Fig. 12d).

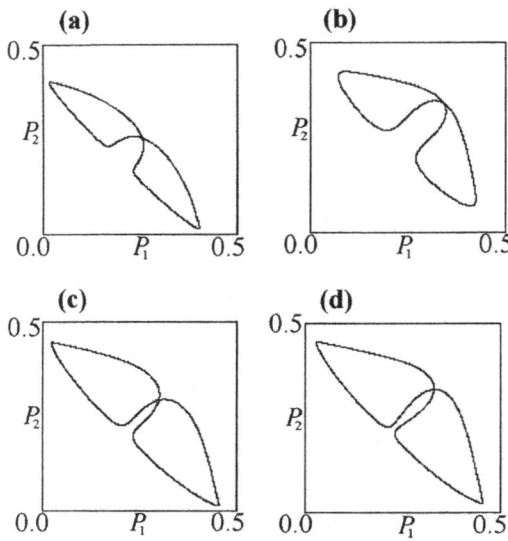

Figure 12. **Robustness of CLB with respect to variation in the breeding system.** Three-locus system was considered with $r_{12} = r_{23} = 0.01$; $d_1=d_2=d_3=1$; $h_1=0.25$, $h_2=-0.125$, $h_3=0.25$; $p=2$; $\tau_1= \tau_2=1$; $\hat{u}_1 =6$, $\hat{u}_2 =0.66$; $\sigma^2 = 2.18$. CLB is displayed under full panmixia and under large deviations from panmixia (e.g., for all levels of apomixis $p_{asex} <0.20$ or selfing $p_{selfing}$ <0.26 this system displays a stable supercycle, but with further increasing of p_{asex} or $p_{selfing}$ only one polymorphic steady state point characterizes the system). The supercycles also proved resistant to a rather large range of simultaneous deviation of p_{asex} and $p_{selfing}$ from zero. **(a)** A stable supercycle in the case of full panmixia $p_{panmixia}=1$. **(b)** A stable supercycle in the case of mixed panmixia and selfing, $p_{panmixia}=0.8$ and $p_{selfing} =0.2$. **(c)** A stable supercycle in the case of mixed panmixia and apomixis, $p_{panmixia}=0.85$ and $p_{asex}=0.15$. **(d)** A stable super-cycle in the case of a mixture, panmixia with selfing and apomixis, $p_{panmixia}=0.8$, $p_{selfing} =0.1$, $p_{asex}=0.1$.

Chaotic-like behavior of systems with fixed recombination rates and CLB with *rec*-modifier can also be robust with respect to deviations from full panmixia. The systems presented in Fig. 13 displayed chaotic behavior under pure panmixia (Fig. 13a). This dynamic pattern was preserved in a mixed breeding system (panmixia with both selfing and apomixis: $p_{panmixia}=0.92$, $p_{selfing} =0.05$, $p_{asex}=0.03$) with Lyapunov exponent $\lambda \approx 0.00045$ (Fig. 13b). CLB in the system with *rec*-modifier presented in Fig. 14 is very robust to deviation from panmixia to asexual reproduction (CLB was observed up to $p_{asex}=0.61$). Note that mean fitness of population in this system is rather high (about 0.34).

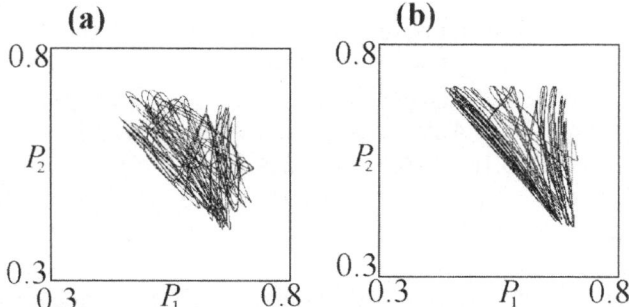

Figure 13. **Robustness of chaotic-like behavior with respect to deviation from panmixia.** Three-locus system was considered with $r_{12} = r_{23} = 0.5$; $d_1 = d_2 = d_3 = 0.5$; $h_1 = 0.06$, $h_2 = 0.1$, $h_3 = 0.06$; $p = 2$; $\tau_1 = \tau_2 = 1$; $\hat{u}_1 = 0.23$, $\hat{u}_2 = 3.03$; $\sigma^2 = 0.77$. (a) Chaotic behavior in the case of full panmixia $p_{\text{panmixia}} = 1$ (Lyapunov exponent $\lambda \approx 0.00045$). (b) Chaotic-like behavior is maintained in the case $p_{\text{panmixia}} = 0.92$, $p_{\text{selfing}} = 0.05$, $p_{\text{asex}} = 0.03$ ($\lambda \approx 0.00045$).

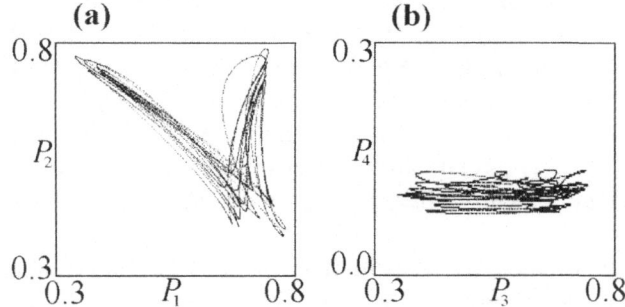

Figure 14. **Robustness of CLB in the system with *rec*-modifier to deviation from panmixia.** Three-locus selected system was considered combined with selectively neutral *rec*-modifier (locus #4): $r_{12} = r_{23} = r$; $r_{MM} = 0.02$, $r_{Mm} = r_{mm} = 0$, $r_{34} = r_c = 0.002$; $d_1 = d_2 = d_3 = 0.5$, $d_4 = 0$; $h_1 = 0.06$, $h_2 = 0.09$, $h_3 = 0.06$, $h_4 = 0$; $p = 2$; $\tau_1 = \tau_2 = 3$, $T = 6$; $\hat{u}_1 = 0.125$, $\hat{u}_2 = 3$; $\sigma^2 = 1$. CLB was observed under a wide range of mixed breeding systems (with parameter intervals: $0 \le p_{\text{selfing}} < 0.6$ or $0 \le p_{\text{asex}} < 0.25$). Here we present projections of CLB on phase planes in the case $p_{\text{panmixia}} = 0.8$, $p_{\text{selfing}} = 0.1$, $p_{\text{asex}} = 0.1$: (a) P_1 and P_2; (b) P_3 and P_4.

Robustness of CLB with respect to variation in recombination rates in population with mixed breeding system: CLB can be displayed by multilocus systems with weak, intermediate, or tight linkage between selective loci. To ascertain that conditions on recombination rates for existence of CLB in the system are not too restrictive, we explored various systems at different recombination rates. For example, a 4-locus selected system with $d_1 = 0.01$, $d_2 = 0.04$, $d_3 = 0.16$, $d_4 = 0.34$; $h_1 = h_2 = h_3 = h_4 = 0$; $p = 2$; $\tau_1 = \tau_2 = 1$; $\hat{u}_1 = 0$, $\hat{u}_2 = 1.1$; $\sigma^2 = 0.015$; $p_{\text{asex}} = 0.2$, and $p_{\text{panmixia}} = 0.8$ displays a stable supercycle for all recombination rates $r_{12} = r_{23} = r_{34} = r \in [0.03, 0.5]$.

System displaying CLB for mixed breeding system but not for pure panmixia: examples can be constructed with CLB displayed by populations with mixed breeding systems, but not with full panmixia. For a 3-locus selected system with $d_1 = d_2 = d_3 = 1$; $h_1=0.25$, $h_2= -0.1$, $h_3=0.25$; $r_{12} = r_{23}= 0.01$; $p=2$; $\tau_1= \tau_2=1$; $\hat{u}_1=6$, $\hat{u}_2=0.8$; $\sigma^2 =1.9$; $p_{\text{selfing}}=0.15$ (panmixia mixed with selfing), and for a 4-locus selected system with $r_{12} = r_{23}= r_{34}=0.5$; $d_1=0.01$, $d_2=0.04$, $d_3=0.16$, $d_4=0.34$; $h_1= h_2= h_3= h_4=0$; $p=2$; $\tau_1= \tau_2=1$; $\hat{u}_1=0$, $\hat{u}_2=1.1$; $\sigma^2 = 0.02$; $p_{\text{asex}}=0.1$ (panmixia mixed with apomixis), stable supercycles were observed, but no CLB was found in both systems in case of full panmixia.

Modifier of breeding system as a source of CLB: by analogy with *rec*-modifier, we considered a selectively neutral modifier of the breeding system. Two types of modifier were considered affecting: (a) the proportion of selfing p_{selfing} (see also: Uyenoyama et al., 1981; Altenberg, 1984; Uyenoyama and Waller, 1991), and (b) the proportion of asexual (apomixictic or vegetative) reproduction p_{asex} (Hamilton, 1980; Eshel and Feldman 1982; Feldman and Otto, 1989), versus random mating. Therefore, $p_{\text{panmixia},ij} + p_{\text{selfing},ij} + p_{\text{asex},ij} =1$. In accordance with the model described in section 5.1, these proportions are genotype-dependent. We considered the simplest case, where the breeding system is controlled by a single diallelic locus. We found systems with CLB of allele frequencies at the selectively neutral modifier locus (recall that the selected trait is subjected to stabilizing selection with periodically changing optimum: see section 3.2). Consider two examples:

(a) a 3-locus selected system combined with selectively neutral modifier of selfing (locus #4): $r_{12}= r_{23} = r_{34} =0.001$; $d_1 = d_2 = d_3 =1$, $d_4=0$; $h_1 = 0.15$, $h_2= -0.1$, $h_3 = 0.15$, $h_4=0$; $p = 2$; $\tau_1 = \tau_2 = 1$; $\hat{u}_1 = 6$, $\hat{u}_2 = 0.76$; $\sigma^2 = 2.3$; $p_{\text{selfing}}(A_4A_4) = 0.1$, $p_{\text{selfing}}(A_4a_4) = 0.05$, $p_{\text{selfing}}(a_4a_4) = 0$, and

(b) a 3-locus selected system combined with selectively neutral modifier of apomixes (locus #4): $r_{12}= r_{23}=0.5$, $r_{34}=0.005$; $d_1=0.9$, $d_2=0.6$, $d_3=2$, $d_4=0$; $h_1= h_2= h_3= h_4=0$; $p=2$; $\tau_1= \tau_2=1$; $\hat{u}_1=0$, $\hat{u}_2=1.1$; $\sigma^2 = 0.92$; $p_{\text{asex}}(A_4A_4) = 0.95$, $p_{\text{asex}}(A_4a_4) = 0.8$, $p_{\text{asex}}(a_4a_4) = 0.8$.

Both these systems display stable supercycle with superoscillations of allele frequencies in all four loci. The last example with modifier of apomixes has an additional interesting feature: the system displays CLB in the case of polymorphism for modifier of apomixis, but no CLB was found for any fixed intermediate level of apomixis. Therefore, similar to *rec*-modifiers (section 4.3), the modifier of apomixis can be considered in this model as a source of CLB.

6. DYNAMIC COMPLEXITY IN POPULATION GENETICS AND EVOLUTION: SOME PERSPECTIVES

6.1 Polymorphic fixed points versus complex trajectories

Temporal environmental fluctuations have long attracted attention as a factor of polymorphism maintenance. Theoretical analysis of single-locus models has set out the conditions of polymorphism (Haldane and Jayakar, 1963; Ewing, 1979; Maynard Smith and Hoekstra, 1980). Much less progress has been made with multilocus systems, due to technical difficulties. Consequently, several attempts have been undertaken to analyze the multilocus problem under some simplified conditions: weak selection and/or loose linkage between the selected loci leading to negligible linkage disequilibria. Under these circumstances, a varying optimum of the selected trait was found not to affect the amount of the maintained genetic variation (Lande, 1976). Our previous studies addressed the problem where no limitation of loose linkage or weak selection is accepted (Korol et al., 1994, 1996; Kirzhner et al., 1995b,c). It was shown that stabilizing selection with a cyclically moving optimum might produce abundant polymorphisms when the selected trait is controlled by additive loci with non-equal effects or by semidominant (dominant) loci. In diploid cyclical selection models, numerous examples were found where polymorphism maintenance under cyclical selection regime is not a phenomenon reminiscent of the constant environment (Korol et al., 1996). Moreover, under haploid constant selection, polymorphic fixed point proved unstable, if it exists at all (Rutschman, 1994; Kirzhner et al., 1995c), whereas in cyclical selection models stable polymorphism can be found *en masse* (Kirzhner et al., 1995c).

Clearly, environmentally determined cyclical selection may result in fixation or in forced oscillation of the population genetic structure with a period equal to that of the environmental changes. In this paper we review a series of our studies that showed an abundance of additional modes of behavior manifested by multilocus systems subjected to cyclical selection: (a) supercycles (auto-oscillations, with a period sometimes comprising dozens or hundreds of external periods) and (b) more complex limiting behavior (a broad spectrum of chaotic-like trajectories). Classical population genetic models resulting in complex behavior usually include some forms of frequency- or density-dependent selection (May and Anderson, 1983; Preygel and Korol, 1990; Altenberg, 1991; Hamilton, 1993; Korol et al., 1994; Kirzhner et al., 1998a; Doebely and de Jong, 1999; Flatt et al., 2001). These models directly or indirectly involve an ecological component. By contrast, in the model of cyclical selection the coefficients in the evolutionary operator do not depend on the system's phase variables. The selection model considered in this paper is a standard one in population genetics (see Nagylaki, 1992). Nevertheless, it manifests *en masse* earlier undetected complex dynamic patterns.

Constant selection at a single locus level, as well as multilocus selection-free regimes, cannot by themselves produce complex limiting population genetic behaviors with an attracting set of a trajectory consisting of more than one point (Geiringer, 1949; Lyubich, 1971, 1992; Kirzhner and Lyubich, 1974; Lyubich et. al., 1976). In a continuous 2-locus model of constant selection, Akin (1979) found some domains of parameter values that can result in auto-oscillations. Hastings (1981) constructed an example demonstrating CLB in a 2-locus discrete-time model. Thus, constant selection can produce CLB in population genetic systems, but only in exceptional cases.

Cyclical selection. Here we describe a wide spectrum of systems with cyclical selection that manifest CLB. These include haploid and diploid selection for non-equal additive loci, diploid selection for semidominant loci, and several types of genomic distribution of selected loci (a block of tightly linked loci, several such unlinked blocks, and a series of independently segregating loci). All these systems appeared to manifest abundant CLBs for a range of the model parameters (recombination rates, individual effects of the involved loci on the selected trait, and dominance effects of the selected loci). Nevertheless, not all combinations of the foregoing types did so (e.g., no such example was found with haploid selection for four unlinked loci). The availability of a broad spectrum of systems manifesting CLB made it possible to check whether these systems share some common features not characteristic of those that do not manifest such behavior. Although we are quite far from a comprehensive answer to this question, two such features may be mentioned here: rather strong selection and substantial linkage disequilibria. Two exactly opposite features were assumed in the first multilocus models with temporarily fluctuating environments (e.g., Lande, 1976; Gillespie and Turelli, 1989).

6.2   Do the conditions of complex dynamics fit biology?

The biological relevance of these findings depends on (**a**) how real are the parameter sets producing CLB and (**b**) whether the required strength of selection and resulting mean fitness are compatible with the reproductive capabilities of real populations. Both questions can be answered affirmatively.

Due to the environmental "initiative" of the revealed unusual dynamic patterns, it was of primary interest to check whether CLBs are robust with respect to disturbances of the parameters characterizing the environmental variation, e.g., of the optimal values of the selected trait along the period, or the period length. The answer is affirmative, that is, the revealed supercyclical movements are, to some extent, resistant to such disturbances. We also affirmatively answer another related question concerning resistance of CLB to random drift: the simulations showed that to a certain extent CLBs are robust to fluctuations caused by finite population size. Likewise, the ranges of the ratios of gene effects, dominance ratios, and rates of recombination in our numerical examples seem quite realistic.

More complex is the question of mean fitness. In the case of purely additive non-equal genes, very low mean fitness appears characteristic of the complex trajectories. However, in the case of dominant gene action, a significant part of situations with CLB lie in the fitness range of 0.1-0.4. Moreover, for the class of more realistic models with the selected trait being controlled by semidominant genes with non-equal effects, it is easy to find CLB regimes with rather high mean fitness, up to 0.4-0.6, which is compatible even with the relatively low reproductive capacities of many reptiles, birds, and mammals, let alone organisms with higher reproduction rates, i.e., most living organisms. An example of a supercycle with quite a high mean fitness (at any generation) is provided in Fig. 14. Here, the mean fitness along the whole supercycle varies in the range 0.33-0.34 at any of the environmental stages.

Ford (1971) was among the first who demonstrated that strong selection could be a common phenomenon in nature. Rich evidence has accumulated in the literature on this subject (reviewed in Nevo, 1999, 2001b). Moreover, the whole concept of the evolution of co-adapted blocks of genes is based on the assumption that strong selection and tight linkage are the major factors maintaining these blocks intact (e.g., Darlington, 1971). Many examples of polymorphic co-adapted gene blocks are well known (reviewed in Ford, 1971; Darlington, 1971; Clegg et al., 1978; Korol et al., 1994). Clearly, if the question is the polymorphism maintenance itself, the stringent requirements of selection can be significantly reduced (see an example with cyclical selection on p.1438 in Korol et al. (1996), where the geometric mean fitness was $\overline{W} = 0.80$ with a nearly global stability of the polymorphism). The number of such examples can easily be increased. However, if the question is preservation of linkage disequilibrium in a polymorphic population, then higher selection intensities are generally required. Finally, existence of polymorphism in the form of CLB may require an even stronger selection. As we could see from the presented examples of CLB (e.g., Fig. 3a,b and d), varying along the trajectories linkage disequilibrium is a characteristic feature of CLB. In the case of non-equal linked loci (Fig. 3a), the highest disequilibria are characteristic of the strongest loci; this is also true for the case of two unlinked blocks (Fig. 3e). Note, too, the existence of linkage disequilibrium between the unlinked blocks.

6.3   Do sex and recombination simplify dynamic patterns?

The foregoing results on CLB were obtained for panmictic models. Therefore, it was interesting to analyze the effect of deviations from "canonical" sex, e.g., in the form of panmixia mixed with selfing or/and apomixis, on CLB. We demonstrated that CLB is compatible with both mixed breeding systems (panmixia combined either with apomixis or selfing) experiencing stabilizing selection with cyclically varying optima. In some examples, CLB manifested remarkable robustness to variation of the breeding system (see section 5.2).

The main source of complex dynamic patterns in population genetics is frequency- or density-dependent selection in single and two- (or multiple-) species interactions (Charlesworth, 1971; Loeschcke and Christiansen, 1984; May and Anderson, 1983; Bell and Maynard Smith, 1987; Preygel and Korol, 1990; Altenberg, 1991; Hamilton, 1993; Korol et al., 1994; Kirzhner et al., 1999; Doebely and de Jong, 1999). One of the generalizations derived from a study of single-locus ecological-genetic models with restricted mixing was that sex <u>reduces</u> the likelihood of complex dynamics and chaos (Ruxton, 1995). Ruxton's single-locus model could deal with only a very restricted aspect of sex, meiotic segregation and syngamy. However, sexual reproduction includes one more major component: genetic recombination. The foregoing conclusion of Ruxton corresponds indirectly to our results presented in previous papers on cyclical selection obtained for purely panmictic multilocus systems (Kirzhner et al., 1996; Korol et al., 1998): CLB were more frequently observed at lower rates of recombination, although intermediate rates or free recombination can also produce CLB.

As for the effect of breeding systems on CLB, it does not seem that more sex corresponds to less complex dynamics. Some of the supercycles observed for low level of apomixis vanished in case of full panmixia. Some of CLBs found for panmictic systems resisted only small deviations from panmixia, whereas others endured at a range of 0-50% of apomixis and 0-20% of selfing. As indicated earlier, recombination also plays an important role. We can conclude that complex dynamics are quite compatible with sexual reproduction, at least within the framework of pure genetic models (not including variations in population density) of multilocus cyclical selection. In general, our results are more compatible with the conclusion of Flatt et al. (2001) derived from studies of another class of models, describing ecological-genetic interactions of antagonistic species, where complex dynamics were characteristic for intermediate sex (see also Korol et al., 1994, for a similar conclusion about the role of recombination in complex dynamics of interacting antagonists). Nevertheless, as follows from our studies, in many examples deviation from panmixia or decreasing recombinations result in simplification of population dynamics: in these cases, less sex corresponds to less complexity (see Fig. 15). Overly broad generalizations seem premature.

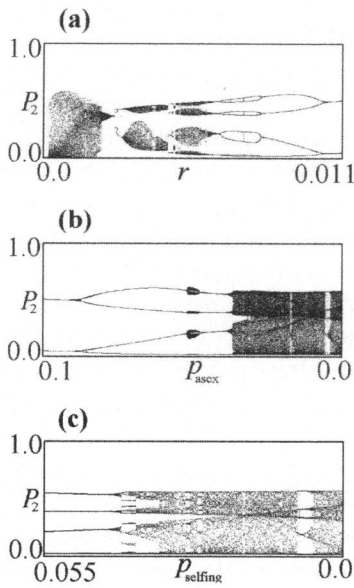

Figure 15. **Bifurcations of limiting attractor, with** r**,** p_{selfing}**, and** p_{asex} **as bifurcation parameters, for a system displaying T-cycle.** 3-locus selected system was considered: $r_{12} = r_{23} = r = 0.011$; $d_1=d_2=d_3=1$; $h_1 = 0.9$, $h_2 = 0.45$, $h_3 = 0.9$; $p = 4$; $\tau_1=\tau_3=3$, $\tau_2=\tau_4=1$, $T=8$; $\hat{u}_1=5.5$, $\hat{u}_2=\hat{u}_4=3.1$, $\hat{u}_3=1.2$; $\sigma^2=0.24$. **(a)** $p_{\text{panmixia}} = 0.9$; $p_{\text{asex}} = 0.1$; $r \in [0, 0.011]$; **(b)** $p_{\text{selfing}} = 0$; $p_{\text{asex}} \in [0, 0.1]$; **(c)** $p_{\text{asex}} = 0$; $p_{\text{selfing}} \in [0, 0.055]$. As one can see, high level of sex does not always simplify the behavior of the system.

As recombination rate proved to be a key factor affecting the mode of CLB and the existence of CLB, a generalized and more natural model was considered: the fixed recombination rate was replaced by a polymorphic recombination modifier (Kirzhner et al., 1995a; Korol et al., 1998). Several types of *rec*-modifiers were considered: **(i)** modification of recombination rates in all intervals of a chromosome; **(ii)** modifiers with an uneven distribution of effects including modification of linkage between blocks of linked loci with no within-block effect, and modification of linkage within blocks of tightly linked loci with no between-block effect. The revealed modifier-dependent changes included **(a)** supercyclical dynamics due to the *rec*-modifier in a system which does not manifest CLB when recombination rate is a fixed parameter; **(b)** appearance of a new level of superoscillations (super-supercycles) in a system which manifests supercycles under a fixed modifier; **(c)** chaotization of the regular supercyclical dynamics. While the attractors of the first two types seem to be rather complex, trajectories starting from neighboring initial points do not diverge. Moreover, the domain of attraction of these movements appears to be quite large, sometimes manifesting a nearly global stability (as in the systems presented in Fig. 3b). This phenomenon is largely due to polymorphism at the modifier locus. The modifier locus is an active participant in the observed complex non-monotonic limiting movements, although its dynamics can differ from that of the selected loci (e.g., Fig. 9). Therefore, a model with genotypic dependence of the rate of sexual recombination, which is more compatible with the real world (natural populations proved polymorphic for recombination rate: see Korol et al., 1994) and better represents the essence of sexually reproducing populations, demonstrates an increased predisposition to and stability of dynamic complexity.

To explain the observed phenomenon of a further complication of CLB with the introduction of a polymorphic *rec*-modifier (birth of a super-supercycle in a system already manifesting super-auto-oscillation), we proposed earlier a new notion of "genetic hysteresis" (see section 2.4 and Kirzhner et al., 1995a). It reflects the fact that under super-supercyclical movement, the characteristics of the system dynamics depend not only on the position of the system's coordinates in the phase space, but also on the direction of this movement (e.g., whether the frequency of allele for higher recombination rate increases or decreases over the specific part of the trajectory). Linkage of the modifier locus to the selected system, a kind of an "inertia factor", proved to be the key component determining the main characteristics or even the very possibility of such a pattern. Clearly, this extension of the relevance of CLB derives from 2-level selection effects: cyclical selection on the trait loci, and changes of recombination rate caused by "induced" selection on the *rec*-modifier (Korol et al., 1994). We can repeat with assurance the speculation of Ferriere and Fox (1995) that selection favors dynamic complexity in (sexual – K.F.K.) genetic systems.

6.4   Evolutionary significance of CLB

Complex (chaotic) dynamics display some important properties that could presumably be utilized in the course of evolution. It is not an easy task to propose hypothetical scenarios ("ecological niches") of such utilization, bearing in mind the controversies on rareness/abundance of complex dynamics (genetic or/and ecological) in nature. Nonetheless, since our discovery of complex dynamics in simple cyclical selection models, Nevo has encouraged our efforts in this direction. Strong selection as a precondition of the CLB phenomenon ideally corroborates Nevo's concepts on strong ecological selection as a major evolutionary factor (e.g., Nevo 1999, 2001a). Here, a few speculations about the potential significance of CLB are briefly presented.

Complex dynamics and evolution of recombination and sex: the described CLB patterns result from high-frequency forced oscillations of the optimum value of the fitness-related trait. In such conditions only very strong selection can provide evolution toward increased recombination (Korol and Preygel, 1989; Charlesworth, 1993; Korol et al., 1994; Kondrashov and Yampolsky, 1996). The high genetic load needed to promote increased recombination is one of the major difficulties when temporal (cyclical or stochastic) environmental fluctuations are to be considered a causal factor (Otto and Michalakis, 1998). Another obstacle was the presumed inability of fluctuating selection to preserve polymorphism, a precondition of recombination evolution (Kondrashov, 1993). We found earlier that this is not really an obstacle: stabilizing selection with a cyclically varying optimum for a quantitative trait that depends on purely additive or semidominant genes with non-equal effects leads to local polymorphism stability with a sufficiently large polymorphism attracting domain (Korol et al., 1994, 1996; see Fig. 3b).

Analysis of this model shows that the demands for selection strength sufficient to promote evolution toward increased recombination can be relaxed, if the effect of linkage of *rec*-modifier to trait loci is taken into account. This allowed us to suggest a simple heuristic explanation of the effects of the modifier on CLB (Korol et al., 1998) based on the "low-pass filter effect" described for evolution of recombination and mutation modifiers (Sasaki and Iwasa, 1987; Ishii et al., 1989). In a system subjected to fluctuating selection with a mixture of oscillations with different period lengths, the modifier fate seems to be determined mainly by the lowest frequency component. Our models have two components of the selected system dynamics: (i) forced oscillations with a relatively short period; (ii) low-frequency CLB movement (e.g., a supercycle consisting of hundreds or even thousands of environmental periods). According to the principle of Sasaki and Iwasa (1987), the fate of the *rec*-modifier should depend mainly on the second component. However, along the trajectory, the modifier itself evolves, which may result in a reduction of the "current" amplitude of the low-frequency movement. In such intervals of system trajectory, the high-frequency component determines the dynamics of the modifier, preparing the conditions for the next phase. Such a dynamic balance may generate different modes of limiting behavior (see Figs. 9-11).

The relevance of the proposed mechanism of CLB effect on recombination evolution could be demonstrated by an <u>artificial construction,</u>

where the observed auto-oscillatory long-periodical movement was replaced by 'external' long-periodical changes in the selected optimum, so that the cyclical selection regime was a superposition of high- and low-frequency oscillations (Korol et al., 1998). The modifier behavior in the initial model and in the artificial example proved to be very similar, confirming the relevance of the proposed mechanism. Therefore, the ability to manifest low-frequency supercyclical dynamics may promote evolution of increased recombination despite the expected fixation of the zero recombination allele at the modifier locus for situations with short-period forced oscillations.

Instability of the population trajectories with respect to small disturbances of allele or haplotype/genotype frequencies: unpredictability of the trajectory due to imprecise information about the initial point in a system undergoing chaotic dynamics can have some interesting biological consequences in situations where such a trajectory is subjected to a short-term external disturbance. From the moment of disturbance, the observed trajectory will exponentially diverge from the one that the system would have followed if the disturbance had not happened. Different genetic processes like mutations, migration, or short-term changes in selection regime can be considered plausible candidates for such a disturbance of population state. The important point is that an extremely small alteration in genotype frequencies can cause drastic changes of trajectory. Such "super-sensitivity" could provide unlimited possibilities of variations in population structure without altering the main factors, incomparable with those of identical populations that have not entered into a complex dynamic regime. One may speculate that the inclination to be affected by a complex dynamic regime could serve as a mechanism for switching between potential scenarios of the evolutionary process. These mechanisms of flexible reactivity displayed by the ability to "correct/modulate" the system trajectory without changing the basic attractor are well known in other fields of biological self-organization, like development or heart oscillations. Under certain combinations of external-internal conditions, this flexible stability can be destroyed resulting in abrupt ontogenetic disturbances, diseases, extinctions, or origins of new evolutionary clades. Such changes of dynamic regimes are referred to as bifurcations. For evolutionary genetics, bifurcation is an important property of system dynamics that can affect population genetic processes (Ferrieri and Fox, 1995).

CLB as a "diversity maker": consider a spatially structured metapopulation that includes a main range and a net of peripheral population islands. One can speculate about evolutionary advantages of CLB as a source of genetic diversity flowing out of the central to peripheral populations, thereby facilitating adaptive radiation. Indeed, we have shown that even cyclical trajectories can have a very complex form, and more so, dynamic chaos. The evolutionary importance of this phenomenon lies in the fact that a rich spectrum of specific population states will be generated along such trajectories, which will increase the range of potential effects of selection. Export of genotypes (emigrants) from such a population to the peripheral ones can increase the amount of the stored genetic diversity as compared with situations of equilibrium dynamics (Holt, 1983). The phenomenon of increased store of variation in peripheral populations is described in the literature (e.g., Nevo, 2001a; Nevo et al., 2002). One more potential effect of CLB in the central population is worth mentioning: the richness of the population states generated along the CLB trajectory may help the peripheral populations to overcome the adaptive valleys and reach new adaptive peaks otherwise unattainable due to epistasis (Whitlock, 1997).

Richness of CLB population dynamics for trajectory time-scale characteristics: one of the outstanding peculiarities of the described phenomenon, CLB caused by cyclical selection, is a wide range of dynamic patterns that a population could display with variation of the system parameters or environmental characteristics, ranging from a few environmental periods (e.g., in T-cycles) up to tens of thousands of periods (e.g., in super-supercyclical oscillations involving *rec*-modifiers linked to the selected system). Especially interesting is the great variation in the rate of changes in phase variables along a trajectory that can be observed in some CLB examples, with regular (supercyclical) or chaotic movements. In particular, some systems display very slow changes, or "quasi steady-states", for very long periods of time, which alternate with short periods of very fast changes. For an external observer interested in long-term (evolutionary) population processes, such changes would recall the scenario referred to as "punctuated equilibrium" (Gould, 1998), at least at first glance. These aspects may also be of interest for studies dealing with temporal variation in population genetic structure. In particular, natural populations subjected to seasonal variation, like *Drosophila* and other insects, sometimes display directed changes across years, and revealing the underlying causal ecological factors seems not a trivial problem (Dobzhansky, 1970).

7.   REFERENCES

Akin E. 1979. The Geometry of Population Genetics, Springer-Verlag, Berlin e. a.
Altenberg L. 1984. A Generalization of Theory on the Evolution of Modifier Genes. Ph.D. thesis, Stanford University.
Altenberg L. 1991. Chaos from linear frequency-dependent selection. Amer Nat. 138, 51-68.
Barton N. H. 1995. A general model for the evolution of recombination. Gene Res. 65, 123-144.
Begon M., Harper J.L. and Townsend C.R. 1996. Ecology: Individuals, Populations, and Communities. Blackwell Science, Oxford.

Bell G. and Maynard Smith J. 1987. Short-term selection for recombination among mutually antagonistic species. Nature 328, 66-68.

Bergman A. and Feldman M.W. 1990. More on selection for and against recombination. Theor Popul Biol. 38, 68-92.

Brooks L.D. 1988. The evolution of recombination rates. In R.E. Michod and B.R. Levin, eds., The Evolution of Sex: An Examination of Current Ideas. Sinauer, Sunderland (Mass.), pp. 87-105.

Bürger R. 1989. Linkage and the maintenance of heritable variation by mutation-selection balance. Genetics 121, 175-184.

Bürger R. 1999. Evolution of genetic variability and the advantage of sex and recombination in changing environments. Genetics 153, 1055-1069.

Burt A. and Bell G. 1987. Mammalian chiasma frequencies as a test of two theories of recombination. Nature 326, 803-805.

Charlesworth B. 1971. Selection in density-regulated populations. Ecology 52, 469-474.

Charlesworth B. 1976. Recombination in a fluctuating environment. Genetics 83, 181-195.

Charlesworth B. 1989. The evolution of sex and recombination. Trends Ecol Evol 4, 264-267.

Charlesworth B. 1993. Directional selection and evolution of sex and recombination. Genet Res. 61, 205-224.

Chinnici J.P. 1971. Modification of recombination frequency in *Drosophila*. II. The polygene control of crossing over. Genetics 69, 85-96.

Clegg M.T., Kahler A.L. and Allard R.W. 1978. Estimation of life cycle components of selection in an experimental plant population. Genetics 89, 765-792.

Darlington C.D. 1971. In Ecological Genetics and Evolution. Blackwell Scientific, Oxford, pp. 1-19.

Dobzhansky Th. 1970. Genetics of the Evolutionary Process, Columbia University Press, New York.

Doebely M. and de Jong G. 1999. Genetic variability in sensitivity to population density affects the dynamics of simple ecological models. Theor Popul Biol. 55, 37-52.

Eshel I. and Feldman M.W. 1982. On the evolution of sex determination and the sex ratio in haplo-diploid populations. Theor Popul Biol. 21, 440-450.

Ewing E.P. 1979. Genetic variation in a heterogeneous environment. VII. Temporal and spatial heterogeneity in infinite population. Amer Nat. 114, 197-212.

Feldman M.W., Christiansen F.B. and Brooks L.D. 1980. Evolution of recombination in a constant environment. Proc Natl Acad Sci USA 77, 4838-4841.

Feldman M.W. and Otto S.P. 1989. More on recombination and selection in the modifier theory of sex-ratio distortion. Theor Popul Biol. 35, 207-225

Feldman M.W., Otto S.P. and Christiansen F.B. 1997. Population genetic perspectives on the evolution of recombination. Ann Rev Genet. 30, 261-295.

Ferrière R. and Fox G.A. 1995. Chaos and evolution. Trends Ecol Evol. 10, 480-485

Fisher R.A. 1930. The Genetical Theory of Natural Selection. Clarendon Press, Oxford.

Flatt T., Maire N. and Doebeli M. 2001. A bit of sex stabilizes host-parasite dynamics. J Theor Biol. 212, 345-354.

Flexon P.B. and Rodell C.F. 1982. Genetic recombination and directional selection for DDT resistance in *Drosophila melanogaster*. Nature 298, 672-675.

Ford E.B. 1971. Ecological Genetics, 3rd edition. Chapman & Hall, London.

Gavrilets S. and Hastings A. 1994a. Maintenance of multilocus variability under strong stabilizing selection. J Math Biol. 32, 287-302.

Gavrilets S. and Hastings A. 1994b. Dynamics of genetic variability in two-locus models of stabilizing selection. Genetics 138, 519-532.

Geiringer H.J. 1949. On some mathematical problems arising in the development of Mendelian genetics. J Amer Stat Assoc. 44, 526-547.

Gillespie J.H. 1978. A general model to account for enzyme variation in natural populations. V. The SAS-CFF model. Theor Popul Biol. 14, 1-45.

Gillespie J.H. and Turelli M. 1989. Genotype-environment interactions and the maintenance of polygenic variation. Genetics 121, 129-138.

Gorlov I.P., Schuler L., Bunger L. and Borodin P.M. 1992. Chiasma frequency in strains of mice selected for litter size and for high body weight. Theor Appl Genet. 84, 640-642.

Gorodetsky V.P., Zhuchenko A.A. and Korol A.B. 1990. Efficiency of feedback selection for recombination in *Drosophila*. Genetika (USSR), 26, 1942-1952 (in Russian).

Gould S.J. 1998. Gulliver's further travels: the necessity and difficulty of a hierarchical theory of selection. Phil Trans Roy Soc Lond B, 353, 307-314.

Grishkan I., Korol A.B., Nevo E. and Wasser S.P. 2003. Ecological stress and sex evolution in soil microfungi. Proc Roy Soc B, 270, 13-18.

Haldane J.B.S. and Jayakar S.D. 1963. Polymorphism due to selection of varying direction. J Genet. 58, 237-242.

Hamilton, W. D. 1980. Sex versus non-sex versus parasite. Oikos, 35, 282–290.

Hamilton, W. D. 1993. Haploid dynamic polymorphism in a host with matching parasites: Effects of mutation/subdivision, linkage, and patterns of selection. J Heredity, 84, 328-338.

Hastings A. 1981. Stable cycling in discrete time genetic models. Proc Natl Acad Sci USA 78, 7224-7225.

Hastings A. 1984. Stable equilibria at two loci in populations with large selfing rates. Genetics 109, 215-228.

Hastings A., Hom C.L., Ellner S., Turchin P. and Godfray H.C.J. 1993. Chaos in ecology - is mother-nature a strange attractor. Ann Rev Ecol Syst. 24, 1-33.

Hedrick P.W. 1998. Maintenance of genetic variation: spatial selection and self-fertilization. Amer Nat. 152, 145-150.

Holt R.D. 1983. In Rhodin A.G.J. and Miyata K., (eds.), Advances in Herpetology and Evolutionary Biology. Harvard University Museum of Comparitive Zoology [Publication Series], pp. 680-694.

Ishii K., Matsuda H., Iwasa I. and Sasaki A. 1989. Evolutionary stable mutation rate in a periodically changing environment. Genetics 121, 163-174.

Kirzhner V.M. and. Lyubich Yu. I. 1974. General evolutionary equation and the limit theorem for genetic systems without selection. Sov Math Doklady, 215, 31-33 (in Russian).

Kirzhner V.M., Korol A.B., Ronin Y.I. and Nevo E. 1994. Cyclical behavior of genotype frequencies in a two-locus population under fluctuating haploid selection. Proc Natl Acad Sci USA 91, 11432-11436.

Kirzhner V., Korol A.B., Ronin Y.I. and Nevo E. 1995a. Genetic super-cycles caused by cyclical selection. Proc Natl Acad Sci USA 92, 7130-7133.

Kirzhner V., Korol A.B. and Ronin Y. 1995b. The dynamics of linkage disequilibrium under temporal environmental fluctuation. Two-locus selection. Theor Popul Biol. 47, 257-276.

Kirzhner V., Korol A.B. and Ronin Y.I. 1995c. Cyclical environmental changes as factor maintaining genetic polymorphism. I. Two-locus haploid selection. J Evol Biol. 8, 93-120.

Kirzhner V., Korol A.B. and Nevo E. 1996. Complex non-stationary trajectories in multilocus genetic systems caused by cyclical selection with moving optimum. Proc Natl Acad Sci USA 93, 6532-6535.

Kirzhner V., Lembrikov B., Korol A.B. and Nevo E. 1998a. Supercycles, strange attractors and chaos in a standard model of population genetics. Physica A 249, 565-570.

Kirzhner V., Korol A.B. and Nevo E. 1998b. Complex limiting behavior of multilocus genetic systems in cyclical environment. J Theor Biol. 190, 215-225.

Kirzhner V., Korol A.B. and Nevo E. 1999. Abundant multilocus polymorphisms caused by genomic interaction between species on trait-for-trait basis. J Theor Biol. 198, 361-370.

Kirzhner V. and Lyubich Yu. 2003. On finiteness of multiplicative selection equilibria. J Appl Math Letters (in press).

Kondrashov A.S. 1993. Classification of hypotheses on the advantage of amphimixis. J Heredity 84, 372-387.

Kondrashov A.S. and Yampolsky L.Yu. 1996. High genetic variability under the balance between symmetric mutation and fluctuating stabilizing selection. Genet Res. 68, 157-164.

Korol A.B. 1999. Selection for adaptive traits as a factor of recombination evolution: Evidence from natural and experimental populations (a review). In S.P. Wasser, ed., Evolutionary Theory and Processes: Modern Perspectives. Kluwer Academic Publishers, Dordrecht. Pp. 31-53.

Korol A.B. 2001. Recombination. In Encyclopedia of Biodiversity. V.5. Academic Press, San Diego. Pp. 53-71.

Korol A.B. and Preygel I.A. 1989. Increase in recombination in a multilocus system under environmental fluctuations. Genetika (USSR) 25, 923-931 (in Russian).

Korol A.B. and Iliadi K.G. 1994. Recombination increase resulting from directional selection for geotaxis in *Drosophila*. Heredity 72, 64-68.

Korol A.B., Preygel I.A. and Preygel S.I. 1994. Recombination Variability and Evolution. Chapman & Hall, London.

Korol A.B., Kirzhner V.M., Ronin Y.I. and Nevo E. 1996. Cyclical environmental changes as factor maintaining genetic polymorphism. II. Two-locus diploid selection. Evolution 50, 1432-1441.

Korol A.B., Kirzhner V. and Nevo E. 1998. Dynamics of recombination modifiers caused by cyclical selection: interaction of forced- and auto-oscillations. Genet Res. 69, 135-147.

Kun L.A. and Lyubich Yu.I. 1979. Convergence to equilibrium under the action of additive selection in multilocus multiallele population. Sov Math Doklady 20, 1380-1382 (in Russian).

Lande R. 1976. The maintenance of genetic variability by mutation in a polygenic character with linked loci. Genet Res. 26, 221-235.

Lewontin R.C. 1974. The Genetic Basis of Evolutionary Change. Columbia University Press, New York.

Loeschcke V. and Christiansen F.B. 1984. Evolution and intraspecific exploitative competition. II. A two-locus model for additive gene effects. Theor Popul Biol. 26, 228-264.

Lyubich Yu.I. 1971. Basic concepts and theorems of evolutionary genetics for free populations. Russian Math Survey 26, 51-123.

Lyubich Yu.I. 1992. Mathematical Structures in Population Genetics. Springer-Verlag, Berlin e. a.

Lyubich Yu.I., Maystrovsky G.D. and Olkhovsky Yu.G. 1976. Selection induced convergence to equilibrium in a single locus autosomal population. Sov Math Doklady 226, 58-60 (in Russian).

Lyubich Yu.I., Kirzhner V. and Ryndin A. 2001. Mathematical theory of phenotypical selection. Adv Appl Math. 26, 330-352.

Marshall D.R. and Weir B.S. 1979. Maintenance of genetic variation in apomictic plant populations. I. Single locus models. Heredity 42,159-172.

May R.M. 1976. Simple mathematical models with very complicated dynamics. Nature 261, 459-467.

May R.M. and Anderson R.M. 1983. Epidemiology and genetics in the coevolution of parasites and host. Proc Roy Soc Lond B 219, 281-313.

Maynard Smith J. 1978. The Evolution of Sex. Cambridge University Press, Cambridge, UK.

Maynard Smith J. 1980. Selection for recombination in a polygenic model. Genet Res. 35, 269-277.

Maynard Smith J. 1988a. Selection for recombination in a polygenic model: the mechanism. Genet Res. 51, 59-63.

Maynard Smith J. 1988b. The evolution of recombination. In R.E. Michod and B.R. Levin, eds., The Evolution of Sex: An Examination of Current Ideas. Sinauer Ass., Sunderland (Mass.). Verlag, Berlin e.a., pp. 106-125.

Maynard Smith J. and Hoekstra R. 1980. Polymorphism in a varied environment: How robust are the models. Genet Res. 35, 45-57.

Moran P.A.P. 1962. The Statistical Processes of Evolutionary Theory. Oxford University Press, Oxford.

Nagylaki T. 1992. Introduction to Theoretical Population genetics. Springer-Verlag, Berlin.

Nagylaki T., Hofbauer J. and Brunovsky P. 1999. Convergence of multilocus systems under weak epistasis or weak selection. J Math Biol. 38, 103-133.

Nevo E. 1997. Evolution in action across phylogeny caused by microclimatic stresses at "Evolution Canyon". Theor Popul Biol. 52, 231-243.

Nevo E. 1999. Mosaic Evolution of Subterranean Mammals: Regression, Progression and Global Convergence. Oxford University Press, Oxford.

Nevo E. 2001a. Evolution of genome-phenome diversity under environmental stress. Proc Natl Acad Sci USA 98, 6233-6240.

Nevo E. 2001b. Genetic diversity. In: Encyclopedia of Biodiversity, 3, 195-213, Academic Press, New York.

Nevo E., Kirzhner V., Beiles A. and Korol A.B. 1997. Selection versus random drift: long-term polymorphism persistence in small populations (evidence and modelling). Phil Trans Roy Soc Lond B 352, 381-391.

Nevo E.,. Korol A.B, Beiles A. and Fahima T. 2002. Evolution of Wild Emmer Wheat: Population genetics, genetic resources, and genome organization of wheat's progenitor, Triticum dicoccoides. Springer-Verlag, Berlin e. a.

Olsen L.F. and Schaffer W.M. 1990. Chaos versus noisy periodicity: Alternative hypotheses for childhood epidemics. Science 249, 499–504.

Otto S.P. and Michalakis Y. 1998. The evolution of recombination in changing environments. Trends Ecol Evol. 13, 145-151.

Otto S.P and Lenormand T. 2002. Resolving the paradox of sex and recombination. Nat Rev Genet. 3, 252-261.

Overath R.D. and Asmussen M.A. 1998. Genetic diversity at a single locus under viability selection and facultative apomixis: Equilibrium structure and deviations from Hardy-Weinberg frequencies. Genetics 148, 2029-2039.

Preygel S.I. and Korol A.B. 1990. Evolution of recombination in systems of "host-parasite" type. Multilocus models. Genetika (USSR) 26, 349-358 (in Russian).

Peng J.H., Ronin Y.I., Fahima T., Roder M.S., Li Y.C., Nevo E. and Korol A.B. 2003. Domestication QTLs in wild emmer wheat, *Triticum dicoccoides*. Proc Natl Acad Sci, USA 100, 2489-2494.

Reiersol O. 1962. Genetic algebras studied recursively and by means of differential operators. Mathematica Scandinavica 10, 25-44.

Rutschman D.H. 1994. Dynamics of the two-locus haploid model. Theor Popul Biol. 45, 167-176.

Ruxton G.D. 1995. Population models with sexual reproduction show a reduced propensity to exhibit chaos. J Theor Biol. 175, 595-601.

Ryndin A., Kirzhner V., Nevo E. and Korol A.B. 2001. Polymorphism maintenance in populations with mixed random mating and apomixis subjected to stabilizing and cyclical selection. J Theor Biol. 212, 168-181.

Sacker R.J. and Bremen H.F. 2003. A new approach to cycling in a 2-locus 2-allele genetic model. J Diff Equat Appl. 9, 441-448.

Saleem M., Lamb B. and Nevo E. 2001. Inherited differences in crossing-over and gene conversion frequencies between wild strains of *Sordaria fimicola* from 'Evolution Canyon'. Genetics 159,1573-1593.

Sasaki A. and Iwasa Y. 1987. Optimal recombination rate in fluctuating environments. Genetics 115, 377-388.

Simchen G. and Stamberg J. 1969. Fine and coarse controls of genetic recombination. Nature 222, 329-332.

Turelli M. and Barton N. 1990. Dynamics of polygenic characters under selection. Theor Popul Biol. 38, 1-57.

Uyenoyama M.K., Feldman M.W. and Mueller L.D. 1981. Population genetic theory of kin selection: Multiple alleles at one locus. Proc Natl Acad Sci USA 78, 5036-5040.

Uyenoyama M.K. and Waller D.M. 1991. Coevolution of self-fertilization and inbreeding depression III. Homozygous lethal mutations at multiple loci. Theor Popul Biol. 40, 173-210.

Whitlock M.C. 1997. Founder effects and peak shifts without genetic drift: Adaptive peak shifts occur easily when environments fluctuate slightly. Evolution 51, 1044-1048.

Wolf H.G., Wohrmann K. and Tomiuk J. 1987. Experimental evidence for the adaptive value of sexual reproduction. Genetica 72, 151-159.

Zhivotovsky L.A. and Feldman M.W. 1992. On models of quantitative genetic variability: a stabilizing selection-balance model. Genetics 50, 947-955.

Zhivotovsky L.A., Feldman M.W. and Christiansen F.B. 1994. Evolution of recombination among multiple selected loci: A generalized reduction principle. Proc Natl Acad Sci USA 91, 1079-1083.

EVOLUTIONARY EPISTEMOLOGY AND INVALIDATION

Yuval Ne'eman
School of Physics and Astronomy, Raymond and Beverly Sackler Faculty of Exact Sciences, Tel Aviv University, Tel Aviv 69978, Israel

Abstract: I show that the two most important conceptual advances in epistemology made by Karl Popper, namely, invalidation (1935) and evolutionary epistemology (1972), though apparently unrelated, fit very precisely together if one applies an improved generalized evolutionary paradigm in which the extinctions are included as chance "mutations" of the environment.

1. INTRODUCTION

It is a pleasure for me to contribute to this Festschrift honoring Eviatar Nevo upon his 75th birthday. As one always interested in evolution – while working mostly in other fields such as Particle Physics, Cosmology, and Philosophy of Science, I was happy to have had the chance to be in contact with Eviatar throughout the years.

The topic I have selected relates to evolutionary epistemology, a philosophical doctrine launched by Karl Popper (1972) and developed by Donald Campbell (1974), with significant contributions by my friend and former student Aharon Kantorovich (1993) including Kantorovich and Ne'eman (1989) and Ne'eman (1999). I do believe in Evolutionary Epistemology. Popper is the most quoted modern philosopher of science and is well known for his work on the concept of *invalidation* (or *falsification*, the term used by Karl Popper – which I believe was a rather

S.P. Wasser (ed.), Evolutionary Theory and Processes: Modern Horizons,
Papers in Honour of Eviatar Nevo
Y. Ne'eman. *Evolutionary Epistemology and Invalidation, 109-112.*
© 2004 *Kluwer Academic Publishers.*

unfortunate choice). I use "invalidation", which fits the concept more precisely, in view of its intended meaning, namely, trying to show that some theory or definition, which had been validated for a certain range of values of some parameters, no longer holds within a range of these parameters previously untested, but is now experimentally within reach. *Falsification* implies trying to show that it is false over the new range. But the term has a different and rather ugly meaning in everyday life (such as *falsifying* a document), whereas *invalidation* is precisely what we mean.

Taking evolutionary epistemology as the correct theory and looking in retrospect at what Popper said about invalidation (Popper, 1935), one tends to wonder: if ideas develop according to the evolutionary paradigm, how does "invalidation" – a most important concept in the history and philosophy of science – fit? What is its evolutionary role? I hope to show that it indeed fits beautifully, provided we first update our evolutionary paradigm.

2.   EXTINCTIONS AND A NEW EVOLUTIONARY PARADIGM

Looking at the history of life on Earth we come across the primordial importance of the extinctions (Becker, 2002) at the transition layers between any two geological eras, as first noted and studied by L. and W. Alvarez in the 1960s. We now know several other cases of massive extinctions which have also given a boost to some otherwise un-evolving species. It is scientifically wrong to leave out these catastrophic developments from evolutionary studies.

In my amended paradigm we shall see how it becomes possible for extinctions to be considered part of the evolutionary processes. Here is a modified schema of evolutionary processes:

A system [S], governed by a program P[S] exists in an Environment [E]. It undergoes some routine R[P] which exposes it to the impact of chance, ("tychic" intervention [T] of random results), causing *mutations* M(S). Dynamical constraints "filter" the mutations; if it is an improvement, it prospers. Up to here we have had only "active" mutations of the system's program.

What is the role of the catastrophes which caused the extinctions (such as a hit by a comet, etc.)? These are *tychic* interventions in a routine exposure of the *Environment* R[E] , or "passive" mutations. The *routine* in this case is the motion of the Sun and Earth through different debris with the passage of time. Here, too, there is a dynamic selection: the Tunguska meteor only set fire to a large area in Siberia - whereas the one that killed the dinosaurs also gave an unexpected chance for the smaller mammals to take over. I have used the term "scalawag" for this kind of improvement by survival.

3. EVOLUTIONARY EPISTEMOLOGY – ACTIVE MUTATIONS

Let me immediately place the various actors in their roles in this case.

The "system" is a conjecture or theory which motivated the present stage. The "environment" is the existing body of theory, the paradigm that it has to fit. The mutation occurs when the system's program is exposed to random developments. The intention was to study X - this is the routine R(P) - and something has happened by chance and either nothing important develops, or B is discovered. Examples abound, such as (Segre, 1980; Thorton et al., 1993) the discovery of x-rays by Roentgen in 1896: he was studying cathode rays when a screen, left on a table nearby from a previous experiment, started to glow. He put out his hand and saw the bones through the flesh. Another such example is Cade's discovery of the psychochemical properties of *lithium* (Lickey and Gordon, 1983). Cade was a physician associated with an asylum. He planned to test *uric acid*, reputed as producing active personalities, and tried it as an energizer for the sufferers of depression. Uric acid was obtained as its *lithium* salt. Having administered it to all inmates, he discovered that the *schizophrenics* calmed down. Let us quote some other examples in other fields out of the multitude, i.e., in experimental physics, the discovery of radioactivity by Becquerel or in theoretical physics the emergence of string theory as a quantum and the theory of gravity; in mathematics, both pure and applied. We have the emergence of the modern computer from the after-shocks of the Russell-Whitehead paradox (Ne'eman, 1998).

4. EVOLUTIONARY EPISTEMOLOGY: INVALIDATION AS AN EXTINCTION

We now come to experiments that have broken an existing paradigm. An example is the Michelson–Morley experiment (1897) (Segre, 1980; Thorton and Rex, 1993), with the resulting extinction of Galilean symmetry and the emergence of relativity. The experiment assumed that the addition of velocities always follows the simple rules of Galilean symmetry. This had been verified over a range of velocities up to 0.1% of the velocity of light. The Michelson-Morley experiment tested it de facto at the velocity of light itself – and it destroyed the basic assumption. Galilean physics became the low-velocity approximation of a new theoretical environment, namely, relativity. Popperian invalidation is thus a *routine* "motion" within the *environment* by changing one parameter (in this case velocity), thereby entering a previously unexplored sector and opening the possibility of finding that this is not at all the theoretical environment we started from.

As to the "willed action" aspect, in Michelson's de facto selection of the new theoretical kinematics environment by his (unconscious) choice of the value of the velocities appearing in the addition equation, the only relevant criterion is his total ignorance of what the kinematical paradigm (the environment) will be like – whether it will be the same Galilean symmetry (i.e., no need to change the current paradigm) or not; neither did he know what it could change into. *Research* is a "blind" activity by definition. Otherwise it becomes *development*. As to the evolutionary role in epistemology of the researcher who, unlike Michelson, consciously initiates a falsification/invalidation experiment, he is the direct analogue to that of the biologist who modified the biological environment by irradiating *Drosophila* flies. Both are in one class with a new predator appearing in a given region, or a dam built on the river home of a species. The falsification/invalidation scientist hitchhikes upon an evolutionary feature and becomes part of it.

5.   REFERENCES

Becker L. 2002. Repeated Blows. Sci Amer. 286, 62-69.

Campbell D.T. 1974. Evolutionary Epistemology, in The Philosophy of Karl Popper, P.A. Schilpp, ed., La Salle: Open Court, v.1, pp. 413-463.

Kantorovich A. and Ne'eman Y. 1989. Serendipity as a Source of Evolutionary Progress in Science. Studies in the History and Philosophy of Science 20, pp. 505-529.

Kantorovich A. 1993. Scientific Discovery – Logic and Tinkering. SUNY Press, Albany, pp. 281

Lickey M.E. and Gordon B. 1983. Drugs for Mental Illness. W.H. Freeman, New York.

Ne'eman Y. 1998. The sophism which ushered in the Age of Information Technology. In Theory and Practice in Mathematics, Science and Technology Education (Hebrew), R. Stavy and D. Tirosh, (Eds.), Tel Aviv University School of Education, Tel Aviv, pp. 15-20.

Ne'eman Y. 1999. Order out of Randomness: Science and Human Society in a Generalized Theory of Evolution (Hebrew). Van Leer Jerusalem Institute and Hakibbutz Hame'uchad, Tel Aviv, 112 pp.

Popper K. 1935. Logik der Forschung. English version, 1959.

Popper K. 1972. Objective Knowledge. Clarendon Press, Oxford.

Segre E. 1980. From X-rays to Quarks. W.H. Freeman, New York.

Serway R.A., Moses C. and Moyer C.A. 1997. Modern Physics. Saunders College and Harcourt, Brace Jovanovich, New York.

Thornton S.T. and Rex A. 1993. Modern Physics for Scientists and Engineers. Saunders College and Harcourt, Brace Jovanovich, New York.

Part Two

Genome Evolution

TUNING FUNCTION OF TANDEMLY REPEATING SEQUENCES: A MOLECULAR DEVICE FOR FAST ADAPTATION

Edward N. Trifonov
Genome Diversity Center, Institute of Evolution, University of Haifa, Mt. Carmel, Haifa 31905, Israel

Abstract: Dispersed and tandemly repeating sequences of various types are the most polymorphic components of genomes, especially of eukaryotic genomes. The tandem repeats are also least stable due to frequent changes in the numbers of repeating units in the runs. One may associate this fluid type of genomic polymorphism with individual phenotypic variations and, perhaps, with fast adaptation to a changing environment. A "modulation (or fast adaptation) code" was suggested in 1989 (by the author) that attributed to tandem repeats the function of tuning gene expression by spontaneous changes in the copy numbers of the repeating elements associated with the gene, with subsequent selection. This potentially powerful mechanism of fast adaptation without any changes in the genes themselves was originally illustrated by only a few examples. Since then several researchers have developed similar thoughts, and numerous new supporting examples have appeared in the literature. The changes in copy numbers of tandem repeats are found to influence nearby gene expression. The variable runs of the repeats may, thus, serve as tuners of the gene expression in response to environmental pressures. The same mechanism could serve for differentiation (adaptation of individual cells to a changing cellular environment) and for gradual changes in quantitative traits.

S.P. Wasser (ed.), Evolutionary Theory and Processes: Modern Horizons,
         Papers in Honour of Eviatar Nevo
E.N. Trifonov. *Tuning Function of Tandemly Repeating Sequences: a Molecular Device for Fast Adaptation, 115-138.*
© 2004 *Kluwer Academic Publishers.*

1.   INTRODUCTION

Amongst many different types of nucleotide sequences, the most conspicuous ones - tandem repeats - are also seemingly least promising and dull. The general structure of the tandemly repeating sequence is $(x_1x_2x_3...x_k)_n$ where $x_1x_2x_3...x_k$ is the repeat unit - either nucleotide or amino acid sequence, generally, of any length k, from one to several thousand bases (amino acids). The subscript *n* (highlighted to reflect its key importance) is the copy number of the repeats in the run. There can be anywhere between two and several hundreds of identical repeats in one run. These are, for an introductory example, $(GCT)_n$ repeats associated with some neurodegenerative diseases (Usdin and Grabczyk, 2000), numerous satellite DNA sequences (Miklos, 1985) or, say, (tyr-ser-pro-thr-ser-pro-ser)$_n$ - sequence at the C-terminus of RNA polymerase II (Allison et al., 1988; Bartolomei et al., 1988).

The tandem repeats constitute 20-40% of the eukaryotic genome (Britten and Kohne, 1968; Jelinek and Schmid, 1982) the rest of which consists, largely, of other "silent" sequence types - dispersed repeats, intergenic and intervening sequences, and pseudogenes leaving only a small percentage to protein-coding sequences. Such a domination of this "junk" (Ohno, 1970), "selfish" (Doolittle and Sapienza, 1980; Orgel and Crick, 1980), "ignorant" (Dover, 1980), but probably "polite" (Zuckerkandl, 1986) DNA, has lead inevitably to the pessimistic thought that "it may be futile to seek functions for the majority of these potentially 'selfish' eukaryotic repetitive DNA sequences since none may exist" (Hardman, 1986), and "however it may ultimately be explained (role of satellites, E.T.), the explanation is unlikely to have general significance" (Macgregor and Sessions, 1986).

Yet a more conservative functionalist view has never been abandoned. As early as 1951, McClintock suggested: "...one should look first to the conspicuous heterochromatic elements in the chromosomes in search of the controlling systems...". It was not even known at that time that the DNA component of the heterochromatin is, largely, tandemly repeating satellite DNA. Mazrimas and Hatch (1972) proposed that "satellite or repetitive DNA are an integral part of one of the mechanisms that have evolved in higher organisms to promote genetic flexibility".

One important property of the repeats could be a clue to the possible mechanism: the tandem repeats represent the most variable part of the eukaryotic genome. Could not this remarkable molecular variability be somehow linked to equally remarkable phenotypic variability? In the review below the thought is advanced that the tandem repeats act as tuners that modulate various cellular functions in a copy-number dependent way. The variability of the copy numbers leads to changes at the molecular level, quantitative variability in patterns of gene expression and, thus, to gradual organismal phenotypic changes. That is, such tuners may cause changes in somatic cells, in individual organisms, and, via germ cells and sexual selection - in species as a whole. The changes in tandem repeats are not major genome changes and, therefore, would not lead to the formation of new taxonomic units. They may, however, play a role in the early stages of speciation.

2. EARLY THOUGHTS AND IDEAS ON THE ADAPTATIONAL ROLE OF TANDEM REPEATS

This section is not intended to review all earlier ideas on the possible function(s) of tandemly repeating sequences, satellites, and heterochromatin (see, e. g., Hsu, 1975). We shall focus on those deliberations which have lead to the idea about the relation of tandem repeats to adaptation.

Various repeating sequences in general and tandem repeats in particular have been frequently thought of as somehow involved in regulation of gene expression (Britten and Davidson, 1969, 1971; Georgiev, 1969). Britten and Davidson were the first to discuss this idea *in extenso*. They suggested that genes-producers or whole batteries of the genes-producers are under complicated control of various "receptor sequences" interacting with diffusable regulatory molecules. Many identical copies of the receptor sequences are needed for producers of the same battery. Saltatory local replication of some sequences, their "diffusion" throughout the genome and involvement in local regulatory contexts, on one hand, and storage of the generated copies to serve as regulatory elements in the future, on the other, would explain the frequent occurrence of the repeats in eukaryotic genomes. They also suggested that the repeats-regulators may well be an important part of the evolutionary process since "major events in evolution require significant changes in patterns of gene regulation" while "the appearance of new structural (producer) genes represents a minor part of the changes involved" (Britten and Davidson, 1971).

Zuckerkandl (1974) suggested that the regulatory functions of the repeats are realized via higher order chromatin structure. Heterochromatin that consists largely of tandemly repeating satellite DNA sequences (e.g., Kurnit, 1979) is viewed as folded into specific quaternary structure. It is, presumably, involved in gene regulation by inducing neighboring sequences to fold in a similar way and by competing with them for binding of certain regulatory molecules.

Broadly discussed speculation on a possible role of the repeats in recombination during meiosis (John and Miklos, 1979) had been abandoned under the weight of much evidence to the contrary (Miklos, 1985).

To the author's knowledge, the potential importance of the *variability of the copy numbers of the repeats,* in general, had not been appreciated in earlier work. In one instance, however, studies on underreplication and amplification of repetitive DNA in plants during differentiation and dedifferentiation lead to the suggestion that changes in the amount of repetitive DNA may somehow result in "new balances of genes in various cells of a multicellular individual" (Nagl, 1976). Also, satellite sequences were considered as responsible for some general function (Zuckerkandl, 1986), which would "depend exclusively on the quantity of nucleotides involved, independent of their nature and sequence". In another study (Schaffner et al., 1988) where enhancer sequences had been truncated, reshuffled, and repeated various numbers of time, a general principle of mammalian gene regulation was proposed based on a "redundancy of information that can be provided either by a combination of different DNA sequence elements or by multiple copies of the same element" ("elements" mean binding sites for various factors present in the cell). Maniatis et al. (1987) also noted: "The additive effect of tandemly reiterated regulatory elements appears to be a general phenomenon".

The general modulatory and adaptational function for the tandem repeats, irrespective of the details of their molecular involvement and the nature of the adjustable function was suggested in 1989 (Trifonov, 1989, 1990): the tandem repeats play a modulatory role, acting as tuners of the functions they are associated with, by increasing or decreasing the repeat copy numbers *n*. A tempting analogy is at hand: the general quantitative function of money vs. the type of currency forms of payment and variety of goods the money can buy. The system of variable copy-number modulators assigns a crucially important quality to the genes and other adjustable entities of the cell - graduality in the changes of their activities during adaptation of species to the changing environment.

An accumulating wealth of highly suggestive experimental data has, inevitably, led other researchers to the same general concept. In 1994-1997, King and his colleagues independently developed the same scheme for involvement of the tandem repeats in the evolutionary process (King, 1994; King et al., 1997; Kashi et al., 1997). Schaffner's group (Künzler et al., 1995), also came to the same general idea on tandem repeats and adaptation. Related, though more specific thoughts were put forward by Holliday (1991), who suggested the variable copy number of repeats associated with developmental genes as the basis for quantitative variation, and by Richards and Sutherland (1992a,b) who introduced the term "dynamic mutation" - change in copy number of tandem repeats involved in expansion diseases. As the latter authors noted (Richards and Sutherland, 1992b), "simple, repetitive DNA sequences have a fundamental role and deserve far more attention than their simplicity might imply".

3. FROM STRESS TO ADAPTATION VIA CHANGES IN REPEAT COPY NUMBERS

The following sections of the review deal largely with experimental evidence and are organized along the following line, which is also a concise description of the main idea of the proposed general molecular mechanism of adaptation:

Environmental challenges cause various **stress** reactions, of which one is **amplification** and/or (partial) **elimination** of the tandem repeats, i.e., changes in the copy number n in the tandem runs. An immediate response is the **retuning** of the expression of many genes and multigene functions influenced by the repeats. The copy numbers n of the variable tuners linked to specific genes most relevant to given environmental changes, are under **selection** pressure. Accordingly, some of the copy-number tuning responses to the stress result in observable **phenotypic changes**. This completes the causal chain of adaptation from environment to phenotype.

3.1 Stress

A certain degree of stress caused by fluctuations in the environment is a necessary starting point for every adaptational change. The intermediate links between environment and phenotype are not clear. A general question is "How can a genome reorganize itself when faced with a difficulty for which it is unprepared?" (B. McClintock, in her Nobel Prize lecture, 1984). One answer to this question is transposable elements, but many other genome rearrangements are caused by the stress as well. The tuning function of the tandemly repeating sequences offers an obvious link: from induced changes in the repeat copy number to retuning of molecular functions under selection pressure and to the organismal phenotype variations.

Changes in the repeating components of the genome, indeed, are the most dramatic genomic consequences of stress and of more gentle pressures. This is best exemplified by the phenomenon of gene amplification: e.g., dihydrofolate reductase genes are amplified in cells treated with methotrexate (Bostock et al., 1979; Kaufman et al., 1979). Changes in satellite DNA are also documented. *Cucumus melo* root tissue culture transiently generates new satellite DNA, 1.719 g/ml, not present in the bulk tissue in culture, in response to phytohormone auxin (Grisvard and Tuffet-Anghileri, 1980). It may well be that numerous individual variations in satellite DNA content or polymorphisms in other tandem repeats are caused, similarly, by environmental pressures. A large survey of genetic diversity of plants and animals at both protein and DNA levels, caused by ecological stress, demonstrates strong correlation of the changes in tandemly repeating sequences with environment (Nevo, 1998). In particular, such correlation is observed between microsatellite diversity in wild emmer wheat and highly localized ecological factors causing aridity stress such as distribution of shadow and soil moisture (Li et al., 2000a,b).

3.2   Amplification/elimination of tandem repeats

The variability of heterochromatin patterns of satellite DNA content and of copy numbers of tandem repeats is well documented. Most of the original observations of this kind failed to make any functional connections of this variability. Craig-Holmes (1977) concludes in his study on inherited heterochromatin variants: "Continued C-band variability observed in the pedigrees question the biological significance of these heterochromatic regions". However, the same author noted that "as methods improve, it is possible that we will each be found to have our own private and unique chromosomal profile". Thus, the C-band patterns, even if not functional, at least could reflect individual genomic (and phenotypic?) differences.

Heterozygosity in C-band intensity and in their very presence is frequently observed in individual karyotypes (John, 1988). C-band size is polymorphic in humans (Cohen et al., 1966; Angell, 1973; Seuanez, 1979; Gosden et al., 1981; Verma, 1988) as well as in other species, e.g., in rye (Appels, 1982; Gustafson et al., 1983) and in the Australian grasshopper (John et al., 1986). In *Drosophila* significant variability in the amounts of different DNA satellites in closely related species is observed (Gall and Atherton, 1974; Holmquist, 1975). Heterochromatin may spontaneously amplify. A classical example - production of megachromosomes in hybrids of *Nicotiana tabacum* with *Nicotiana tomentosiformis* is up to 15 times the normal length due to amplification of heterochromatin blocks (Burns and Gerstel, 1973). Amplification of various parts of chromosomes with the appearance of homogeneously stained regions under the action of various drugs (colchicine, adriablastin, methotrexate) is a broadly studied phenomenon (e.g., Kopnin et al., 1985, and references therein). Heterochromatin is also frequently lost (eliminated), sometimes completely, during ontogeny, e.g., in nematodes (John, 1988). Somatic cells contain only euchromatin in this case and substantially less satellite DNA. Elimination of heterochromatin also occurs in cell cultures. When *Brassica nigra* is subjected to heat shock conditions, some of the rRNA coding repeats in its genome are eliminated (Waters and Schaal, 1996). Large megabase scale individual variations in numbers of alphoid DNA repeats (170 bp) in centromeric regions of human chromosomes are documented (Jabs et al., 1989).

Massive amplification of the repeating sequences results in very large differences in haploid DNA content of taxonomically similar species that do not have any substantially different gene numbers - C-value paradox. In eukaryotes, this difference could be up to three orders of magnitude (Hinegardner, 1976). Excessive C-value is basically due to repetitive sequences, e.g., in amphibians (Britten and Davidson, 1971). It is proportional to heterochromatin content, e.g., in Cricetidae and Muridae (Birshtein, 1987). Interestingly, in specialized forms with a narrow habitat C-value is lower (Hinegardner, 1976), as has been observed for fish, insects, amphibians, mollusks, sea cucumbers, mammals, and plants. This indicates that the excessive heterochromatin and satellites have something to do with adaptation (see below).

The same variability and amplification/elimination phenomena are typical also for simpler sequences. A somewhat different terminology is used in this case: *expansion* (increase in *n*) instead of *amplification*, and *contraction* (decrease in *n*) instead of *elimination*. Low resolution data on heterochromatin and satellite DNA do not allow a clear distinction between appearance/disappearance of new repeat sites and changes in the lengths of already existing ones. There is some evidence, however, that the low-resolution changes are rather of an expansion/contraction type as well (John and King, 1983; Kato and Tanifuji, 1986; Verma, 1988).

There are hypervariable regions in the human genome containing simple tandemly repeating sequences ("minisatellites") with copy number hetero-zygosity (Jeffreys et al., 1985, and references therein). "These variants were transmitted apparently in a Mendelian fashion, in that each polymorphic band in the daughter could be identified within one or the other (but not both) parents" (Jeffreys et al., 1985). Heterozygosity of some repeats is close to 100%. Mendelian inheritance indicates that these repeats are autosomal in origin. In one of 27 individuals, one of 240 (total for 27) resolvable bands is newly formed, not inherited. Thus the copy number change rate is ~ 0.004 (Jeffreys et al., 1985). The tandem repeat $(ACAGGGGTGTGGGG)_n$ is found 5' to human insulin gene, with *n* varying between 26 and 209. Of the individuals examined, 63% are heterozygous at this site (Bell et al., 1982). The variable number of tandem repeats (VNTR) in many loci of the human genome serves as a convenient marker for genome mapping and forensics (Nakamura et al., 1987). A highly heterozygous 3'-region of human apolipoprotein B gene contains tandemly repeating 15 bp long A+T rich sequence units usually repeated 30 to 50 times in at least 12 different alleles (Boerwinkle et al., 1989). The respective allele numbering system (3'-beta-30 to 3'-beta-50) is suggested by the authors.

Several possible mechanisms for the repeat copy number variations are discussed in the literature. As Britten and Davidson (1971) suggested, the repeated sequences may originate accidentally by saltatory replication or by unequal crossing over (Smith, 1976). Unequal crossing over can cause a dramatic drop in, e.g., copy number of ribosomal RNA genes in *Drosophila melanogaster*, leading to slow growing bobbed mutant flies (Stern, 1927). The same process occasionally brings the copy number back to normal. The repeats may also be formed by recombination via short, highly recombinogenic sequences, or by (retro) transposition (Hardman, 1986). Saltatory replication may occur due to an intrinsic property of DNA polymerase to synthesize spontaneously long stretches of tandemly repeating sequences, even in the absence of any initial template. For example, $(AT)_n$ is synthesized by DNA polymerase of *Esherichia coli* (Schachman et al., 1960); $(TATCCGCA)_n$, $(TATAGTTATAAC)_n$ and repeats of other units, 4 to 18 bases long, are synthesized by DNA polymerase of the archaeon *Thermococcus litoralis* (Ogata and Miura, 1998a); repeats $(CATGTATA)_n$, $(TATACGTA)_n$, and $(TGTATGTATACATACATA)_n$ are spontaneously generated by DNA polymerase of *Thermus thermophilus* (Ogata and Miura, 1998b). In the presence of DNA such tandem replication is likely to happen at sites of pre-existing repeats. In both cases the expansion of the (pre)formed repeats is mediated, probably, by slippage structure (Chamberlin and Berg, 1962; Schlötterer and Tautz, 1992). Its formation would result in repeated replication of the same unit and, thus, extension of the run of the repeats. Expansions and contractions of the tandem repeats may also be induced by the double-strand break repair system, as observed in yeast (Pâques et al., 1998).

One possible way to causally link the stress system to expansion/contraction of the tandem repeats would be modulation of fidelity of the enzymes involved in DNA replication. Stress-inducible lower fidelity enzymes may exist that are directly involved in the copy-number tuning process. For example, some genes of the eukaryotic DNA mismatch repair systems are found to contain unusually high numbers of variable exonic mononucleotide repeats (Chang et al., 2001). The authors note that "it is a paradox that the MMR system, which limits mutations in microsatellite sequences, would be particularly vulnerable to mutation by virtue of having microsatellites in its own coding regions". An interesting speculation is that the microsatellites of the genes may "act as an evolutionary switch that modulates the mutation rate under conditions that require rapid evolution".

3.3   Retuning

The link between the changing copy number of tandem repeats and phenotype is provided by an accumulating number of experimental observations showing a dependence of gene expression and other functions on the copy numbers of the associated repeats. Virtually every eukaryotic gene is accompanied by tandem repeats of one kind or another. Even in the compact genome of the *Fugu* fish microsatellites are found every 1.9 kb in average (Edwards et al., 1998), i.e., 2 to 4 runs of the repeats per gene. It is quite likely that some of these, if not all, serve as tuners. The available data on the dependence of the gene's expression and other cellular functions on the repeat copy number n are summarized in Table 1, where mostly eukaryotic, but also several prokaryotic examples are listed. For convenience each section of the Table is split in two: nucleic acid sequence repeats and amino acid sequence repeats. Many of the examples given in Table 1 are provided by *in vitro* studies. The readiness of the variety of the systems to retuning suggests, however, that tuning by the repeat copy number is rather a general phenomenon. Still, *in vivo* studies are clearly needed to make the outlined theory of tandem tuners sounder. The Table illustrates the tuning function of the tandem repeats. They may act as "tuning knobs" (King et al., 1997; Trifonov, 1999) for modulation of gene expression or other functions as gradually as discrete numbers of the repeats in the tandem runs would allow. The larger the number of the repeats in the run and the weaker the influence of any individual repeat: the finer the tuning. The data actually, represent the middle part of the causal route from environmental changes to gross phenotypic response - expansion/contraction of the repeats and respective changes in gene expression patterns, leaving out the initial stress component and the final stage of the organismal phenotype change. The genes themselves (classical genotype) remain unchanged.

The graduality is an important feature of the tuning mechanism since it provides quantitative character to the modulated molecular traits. Studies on heterochromatin have already shown that the "variations in heteromorphisms are continuous rather than discrete" (Verma, 1988). "The variants of each chromosome can be arranged into a graded series of increasing C-band size" (John and King, 1983). The graduality dimension characteristic for adaptation is especially important for development of quantitative traits (see below).

3.4 Selection, setting of tuners

The initial changes in the repeat copy numbers caused by environmental pressures, due to the very nature of random saltatory generation of the repeats, are not directed. Many tuners, if not all, presumably change their copy numbers under duress. The specific adaptive response can only be established by a selection process spanning several sexual generations or, at the cellular level - several cellular divisions. The possible scenario is rather straight forward. Initially, with the onset of the stress condition, all or many tandem tuners, depending on the magnitude of the stress, change randomly to somewhat larger or smaller values of n. Since not every change in this or another gene's expression is adaptive, the tuners of adaptively incompetent genes keep changing, both loosing and acquiring a few copies of the respective repeats, and staying, essentially, close to their pre-stress activities and *n*-values. The changes of tuners for adaptationally competent genes, however, cause the adaptive response and are, thus, under selection pressure. In other words, the changes in the tuners of relevant genes in desirable direction and respective changes of the gene activities towards relaxation of the stress are selected for. The selection pressure, thus, results in systematic *directional* changes of respective *n* values that lead, finally, to desirable activity levels of the adaptationally relevant genes and relaxation of the stress, i. e., adaptation.

3.5   Phenotypic changes

Since most of the phenotype characters are pleiotropic with many genes involved in their expression, and since effects of many subtle chromosomal changes can be compensated by cellular or organismal homeostasis, one should not probably expect a given specific C-band or run of repeats to be necessarily detectably connected with a specific phenotypic trait. In a few cases, however, such a connection can be established. In particular, climatic adaptations are often reflected in the heterochromatin. The heterochromatin content of the grasshopper, *Atractomorpha similis,* shows a clinal increase southwards within Australia (John et al., 1986). *A. similis* is believed to have entered northern Australia from Eurasia via the land bridge some 10 000 years ago. The repeat (thr-gly)$_n$ in the circadian rhythm gene shows dependence of *n* on latitude (Sawyer et al., 1997). Twelve species of kangaroo rat (*Dipodomys*), one of the most successful residents of arid lands in North America, all contain different amounts of 1.702, 1.707, and 1.713 g/ml satellite DNA from 36% to 74%. The total content of the satellites has an inverse correlation with the degree of specialization toward a saltatorial bipedal type of locomotion according to quantitative osteological measurements (Setzer, 1949; Mazrimas and Hatch, 1972). The bipedal versus quadripedal locomotion gives an advantage in arid areas where the animal has to cover large distances in search for food. The satellite content also correlates with the number of subspecies in each species, and with intermembral index (body parameter describing the shift towards bipedal mode) (Howell, 1944). As Mazrimas and Hatch (1972) noted: "satellite DNA may have supragenic functions that are reflected in relationships with morphological and ecological specialization". Similarly, from studies on metal tolerance in plants, Antonovics (1971) concluded that "genetic differentiation within species in response to habitat differences has now been repeatedly observed and has become an established phenomenon". One "can no longer assume that members of a species from one habitat are identical to those from another". Plants, *Festuca ovina* and *Agrostis canina,* during 30 years of growth under a zinc-coated fence developed tolerance to zinc. A few inches from the fence the same plants were not tolerant (Snaydon R.W., as referred to in Bradshaw et al., 1965). Human chromosome studies (Halbrecht and Shabtay, 1976) show that although most of heterochromatin polymorphisms are not clearly connected with any phenotypic abnormalities, the individuals with a higher number of C-band variants usually display one or another kind of malformations. Perhaps, rather normal phenotypic variations should be scrutinized to trace the connection between tandem repeats and phenotype. *Homo sapiens* would be the best object for such studies considering the ability of humans to recognize even very subtle differences between human individuals.

Copy number variation of tandem repeats could influence the phenotype both at the moment of the change and after sexual spread of the new change. A particular mechanism of the molecular change could be any intracellular DNA turnover mechanism including the tuning by tandem repeats. The molecular drive suggested for multigene families is applicable as well to any other repeating sequences involved in regulation of gene expression (Dover, 1982, 1986). The DNA turnover, however, can be viewed as another kind of mutation (resulting in deletions and insertions of the repeats), in which case the copy number - phenotype relationship would be a simple case of Darwinian selection.

Phenotypic variations that correlate with the copy number of variable tandem repeats are typical of eukaryotic viruses. The AS virus, antigenic variant of BK virus with distinct antigenicity of its capsid is sequence-wise 94.9% homologous to its prototype. One of the major differences is the absence of the first and second tandemly repeated enhancers in the AS virus (Tavis et al., 1989). A deletion mutant of the wild type BK virus, lacking one copy of the 68 bp enhancer demonstrates higher transforming capacity (Hara et al., 1985; Watanabe and Yoshiike, 1985). Repetition 2-4 times of various 10 to 90 bp fragments in the promoter region of BK virus is associated with adaptation of the virus to cell culture (Rubinstein et al., 1991).

At high levels of fertilizers the flax plant dramatically changes its morphology to a bushy type, which is accompanied by an increase in copy number of several types of tandem repeats of unit sizes 320 to 600 bp (Cullis, 1986). Under pressure for surface changes in *Haemophilus influenzae*, the copy number n of $(GACA)_n$ repeats is changed causing modulation of downstream glycosyltransferase gene and phenotypic switch (Hood et al., 1996). The authors comment: "This genetic potential to generate a repertoire of variant antigens is one of the mechanisms by which pathogenic microbes can adapt to the differing microenvironments of the host and evade host immune responses".

4.   RELATED ISSUES

In the following sections, several issues are discussed related to the concept of the tuning function of tandemly repeating sequences, and respective literature are briefly reviewed.

4.1   Repeat expansion diseases

Malfunction of tandemly repeating tuners of gene expression would be expected to result in the malfunction of the gene itself. For example, over- or undertuning beyond normal limits of the respective copy numbers may switch some genes off or on, thus causing major disorders. Indeed, many examples are known today of such overtuning (Usdin and Grabczyk, 2000), or undertuning (Delot et al., 1999) - so-called triplet expansion diseases, where both noncoding and (mostly) protein coding sequences are involved. The very nature of these diseases, caused by spontaneous excessive change of the copy numbers of the tandem repeats-tuners, suggests that the unit size *per se* (3 in the examples above) is not essential, and that expansion diseases should exist with repeats of different unit sizes. Indeed, as Table 2 illustrates, the over(under)-repeating unit may be virtually any size, from 3 to 3200 bp. One may predict that the tandem repeat expansion/contraction diseases with unit sizes 1 and 2 as well as over 3200 bp will also be discovered.

4.2   Cancer

Cancer cells are characterized by an immense spectrum of differences in their molecular patterns compared to normal cells including numerous changes in the tandemly repeating sequences. It may well be that the original changes that ignite the process are of the same nature as in the case of expansion diseases - spontaneous expansion/contraction of some repeats. For example, the ovarian cancer risk is, indeed, associated with a certain run of tandem repeats downstream of the *HRAS1* proto-oncogene (Phelan et al., 1996, see Table 2). The early carcinogenesis may be considered, perhaps, as an aberrant form of cellular differentiation (see below), presumably, directed by the copy-number tuners.

The tumors display various changes in the chromosomes including changes in heterochromatin distribution (Sinha and Pathak, 1973; Chen and Shaw, 1973). Gene losses in cancer cells and large deletions in their chromosomes occur frequently, in a tissue-specific manner (Ponder, 1988, and references therein). The same holds true for repetitive sequences. "The length of repetitive sequences is (relatively, E. T.) stable in normal cells, but increases and decreases in the number of repeats can occur in tumor cells" (Jackson et al., 1998).

4.3   Differentiation

Since events of amplification/elimination (expansion/contraction) of tandem repeats are rather frequent, they may happen many times in the ontogenetic chain of individual cell divisions. That would cause, respectively, retuning of the genes involved, and "adaptation" of the dividing cells to the changing *cellular* environment. In other words, the retuning mechanism described above applied to organismal adaptation, is equally applicable to cell differentiation. The cellular reactions of the differentiated cells to tuning by tandem repeats are generally expected to be different compared to other tissues and to the entire organism. The specific tuning changes may be advantageous to only one type of cells. The possible role of heterochromatin in cell differentiation was appreciated as early as 1955 (Goldschmidt, 1955): "The functions of heterochromatin must be involved in the differences between somatic and germinal cells". Craig-Holmes et al. (1973) observed variations in C-bands within individuals, though these are less frequent than between individuals. That is, the heterochromatin pattern may, indeed, change during ontogenesis. Zuckerkandl (1974) also noted: "the participation of heterochromatin in achieving differentiation and morphogenesis may well be important", and "heterochromatin is considered here as an active entity that participates in the control of eukaryote-type cell differentiation". On the basis of numerous early data on under- and over-replication (amplification) of repetitive DNA during differentiation Nagl (1976) concluded: "the development of a high number of repetitive, noninformative DNA sequences evidently is prerequisite for the complex control of differential gene activity".

The connection between tandem repeats and differentiation can be illustrated by several examples of higher resolution. In dedifferentiating cells of carrot explants placed on agar with phytohormones, certain fractions of satellite DNA start to replicate earlier than the bulk DNA (1.711 g/ml and, minor, 1.695 g/ml) (Kato and Tanifuji, 1986). As the authors commented: "it is possible that...early replication of some types of highly repetitive DNA sequences is related to regulation of genome activity to induce dedifferentiation". It is worth noting that carrot cells are totipotent, i. e., their differentiation and dedifferentiation are reversible. The retuning function of the tandem repeats may, perhaps, constitute the very basis of the phenomenon of totipotency. In the study by Parenti et al. (1973), during callus formation from *Nicotiana glanca* pith tissue (i.e., dedifferentiation) transient satellite DNA (1.722 g/ml) amplification was observed. The amount of satellite DNA in the larval salivary gland of *Drosophila nasutoides* drops from 60% in mitotic chromosomes to only trace amounts (Zacharias, 1986, and references therein). John (1988) describes other similar examples. In the yeast differentiation model, copy number-dependent activation of the reporter *CYC7* (iso-2-cytochrome C) gene by the repeat (57 bp)$_{1-4}$ in its 5'-region is cell type-dependent. The gene is not expressed in a/α cells (Company and Errede, 1988). It is not clear, however, that this change of **n** is the sole factor in the differentiation.

4.4   Quantitative traits

The puzzles of inheritance and development of quantitative traits are best illustrated by classical experiments with artificial selection for the abdominal bristle number in *D. melanogaster* (Mather and Harrison, 1949a,b; Mayr, 1963), from normal 36 to 63, or down to 24. The effect of gradual growth of the bristle number during 20 generations is thought to be pleiotropic, i.e., involving many different genes. However, on several occasions the bristle number stopped changing for more than 50 generations and then growth continued. This is hard to reconcile with pleiotropy. Apparently, a new mutation was needed for a new important allele to appear, in which case further change would not be expected to be gradual. Or, as we would like to interpret it - the graduality is provided by copy-number tuner(s) of one or a few genes only. This copy number might stay below some threshold during these 50 generations, or an additional tuner might take over with its copy number originally insufficiently large (or small). Remarkably, when the selection is interrupted, the bristle number rapidly comes back to the norm (2-3 generations), which is hard to explain by the quick loss of the presumed selected alleles abundant in the gene pool just a few generations before. It can be explained easily, however, by quick retuning of a few tandemly repeating sequences presumably involved in the selection process. This phenomenon of reversal to the original condition is called genetic homeostasis (Lerner, 1954) and is observed in many other artificial selection cases. Its usual interpretation is the superiority of the original genotype under the original conditions and, therefore, reversal to that genotype (Mayr, 1963).

Another indication of quantitative changes at the single gene level is the effect of correlated responses known to animal and plant breeders (Mayr, 1963). During the selection for a given phenotype, occasionally some other independent aspect of phenotype shows concomitant change. Obviously, only a very small number of common genes could be involved in expression of both responses. And yet, their correlated changes are gradual, rather than stepwise.

When non-identical stocks are initially chosen for artificial selection, then usually different rates and final levels of response are observed as well as a difference in correlated responses (Dobzhansky and Spassky, 1962). This also indicates that, despite the assumed pleiotropy, only a few key genes are predominantly involved in the selection in each case. The changes, however, are always gradual.

Out of all these major features of the quantitative traits, graduality and reversibility, in particular, may well be explained by the tuning capacity of the tandemly repeating sequences. More attention, perhaps, should be paid to the copy numbers of the tandem repeats than just to their presence as markers for the mapping of quantitative traits.

5. PREDICTIONS

The concept outlined in the review may become a theory, a molecular theory of adaptation, if only it would offer some predictions that, hopefully, would turn out to be correct. Some possible predictions are listed below.

If, indeed, the tandemly repeating tuners are a general device for adaptation of species, then there should also be a universal molecular mechanism(s) that would boost the expansion/contraction of the repeats under conditions of stress. It may actually be a part of already known stress systems, which has not yet been revealed.

Stressed populations, e.g., human populations in conditions of famine, war, emigration, etc. would be expected to display higher variability in the tandemly repeating sequences and be at higher risk of repeat expansion disorders.

Early carcinogenesis may involve expansions/contractions of some specific repeats.

Possible involvement of the tandem repeats in differentiation suggests that there should be tissue-specific differences in the patterns of the repeats.

Certain variable repeats should be linked to at least some quantitative traits so that the gradually changing copy numbers of the repeats would provide the quantitative character to the traits.

6. ACKNOWLEDGMENTS

Discussions with many colleagues have been enlightening and often crucial in the development of the idea on the tuning role of the tandem repeats in adaptation. Special thanks to J. Beckmann, M. Bellgard, G. Bernardi, T. Bettecken, H. Bünemann, M. Burmeister, U. Grossbach, H. Hoehn, I. Irlin, A. Konopka, E. Nevo, D. Raveh, C. Schmid, E. Schmidt, J. Shapiro, M. Singer, L. Ulanovsky, J. Wang, and E. Zuckerkandl. The author

Edward N. Trifonov

is grateful to his colleagues for the intellectual pleasure of discussing and absorbing their comments. Opinions and suggestions of anonymous reviewers of the paper are highly appreciated as well.

Table 1. Tuning function of tandemly repeating sequences. Experimental data.

MODULATION OF TRANSCRIPTION

Unit	No. of repeats		
A	20-55	upstream of *ADR2* gene of *S. cerevisiae*	Nature 304, 652, 1983
T	11-45	upstream of *Dictyostellium* actin genes	NAR 22, 5099, 1994
T	9-42	Gcn4-activated transcription, yeast *his3* gene	EMBO J 14, 2570, 1995
T	10-80	upstream, vaccinia virus late promoters	JMB 210, 771, 1989
GT	30-130	*CAT* constructs, monkey, human cells	MCB 4, 2622, 1984
RY	94,144	mouse *ADH1* gene, first intron	Gene 57, 27, 1987
ACCGA	5-12	UAS1 site of yeast *CYC1* gene	MCB 6, 4690, 1986
CTTCC	2,3	upstream activator of yeast *PGK* gene	NAR 16, 8245, 1988
AARKGA	2-8	human IFN beta gene, PRDI element	Science 236, 1237, 1987 EMBO J 8, 101, 1989
ATCTTTC	15-28	Between promoters P2 and P1 of adhesin genes of *H. influenzae*	PNAS 96, 1077, 1999
AGGGCAGAGC	1-3	mouse βDRE element, β-globin promoter	MCB 10, 972, 1990
GGGGCGGGGC	1,2	Sp1 sites, adenovirus early promoter	JBC 266, 20406, 1991
CAAAAATGCC	9-35	transient expression of galactokinase	BBRC 180, 1273, 1991
11 bp	1-4	mouse metallothionein I gene, MREa element	MCB 5, 1480, 1985
12 bp	1,3	bovine papilloma virus, E2 site	EMBO J 7, 525, 1988
12 bp	1-4	human IFN beta gene, PRDII element	EMBO J 8, 101, 1989
12 bp	1-6	MRE element of mouse metallothionein-1 promoter	Nature 317, 828, 1985
14 bp	1-4	soybean heat shock promoter element	JMB 199, 549, 1988
14 bp	1-4	*C. elegans* HS element in mouse cells	MCB 6, 3134, 1986
14 bp	1-4	*Drosophila* HS element in yeast cells	NAR 14, 8183, 1986
14 bp	1-5	cell-cycle dependent transcription of the yeast *HO* gene	Cell 42, 225, 1985
16 bp	1,5	human oligoA synthetase gene	EMBO J 7, 411, 1988
17 bp	1,3	yeast allantoate permease gene, GATAA containing element	MCB 9, 602, 1989
17 bp	1-8	SV40-rat construct, preproinsulin gene	MCB 8, 2737, 1988
17 bp	1,5	yeast allantoate permease gene	MCB 9, 602, 1989
18 bp	1-5	immediately early genes, human cytomegalovirus	JV 63, 1435, 1989
31 bp	1-8	NF-κB factor binding site upstream of mouse beta-globin gene	JMB 214, 373, 1990
32 bp	1,2	yeast allantoate permease gene	MCB 9, 602, 1989
32 bp	1,2	immediately early genes, human cytomegalovirus	JV 63, 1435, 1989
32 bp	1-4	upstream of the *SUC2* gene of *S. cerevisiae*	MCB 6, 2324, 1986
39 bp	1,2	copper-induced transcription of yeast copper-metallothionein gene	MCB 6, 1158, 1986
57 bp	1-4	H element, Ty1 transposon, yeast *CYC7*	MCB 8, 5299, 1988
60 bp	1-3	cauliflower mosaic virus activator	EMBO J 7, 1589, 1988

113 bp	n	expression of a reporter gene	Gene 189, 13, 1997
122 bp	1-4	maize streak virus activator element	EMBO J 7, 1589, 1988
240 bp	n	rDNA spacer in *Drosophila*	NAR 10, 7017, 1982
			PNAS 85, 5508, 1988
			MCB 10, 4667, 1990
———			
Q	11-38	activity of transcription factors harboring the repeats	Science 263, 808, 1994
Q	25-77	efficiency of transactivation of androgen receptor	NAR 22, 3181, 1994

ENHANCERS

Unit	No. of repeats		
12 bp	1-3	SV40 constructs expressing E2 peptide of bovine papilloma virus	EMBO J 7, 525, 1988
12 bp	2-6	ftz-dependent enhancer, *Drosophila*	Nature 336, 744, 1988
14 bp	1,2	phorbol ester induction, HIV, R region	MCB 7, 3994, 1987
16 bp	1,5	interferon-responsive, *tk* gene constructs, transfected monkey cells	EMBO J 7, 1411, 1988
17 bp	1,2	yeast upstream activator sequence, in HeLa cells	Cell 52, 169, 1988
17 bp	1,4	CRE enhancer of human vasoactive intestinal peptide gene	PNAS 85, 6662, 1988
18 bp	1,2	cAMP responsive, human glycoprotein hormone	MCB 7, 3759, 1987
20 bp	4,8	core of SV40 enhancer, constructs	JMB 201, 81, 1988
30 bp	11-21	EBV transcription and replication	MCB 6, 3838, 1986
50 bp	1-6	herpes virus saimiri	JMB 201, 81, 1988
57 bp	1-4	H element of Ty1 transposon, *CYC7* gene	MCB 8, 5299, 1988
60 bp	n	rDNA spacer, *X. laevis*	Cell 35, 449, 1983
68 bp	1-3	BKV transcription	Science 222, 749, 1983
72 bp	1-3	SV40, constructs	JV 55, 823, 1981
81 bp	n	rDNA spacer, *X. laevis*	Cell 35, 449, 1983
99 bp	1,2	murine Akv retrovirus	JV 64, 3185, 1990
109 bp	1,2	MCF virus, oncogenicity	JV 63, 1284, 1989
140 bp	1-13	mouse rRNA gene spacer	PNAS 87, 7527, 1990

OTHER ACTIVITIES

Unit	No. of repeats		
A	17-20	promoter region, *Mycoplasma* surface antigen variation	EMBO J 10, 4069, 1991
C	8-44	5'-UTR, virulence of mengovirus	JV 70, 2027, 1996
GT	n	recombination, mouse somatic cells	MCB 6, 3948, 1986
GT	n	recombination, Rec A binding	JMB 273, 105, 1997
GT	n	meiosis, yeast	MCB 6, 3934, 1986
CG	n	recombination, mouse somatic cells	MCB 6, 3948, 1986
AAG	2-8	exon M2 of mouse IGQgene, enhancement of splicing	MCB 14, 1347, 1994
GACA	22-35	phenotypic switching of a lypopolysaccharide	PNAS 93, 11121, 1996

		epitope	
AAGTGA	4-8	upstream inducible element, human beta interferon gene	JV 64, 3063, 1990
GAAAGT	2,4	mediates virus-inducible transcription of human interferon genes	PNAS 88, 1369, 1991
ATAGTAAA	13,17	iteron in plasmid pAD1 of *E. faecalis*, mating response to sex pheromone	J J Bact 177, 5453, 1995
CTGAGGTCAA	1-5	F2 half-element of chicken lysozyme silencer S-2.4 kb	Cell 61, 505, 1990
14 bp	1-5	3'-terminal UTR, tobacco vein mottling virus, disease symptom severity	PNAS 88, 9863, 1991
17 bp	1-8	modulation of translation, rat preproinsulin	MCB 8, 2737, 1988
31 bp	1-6	packaging of Adenovirus Type 5 DNA	JV 64, 2047, 1990
40 bp	1,2	polyoma virus expression	JV 62, 3896, 1988
46 bp	1-4	virus-responsive element of IFNα1 promoter, induced expression	Cell 50, 1057, 1987
48 bp	2,5	transforming activity of a retrovirus	NAR 26, 4868, 1998
68 bp	1-3	BK virus, transforming activity	JV 55, 867 & 823, 1985
240 bp	13-350	modulation of meiotic drive, Rsp of SD system of *Drosophila*	Cell 54, 179, 1988 / Nature 332, 394, 1988
————			
TG	20-30	regulation of period in circadian rhythm	Science 278, 2117, 1997
SKQPFRK	2-7	chloroplast ribosomal protein S18	FEBS Let 279, 190, 1991
YSPTSPS	9-26	yeast RNApolII, modulation, response to enhancer signals	MCB 8, 321, 1988 / Nature 347, 491, 1990
YSPTSPS	3-78	mouse RNApolII, modulation	MCB 8, 330, 1988
12 aa	7-11	Mycoplasma surface antigen variation	EMBO J 10, 4069, 1991
31 aa	3,4	stage- and tissue specificity of human microtubule-associated protein tau	EMBO J 8, 393, 1989
34 aa	0-17	plant resistance to bacterial spot disease	Nature 356, 172, 1992
42 aa	3-13	segment polarity armadillo gene, *Drosophila*, phenotypic series	Cell 63, 1167, 1990
53 aa	11-50	kringle IV, processing and secretion of apolipoprotein (a)	JBC 271, 32403, 1996
82 aa	1-9	alpha C protein, *Streptococci*, modulation of host immunity	PNAS 93, 4131, 1996

The references are given in abbreviated form and then its full name of the Journal. Journal abbreviations: BBRC - Biochemical and Biophysical Research Communications; EMBO J - EMBO Journal; FEBS Let - FEBS Letters; J Bact - Journal of Bacteriology; JBC - Journal of Biological Chemistry; JMB - Journal of Molecular Biology; JV - Journal of Virology; MCB - Molecular and Cellular Biology; NAR - Nucleic Acid Research; PNAS - Proceedings of National Academy of Sciences, USA.

Table 2. Tandem repeat expansion diseases and disorders.

Repeat	Copy number *n* range	Location	Disease or disorder	References
(3 bp/1 aa)$_n$	5 to over 200	5'-, 3'- and coding regions	over 10 different neurodegenerative and other diseases	Usdin and Grabczyk, 2000 Brais et al., 1998 Delot et al., 1999
(4 bp)$_n$	75 to 11.000	intron 1 of *NF9* gene	myotonic dystrophy type 2	Liquori et al., 2001
(5 bp)$_n$	10 to 4.500	intron 9 of *SCA10* gene	spinocerebellar ataxia type 10	Matsuura et al., 2000
(12 bp)$_n$	2 to over 60	5' from cystatin B gene	progressive myoclonus epilepsy	Lalioti et al., 1997
(14 bp)$_n$	40 to 150	5' from insulin gene	susceptibility to type 1 diabetes	Bennett et al., 1995 Kennedy et al., 1995
(15 bp)$_n$ and (18 bp)$_n$	few to 90	5' from cystatin B gene	progressive myoclonus epilepsy	Virtaneva et al., 1997
(24 bp/8 a)$_n$	5 to 34	coding region of the prion protein gene	Creutzfeldt-Jakob disease	Cochran et al., 1996
(28 bp)$_n$	30 to 100	3' from *HRAS1* proto-ncogene	ovarian cancer risk	Phelan et al., 1996
(342 bp/114 aa)$_n$	15 to 37	apo(a) coding region	Lp(a) level, susceptibility to atherosclerosis and thrombosis	Lindahl et al., 1990 Koschinsky et al., 1990
(3200 bp)$_n$	2 to 100	*FSHD* gene region	FSHD muscular dystrophy	van Deutekom et al., 1993

7.   REFERENCES

Allison L.A., Wong J.K.C., Fitzpatrick V.D., Moyle M. and Ingles C.J. 1988. The C-terminal domain of the largest subunit of RNA polymerase II of *Saccharomyces cerevisiae*, *Drosophila melanogaster*, and mammals: A conserved structure with an essential function. Mol Cell Biol. 8, 321-329.

Angell R.R. 1973. The chromosomes of Australian aborigines. In: The Human Biology Of Aborigines in Cape York, Kirk R.L. (Ed.), A.C.T., Canberra, pp. 103-109.

Antonovics J. 1971. The effects of a heterogeneous environment on the genetics of natural populations. Amer Scient. 59, 593-599.

Appels R. 1982. The molecular cytology of wheat-rye hybrids. Int Rev Cytol. 80, 93-132.

Bartolomei M.S., Halden N.F., Cullen C.R. and Corden J.L. 1988. Genetic analysis of the repetitive carboxyl-terminal domain of the largest subunit of mouse RNA polymerase II. Mol Cell Biol. 8, 330-339.

Bell G.I., Selby M.J. and Rutter W.J. 1982. The highly polymorphic region near the human insulin gene is composed of simple tandemly repeating sequences. Nature 295, 31-35.

Bennett S.T., Lucassen A.M., Gough S.C.L., Powell E.E., Undlien D.E., Pritchard L.E., Merriman M.E., Kawaguchi Y., Dronsfield M.J., Pociot F., Nerup J., Bouzekri N., Cambon-Thomsen A., Ronningen K.S., Barnett A.H., Bain S.C. and Todd J.A. 1995.

Susceptibility to human type 1 diabetes at IDDM2 is determined by tandem repeat variation at the insulin gene minisatellite locus. Nat Gene. 9, 284-291.

Birshtein V.Y. 1987. Cytogenetic and molecular aspects of evolution of vertebrates. Nauka, Moscow (in Russian).

Boerwinkle E., Xiong W., Fourest E. and Chan L. 1989. Rapid typing of tandemly repeated hypervariable loci by the polymerase chain reaction: application to the apolipoprotein B 3' hypervariable region. Proc Natl Acad Sci USA 86, 212-216.

Bostock C.J., Clark E.M., Harding N.G.L., Mounts P.M., Tyler-Smith C., van Heyningen V. and Walker P.M.B. 1979. The development of resistance to methotrexate in a mouse melanoma cell line. I. Characterization of the dihydrofolate reductases and chromosomes in sensitive and resistant cells. Chromosoma 74, 153-177.

Bradshaw A.D., McNeilly T.S. and Gregory R.P.G. 1965. Industrialization, evolution and the development of heavy metal tolerance in plants. Brit Ecol Soc Symp. 6, 327-343.

Brais B., Bouchard J-P., Xie Y-G., Rochefort D.L., Chretien N., Tome F.M.S., Lafreniere R.G., Rommens J.M., Uyama E., Nohira O., Blumen S., Korcyn A.D., Heutink P., Mathieu J., Duranceau A., Codere F., Fardeau M. and Rouleau G.A. 1998. Short GCG expansions in the *PABP2* gene cause oculopharyngeal muscular dystrophy. Nat Gene. 18, 164-167.

Britten R.J. and Davidson E.H. 1969. Gene regulation for higher cells: a theory. Science 165, 349-357.

Britten R.J. and Davidson E.H. 1971. Repetitive and non-repetitive DNA sequences and a speculation on the origins of evolutionary novelty. Quart Rev Biol. 46, 111-133.

Britten R.J. and Kohne D.E. 1968. Repeated sequences in DNA. Science 161, 529-540.

Burns J.A. and Gerstel D.U. 1973. The formation of megachromosomes from heterochromatic blocks of *Nicotiana tomentosiformis*. Gene. 75, 497-502.

Chamberlin M. and Berg P. 1962. Deoxyribonucleic acid-directed synthesis of ribonucleic acid by an enzyme from *Escherichia coli*. Proc Natl Acad Sci, USA 48, 81-94.

Chang D.K., Metzgar D., Wills C. and Boland C.R. 2001. Microsatellites in the eukaryotic DNA mismatch repair genes as modulators of evolutionary mutation rate. Genome Res. 11, 1145-1146.

Chen T.R. and Shaw M.W. 1973. Stable chromosome changes in a human malignant melanoma. Cancer Res. 33, 2042-2047.

Cohen M.M., Shaw M. and MacCluer J.W. 1966. Racial differences in the length of the human Y chromosomes. Cytogenetics 5, 34-52.

Cochran E.J., Bennett D.A., Cervenakova L., Kenney K., Bernard B., Foster N.L., Benson D.F., Goldfarb L.G. and Brown P. 1996. Familial Creutzfeldt-Jakob disease with a five-repeat octapeptide insert mutation. Neurol. 47, 727-733.

Company M. and Errede B. 1988. A Ty1 cell-type specific regulatory sequence is a recognition element for a constitutive binding factor. Mol Cell Biol. 8, 5299-5309.

Craig-Holmes A.P. 1977. C-band polymorphism in human populations. In: Population Cytogenetics, Studies in Humans, Hook E.B. and Porter I.H. (Eds.), Acad. Press, New York.

Craig-Holmes A.P., Moore F.B. and Shaw M.W. 1973. Polymorphism of human C-band heterochromatin. I. Frequency of variants. Amer J Hum Gene. 25, 181-192.

Cullis C.A. 1986. Phenotypic consequences of environmentally induced changes in plant DNA. Trends Gene. 2, 307-309.

Delot E., King L.M., Briggs M.D., Wilcox W.R. and Cohn D.H. 1999. Trinucleotide expansion mutations in the cartilage oligomeric matrix protein (COMP) gene. Hum Mol Gene. 8, 123-128.

Dobzhansky T. and Spassky B. 1962. Genetic drift and natural selection in experimental populations of *Drosophila pseudoobscura*. Proc Natl Acad Sci, USA 48, 148-156.

Doolittle W.F. and Sapienza C. 1980. Selfish genes, the phenotype paradigm and genome evolution. Nature 284, 601-603.

Dover G .1980. Ignorant DNA? Nature 285, 618-620.

Dover G. 1982. Molecular drive: a cohesive mode of species evolution. Nature 299, 111-117.

Dover G. 1986. Molecular drive in multigene families: how biological novelties arise, spread and are assimilated. Trends Gene. 2, 159-165.

Edwards Y.J.K., Elgar G., Clark M.C. and Bishop M.J. 1998. The identification and characterization of microsatellites in the compact genome of the Japanese pufferfish, Fugu rubripes: perspectives in functional and comparative genomic analyses. J Mol Biol. 278, 843-854.

Gall J.G. and Atherton D.D. 1974. Satellite DNA sequences in *Drosophila virilis*. J Mol Biol. 85, 633-664.

Georgiev G.P. 1969. On the structural organization of operon and the regulation of RNA synthesis in animal cells. J Theor Biol. 25, 473-490.

Goldschmidt R.B. 1955. Theoretical Genetics. Univ. California Press, Berkeley.

Gosden J.R., Laurie S.S. and Cooke H.J. 1981. A cloned repeat DNA sequence in human chromosome heteromorphisms. Cytogenet Cell Gene. 29, 32-39.

Grisvard J. and Tuffet-Anghileri A. 1980. Variations in the satellite DNA content of *Cucumis melo* in relation to dedifferentiation and hormone concentration. Nucl Acids Res. 8, 2843-2858.

Gustafson J.P., Lukaszewski A.J. and Bennet M.D. 1983. Somatic deletion and redistribution of telomeric heterochromatin in the genus *Secale* and *Triticale*. Chromosoma 88, 293-298.

Halbrecht I. and Shabtay F. 1976. Human chromosome polymorphism and congenital malformations. Clin Genetm. 10, 113-122.

Hamada H., Seidman M., Howard B.H. and Gorman C.M. 1984. Enhanced gene expression by the poly(dT-dG)·poly(dC-dA) sequence. Mol Cell Biol. 4, 2622-2630.

Hara K., Oya Y. and Yogo Y. 1985. Enhancement of the transforming capacity of BK virus by partial deletion of the 68-base-pair tandem repeats. J Virol. 55, 867-869.

Hardman N. 1986. Structure and function of repetitive DNA in eukaryotes. Biochem. J 234, 1-11.

Hinegardner R. 1976. Evolution of genome size. In: Molecular Evolution, Ayala F.J. (Ed.), Sinauer Assoc Inc, Sunderland, pp. 179-199.

Holliday R. 1991. Quantitative genetic variation and developmental clocks. J Theor Biol. 151, 351-358.

Holmquist G. 1975. Organization and evolution of *Drosophila virilis* heterochromatin. Naturem. 257, 503-505.

Hood D.W, Deadman M.E., Jennings M.P., Bisercic M., Fleischmann R.D., Venter J.C. and Moxon E.R. 1996. DNA repeats identify novel virulence genes in *Haemophilus influenzae*. Proc Natl Acad Sci USA 93, 11121-11125.

Howell A.B. 1944. Speed in Animals. Univ. Chicago Press, Chicago.

Hsu T.C. 1975. A possible function of constitutive heterochromatin: the bodyguard hypothesis. Gene Suppl. 79, 137-150.

Jabs E.W., Goble C.A. and Cutting G.R. 1989. Macromolecular organization of human centromeric regions reveals high-frequency, polymorphic macro DNA repeats. Proc Natl Acad Sci, USA 86, 202-206.

Jackson A.L., Chen R. and Loeb L.A. 1998. Induction of microsatellite instability by oxidative DNA damage. Proc Natl Acad Sci USA 95, 12468-12473.

Jeffreys A.J., Wilson V. and Thein S.L. 1985. Hypervariable "minisatellite" regions in human DNA. Nature 314, 67-73.

Jelinek W.R. and Schmid C.W. 1982. Repetitive sequences in eukaryotic DNA and their expression. Ann Rev Bioch. 51, 813-844.

John B. 1988. In: Heterochromatin, Molecular and Structural Aspects, Verma R.S. (Ed.), Cambridge Univ. Press, Cambridge.

John B., Appels R. and Contreras N. 1986. Population cytogenetics of *Atractomorpha similis*. II. Molecular characterization of the distal C-band polymorphisms. Chromosoma 94, 45-58.

John B. and King M. 1983. Population cytogenetics of *Atractomorpha similis*. I. C-band variations. Chromosoma 88, 57-68.

John B. and Miklos G.L. 1979. Functional aspects of satellite DNA and heterochromatin. Int Rev Cytol. 58, 1-114.

Kashi Y., King D. and Soller M. 1997. Simple sequence repeats as a source of quantitative genetic variation. Trends Gene. 13, 74-78.

Kato A. and Tanifuji S. 1986. Early replication of highly repeated DNA sequences during dedifferentiation of carrot. Plant Cell Physiol. 27, 261-264.

Kaufman R.J., Brown P.C. and Schimke R.T. 1979. Amplified dihydrofolate reductase genes in unstable methhotrexate-resistant cells are associated with double minute chromosomes. Proc Natl Acad Sci USA 76, 5669-5673.

Kennedy G.C., German M.S. and Rutter W.J. 1995. The minisatellite in the diabetes susceptibility locus IDDM2 regulates insulin transcription. Nat Gene. 9, 293-298.

King D.G. 1994. Triple repeat DNA as a highly mutable regulatory mechanism. Science 263, 595-596.

King D.G., Soller M. and Kashi Y. 1997. Evolutionary tuning knobs. Endeavor 21, 36-40.

Kopnin B.P., Massino J.S. and Gudkov A.V. 1985. Regular pattern of karyotypic alterations accompanying gene amplification in Djungarian hamster cells: study of colchicine, adriablastin, and methotrexate resistance. Chromosoma 92, 25-36.

Koschinsky M.L., Beisiegel U., Henne-Bruns D., Eaton D.L. and Lawn R.M. 1990. Apolipoprotein(a) size heterogeneity is related to variable number of repeat sequences in its mRNA. Biochem. 29, 640-644.

Künzler P., Matsuo K. and Schaffner W. 1995. Pathological, physiological, and evolutionary aspects of short unstable DNA repeats in the human genome. Biol Chem Hoppe-Seyler 376, 201-211.

Kurnit D.M. 1979. Satellite DNA and heterochromatin variants: the case for unequal mitotic crossing over. Hum Gene. 47,169-186.

Lalioti M.D., Scott H.S., Buresi C., Rossier C., Bottani A., Morris M.A., Malafosse A. and Antonarakis S.E. 1997. Dodecamer repeat expansion in cystatin B gene in progressive myoclonus epilepsy. Nature 386, 847-851.

Lerner I.M. 1954. Genetic Homeostasis. Oliver and Boyd, Edinburgh.

Li Y.-C., Röder M.S., Fahima T., Kirzhner V.M., Beiles A., Korol A.B. and Nevo E. 2000a. Natural selection causing microsatellite divergence in wild emmer wheat at the ecologically variable microsite at Ammiad, Israel. Theor Appl Gene. 100, 985-999.

Li Y., Fahima T., Korol A.B., Peng J., Röder M.S., Kirzhner V., Beiles A. and Nevo E. 2000b. Microsatellite diversity correlated with ecological-edaphic and genetic factors in three microsites of wild emmer wheat in North Israel. Mol Biol Evol. 17, 851-862.

Lindahl G., Gersdorf E., Menzel H.J., Seed M., Humphries S. and Utermann G. 1990.Variation in the size of human apolipoprotein(a) is due to a hypervariable region in the gene. Hum Gene. 84, 563-567.

Liquori C.L., Ricker K., Moseley M.L., Jacobsen J.F., Kress W., Naylor S.L., Day J.W. and Ranum L.P.W. 2001. Myotonic dystrophy type 2 caused by a CCTG expansion in intron 1 of *ZNF9*. Science 293, 864-867.

Macgregor H.C. and Sessions S.K. 1986. The biological significance of variation in satellite DNA and heterochromatin in newts of the genus *Triturus*: an evolutionary perspective. Phil Trans Roy Soc Lond. B. 312, 243-259.

Maniatis T., Goodbourn S. and Fischer J.A. 1987. Regulation of inducible and tissue-specific gene expression. Science 236, 1237-1245.

Mather K. and Harrison B.J. 1949a. The manifold effect of selection. Part I. Hered. 3, 1-52.

Mather K. and Harrison B.J. 1949b. The manifold effect of selection. Part II. Hered. 3, 131-162.

Matsuura T., Yamagata T., Burgess D.L., Rasmussen A., Grewal R.P., Watase K., Khajavi M., McCall A.E., Davis C.F., Zu L., Achari M., Pulst S.M., Alonso E., Noebels J.L., Nelson D.L., Zoghbi H.Y. and Ashizawa T. 2000. Large expansion of the ATTCT pentanucleotide repeat in spinocerebellar ataxia type 10. Nat Gene. 26, 191-194.

Mayr E. 1963. Animal Species and Evolution. Harvard Univ. Press, Cambridge.

Mazrimas J.A. and Hatch F.T. 1972. A possible relationship between satellite DNA and the evolution of kangaroo rat species (genus *Dipodomys*). Nature New Biol. 240, 102-105.

McClintock B. 1951. Chromosome organization and genic expression, Cold Spring Harbor Symp Quant Biol. 16, 13-47.

McClintock B. 1984. The significance of responses of the genome to challenge. Science 226, 792-801.

Miklos G.L.G. 1985. Localized highly repetitive DNA sequences in vertebrate and invertebrate genomes. In: Molecular Evolutionary Genetics, MacIntyre R.J. (Ed.), Plenum Press, New York, pp. 241-321.

Nagl W. 1976. Nuclear organization. Ann Rev Plant Physiol. 27, 39-69.

Nakamura Y., Leppert M., O'Connell P., Wolff R., Holm T., Culver M., Martin C., Fujimoto E., Hoff M., Kumlin E. and White R. 1987. Variable number of tandem repeat (VNTR) markers for human gene mapping. Science 235, 1616-1622.

Nevo E. 1998. Molecular evolution and ecological stress at global, regional and local scales: the Israeli perspective. J Exp Zool. 28, 95-119.

Ogata N. and Miura T. 1998a. Creation of genetic information by DNA polymerase of the archaeon *Thermococcus litoralis*: influences of temperature and ionic strength. Nucl Acid Res. 26, 4652-4656.

Ogata N. and Miura T. 1998b. Creation of genetic information by DNA polymerase of the thermophilic bacterium *Thermus thermophilus*. Nucl Acid Res. 26, 4657-4661.

Ohno S. 1970. So much "junk" DNA in our genome. In Evolution of Genetic Systems, Smith H.H. (Ed.), Gordon and Breach, New York, v. 23. pp. 366-370.

Orgel L.E. and Crick F.H.C. 1980. Selfish DNA: the ultimate parasite. Nature 284, 604-607.

Pâques F., Leung W-Y. and Haber J.E. 1998. Expansions and contractions in a tandem repeat induced by double-strand break repair. Mol Cell Biol. 18, 2045-2054.

Parenti R., Guille E., Grisvard J., Durante M., Giorgi L. and Buiatti M. 1973. Transient DNA satellite in dedifferentiating pith tissue. Nature New Biol. 246, 237-239.

Phelan C.M., Rebbeck T.R., Weber B.L., Devilee P., Ruttledge M.H., Lynch H.T., Lenoir G.M., Stratton M.R., Easton D.F., Ponder B.A.J., Cannon-Albright L., Larsson C., Goldgar D.E. and Narod S.A. 1996. Ovarian cancer risk in BRCA1 carriers is modified by the HRAS1 variable number of tandem repeat (VNTR) locus. Nat Gene. 12, 309-311.

Ponder B.1988. Gene losses in human tumors. Nature 335, 400-402.

Richards R.I. and Sutherland G.R. 1992a. Dynamic mutations: a new class of mutations causing human disease. Cell 70, 709-712.

Richards R.I. and Sutherland G.R. 1992b. Heritable unstable DNA sequences. Nat Gene. 1, 7-9.

Rubinstein R., Schoonakker B.C.A. and Harley E.H. 1991. Recurring theme of changes in the transcriptional control region of BK virus during adaptation to cell culture. J Virol. 65, 1600-1604.

Sawyer L.A., Hennessy J.M., Peixoto A.A., Rosato E., Parkinson H., Costa R. and Kyriacou C.P.1997. Natural variation in a *Drosofila* clock gene and temperature compensation. Science 278, 2117-2120.

Schachman H.K., Adler J., Radding C.M., Lehman I.R. and Kornberg A. 1960. Enzymatic synthesis of deoxyribonucleic acid. VII. Synthesis of a polymer of deoxyadenylate and deoxythymidilate. J Biol Chem. 235, 3242-3249.

Schaffner G., Schirm S., Muller-Baden B., Weber F. and Schaffner W. 1988. Redundancy of information in enhancers as a principle of mammalian transcription control. J Mol Biol. 201, 81-90.

Schlötterer C. and Tautz D. 1992. Slippage synthesis of simple sequence DNA. Nucleic Acid Res. 20, 211-215.

Setzer H.W. 1949. Univ Kansas Publ Museum Nat Hist. 1, pp. 473.

Seuanez H.N. 1979. The phylogeny of human chromosomes. Springer-Verlag, Berlin.

Sinha A.K. and Pathak S. 1973. Distribution of constitutive heterochromatin in HeLa and HEp-2 cell lines. Humangenetik 18, 47-54.

Smith G.P. 1976. Evolution of repeated DNA sequences by unequal crossover. Science 191, 528-535.

Stern C. 1927. Ein genetischer und zytologischer Beweis für Verebung in Y-Chromosome von *Drosophila melanogaster*. Z Induktive Abst Vererbungslehre 44, 187-231.

Tavis J.E., Walker D.L., Gardner S.D. and Frisque R.J. 1989. Nucleotide sequence of the human polyomavirus AS virus, an antigenic variant of BK virus. J Virol 63, 901-911.

Trifonov E.N. 1989. The multiple codes of nucleotide sequences. Bull Math Biol. 51, 417-432.

Trifonov E.N. 1990. Making sense of the human genome. In: Structure and Methods, vol. 1, Human Genome Initiative and DNA Recombination, Sarma R.H. and Sarma M.H. (Eds.), Adenine Press, New York, pp. 69-77.

Trifonov E.N. 1999. Elucidating sequence codes: three codes for evolution. Ann NY Acad Sci. 870, 330-338.

Usdin K. and Grabczyk E. 2000. DNA repeat expansions and human disease. Cell Mol Life Sci. 57, 914-931.

van Deutekom J.C.T., Mijmenga C., van Tienhofen E.A.E., Gruter A.M., Hewitt J.E., Padberg G.W., van Ommen G-J.B., Hofker M.H. and Frants R.R. 1993. FSHD associated DNA rearrangements are due to deletions of integral copies of a 3.2 kb tandemly repeating unit. Hum Mol Gene. 2, 2037-2042.

Verma R.S. 1988. Heteromorphisms of heterochromatin. In: Heterochromatin. Molecular and structural aspects, Verma R.S. (Ed.), Cambridge Univ Press, Cambridge. Pp. 276-292.

Virtaneva K., D'Amato E., Miao J., Koskiniemi M., Norio R., Avanzini G., Franceschetti S., Michelucci R., Tassinari C.A., Omer S., Pennacchio L.A., Myers R.M., Dieguez-Lucena J.L., Krahe R., de la Chapelle A. and Lehesjoki A-E. 1997. Unstable minisatellite expansion causing recessively inherited myoclonus epilepsy, EPM1. Nat Gene. 15, 393-396.

Watanabe S. and Yoshiike K. 1985. Decreasing the number of 68-base-pair tandem repeats in the BK-virus transcriptional control region reduces plaque size and enhances transforming capacity. J Virol. 55, 823-825.

Waters E.R. and Schaal B.A. 1996. Heat shock induces a loss of rRNA-encoding DNA repeats in *Brassica nigra*. Proc Natl Acad Sci, USA 93, 1449-1452.

Zacharias H. 1986. Tissue-specific schedule of selective replication in *Drosophila nasutoides*. Roux's Arch Dev Biol. 195, 378-388.

Zuckerkandl E. 1974. A possible role of "inert" heterochromatin in cell differentiation. Action of and competition for "locking" molecules. Biochem. 56, 937-954.

Zuckerkandl E. 1986. Polite DNA: Functional density and functional compatibility in genomes. J Mol Evol. 24, 12-27.

COMPARATIVE MAMMALIAN GENOMICS AND ADAPTIVE EVOLUTION: DIVERGENT HOMOLOGS AND NOVEL GENES IN THE CATTLE GENOME

Harris A. Lewin, Joshua H. Larson, and Charu G. Kumar
The Department of Animal Sciences and The W. M. Keck Center for Comparative and Functional Genomics, The University of Illinois at Urbana-Champaign, Urbana, IL 6180, USA

Abstract: The convergence of two major currents in biology, comparative genomics and evolutionary biology, provides the pretext for greater understanding of the molecular mechanisms of adaptive evolutionary change. Our hypothesis is that rapidly evolving and novel genes contribute to the broad range of adaptive phenotypes observed among the mammalian orders. We have implemented an efficient scheme for the identification and characterization of divergent homologs and novel mammalian genes in order to obtain the necessary resources to test this hypothesis. The scheme is applicable to any species, particularly those whose genomes have not been completely sequenced. Herein, we discuss *Bos taurus* (domesticated cattle) as a model organism for studies in evolutionary biology and present a summary of results obtained using our scheme that led to the discovery of the *MHCLA* genes, a rapidly evolving gene family in cattle that is related to human *ULBP/RAET1*. The rapid evolution of these genes appears to be related to their function as ligands for receptors present on natural killer cells of the immune system. In addition to *MHCLA*, 13 divergent homologs and 257 novel transcripts have been identified and are currently being characterized in detail for their chromosome map location and expression profiles in different tissues. The essential question to be answered is what roles these genes play in generating the adaptive phenotypes that arose among the Ruminantia during the past 85 million years.

S.P. Wasser (ed.), Evolutionary Theory and Processes: Modern Horizons,
  Papers in Honour of Eviatar Nevo
H.A. Lewin, J.H. Larson, C.G. Kumar. *Comparative Mammalian Genomics and Adaptive Evolution: Divergent Homologs and Novel Genes in the Cattle Genome, 139-152.*
© 2004 *Kluwer Academic Publishers.*

1.   INTRODUCTION

My relationship with Eibi Nevo began after a symposium talk that I (HAL) gave on the subject of this paper at the 2001 Plant and Animal Genome Meeting. In his inimitable style, Eibi directly approached me after the talk to share his enthusiasm for our results and approach. In a taxi on the way to the airport we agreed to stay in contact, and during the following months I was flooded by an awesome volume of fascinating reprints and books from Eibi's archives. I am sure that everyone who knows Eibi has had a similar experience! Reading Eibi's main works on adaptive evolution in *Spalax* compelled me to think about our recent research findings in an even broader evolutionary context. During the summer of 2001 I visited Eibi at the University of Haifa, cementing our friendship and common scientific pursuit. Eibi's lifelong interest in understanding the molecular mechanisms of the adaptive evolutionary features of his beloved *Spalax* and my interest in elucidating the molecular bases for the unique adaptations found in ruminants (collectively called the ruminant adaptation) had found unexpected common ground in the path of genomic biology. The following brief review outlines the intellectual framework and experimental approach we have taken to the problem of identifying rapidly evolving genes in domesticated cattle (*Bos taurus*). The general approach described is applicable to any species that has a biologically interesting adaptation and is, at present, the only feasible alternative to genome-wide comparisons based on whole genome sequencing.

Evolutionary biologists have long puzzled over the molecular basis of adaptation and speciation (Raff, 1996; Nevo, 1999). As Nevo (1999) pointed out "Major evolutionary patterns and processes of speciation, adaptation, community ecology, biodiversity, and biota evolution, as well as extinction, are either unresolved or await theoretical and empirical elucidation". Although it is beyond the scope of this article to review all pertinent theories on the molecular mechanisms of adaptive evolutionary change, any discussion of the origins of phenotypic diversity must consider the current dogma. A widely accepted hypothesis in the field of evolutionary developmental biology is that regulatory changes affecting the temporal, spatial, and quantitative levels of expression of conserved genes that control developmental processes are responsible for the morphological differences between higher organisms (Raff, 1996). While we do not discount the experimental evidence in support of this and other "classical" hypotheses, the current dogma has not yet fully embraced the growing body of data arising from the field of comparative genomics (Kirshner and Gerhart, 1998). In our view, it is important for evolutionary biologists to consider the following recent results obtained from genomic studies: (*i*) a relatively large number of genes are under positive Darwinian selection in mammalian genomes (Liberles et al., 2001), (*ii*) a small but significant number of all genes are novel in any particular pair-wise comparison of mammalian genomes (Waterston et al., 2002), and (*iii*) a large number of non-conserved, noncoding regulatory RNAs can be found in EST databases and RNA pools (Hüttenhofer et al., 2001; Yelin et al., 2003). An essential problem experimentalists must now address is to determine the role(s) novel genes and regulatory RNAs play in generating adaptive, metabolic, morphologic, and physiologic change. The work we describe herein summarizes our approach for identifying highly divergent and novel genes in cattle, which is the first necessary step in addressing their role in adaptive evolutionary change.

2. ORFans, DIVERGENT HOMOLOGS, AND NOVEL GENES

In bacterial genomes, open reading frames (ORFs) which have no detectable sequence similarity to proteins in other genomes (termed *ORFans*) have been identified in large numbers (Fischer and Eisenberg, 1999). Fischer (1999) proposed that ORFans encode either "unique proteins with novel function" or "distant members of known gene families." In our scheme the term ORFan was borrowed from bacterial genomics to provide a consistent descriptor for similar mammalian DNA sequences. ORFans can be divided into two general categories: divergent homologs and novel genes. A *divergent homolog* is defined as a gene of common evolutionary origin in two or more species whose DNA sequence has diverged to the extent that homology cannot be detected using BLASTN search with default parameters (e.g., *MHCLA*, Larson et al., 2003). More sensitive search methods that use conceptual translations, such as BLASTX and TBLASTX, are required to identify divergent homologs. Mapping of a putative divergent homolog to an orthologous position in two or more mammalian genomes is one necessary step to confirm that the gene is indeed a divergent homolog. Although divergent homologs may be shown to be evolutionarily related and map to the predicted orthologous position, the proteins they encode may vary in function (Olson, 1999). Gene duplication and positive selection appear to be major mechanisms for generating divergent homologs (Hughes, 1999).

We define the term *novel gene* as a DNA sequence or its conceptually-translated protein(s) that has no detectable homology to any DNA or protein sequence in the public domain including draft or finished human and mouse genome/proteome sequences. This definition does not include genes that arise by exon fusion or genes that give rise to alternative splicing variants. Possible models to account for the presence of these novel genes include horizontal gene transfer (Lander et al., 2001), gene loss (Olson, 1999), and *de novo* formation (Yamauchi et al., 2002; Kondrashov and Koonin, 2003).

The most direct approach now available for identifying divergent homologs and novel genes is genome-to-genome comparison. This approach, used extensively in bacterial genomics, was recently used for comparing gene content in mice and humans (Waterston et al., 2002). Approximately 80 percent of all mouse genes were found to have a clearly identifiable ortholog in the human genome. The remaining 20 percent consist largely of species- or lineage-specific paralogs (Waterson et al., 2002). Many of these genes and gene families are known to be involved in reproduction, olfaction, and immunity and are under positive selection, as judged by the rate of nonsynonymous to synonymous amino acid substitutions (K_a/K_s). A substantial fraction of the proteins encoded by these genes can also be found in The Adaptive Evolution Database (Liberles et al., 2001), which contains data that suggest 10 to 20 percent of protein families are under positive selection. A particularly interesting example is *morpheus,* an extensively duplicated gene family in humans that has evolved rapidly in hominids after divergence from the great apes (Johnson et al., 2001). These results suggest that rapid divergence of homologous genes and gene families by gene duplication, deletion, and mutation may play an important role in adaptive evolution even among closely related mammalian species.

The best evidence that a relatively large number of novel genes exist in mammals comes from direct comparison of the human and mouse genomes. In total, 118 mouse ORFs could not be found in the human genome (Waterston et al., 2002). In whole genome comparisons it is possible that at least part of a "novel" set of genes is due to false gene predictions or missing DNA sequence. Although ESTs and full-length cDNA sequencing help to appropriately classify putatively novel genes, there are still a large number of predicted genes for which no mRNA has been identified, leaving the question as to whether they really are genes. Nevertheless, many of the novel genes identified in the mouse are likely to be authentic. The most plausible explanations for the presence of novel genes in the mouse are that their orthologs were deleted in an ancestral species (Olson, 1999) or that they have lost their phylogenetic signals due to strong positive selection (Johnson et al., 2001). An alternative although less likely possibility is that novel genes were acquired by horizontal transfer after the divergence of the two species (Stanhope et al., 2001). Large EST collections in other mammalian species suggest a high frequency of novel genes, particularly in tissues of reproductive and immunological importance. However, BLASTN was often the only method used in these studies. Thus, other factors that might affect classification of a sequence as novel were not accounted for or controlled (e.g., long 3' UTRs and genomic contaminants).

3. CATTLE AS A MODEL SPECIES FOR EVOLUTIONARY BIOLOGY

At this juncture it is important to justify why cattle are an excellent model to study adaptive evolution in mammals, and more specifically, why we chose a placenta cDNA library for the identification of novel genes and divergent homologs. *Bos taurus* is a species within the mammalian order Cetartiodactyla, which diverged from a common ancestor of the Rodentia and Primates approximately 94 million years ago at the Laurasiatheria-Euarchontoglires split (Springer et al., 2003). In contrast, the Rodentia and Primates diverged ~87 million years ago. The deep divergence time of Cetartiodactyla has several important consequences for comparative evolutionary studies. Firstly, the cattle genome will be useful for annotating the human genome for genes and conserved noncoding regulatory elements (Thomas and Touchman, 2002). Secondly, the accumulated adaptive phenotypes in the Cetartiodactyl and Primate orders make cattle an ideal species to explore the molecular bases for adaptive evolution. For example, ruminants have distinctive features of their digestive system (four chambered stomach), possess a cotyledonary, synepitheliochorial placenta, and exhibit variable structures in their immune system (e.g., hemal lymph nodes and high frequency of γ/δ T cells) as compared to humans. These features provide a strong justification for using cattle in comparative evolutionary studies and were thus essential criteria used by NIH to assign high priority to sequencing the cattle genome (Gibbs et al., 2002). Indeed, the cattle genomics community is very well organized for the sequencing effort, which is likely to begin late 2003. There are detailed linkage, radiation hybrid and comparative maps (Kappes et al., 1997; Band et al., 2000) as well as bacterial artificial chromosome (BAC) libraries and clone maps (Larkin et al., 2003) to facilitate genome assembly. Thus, despite the relatively long generation interval and high *per diem* costs to maintain experimental animals, the detailed knowledge of the biology of the organism due to its agricultural importance and full complement of genomic resources will add further to the attractiveness of cattle as a model species for evolutionary biology.

The placenta, with its very significant architectural differences in cattle, humans, and other mammals (Mossman, 1987) provides an ideal living laboratory to explore a group of adaptations that was necessary to exploit the appearance of extensive plains and grasslands during the late Miocene. Even prior to the development of genomic approaches, novel gene families were identified in ruminant placentae including the *pregnancy associated glycoprotein (PAG)* genes. The *PAG*s represent a seminal example of extensive duplication and divergence in ruminants of an ancestral gene encoding an aspartyl protease (Xie et al., 1997). However, the PAGs appear to be inactive as proteinases. There may be more than 100 *PAG* genes in ruminant genomes, many of which are expressed exclusively in the placental epithelium. Although related in amino acid sequence to human chymosin, cathepsin, and renin the *PAG* gene family has no clear ortholog in primates. However, the analysis of K_a/K_s suggests that the *PAG* genes are under positive selection (Hughes et al., 2000). These findings suggest that the cattle placenta will be a rich source of genes to explore the comparative biology of placentation in mammals and are the primary justification for our choice of a placenta cDNA library as a resource to search for divergent and novel genes.

4.   EXPERIMENTAL APPROACH

Our work in the area of divergent orthologs and novel genes as agents of evolutionary change began with the then unexpected observation that a large number of expressed sequence tags (ESTs) from cattle ovary and spleen cDNA libraries had no similarity to any sequences in the public domain DNA databases *circa* 1997-1999 (Ma et al., 1998; Band et al., 2000). In those studies we speculated that the mammalian transcripts with no significant similarity to DNA sequences in the public domain databases could represent genes that are important for species-specific adaptations. However, prior to complete sequencing of the human genome, it was not possible to know whether these ESTs represented genes that were as yet undiscovered in the human genome or cattle-specific transcripts. Also, these "novel" ESTs may have contained cloned genomic DNA sequence contaminants containing low-complexity adenine-rich elements (cloning artifacts during cDNA library construction), alternate or immature splice variants of known genes (Wolfsberg and Landsman, 1997; Mironov et al., 1999), previously unsequenced central portions of transcripts with 5' or 3' extremities not present in dbEST (Neto et al., 2000) and ESTs with divergent 3' untranslated regions. Subsequent large-scale cattle EST sequencing and the completion of the draft human and mouse genome sequences permitted us to approach the problem of ESTs as sentinels for divergent homologs and novel genes in a more systematic manner using computational genomics and phylogenetics.

The rapid expansion of public domain sequence databases and the easy accessibility of sequence analysis tools have permitted identification of orthologous genes on a massive scale (e.g., Rebeiz and Lewin, 2000). In addition, extensive web-based resources are now available for locating homologs across the phylogenetic spectrum including NCBI's HomoloGene and TIGR's Gene Indices. With more than 300,000 cattle ESTs now in the public domain and the relatively low rate of redundancy (as compared with the human collection), it is likely that representative sequences of more than half of all cattle genes have been identified (there are currently 18,000 cattle UniGenes). The general scheme we have used to identify divergent homologs and novel genes from a collection of ESTs is outlined in Fig. 1. A cattle placenta-specific EST database created by our laboratory was selected as an EST resource (http://titan.biotec.uiuc.edu/cattle/cattle_project.htm). The ESTs were cleansed of vector sequences, masked for repeats, and run through an in-house BLASTN pipeline (PipeBlastn) that permitted identification and digital subtraction of homologs (orthologs and paralogs) in other species, using $E < e^{-5}$ as a threshold. In this step of the analysis, it is critical that all nucleotide sequences from the organism under investigation be removed from the target database of the BLASTN search as well as sequences from other members of the same order (in our experiment, Cetartiodactyl sequences). The remaining set of ESTs was then extended from their 5' and 3' ends *in silico* using BLAST against other cattle ESTs and cDNAs in the public domain. For all candidate ORFans, full-length clone sequences were obtained by primer walking using the cDNA source clones as templates. PipeBlastn was then run again, subtracting any newly detected non-Cetartiodactyl homologs in genome space. A set of high probability ORFans was created by ORF scanning in the three forward reading frames. The largest complete ORF was derived for each sequence (beginning with a methionine initiation codon and ending with a stop codon) as well as the longest partial ORF (reading frame lacking N-terminal initiation codon to account for partial transcripts). The putative ORFan gene set was also interrogated for distant homology in other species using BLASTX. ORFans that aligned to database sequences with the empirically chosen threshold of $E<e^{-10}$ were designated as putative divergent homologs. ORFans with $E\geq e^{-10}$ were designated as putative novel genes. The working set of divergent homologs was then subjected to radiation hybrid mapping followed by multiple sequence alignment and phylogenetic analysis. These two experimental approaches permitted the identification of orthologous and paralogous relationships. For the divergent homologs, comparison of K_a/K_s was carried out with aligned homologous sequences. The working set of divergent homologs and novel genes was then subjected to a battery of bioinformatics tools that enabled identification of structural features in the encoded proteins and motifs that might provide clues to their molecular function. Both sets of sequences have been spotted on a high-density microarray (Band et al., 2002; Everts et al., 2002) for transcript profiling of different cattle tissues, including placenta (Band et al., 2000, 2003). Biological inference was then drawn from bioinformatic characterization and transcript profiling experiments. Additional methods for functional charac-terization, such as *in situ* hybridization and laser capture microdissection coupled with microarray analysis, may be useful in the future for elucidating the function(s) of the ORFan set.

5.   DISCOVERY OF A RAPIDLY EVOLVING GENE FAMILY IN THE CATTLE GENOME

The subtractive comparative genomics approach described above was used in our recent discovery of the cattle *major histocompatibility complex class I-like family A* (*MHCLA*) genes (Larson et al., 2003). Using a starting set of 888 3' cattle ovary and spleen ESTs, we removed sequences with significant BLASTN nucleotide alignments ($E < e^{-5}$) to sequences in the public domain and fully sequenced 14 novel clones containing polyadenylation signals and polyadenylated tails. Upon conceptual translation and protein alignment, one clone was found to share 24 percent sequence identity with the cattle MHC class I molecule, BoLA-BL3-7. We subsequently used this clone, designated *MHCLA1*, to retrieve an additional paralogous cattle EST and five distantly related human ESTs from GenBank. A cDNA clone containing the paralogous cattle gene was fully sequenced and named *MHCLA2*. The five human ESTs/genes were published soon afterwards as *ULBP1*, *ULBP2*, *ULBP3* (Cosman et al., 2001), *RAET1G*, and *RAET1K* (Radosavljevic et al., 2002).

Homology modeling revealed that the cattle MHCLA molecules exhibit the same unique domain architecture reported for the human ULBP/RAET1 and mouse H60/RAE-1 molecules except that their cytoplasmic domains are different. Multiple sequence alignment and phylogenetic analysis demonstrated that the cattle MHCLA molecules were more related to the mouse RAE-1/H60 and human ULBP/RAET1 molecules than they were to other classical and nonclassical MHC class I molecules. However, orthology could not be assigned to the cattle sequences as they were a great deal more similar to one another than to their human and mouse homologs. Comparative mapping placed cattle *MHCLA1* on BTA9 in a region of conserved synteny in the mouse and human genomes, thus confirming the homologous relationship between the cattle, human, and mouse genes. A survey of gene expression in different tissues using RT-PCR showed that *MHCLA1* is expressed constitutively in the 17 tissues examined, similar to the *ULBP* and classical MHC class I genes (Klein, 1986; Cosman et al., 2001). Southern blotting using an *MHCLA1*-derived sequence to probe a panel of genomic DNA from 14 species representing five mammalian orders showed evidence of many *MHCLA* genes, but hybridization was restricted to the suborder Ruminantia. These results were consistent with the sequence divergence detected by BLAST and suggest that the *MHCLA* genes are rapidly evolving.

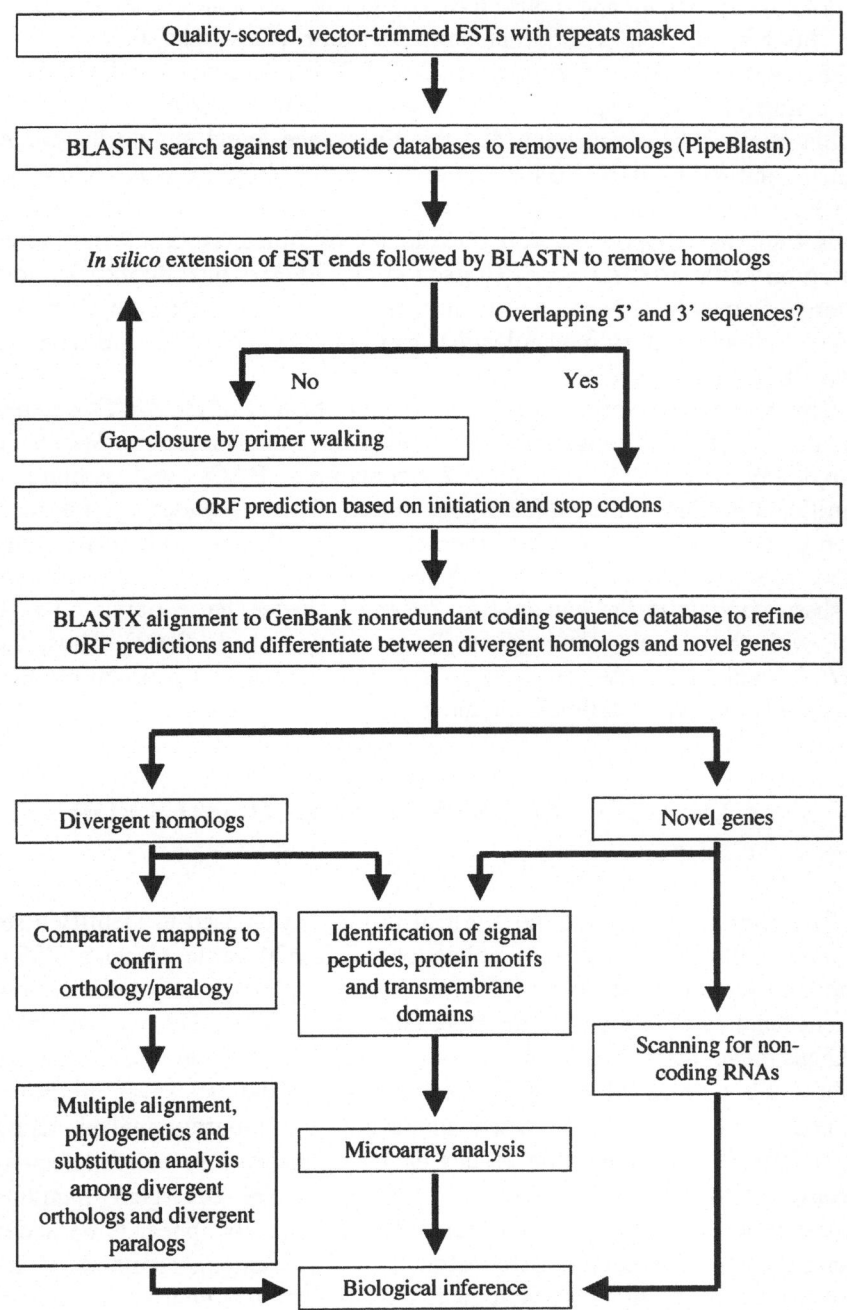

Figure 1. Scheme for identification and characterization of divergent homologs, novel genes, and noncoding RNAs. See text for further details.

The human ULBP and mouse RAE-1 and H60 molecules serve as ligands for the NK cell NKG2D stimulatory receptor (Cerwenka et al., 2000; Diefenbach et al., 2000; Steinle et al., 2001). This interaction can be blocked by a human cytomegalovirus UL16 glycoprotein (Cosman et al., 2001; Kubin et al., 2001) suggesting that the divergence found among the cattle, human, and mouse homologs is due to selection exerted by viral pathogens. We have observed a UL16-like molecule in the genome of bovine herpes virus-4 and have hypothesized that cattle herpes viruses are a selective force acting on cattle *MHCLA* genes (Larson et al., 2003). Thus, duplication and divergence may have been a mechanism to effectively increase the repertoire of NK cell stimulatory molecules that can bind NKG2D in the presence of viral UL16-like molecules.

The genomic organization and evolutionary history of the *MHCLA* genes are the subjects of ongoing investigation in our laboratory. We have completely sequenced four *MHCLA*-containing BACs that contain a significant portion of the *MHCLA* gene cluster. There is evidence for at least twenty *MHCLA* genes or gene fragments in the cluster, with many more likely to be identified in the gap in the current BAC contig. These results are in sharp contrast to the human *ULBP/RAET1* cluster that contains only 10 loci, with four being pseudogenes (Radosavljevic et al., 2002). Thus, the *MHCLA* genes of cattle appear to be another example of a gene family that has greatly expanded in the Ruminantia.

6.   CURRENT EFFORTS TO FIND HOMES FOR ORFans

The bioinformatics pipeline outlined in Fig. 1 was used to identify a set of unique transcripts within a collection of 12,620 cattle placenta ESTs. Although we are currently in the process of completing our studies, a few salient features of our results are presented here. To date, 373 putative cattle ORFans have been identified. Among these, 13 have been identified as divergent homologs, of which four are divergent orthologs, seven are novel paralogs, and two are unclassified. Interestingly, microarray analysis using RNA from 18 different cattle tissues including placenta revealed placenta-specific expression of the novel paralogs. Three divergent paralogs belonging to the bovine *prolactin related protein* family appear to be under strong positive selection as demonstrated by K_a/K_s among members of the family. One divergent ortholog appears to have distant similarity to a human receptor for a hepatitis virus and shows very strong expression in the liver. We speculate that the rapid diversification of the protein encoded by this gene may be a result of selection exerted by a viral pathogen. Other divergent homologs were found to be expressed differentially in the different tissues. At the present time the incomplete functional annotation of the human homologs of these divergent genes makes it difficult to ascribe functional significance to their apparent rapid rate of evolution. However, some general features were found among them, including the presence of signal peptides and domains encoding protease inhibitors.

The novel gene set, consisting of 257 sequence contigs, has been considerably more difficult to characterize. Protein motif searching tools have turned up only a few convincing results (e.g., transmembrane domains, signal peptides, and Kunitz domains). However, a substantial fraction of them cannot be excluded as coding sequences. If substantiated, these would represent novel genes of immense biological interest. As discussed above, the mechanisms by which novel genes may appear are gene loss, gene duplication followed by rapid divergence, and horizontal transfer (for other mechanisms see Long, 2001).

We are intrigued by the possibility that at least some of the novel "transcripts" we found in the placenta cDNA library might represent antisense transcripts and/or a subset of small non-messenger RNAs (snmRNAs). To date, we have identified several sequences in our "novel set" that have the characteristics of snmRNA using a specialized algorithm (Rivas and Eddy, 2001). The noncoding RNAs have been shown to affect transcription and translation in a variety of ways, including gene silencing and binding directly to the ribosome (Hüttenhofer et al., 2001). Numerous candidates for snmRNAs have been found in representative organisms from all three domains of life, including the archaeon *Archaeglobus fulgidus*, the bacterium *E. coli*, and in Eukarya, such as mouse and *A. thaliana* (Hüttenhofer et al., 2001). To what extent noncoding RNAs play a role in adaptive evolution we do not yet know. However, given their tissue and developmental-stage specificity and ubiquity (Yelin et al., 2003), it is plausible that these RNAs play an important role in generating adaptive phenotypes. Thus, our collection of "novel genes" derived from cattle placenta may also contain antisense and snmRNA, which provides a unique resource to explore their possible roles in the ruminant adaptation.

7. CONCLUSIONS

Genomics provides a powerful approach for exploring the genetic mechanisms underlying the evolutionary processes of adaptation and speciation. We have developed a combined bioinformatics and experimental strategy for identifying genes and noncoding RNAs that might play a role in adaptive evolution. In the absence of complete genome sequence information for most extant species, each with its own unique evolutionary adaptations, an EST-based strategy like the one we have developed may be the only viable means of identifying divergent homologs, novel genes, and snmRNAs on a genomic scale. With the appropriate resources, such as tissue and development stage-specific cDNA libraries, it will be possible to catalog rapidly evolving and novel genetic elements in a cost effective manner for any species. The greatest need is for suitable experimental tools to address the functions of these genes and RNAs. While microarray analysis can provide important information on tissue distribution and timing of expression of these sequences, expression profiles are not a substitute for biochemical data, cell or organismal biology. The true test of the functions of these genes/RNAs will lie in their ability to transform mammalian phenotypes. When this can be achieved, perhaps using transfection and transgenesis, comparative genomics will be viewed as contributing significantly to the field of evolutionary developmental biology.

8.   ACKNOWLEDGMENTS

We gratefully acknowledge contributions by Drs. Mark Band and Lei Liu in the W. M. Keck Center for Comparative and Functional Genomics at the University of Illinois, and Dr. Robin Everts. We also thank Professor Gene Robinson for his helpful suggestions on the manuscript. Funding for this work was provided in part by grants from the USDA National Research Initiative, Project No. 98-32205-6644 and 99-35205-8534.

9.   REFERENCES

Band M.R., Everts R.E., Liu Z.L., Morin D.E., Peled J.U., Rodriguez-Zas S.L. and Lewin H.A. 2003. Gene expression profiling of 17 cattle tissues reveals unique patterns related to tissue function. Plant and Animal Genome XI, San Diego, CA (abstract p. 745).

Band M.R., Larson J.H., Rebeiz M., Green C.A., Heyen D.W., Donovan J., Windish R., Steining C., Mahyuddin P., Womack J.E. and Lewin H.A. 2000. An ordered comparative map of the cattle and human genomes. Genome Res. 10, 1359-1368.

Band M.R., Olmstead C., Everts R.E., Liu Z.L. and Lewin H.A. 2002. A 3800 gene microarray for cattle functional genomics: comparison of gene expression in spleen, placenta, and brain. Anim Biotechnol. 13, 163-172.

Cerwenka A., Bakker A.B., McClanahan T., Wagner J., Wu J., Phillips J.H. and Lanier L.L. 2000. Retinoic acid early inducible genes define a ligand family for the activating NKG2D receptor in mice. Immun. 12, 721-727.

Cosman D., Mullberg J., Sutherland C.L., Chin W., Armitage R., Fanslow W., Kubin M. and Chalupny N.J. 2001. ULBPs, novel MHC class I-related molecules, bind to CMV glycoprotein UL16 and stimulate NK cytotoxicity through the NKG2D receptor. Immun. 14, 123-133.

Diefenbach A., Jamieson A.M., Liu S.D., Shastri N. and Raulet D.H. 2000. Ligands for the murine NKG2D receptor: expression by tumor cells and activation of NK cells and macrophages. Nat Immunol. 1, 119-126.

Everts R.E., Band M.R., Liu Z.L., Peled J.U., Kumar C.G., Bari A., Liu L. and Lewin H.A. 2002. Transcript profiling of bovine leukemia virus (BLV) infected and uninfected cell lines using a cDNA microarray containing 7653 cattle genes. Proceedings of the XXVIIIth International Conference on Animal Genetics, Göttingen, Germany, p. 55 (abstract).

Fischer D. and Eisenberg D. 1999. Finding families for genomic ORFans. Bioinform. 15, 759-762.

Fischer D. 1999. Rational structural genomics: affirmative action for ORFans and the growth in our structural knowledge. Protein Eng. 12, 1029-1030.

Gibbs R., Weinstock G., Kappes S., Schook L., Skow L. and Womack J. 2002. Bovine genomic sequencing initiative. http://www.genome.gov/Pages/Research/Sequencing/ /SeqProposals/BovineSEQ.pdf

Hughes A.L. 1999. Adaptive evolution of genes and genomes. Oxford University Press, New York.

Hughes A.L., Green J.A., Garbayo J.M. and Roberts R.M. 2000. Adaptive diversification within a large family of recently duplicated, placentally expressed genes. Proc Natl Acad Sci, USA 97, 3319-3323.

Hüttenhofer A., Kiefmann M., Meier-Ewert S., O'Brien J., Lehrach H., Bachellerie J.P. and Brosius J. 2001. RNomics: an experimental approach that identifies 201 candidates for novel, small, non-messenger RNAs in mouse. EMBO J. 20, 2943-2953.

Johnson M.E., Viggiano L., Bailey J.A., Abdul-Rauf M., Goodwin G., Rocchi M. and Eichler E.E. 2001. Positive selection of a gene family during the emergence of humans and African apes. Nature 413, 514-519.

Kappes S.M., Keele J.W., Stone R.T., McGraw R.A., Sonstegard T.S., Smith T.P., Lopez-Corrales N.L. and Beattie C.W. 1997. A second-generation linkage map of the bovine genome. Genome Res. 7, 235-249.

Kirschner M. and Gerhart J. 1998. Evolvability. Proc Natl Acad Sci USA 95, 8420-8427.

Klein J. 1986. Natural history of the major histocompatibility complex. Wiley-Interscience, New York.

Kondrashov F.A. and Koonin E.V. 2003. Evolution of alternative splicing: deletions, insertions and origin of functional parts of proteins from intron sequences. Trends Genet. 19, 115-119.

Kubin M., Cassiano L., Chalupny J., Chin W., Cosman D., Fanslow W., Mullberg J., Rousseau A.M., Ulrich D. and Armitage R. 2001. ULBP1, 2, 3: novel MHC class I-related molecules that bind to human cytomegalovirus glycoprotein UL16, activate NK cells. Eur J Immunol. 31, 1428-1437.

Lander E.S., Linton L.M., Birren B., Nusbaum C., Zody M.C., Baldwin J., Devon K., Dewar K., Doyle M., FitzHugh W., et al. 2001. Initial sequencing and analysis of the human genome. Nature 409, 860-921.

Larkin D.M., Everts-van der Wind A., Rebeiz M., Schweitzer P.A., Bachman S., Green C., Wright C.L., Campos E.J., Benson L.D., Edwards J., Liu L., Osoegawa K., Womack J.E., de Jong P.J. and Lewin H.A. 2003. A cattle-human comparative map built with cattle BAC-ends and human genome sequence. Genome Res. 13, 1966-1972.

Larson J.H., Rebeiz M.J., Stiening C.M., Windish R.L., Beever J.E. and Lewin H.A. 2003. MHC class I-like genes in cattle, MHCLA, with similarity to genes encoding NK cell stimulatory ligands. Immunogene. 55, 16-22.

Liberles D.A., Schreiber D.R., Govindarajan S., Chamberlin S.G. and Benner S.A. 2001. The adaptive evolution database (TAED). Genome Biol. 2, 1-6.

Long M. 2001. Evolution of novel genes. Curr Opin Genet Dev. 11, 673-680.

Ma R.Z., van Eijk M.J., Beever J.E., Guérin G., Mummery C.L. and Lewin H.A. 1998. Comparative analysis of 82 expressed sequence tags from a cattle ovary cDNA library. Mamm Genome. 9, 545-549.

Mironov A.A., Fickett J.W. and Gelfand M.S. 1999. Frequent alternative splicing of human genes. Genome Res. 9, 1288-1293.

Mossman H.W. 1987. Vertebrate Fetal Membranes. MacMillan, Houndmills.

Neto E.D., Correa R.G., Verjovski-Almeida S., Briones M.R.S., Nagai M.A., da Silva W. Jr., Zago M.A., Bordin S., Costa F.F., Goldman G.H., et al. 2000. Shotgun sequencing of the human transcriptome with ORF expressed sequence tags. Proc Natl Acad Sci, USA 97, 3491-3496.

Nevo E. 1999. Mosaic evolution of subterranean mammals. Regression, progression and global convergence. Oxford Press, UK.

Olson M.V. 1999. When less is more: gene loss as an engine of evolutionary change. Am J Hum Genet. 64, 18-23.

Raff R.A. 1996. The shape of life: genes, development, and the evolution of animal form. The University of Chicago Press, Chicago.

Radosavljevic M., Cuillerier B., Wilson M.J., Clement O., Wicker S., Gilfillan S., Beck S., Trowsdale J. and Bahram S. 2002. A cluster of ten novel MHC class I related genes on human chromosome 6q24.2-q25.3. Genom. 79, 114-123.

Rebeiz M. and Lewin H.A. 2000. COMPASS of 47,787 cattle ESTs. Anim Biotechnol. 11, 75-241.

Rivas E. and Eddy S.R. 2001. Noncoding RNA gene detection using comparative sequence analysis. BMC Bioinform. 2, 8.

Springer M.S., Murphy W.J., Eizirik E. and O'Brien S.J. 2003. Placental mammal diversification and the Cretaceous-Tertiary boundary. Proc Natl Acad Sci, USA 100, 1056-1061.

Stanhope M.J., Lupas A., Italia M.J., Koretke K.K., Volker C. and Brown J.R. 2001. Phylogenetic analyses do not support horizontal gene transfers from bacteria to vertebrates. Nature 411, 940-944.

Steinle A., Li P., Morris D.L., Groh V., Lanier L.L., Strong R.K. and Spies T. 2001. Interactions of human NKG2D with its ligands MICA, MICB, and homologs of the mouse RAE-1 protein family. Immunogene. 53, 279-87.

Thomas J.W. and Touchman J.W. 2002. Vertebrate genome sequencing: building a backbone for comparative genomics. Trends Genet. 18, 104-108.

Waterston R.H., Lindblad-Toh K., Birney E., Rogers J., Abril J.F., Agarwal P., Agarwala R., Ainscough R., Alexandersson M., An P., et al. 2002. Initial sequencing and comparative analysis of the mouse genome. Nature 420, 520-562.

Wolfsberg T.G. and Landsman D. 1997. A comparison of expressed sequence tags (ESTs) to human genomic sequences. Nucl Acids Res. 25, 1626-1632.

Xie S., Green J., Bixby J.B., Szafranska B., DeMartini J.C., Hecht S. and Roberts R.M. 1997. The diversity and evolutionary relationships of the pregnancy-associated glycoproteins, an aspartic proteinase subfamily consisting of many trophoblast-expressed genes. Proc Natl Acad Sci, USA 94, 12809-12816.

Yamauchi A., Nakashima T., Tokuriki N., Hosokawa M., Nogami H., Arioka S., Urabe I. and Yomo T. 2002. Evolvability of random polypeptides through functional selection within a small library. Protein Eng. 15, 619-626.

Yelin R., Dahary D., Sorek R., Levanon E.Y., Goldstein O., Shoshan A., Diber A., Biton S., Tamir Y., Khosravi R., Nemzer S., Pinner E., Walach S., Bernstein J., Savitsky K. and Rotman G. 2003. Widespread occurrence of antisense transcription in the human genome. Nat Biotechnol. 21, 379-386.

SELECTION INDUCED GENETIC VARIATION:
A New Model to Explain Direct and Indirect Effects of Sixty Years of Commercial Selection for Juvenile Growth Rate in Broiler Chickens, with Implications for Episodes of Rapid Evolutionary Change

Yoav Eitan and Morris Soller
Department of Genetics, The Hebrew University of Jerusalem, 91904 Jerusalem, Israel

Abstract: Selection of broiler chickens for juvenile growth rate and proportion of breast weight is an ongoing instance of intense long-term directional selection. The response was accompanied by correlated effects on reproductive performance and livability. Thus, the modern broiler is a unique resource for studies of the genetics of response to long term selection. Two remarkable aspects characterize this response. (1) After more than sixty generations of selection, genetic progress continues with no indication of a plateau, and (2) Correlated effects appeared in a punctuated and coordinated manner; that is, they appeared sequentially in time and simultaneously in the stocks of all mainline breeders. We present two models to explain these aspects: (1) A combination of Island Model and group selection together with internal changes caused by selection (endo-environmental effects) and external changes due to changed management practices introduced to ameliorate the correlated effects of selection (exo-environmental effects). (2) A new model: Selection Induced Genetic Variation, which explains the genetic variation required for the long continued response as resulting from sequential changes in genetic background by selection bringing new cohorts of genes into play through epistatic interactions, in a programmed manner.

S.P. Wasser (ed.), Evolutionary Theory and Processes: Modern Horizons,
Papers in Honour of Eviatar Nevo
Y. Eitan and M. Soller. Selection Induced Genetic Variation, 153-176.
© 2004 *Kluwer Academic Publishers.*

1.   INTRODUCTION

1.1   Origins of the modern broiler industry and modern broiler chicken

Until the mid-1940s, eggs were the main product of the poultry industry. Meat was a secondary product produced primarily in the spring of the year when the layer flocks were reproduced. At this time, almost all male chicks and excess female chicks were reared and sold for meat as young "tender" chickens (hence, the term "Spring Chicken"). Thus, chicken was an expensive meat product, not generally available. In addition, "tough" culled mature females from the layer flocks for making soup were available at low cost throughout the year. This is the source of the famous Jewish chicken soup. Within this constellation, there were "layer" breeds, with high-egg production and small body size which produced a smaller and, hence, less profitable "spring" chicken; and "dual purpose" breeds, with somewhat lower-egg production but larger body size that produced a more profitable "spring" chicken. Both types were reared primarily as egg layers, and overall economic value of the two breed types was similar, but varied according to year and location, depending on relative prices for eggs and meat.

The modern broiler industry had its start in the United States in the midst of WWII. There was a shortage of meat, and, the existing large meat animals (beef, sheep, and swine) because of their limited reproductive ability and long-rearing period could not easily expand production to meet the demand, while the German submarine blockade prevented imports from Argentina. After intense discussions, a government committee appointed to provide a solution, came up with the suggestion that chickens, because of their high reproductive rate (over 200 chicks a year per mother hen) and relatively short growing period (four to five months to market at the time), could rapidly provide large quantities of meat. All that was needed was to use existing facilities to allow the males to mate with the females all through the year (not only when the flock is reproduced) and to hatch the eggs collected for chick production instead of using them as table eggs. The program was a great success, proved economically viable, and continued in full force even when the war was over and shortages a thing of the past.

The broiler industry program was initially based on the dual purpose breeds. However, it was soon found that a cross between a dual purpose female of the White Rock breed, and a male of the somewhat exotic "Cornish" breed, known for its broad breast, provided a chick with a higher proportion of breast meat, and this cross (male of a "Cornish" male line x female of a "White Rock" female line) became the standard. (The modern

"Cornish" and "White Rock" broiler breeds have been extensively modified by introgression and hybridization, hence, the quotation marks).

1.2 Selection for juvenile growth rate

With the introduction of the concept of raising chickens primarily for meat production, it soon became evident that "days to market weight" (about 1.5 to 2.0 kg) was the main determinant of profitability. Days to market weight, in turn, was almost completely determined by juvenile growth rate, measured as body weight at a given age. Thus, intensive selection for rapid juvenile growth rate was implemented in both the White Rock female and Cornish male lines. This trait turned out to be ideal for mass selection. It possessed a high heritability, came to expression at an early age in both males and females, and was easily measured with high accuracy and low cost. Furthermore, generation interval in chickens is short (about one year) allowing rapid turnover of the breeding nucleus; and the high reproduction rate of the female chicken (200 chicks per year per female, and one male per ten females) allowed enormous selection intensity to be applied. In the early years, the best 1% of males and 5% of females with respect to juvenile growth rate were chosen to reproduce the breeding nuclei.

1.3 Response to selection for juvenile growth rate

In the initial generations of selection, the intense and effective selection for juvenile growth rate resulted in a decrease in days to market of three to four days per generation; from 140 days in 1942 to 84 days in 1957. In later years it became necessary to apply selection to additional traits: proportion of breast meat out of total carcass weight, efficiency of feed conversion (which acted primarily to reduce body fat content); and female reproductive performance in the female lines. Still, progress in juvenile growth rate continued (Havenstein et al., 1994a) achieving a further reduction in days to market, of one or two days per generation, bringing days to market from 84 days in 1957 to 35 days in 1991 and to 28 days today (Havenstein et al., 1994b; Havenstein, 2002). Thus, over the course of sixty years there was a remarkable continued response to selection for juvenile growth rate, with a total reduction in days to market weight from 140 days to 28 days. At the same time, the amount of feed required to produce a kg of live chicken has been reduced from 5.0 kg to 1.5 kg., and the proportion of breast meat has increased from 15% to 21% of live body weight. The proportion of body fat increased greatly in the initial years of selection, but with increased attention to feed efficiency, the proportion of body fat has been progressively reduced to about half of its maximum. Taken together, these changes have made the chicken the most efficient converter of grain into meat, and the cheapest source of meat for the consumer.

Remarkably, in spite of the long continued and intense selection for juvenile growth rate exerted over the course of 60 generations, there is no indication of a plateau. Realized heritability of juvenile growth rate in the broiler breeding nuclei remains moderate and a further increase in juvenile growth rate and reduction in days to market appears quite feasible.

The enormous and continued response in juvenile growth rate, and proportion of breast weight, has been accompanied by deleterious effects on reproductive performance of the parent flocks, hatchability of the broiler embryo, and health and liveability of the broiler chick. Indeed, these deleterious effects may render problematic the net economic value of any additional gains in juvenile growth rate and proportion of breast meat (Middlekoop et al., 2002).

A remarkable aspect of the correlated effects is their punctuated and coordinated appearance in time. That is, the various correlated effects appeared sequentially, each within a rather short period of time, yet appeared simultaneously in the stocks of all main line breeders.

In the following section, we will describe in detail some of the correlated effects of selection for juvenile growth rate. In later sections, we will propose some explanations for the long continued response to intense selection for juvenile growth rate, without loss of genetic variation in the trait; and for the punctuated and coordinated appearance of the correlated effects of the selection.

2.   CORRELATED EFFECTS OF SELECTION FOR JUVENILE GROWTH RATE

In the interests of brevity, we concentrate, in the present section, on effects of selection for juvenile growth rate on female and male reproductive performance, including effects on chick hatchability; we also describe some of the management practices that evolved to ameliorate these effects. Although not discussed in detail, deleterious effects at the broiler level are also many and also appeared in a punctuated and coordinated manner. Briefly, broiler level effects include skeletal problems expressed as trembling and wobbly gait and perosis (failure of the leg tendons); heart and circulation failure leading to ascites (accumulation of fluid in the heart cavity), and heart failure syndrome (sudden death due to heart failure); increased sensitivity to heat stress so that all broilers today must be reared with tunnel ventilation to provide cooling and maximum air circulation; reduced immunocompetence expressed as respiratory problems, *Escherichia coli* infection, necessity for a high level of sanitation, and a complex vaccination schedule. Further details are given in Table 1.

Table 1. Correlated effects of selection for juvenile growth rate and breast weight, their proposed endo/exo-physiological mechanisms, and their management solutions

Line	Associated effect	Type	Mechanism	Management solution
Female	Obesity (1960)	Endo	Lack of appetite control caused by selection for JGR	
	Reduced lay (1965)	Endo	Hormonal imbalance caused by obesity	Quantitative feed restriction1
	Increase in minimum weight for sexual maturity (1972)	Genetic	Allometric change in weight for onset of sexual maturity as a result of change in growth curve	Adjustment of recommended growth curves
	Reduced lay (1980)	Exo	Physiological hunger state caused by feed restriction (?)	Stimulatory lighting
	Prolapse/internal lay (2000)	Exo	Stimulatory lighting	Slower release from feed restriction at onset of lay
Male	Reduced fertility (1985)	Endo	Hormonal imbalance induced by obesity	Quantitative feed restriction
	Reduced fertility (1995)	Exo	Stimulatory lighting: simultaneous induction of sexual maturity in males and females (?)	Males transferred from darkout to laying house before the females.
Broiler	Skeletal defects (1980)	Endo	Developmental imbalance, Skeleton: body mass	Mild feed restriction to slow growth
	Reduced immune response (1980)	Endo	Muscle growth competes with immune response	Vaccination programs, pathogen free parent flocks
	Ascites (1990)	Endo	Developmental imbalance, Heart: body mass, exacerbated by cold	Maintain optimal growing temperature; reduce feed intake.
Embryo	Reduced hatchability (2001)	Endo	Increased muscle mass, more embryo heat production	Lower incubator temperature

2.1 Reduced female peak egg production (1960)

Chicken egg production peaks very rapidly, reaching a maximum of an egg a day (100% production) two weeks after the first egg is produced. Egg production drops off slowly thereafter, until the end of the reproductive period of about 10 to 12 months. The first hint of the troubles to come appeared in the early 1960s, which saw a sudden and precipitous drop in female reproductive performance. From a level of 180 to 200 chicks per female bird in the 1950s performance dropped rapidly to less than 120 chicks in the mid 1960s. The problem appeared to result from the fact that females in the hatchery flocks were reaching a much lower level of peak production at entry into lay, only 70 to 80%, with consequent reduced egg production over the entire reproductive period. Furthermore, fertility of the eggs was also reduced. It seemed evident that this was due to the extreme obesity of the mature females. Apparently selection for rapid juvenile growth rate was in large part selection for breakdown in normal appetite control, so that the birds ate without an internal limit (Soller and Eitan, 1984). Experiments showed that under free feeding (ad libitum feeding) conditions, it was possible to force feed additional food to birds of the original dual-purpose or layer breeds, but not to the new broiler lines (Nir et al., 1978). This lack of appetite control led to accumulation of fat during the growth period and to marked obesity in the female entering lay. It is known that adipose tissue is active in producing estrogens and other hormones, and it is these that apparently interfered with female reproductive performance. The reduced fertility resulted from impairment of male mating performance, perhaps due to sheer physical difficulty of the obese males and females to handle the mechanics of courting and mating (Soller and Rappaport, 1966) or possibly due to hormonal effects of the excess adipose tissue in the male affecting semen quality (Soller et al., 1965).

The solution was elegant but radical: introduction of quantitative feed restriction during the rearing of the reproductive males and females. That is, after a few weeks of ad libitum feeding to get the chicks started, the amount of feed per bird per day was limited to an amount that would allow the birds to gain lean body weight at a slow steady pace, but not overeat and become obese. This was achieved by rapidly introducing the desired quantity of feed per bird to the feeding troughs (by means of a mechanical feeder) and allowing sufficient space along the troughs so that all birds could access the feed simultaneously, reducing inter-bird competition.

The introduction of quantitative feed restriction returned the birds to high peak production, although not to levels attained by layers. Broiler chicks turned out to be amazingly tolerant of feed restriction. By 1979, reproductive males and females were fed 50% of ad libitum levels during the rearing period; while, at present, females are fed approximately 30% and males 35% of their ad libitum intake during the rearing period without any untoward effects (Havenstein et al., 1994b). However, tolerance of feed restriction is not unlimited. Work in our laboratory showed that there is a minimum body weight below which females will not enter lay, although they are otherwise healthy (Brody et al., 1980). A similar minimum body weight threshold for onset of female sexual maturity exists over a wide range of vertebrates, from fish to man. Thus, in order to bring the birds into lay, it is necessary to relax feed restriction and increase the amount of feed to allow the female birds to reach their threshold weights for onset of sexual maturity.

2.2 Reduced female post peak egg production (1970)

Initially, birds of the reproductive flock were shifted to ad libitum feeding to induce entry into lay, and remained on ad libitum feeding during the entire reproductive period. However, in the 1970s while peak production remained high, post-peak production dropped rapidly. This was apparently due to continued loss of appetite control which led to overeating and fat accumulation during the laying period, by both females and males. As a consequence, feed restriction was extended into the laying period itself.

2.3 Delay in onset of lay in the female (1980)

Under natural conditions, onset of sexual maturity in chickens is controlled by day length (photoperiod). Male and female birds enter sexual maturity (onset of lay in the females, and onset of mature semen production in males) in the spring, under the stimulus of gradually increasing photoperiod. Normally, chickens will not enter lay in the fall, under the negative stimulus of the naturally decreasing photoperiod at this season. In order to have a steady supply of broiler chicks throughout the year, however, it is necessary to have males and females enter sexual maturity and lay throughout the year. For birds maturing in the fall or winter, this was achieved by extending the photoperiod through use of supplemental artificial light, added at both ends of the natural day. This produces an artificial Spring-type light pattern, and brought the birds into sexual maturity in a reliable manner throughout the year.

Beginning in the 1980s, however, a new problem in reproductive performance became apparent. This was a delay in the onset of sexual maturity for birds maturing in the fall season of the year, even under the stimulus of supplemental artificial light. Some of the females did not enter lay at all; in others, onset of lay was delayed and peak lay was lower. A management solution was found in the form of "stimulatory lighting". For this, the chicks are reared under strict photoperiod control until it is time for them to enter sexual maturity. Photoperiod control is achieved by rearing the chicks in so-called "dark-out houses", in which photoperiod is under total artificial control. The birds are reared under a photoperiod of 16 hours total darkness and 8 hours dim light until about six months of age. During this period they are also under quantitative feed restriction. At this age, they are moved to the laying pens, and exposed to a photoperiod of 14 hours daylight while at the same time feed quantities are increased rapidly (within a few weeks from 70 g/day to over 140 g/day). Under the combined stimulus of the simultaneous increase in photoperiod and feed quantity, the birds are "forced" into sexual maturity and lay.

Initially, stimulatory lighting and dark-out rearing was required only for birds entering lay in the fall. However, with the passage of generations, and continued selection for juvenile growth rate, problems in entering lay appeared even in birds coming to sexual maturity in the spring of the year, under optimal natural lighting. At present, therefore, stimulatory lighting is required at all seasons of the year.

A series of studies in our laboratory showed that need for stimulatory lighting was due to reduced photosensitivity of the broiler chicks, as compared to layers, or to broiler chicks of earlier generations (Eitan and Soller, 1994). The reduced photosensitivity was expressed as a requirement for a longer photoperiod to stimulate the onset of sexual maturity. Thus, under natural decreasing fall light, birds of a layer stock entered lay at 172 days of age, broiler stocks entered lay at 237 days of age (a difference of 65 days). Under stimulatory lighting, the layers entered lay at 174 days, and the broilers at 216 days (a difference of only 42 days). In addition, we found that quantitative feed restriction per se, brought a reduction in photosensitivity in broiler breeder females. Birds fed ad libitum with light stimulation entered lay 21 days earlier than feed restricted birds. This difference increased to 27 days in the absence of light stimulation (Eitan and Soller, 1991). The combination of reduced photosensitivity due to selection for juvenile growth rate and reduced photosensitivity due to feed restriction was apparently responsible for the delay in onset of lay in the broiler females, which could not be overcome by simple increase in photoperiod alone.

2.4   Reduced male fertility (1985)

In the mid 1980's, growth and appetite patterns of males and females began to diverge sufficiently, so that the males became obese and lost fertility when fed together with the females. This made it necessary to use special feeders and feed composition for the males while preventing the males from having access to the female feeders.

2.5   Reduced male fertility in the second half of the laying cycle (1995)

In the early 1990s male fertility began to decline precipitously in the second half of the laying cycle (Creel et al., 1998). In order to deal with this phenomenon, the industry simply reinforced the older males with younger males in the second half of the laying cycle, a procedure termed "spiking" the flock with younger males.

2.6   Increased mortality of females on onset of lay (1995)

Beginning in the mid 1990s females began to show increased mortality at onset of lay. This was due to cloacal prolapse and eversion of the oviduct, internal lay (in which the egg is deposited in the abdominal cavity instead of the oviduct); and sudden death syndrome, in which the bird suddenly dies as a result of accumulation of fluid in the heart cavity. The definitive reasons for these pathological manifestations have not yet been established, and optimal management methods for control of the problem are not yet available.

2.7   Reduced chick hatchability (2000)

The most recent set of problems affecting the broiler industry relate to reduced hatchability and quality of the hatched chicks. Hatchability has fallen from over 90% to 88% or less. There is an increase in the proportion of chicks born with incomplete naval closure, and the sex ratio has changed, from a slight excess of males, to a slight deficiency of males (Taylor, 1999). The hatching eggs are very "unforgiving" and the slightest deviation from optimal conditions results in embryo death and reduced hatchability.

Dealing with this problem required a major change in hatchery practice. During the first nine days of incubation, the incubating egg in the hatchery requires an external heat source to maintain appropriate temperature. During the next twelve days of incubation, the egg produces excess heat, which must be removed. Thus, the traditional hatchery used a multistage incubation system, in which hatching eggs at different stages of development were incubated together, so that eggs at later stages of incubation provided heat for the eggs at the younger stages; and relatively little external adjustment of conditions within the incubator was required. The hatching eggs managed themselves. This has now been replaced by single-stage incubation, with all eggs at the same stage of incubation, and with very tightly controlled external monitoring and continuous adjustment of incubation conditions to maintain optimal temperature, humidity, and CO_2 content across the incubation period.

3.   TWO GENETIC CONUNDRUMS

With this list of afflictions, we close for the time being at least, the litany of troubles that have accompanied the intense and highly successful selection for juvenile growth rate and breast proportion in the modern broiler chicken. Disruption of appetite control resulting from selection for juvenile growth rate, plausibly explain the reduced reproductive performance of modern broiler stocks under ad libitum feeding; while increased heat production by the modern "high breast proportion" embryo explain the recent problems in hatchability. In contrast, there is no clear connection between selection for juvenile growth rate/breast proportion and reduced photo-sensitivity, second half collapse of male fertility, or increased incidence of prolapse, internal lay, and sudden death syndrome at entry into lay. Consequently, the punctuated and simultaneous appearance of these deleterious effects in all breeding stocks in the last two decades of the previous century, presents a companion genetic conundrum to the source of the genetic variation permitting the long continued response to selection for juvenile growth rate. We now detail the two question points in this history that require explanation and consider various speculative solutions to these conundrums.

3.1   The long continued response to selection

Intense directional selection would have been expected to rapidly exhaust the additive genetic variation in the broiler populations: "The simple theoretical expectation is that selection should lead to fixation with the loss of genetic variance... sooner or later it is to be expected that all favorable alleles will be brought to fixation" (Falconer, 1972).

What then, are the sources of the genetic variation that allowed for the extended response to selection over the course of 60 generations, with little if any loss of additive genetic variation?

3.2   The punctuated and coordinated appearance of secondary effects of selection

Correlated effects of selection, whether due to pleiotropy or linkage, are expected to be linear on changes in allele frequency of the loci under selection. Thus, we would expect the various deleterious effects noted above, to have appeared in more or less smoothly increasing degree, with the passage of the generations of selection. Alternatively, if due to pleiotropic or linkage effects of new mutations or recombinants, the deleterious effects would have been expected to appear in a random isolated order, among the stocks of the various breeders. Instead, as shown by the above history, deleterious effects appeared in a punctuated and coordinated fashion.

How, then, do we explain the *punctuated* and *coordinated* appearance of the various deleterious associated effects; *punctuated*, in that the deleterious effects did not develop gradually in a smoothly increasing degree from the initial start of selection, but instead appeared rather suddenly at specific point of selection and; *Coordinated*, in that the deleterious effects appeared in the stocks of all the international breeders, more or less simultaneously, at the same point of time.

4. ISLAND MODEL, GROUP SELECTION AND ENDO/EXO ENVIRONMENTAL EFFECTS

Our initial attempt to explain the two unexpected genetic attributes of the long term selection for juvenile growth rate, was a two part model, which provided separate explanations for the long continued response to selection on the one hand, and for the punctuated and coordinated appearance of the associated deleterious effects on the other. Our explanation for the long continued response to selection was based on Sewall Wrigth's Island model for evolution. Our explanation for the punctuated and coordinated appearance of the associated deleterious effects was based on an extension of the endo-environmental model for correlated effects of selection proposed by Reuven Bar Anan and his student, Micha Ron (1987).

4.1 The island/group selection for the long continued response to selection

With respect to the long continued response to selection for juvenile growth rate, while maintaining additive genetic variation, it is useful to consider the various independent poultry breeders as semi-isolated reproductive islands. Initially, there were hundreds of such independent breeders. In Israel alone, in the 1960s there were over a dozen farms engaged in broiler breeding; all competing within the Israeli market, some with overseas sales as well. Although these numbers were rapidly reduced, the number of international and large national breeders remained quite high until the last round of consolidation and acquisition, after which there remain only three major international broiler breeders.

Initially, the different breeders (the "islands") began with somewhat different stocks, and many crossed their stocks to various large body commercial and show breeds, to increase genetic variation. Thus, taking all breeders together, a very wide net was cast to gather existing genetic variation affecting juvenile growth rate. The effective population size of any of the individual breeding nuclei was not large, ranging up to the high hundreds; but across the large number of independent breeders, the total effective population size of the broiler breeding population numbered many thousands. Nevertheless, because of the island structure of the overall breeding population, by founder effect, any rare allele would be present in one or more of the subpopulations at an appreciable frequency. In any of the breeding nuclei, a single male carrier of a favorable novelty, if selected as a parent animal, would single handedly bring the new variant to a frequency of 0.02 to 0.05 in the breeding nuclei. At this point, intense selection would rapidly move it to moderate and high frequencies. In the same manner, the very large effective population size of the broiler population as a whole, provided a wide field within which mutation, or recombination among tightly linked loci in repulsion phase in balanced blocks, could release novel genetic variation. Here too, since the individual breeding nuclei were not large, any novel mutation or recombination in one of the breeding nuclei appeared initially at non-negligible frequency and could eventually be brought to high frequencies by the intense selection applied.

All breeders need to sell their stocks, and indeed all commercial breeders continually test their stocks against those of their competitors. This provides ready opportunity for introgression of desired genetic material from one breeding nucleus to another. Thus, the breeding nuclei of the various breeders are interconnected by a web of migration. This rapidly spread any novel favorable alleles or allele blocks, originating and reaching high frequency in any one of the breeding nuclei, throughout the entire broiler breeding population.

Thus, in this model we envision three consecutive sources of genetic variation that in principle at least, could underlie the long continued response to selection. The first source consisted of existing genetic variation within each of the breeding nuclei. While this was being utilized, rare favorable alleles present in one or more of the breeding nuclei by founder effects, came to moderate to high frequency, and spread by migration throughout the broiler breeding population. These rare alleles provided the second source of genetic variation. The third source consisted of novel variants occurring by mutation or recombination in individual breeding nuclei that came to sufficiently high frequency to be caught up by selection and provide new substrates for response. Migration of breeding material among breeding nuclei then spread the genetic innovations throughout the entire breeding population.

4.2   Breakdown of development homeostasis and group selection

Two additional factors may have contributed to the strong response to selection. Selection for extreme phenotypic variants is also selection for breakdown of developmental homeostasis. This can release cryptic genetic variation, and may be responsible for the increased sensitivity of the broiler embryo and chick and mature bird to environmental stresses of all sorts. The second factor is the strong selection among breeding nuclei, which was a consequence of the intense competition for market share among breeders. This competition is a consequence of the organization of the broiler industry. The breeders sell their stock to hatchery flocks; the hatcheries sell the day-old broiler chicks to the grow-out farms. The grow-out farm is the final consumer, and is interested in minimizing hatching chick costs, while maximizing chick growth performance. In the initial stages, there were government test farms, which provided unbiased information as to the relative performance of the various breeders. These guided the hatcheries in their purchase of breeding stocks. Later, the broiler growers were organized into large integrations (hatchery, grow out farms, and processor) that carried out their own internal tests of the performance of the stocks of the various breeders. Thus, the best of the breeders were able to maintain and increase market share; the others rapidly fell by the wayside. This very effective "group selection" surely contributed in some appreciable manner to overall genetic progress in juvenile growth rate.

4.3   The endo-/exo-environmental explanation for the punctuated and coordinated appearance of secondary effects of selection

The term "endo-environmental correlation" was coined by Reuven Bar Anan, to describe an indirect effect of one trait (the independent trait) on another (the dependent trait), produced by a causal chain in which a change in the independent trait, changes the internal molecular/physiological environment of the animal in a manner that causes in turn a change in the expression of the dependent trait. Bar Anan proposed this concept to explain the paradoxical finding that the genetic correlation between milk production and fertility in dairy cattle was unequivocally negative, yet major genetic gains in milk production were being made without any concurrent change in cow fertility. Bar Anan pointed out that in high producing cows at the beginning of lactation, milk production is so high and intense that feed requirements to maintain body weight and milk production are beyond the feeding capacity of the cow. As a result the cow loses weight, literally converting her body protein and fat into milk. This reduces body weight below the minimum threshold required for normal ovarian cycling, with consequent reduced fertility in the initial months of lactation. This resulted in a clear negative genetic correlation of high milk production with female fertility; yet the negative genetic correlation was not an intrinsic pleiotropic effect of the alleles producing high milk production. It was secondary to the effect of high milk production on nutritional status of the cow and could, therefore, be addressed by measures to improve nutritional status. Indeed, in the specific case of milk production and fertility, as overall genetic level of milk production increased, average feeding levels increased as well. Thus, the fertility differential among high- and low-producing cows within a given herd environment remains, but average fertility did not decrease although genetic levels of performance have almost doubled.

4.3.1   Obesity related effects

In the case of selection for juvenile growth rate in broilers, it seemed clear that the initial constellation of deleterious effects of selection for juvenile growth rate on reproductive performance of females and males was also due to a clear endo-environmental effect. As noted, selection for juvenile growth rate was in large part selection for breakdown of appetite control, with resultant obesity of the mature animal. It is commonly accepted that it was the obesity per se, which resulted in reduced egg production. Thus, as selection progressed, the mature animals became progressively more obese at entry into lay. Apparently, moderate obesity did not interfere with reproductive performance, so the deleterious effects were noted only when a certain obesity threshold was reached. Once the threshold was reached, effects were strong and cumulative. Since all breeder stocks had to achieve about the same level of juvenile growth rate to remain competitive, they all reached the tipping point at which obesity interfered with reproductive performance at about the same time. In this way, the appearance of reduced female reproductive performance appeared world wide, in a punctuated and coordinated manner among all breeder stocks.

As noted, these deleterious effects were controlled to a large extent by the introduction of quantitative feed restriction. Continued selection for juvenile growth rate led to further disruption of appetite control with deleterious effects among females and males, all controlled by extending feed restriction into the laying period, and then by separate restricted feeding for males and females. With continued selection for rapid juvenile growth rate, and breakdown in natural appetite controls, it has become more and more critical very tightly to monitor feed intake and body weight of the birds during the laying period to allow a slight gain in body weight, without permitting overt obesity.

4.3.2 Photoperiod related effects

The second constellation of deleterious effects, appearing in the 1980s related to reduced photosensitivity, which was countered by the introduction of dark-out housing during the rearing period, was followed by stimulatory lighting and feeding to force the transition to lay. We have speculated (Eitan and Soller, 2001) that the reduced photosensitivity of the birds during the transition to lay was a result of the increasing physiological hunger state induced by the increasing gap between the intake of the broiler bird under ad libitum feeding and the feed amount allotted under quantitative feed restriction. We proposed that the bird interprets this as signaling that the "season" is a bad one for reproduction, and hence delays onset of lay hoping for better times. On this hypothesis, the reduced photosensitivity is caused by an endo-environmental effect of the feed restriction implemented to counter the obesity effect of the selection. Feed restriction is an externally imposed environmental effect, and hence we term this an exo-environmental correlation.

A hypothesis first proposed by Brake (1990) can be adapted to provide an exo-environmental explanation for the more recent collapse of male fertility in the second half of the reproductive cycle. He relates this to abnormal timing of onset of male and female sexual maturity due to stimulatory lighting. Under natural conditions, males are somewhat more photosensitive than females. As a result, in the spring season, or under supplemental lighting, the males begin to enter sexual maturity about two weeks earlier than the females. This enables the males to settle dominance relationships among them, before the females mature. Thus, under "normal" circum-stances when the females mature, the males are ready to service them and are not distracted by male to male interactions. Under stimulatory lighting, males and females are brought to sexual maturity simultaneously. As a result, male-to-male and male-to-female interactions take place at the same time, disrupting the normal behavioral patterns in the flock. Brake's hypothesis holds that when males and females are young and sexually active, they are able to compensate for this, but as they grow older, male mating frequency declines, leading to increased fat accumulation in the males with a corresponding accelerating decline in male fertility. In support of this hypothesis, limited experiments showed that bringing the males into sexual maturity two weeks ahead of the females could improve male fertility (Brake, 1990).

4.3.3 Onset of lay effects

We believe that a combination of endo- and exo-environmental effects may also be responsible for the more recent pathologies associated with entry into lay (prolapse, internal lay, sudden death syndrome). Based in part on suggestions by Brake (Walsh and Brake, 1997), we propose to relate this set of problems to the strong selection for increased breast proportion and improved feed efficiency that has taken center stage of broiler breeding in the last decade. This has created a strong drive for breast muscle deposition in the modern broiler chick, and possibly a further increase in the minimum lean body weight needed for female sexual maturation. In addition, the selection for increased feed efficiency has produced a bird that will grow and gain weight during rearing, even when kept under very severe feed restriction. Results in our laboratory show that cage housed females of a broiler stock retained without selection from 1985 required 70 g/d to achieve weight gains comparable to those achieved by females of a modern broiler stock on 50 g/d (unpublished data, our laboratory). Feed composition has not been adjusted to take this difference in efficiency into account. As a result, the birds grow and gain weight in an apparently normal fashion, but may be suffering a sub-clinical deficiency in protein, and in minerals and other micronutrients. In this condition, the birds are provided with powerful food and light stimulation to bring them into lay. In a short period of time they are expected to engage the hypothalamic-pituitary-gonadal (HPG) axis, build the ovary and oviduct, and simultaneously build their body mass to make up for the reduced growth under feed restriction. In the "natural" (unselected) bird, development of the reproductive system has first call on nutrients relative to muscle mass. In the modern broiler, however, because of the strong selection for breast muscle deposition, breast muscle during compensatory growth may compete more strongly than in the past with the reproductive system (HPG axis, ovary, and oviduct) for protein. This may cause the female reproductive system to lag behind muscle deposition. Under the powerful impetus of stimulatory light and additional feed, the birds enter lay before the reproductive system and its regulatory system are fully developed. In addition, egg production, with its strong call on body mineral stores, begins under conditions of borderline mineral deficiency. One result is an increase in the number of double ovulations (the bird attempts to lay two eggs on the same day). This can lead to internal lay, when the first egg interferes with entrance of the second egg into the oviduct, and to prolapse when both eggs are encapsulated together forming a very large "double yolk egg". The net result of this overall imbalance may be the constellation of syndromes accompanying present day entry into lay: prolapse, internal lay, and sudden death syndrome. A possible management solution derives from the hypothesis: a slower and more gradual entry into lay, using an enriched diet with increased mineral content, giving the entire reproductive system time to build and mature before onset of lay. Whether these adjustments will serve to control the new conditions remains to be seen. Initial experimental and field results are encouraging.

4.3.4 Hatchability effects

A plausible explanation for the recent problems in hatchability attributes this to an endo-environmental effect, resulting from the intense selection for increased proportion of breast meat, particularly in the past ten years (Meijerhof, 2002). This change in body proportions extends into the embryo, and the modern broiler embryo, particularly in the male, has a higher proportion of breast weight than the traditional broiler embryo. Muscle is a metabolically active tissue, leading to excessive heat production and oxygen consumption, particularly by the male chicks. This excess heat production by the embryo is thought to be the major source of the recent problems in the traditional hatchery.

The punctuated appearance of the various correlated effects on reproductive performance, their proposed endo-/exo-environmental origin, and their management solutions, are summarized in Table 1.

4.3.5 Felicities and difficulties for the endo-/exo-environmental explanation

Thus, on these hypotheses, the punctuated and coordinated appearance of the deleterious effects accompanying selection for juvenile growth rate and breast proportion, may be explained, in part, due to nonlinear nature of endo-environmental effects, which therefore appear at specific stages of selection for juvenile growth rate; and, in part, as exo-environmental effects due to the stepwise introduction of management solutions (quantitative feed restriction and stimulatory lighting) to previous problems.

The endo-/exo-environmental explanation for the punctuated and coordinated appearance of deleterious effects provides a highly plausible explanation for some of the effects noted (e.g., reduced reproductive performance of females on ad libitum feeding); a moderately plausible explanation for others (e.g., current problems in hatchability); and highly speculative explanations for yet others (e.g., reduced photosensivity). Indeed, experiments in our laboratory have shown that reduced photosensitivity is also shown by female broiler chickens fed ad libitum (Eitan and Soller, 2001). Also, reduced photosensitvity and need for dark-out housing are found in turkeys, although turkeys do not tolerate quantitative feed restriction. Neither of these phenomena is expected on the hypothesis that loss of photosensitivity derives from the physiological hunger state caused by stringent quantitative feed restriction.

5.   A NEW GENETIC EXPLANATION: SELECTION INDUCED GENETIC VARIATION

These considerations led one of us (Y. Eitan) to propose an alternative hypothesis to account for the punctuated and coordinated appearance of at least some of the deleterious effects of selection for juvenile growth rate. The proposed hypothesis also purports to provide a source for at least part of the new additive genetic variation required for the long continued response to selection for this trait.

The new hypothesis is based on the following argument. We start with the observation that many QTL (quantitative trait loci), while having additive effects, also exhibit strong epistatic interactions with other QTL or with the genetic background. This implies that a locus with minor or even neutral effects in a particular genetic background might have strong effects in some other genetic background. In essence, a neutral locus, transferred to a different genetic background, would transform into a QTL. With this possibility in mind, Eitan notes that directional selection changes allele frequency at numerous QTL eventually bringing positive alleles to fixation or close to fixation. This change in allele frequency at numerous QTL, changes the genetic background and internal "molecular environment" of the cell/organism. On the argument above, some loci that were previously present, but at which allelic variation was neutral with respect to the trait under selection, now operate in a changed genetic environment and come to expression as full fledged QTL. These QTL exhibit additive effects in the new genetic background, but the expression of these additive effects are dependent on the background.

Selection on these QTL produces further changes in the genetic background and in the molecular environment of the cell and organism. Against the changed background and internal environment, new classes of genes, which were previously neutral with respect to the trait under selection, come into play. This continues in an iterative manner, with no obvious limit. On this hypothesis, long-term selection itself acts as a "bootstrap" continuously generating at least part of the genetic variation required for its' own continued response.

An important consequence of the hypothesis is that as selection proceeds to change the genetic background, new sets of genes would continually come into play as sources of variation in a "programmed" manner. That is, when a particular stage of selection is reached, a specific group of genes that were previously neutral, would become active. This would take place at the same stage of selection, even in otherwise separated breeding nuclei. Thus, a parallel series of allele substitutions would be taking place at more or less the same time and rate in all breeding nuclei that were achieving the same level of genetic progress in their breeding programs. Pleiotropic effects of allele substitution at these "new" genes might then be responsible for the punctuated and coordinated appearance of new deleterious effects with continued selection. The same deleterious effects would appear in the stocks of different breeders since the same genes would be undergoing allele substitutions in all stocks.

Thus, this hypothesis, which we term "Selection Induced Genetic Variation" (SIGV), can explain both the long continued response to selection, and the punctuated and coordinated appearance of at least some of the deleterious associated effects.

5.1   An experimental test of the SIGV hypothesis

A strong experimental test of the SIGV hypothesis would be to cross the modern broiler to early broiler lines or to a layer line. In the F1 and F2 of such a cross, the genetic background will revert to a prior situation. If indeed parts of the genetic gains in juvenile growth rate of the modern broiler derive from epistatic interactions with genetic background, then some of the positive alleles contributing to selection effect in the later generations should lose their primary effect in the more "primitive" genetic background of the cross. Consequently, the F1 and F2 of the cross between the modern broiler and stocks representing previous generations, should show a strong negative heterosis for growth rate. In addition, if alleles contributing to the selection effect on later generations also lose their pleiotropic effects, in the genetic background of the F1 / F2 cross, then positive heterosis for male and female reproductive traits should also be observed. Just such a cross has recently been carried out (Deeb and Lamont, 2002). The results conform to expectation with respect to juvenile growth rate. Daily weight gains of the modern broiler parent, layer parent, and F2 progeny to eight weeks of age were 85 g/d, 10 g/d, and 30 g/d, respectively. If the difference between modern broiler and layer were solely additive the F2 would be expected to perform at the midpoint, 47.5 g/d. Thus, strong negative heterosis for growth rate was found, as predicted by the SIGV hypothesis. Results on reproductive performance were not presented. It will be interesting to see if positive heterosis for reproductive performance was indeed obtained. One imagines that under conditions of ad libitum feeding and non-stimulatory photoperiod, the modern broiler would barely manage to enter lay, while the F2 would perform quite respectably.

6.   DISCUSSION

The history and hypotheses presented here in relation to the broiler chicken are of interest in a number of broader contexts. We will briefly discuss some of these here. These broader contexts are of particular importance in view of the fact that the chicken genome is currently being sequenced, with a complete genomic sequence expected by the end of the year. Thus, many of the genetic phenomena exemplified by the broiler model can now be taken to the genomic level for deep understanding of their causes and structure.

6.1   The broiler chicken as an experimental model for QTL mapping

Broiler and layer chickens differ widely in almost all physiological traits. As noted, they are fully fertile using artificial insemination, and their products are fully fertile as well. Thus, a cross between broiler and layer chickens provides excellent material for QTL mapping of a host of traits of basic physiological importance, among them appetite control, body composition, obesity, muscle development, reproductive performance, and immune response. The broiler chicken represents a wide collection of QTL affecting these traits. At most of the relevant sites, allele frequencies in broiler and layer can be expected to differ markedly. Thus, broiler x layer crosses can be analyzed as F2 populations, providing high statistical power per genotyping data point. Broilers x layer crosses are also amenable to analysis by full-sib intercross design (Song et al., 1998). When extended across a number of generations, such a design offers the statistical power for high resolution mapping that is provided by the advanced intercross line design. Dense marker maps are available for mapping in the chicken. Taken together with the genomic infrastructure that is already present or anticipated in the near future, it should be possible to move rapidly from QTL mapping to identifying the gene behind the QTL. Thus, the chicken model has much to offer in the way of functional analysis of genes affecting a wide spectrum of genes influencing very basic physiological processes.

6.2   The broiler chicken as an experimental model for long-term selection

The modern broiler represents unique and unprecedented model of long-term directional artificial selection occurring under virtually optimal conditions. Effective population size was very large and organized in an island model structure with migration. Selection was intense and operated at both the individual and group levels. The products of selection had to make their way in the real world, grow effectively, survive, and reproduce in a variety of climatic and management environments across the entire globe. Thus, the broiler model closely reproduces many aspects of what has been proposed to drive evolution at the biological level. The most remarkable conclusion of this model is that under evolutionary circumstances, the response to selection is apparently open-ended. There is no evidence that genetic variation in juvenile growth rate has been exhausted. This supports the accepted view that natural selection has almost a free hand in molding the physical nature of life.

Founder stocks from which the modern lines were developed have been preserved as unselected control strains, enabling detailed analysis of the genetic structure of this response, the nature and source of the genetic variation that contributed to the response, and which remains in the population.

Although broiler and layer are fully fertile on artificial insemination, it would be of interest to see if reproductive behavioral separation has begun. One imagines that it would be easy to achieve complete reproductive isolation of broiler and layer stocks by a few generations of disruptive selection for juvenile growth rate in a mixed broiler/layer population. Special feeding arrangements to provide restricted food intake to the broilers and ad libitum feeding for the layers could be easily provided.

6.3   The broiler chicken as a parable of man and civilization

At the start of the process that led to development of the modern broiler chicken, the bird was a healthy, robust animal. It regulated its body weight and composition throughout rearing and maturity under ad libitum feeding; came into egg production at the proper age under signals provided by the natural change in the seasons; maintained a proper social order of males and females, remaining highly fertile under natural mating, throughout a long reproductive period. All of the complex regulation and feedback required for this normal functioning took place under natural signals from the external and internal environment, without the conscious awareness on the part of the scientific and farming community as to what these signals were and what was being regulated. It was only as the animal became progressively changed genetically, and as flock management changed relative to that which had been obtained since domestication that previously normal functions turned out to be highly dependent on appropriate signals from the internal and external environment. In each case, a new signal was uncovered by a breakdown in normal functioning and had to be artificially supplied. In some cases, the artificial environment constructed to make up for the missing signal, itself caused further disruption in normal signaling and functioning. Thus, progressively severe feed restriction may have caused the breakdown in normal photoperiod control; stimulatory lighting may have brought about the recent problems at onset of lay and fertility problems in the second half of lay. The management solutions for these problems may produce their own train of new problems. Thus, as the broiler chicken enters an increasingly artificial environment, more and more of the hidden developmental, physiological and behavioral signaling ceases to function causing a crisis, which must be specifically addressed by yet further changes in management practices and deviation of the farm environment from the historical norm.

This process of alienation from the natural environment and progressive disengagement from historical environmental signaling may hold a warning message for mankind as well. We are in the process of constructing a more and more artificial environment for ourselves and our children. We do not know what unknown and unrecognized signals we customarily received from our historical environment that caused us to develop along acceptable social, psychological, and biological paths. We may only recognize the absence or abnormal functioning of these signals when new social, psychological, and biological pathologies become evident. The current obesity epidemic may be a case in point. So, let us keep at least some oases of historical environment safe for our children; Sabbath and holidays, summer camp and vacation time, when we turn off the paraphernalia of modern life; our computers, TV, videos, compact disc players, and email, and return, if only for a brief period, to the environment that historically nurtured mankind.

6.4   The SIGV hypothesis and episodes of rapid evolutionary change

The hypothesis presented here proposes that the genetic changes resulting from selection, can act as a reiterative internal bootstrap converting previously neutral genetic variation to additive genetic variation that can provide a substrate for selection response. Thus, this hypothesis implies that the amount of genetic variation already present in a population, and that is potentially accessible for selection, is much greater than the genetic variation that is immediately available for selection. Admittedly, even on the SIGV hypothesis, new mutations would be required at some point to enable the selection process to continue, but this point might be further down the line than customarily believed. Thus, the first sweep of selection might be sufficient to generate enough change to provide the reproductive isolation required for species generation. In addition, the associated pleiotropic effects come into play as new cohorts of genes enter the selection process, although primarily deleterious as in the broiler model, might also include effects that make a positive contribution to fitness and can be seized upon as substrates for selection in their own right. This would engender its own series of associated effects and so on. Thus, the SIGV model implies that biological populations contain enormous potential for multidimensional genetic change under selection, before entering a phase where further progress depends on new mutation. Considering the changes wrought in the modern broiler after only 60 generations of selection, the potential of 1,000 generations, a blink in time on the scale of evolution, would seem virtually unlimited.

7.   REFERENCES

Brake J. 1990. The effect of a 2-hour increase in photoperiod at 18 weeks of age on broiler breeder performance. Poultry Sci. 69, 910-914.

Brody T., Eitan Y., Soller M., Nir I. and Nitsan Z. 1980. Compensatory growth and sexual maturity in broiler females reared under severe food restriction from day of hatching. Br Poultry Sci. 21, 437-446.

Creel L.H., Denzil M., Bridges W.C. and Grimes L.W. 1998. A model to describe and predict post-peak changes in broiler hatchability. J Appl Poultry Res. 7, 85-89.

Darvasi A. and Soller M. 1995. Advanced intercross lines, an experimental population for fine genetic mapping. Genetics 141, 1199-1207.

Deeb N. and Lamont S.J. 2002 Genetic architecture of growth and body composition in unique chicken populations. J Heredity 93, 107-118.

Eitan Y. and Soller M. 1991. Two-way selection for threshold body weight at first egg in broiler strain females. II. Effect of supplemental light on weight and age at first egg. Poultry Sci. 70, 2017-2022.

Eitan Y. and Soller M. 1994. Selection for high and low threshold body weight at first egg in broiler strain females. 4. Photoperiodic drive in the selection lines and in commercial layers and broiler breeders. Poultry Sci. 73, 769-780.

Eitan Y. and Soller M. 2001. Effect of photoperiod and quantitative feed restriction in a broiler strain on onset of lay in females and onset of semen production in males: A genetic hypothesis. Poultry Sci. 80, 1397-1405.

Falconer D.S. 1972. Introduction to Quantitative Genetics. Ronald Press, N.Y.

Havenstein G.B. 2002. Time trends in the broiler industry. World Poultry 18, 20.

Havenstein G.B., Ferket P.R., Scheideler S.E. and Larson B.T. 1994. Growth, livability and feed conversion of 1957 vs. 1991 broilers when fed 'typical" 1957 and 1991 broiler diets. Poultry Sci. 73, 1785-1794.

Havenstein G.B., Ferket P.R., Scheideler S.E. and Rives D.V. 1994. Carcass composition and yield of 1991 vs. 1957 broilers when fed 1957 and 1991 broiler diets. Poultry Sci. 73, 1795-1804.

Meijerhof R. 2002. Incubation by embryo temperature. World Poultry 18, 36-37.

Nir I., Nitsan Z., Dror Y. and Shapira N. 1978 Influence of over feeding on growth, obesity and intestinal tract in young chicks of light and heavy breeds. Br J Nutrition 39, 27-35.

Robinson R.E., Renema R.A., Bouvier L., Feddes J.J.R., Wilson J.L., Newcombe M. and McKay R.I. 1998. Effect of photostimulatatory lighting and feed allocation in female broiler breeders 1. Reproductive development. Can J Anim Sci. 78, 603-613.

Ron M. 1987. Genetic and physiological associations between milk production and fertility traits in rats and dairy cattle. Ph.D. Thesis, The Hebrew University of Jerusalem.

Soller M. and Eitan Y. 1984. Why does selection for live-weight gain increase fat deposition? A model. World's Poultry Sci J. 40, 5-9.

Soller M. and Rappaport S. 1966. Mating behavior, fertility and rate-of-gain in Cornish males. Poultry Sci. 45, 997-1003.

Soller M., Snapir N. and Shindler H. 1965. Heritability of semen characteristics in White Rock roosters and their genetic correlation with rate of gain. Poultry Sci. 44, 1527-1529.

Song J.Z., Soller M. and Genizi A. 1999. The full-sib intercross line (FSIL) design: A QTL mapping design for outcrossing species. Genet Res. 73, 61-73.

Taylor G. 1999. High yield breeds require special incubation. World Poultry 15, 27-29.

van Middlekoop K., van Harn J., Wiers W.U. and van Horne P. 2002. Slower growing broilers pose lower welfare risks. World Poultry 18, 20-21.

Walsh T.J. and Brake J. 1997. The effect of nutrient intake during rearing of broiler breeder females on subsequent fertility. Poultry Sci. 76, 297-305.

PSEUDOGENES ARE NOT JUNK DNA

Evgeniy S. Balakirev[1,2] and Francisco J. Ayala[1]

[1]Department of Ecology and Evolutionary Biology, University of California, Irvine, CA 92697-2525, USA; [2]Institute of Marine Biology, Vladivostok 690041, Russia and Academy of Ecology, Marine Biology, and Biotechnology, Far Eastern State University, Vladivostok 690600, Russia

Abstract: We describe some unexpected features of pseudogenes in diverse organisms that are inconsistent with the traditional view that considers pseudogenes as nonfunctional sequences of genomic DNA ("junk" DNA) not subject to natural selection. Pseudogenes are often evolutionarily conserved and transcriptionally active. Moreover, there is evidence indicating that some pseudogenes engage in regulation of gene expression and generation of genetic diversity. Pseudogene patterns of nucleotide variability evince that not all pseudogene mutations are selectively neutral and, thus, do not all have equal probability of becoming fixed in the population. A pseudogene that has arisen by duplication or retroposition may, at first, not be subject to natural selection, if the source gene remains functional. Therefore, alleles of the pseudogene will accumulate mutations, including disabling mutations, over time. But a mutant allele that incorporates a new function may be favored by natural selection and will have enhanced probability of becoming fixed in the population. Thus, some pseudogenes that have lost the original function may have acquired new ones. A review of the evidence leads to the conclusion that pseudogenes are important components of genomes, representing a repertoire of sequences available for functional evolution and subject to non-neutral evolutionary changes. Pseudogenes might be considered as potogenes (following the terminology of J. Brosius and S.J. Gould, 1992), i.e., DNA sequences with a potentiality for becoming new genes (Balakirev and Ayala, 2003). Furthermore, we conjecture that some pseudogenes along with their parental sequences may constitute sets of indivisible functionally interacting entities (intergenic complexes or "intergenes"), in which all the component elements are required in order to fulfill a collective functional role.

S.P. Wasser (ed.), Evolutionary Theory and Processes: Modern Horizons,
Papers in Honour of Eviatar Nevo
E. S. Balakirev and F. J. Ayala. Pseudogenes Are Not Junk DNA, 177-193.

1.   INTRODUCTION

Traditionally pseudogenes are defined as nonfunctional sequences of genomic DNA, originally derived from functional genes, which exhibit degenerative features such as premature stop codons and frameshift mutations preventing their expression (Jacq et al., 1977; Proudfoot, 1980; Little, 1982; Wilde, 1986). Pseudogenes with introns arise as a result of tandem duplication of genes with a gradual loss of function because one copy of the gene may suffice for the needs of the organism. "Processed" pseudogenes lack introns and arise from reverse transcription of processed mRNA followed by integration into the genome (Vanin, 1985; Weiner et al., 1986). Partially processed and chimeric pseudogenes also have been described in various organisms (see Balakirev and Ayala, 2003).

Pseudogenes are common in vertebrate species (Mighell et al., 2000) but rare in bacteria (Andersson and Andersson, 2001; Lawrence et al., 2001), yeast (Harrison et al., 2002), *Drosophila* (Powell, 1997; Harrison et al., 2003), and polyploid animal species such as carp, goldfish, *Xenopus laevis*, and salmonids (Larhammar and Risinger, 1994). The distribution of pseudogenes is often not uniform within genomes. In humans, chromosomes 21 and 22 exhibit notable excesses of pseudogenes near the centromeres of both chromosomes (Harrison et al., 2002). In *Caenorhabditis elegans* the density of pseudogenes is higher toward the ends of the chromosomes than in the middle, where the density of genes is highest (Harrison et al., 2001). In the yeast *Saccharomyces cerevisiae* there is a highly increased density of pseudogenes at or near the telomeres of the chromosomes (Harrison et al., 2001).

Pseudogenes have been thought to be regions free of selection (Li et al., 1981), which would degenerate due to recurrent deleterious mutations and absence of purifying selection. However, eukaryotic genomes contain many pseudogenes that appear to have avoided strong degeneration (Mighell et al., 2000; Harrison and Gerstein 2002; Balakirev and Ayala, 2003). Moreover, pseudogenes that have been suitably investigated often exhibit functional roles such as gene expression, gene regulation, and generation of genetic diversity. Here, we describe some of these pseudogene features detected in diverse organisms. We conclude that the sharp division between genes and pseudogenes is not appropriate and that the generally accepted definition of pseudogenes as nonfunctional sequences of genomic DNA may not be appropriate. We have proposed that pseudogenes be considered as *poto*genes (following the terminology of Brosius and Gould, 1992), i.e., DNA sequences with a *pot*entiality for becoming new genes (Balakirev and Ayala, 2003). Herein, we propose that pseudogenes along with their parental sequences may, in some cases, represent indivisible functionally interacting entities (intergenic complexes or "intergenes") in which no single part can separately accomplish the final functional role.

2. SEQUENCE CONSERVATION AND NON-NEUTRAL EVOLUTION OF PSEUDOGENES

It is generally assumed that all pseudogene mutations are selectively neutral and have equal probability to become fixed in the population (Kimura, 1980; Li et al., 1981; Gojobori et al., 1982; Li, 1983; Graur and Li, 2000, p. 124). However, the sequence evolution of many pseudogenes has characteristics of non-neutral evolution. For instance, for a majority of *Drosophila* pseudogenes there are clear indications of functional constraints, such as: lower than expected intraspecific variability and interspecific divergence; significant heterogeneity of nucleotide variability and divergence along the sequences; higher rate of substitution at synonymous than nonsynonymous nucleotide positions; conservation of important functional regions; transcriptional activity; and codon bias (Jeffs and Ashburner, 1991; Currie and Sullivan, 1994; Jeffs et al., 1994; Sullivan et al., 1994; Balakirev and Ayala, 1996; Pritchard and Schaeffer 1997; Ramos-Onsins and Aguadé 1998; Balakirev et al., 2003).

Non-neutral patterns of pseudogene evolution have been revealed in many organisms. In chicken, *IglV* and *IghV* pseudogenes show a nonrandom distribution of variability along the complementarity-determining regions (CDR) with a peak of nucleotide and amino acid variability at the 3' end of the pseudogenes. In contrast, the framework regions (FR) appear to have been highly conserved. The chicken *IglV* and *IghV* pseudogenes display the hallmarks of expressed *IgV* genes: selection for open reading frames, diversification of the CDRs, and conservation of the FRs (Rothenfluh et al., 1995). Only a minority of the *IglV* and *IghV* pseudogenes contains more than one crippling mutation and, remarkably, very few are crippled because of stop codons or frameshift mutations. Moreover, the *IglV* and *IghV* pseudogenes contain significantly fewer stop codons generated by point mutation than would be expected under a model of random nucleotide changes. Most nucleotide substitution-generated stop codons are rescued by additional nucleotide changes within the same codon. The majority of insertions and deletions are in frame. Furthermore, the majority of *IgV* pseudogenes have not diverged significantly from functional genes (Rothenfluh et al., 1995). Thus, these chicken pseudogenes exhibit evidence of selection for protein function. It might appear that selection for open reading frames occurs in these genes because they donate DNA sequence during somatic gene conversion. However, an analysis of 217 gene conversion events at the chicken *IglVI* locus (McCormack and Thompson, 1990) reveals that pseudogenes were rarely utilized as gene conversion sequence donors; three of the pseudogenes were never used. Most gene conversion events involve a small segment of the pseudogene (mean size approximately 27 bp; McCormack and Thompson, 1990). An earlier comparison of seven functional mouse V_H genes with nine V_H pseudogenes also led to the conclusion that *IgV* pseudogenes seem to be subject to similar selection pressure as the functional genes (Schiff et al., 1985).

A small number of detrimental alterations is a common feature of pseudogenes; there are many examples of extremely conserved pseudogenes (see below). A single point mutation is responsible by itself for the non-functionality of the human *TCRG-V10* gene (Zhang et al., 1996). Immuno-globulin variable-region pseudogenes have relatively minor defects, most frequently single nucleotide substitutions or the insertion or deletion of single nucleotides leading to stop codons or loss of an initiation codon. The relatively "intact" status of most immunoglobulin pseudogenes has been noted (Givol et al., 1981; Schiff et al., 1985; Blankenstein et al., 1987; Ferguson et al., 1988). An interesting example of "preserved" pseudogene, ψ*PgiC,* has been described in the wildflower genus *Clarkia* (Gottlieb and Ford, 1997). Relative-rate tests show that exon nucleotides have not diverged faster in the pseudogene than in the functional counterpart. Moreover, the ratio of synonymous to replacement substitutions is low in the pseudogene. The observed pattern of ψ*PgiC2* cannot be accounted for by gene conversion from *PgiC1*. There is a possibility that the pseudogene may have been preserved by selection against mutants causing defective PGIC1-PGIC2 heterodimers. Experimental evidence of such deleterious subunit interactions has recently been shown for the Cu/Zn *Sod* gene in *Drosophila* (Phillips et al., 1995).

Many pseudogenes retain 90% or higher sequence similarity with their functional counterparts. The degree of similarity is often not uniform along the sequences. In the human ferrochelatase gene and pseudogene, nucleotide identity is 82-93% in exons 2-11, but only 63% in the 5'-flanking region (Whitcombe et al., 1994). Between a human lactate dehydrogenase gene and its pseudogene, the 5'-flanking region has twice as many differences as the coding and 3'-flanking regions (Sudo et al., 1990). In salmon, there is high conservation (up to 80-90%) between the intron sequences of the growth hormone pseudogene and the functional gene, but much less homology in the coding regions (Kavsan et al., 1994). Similarly, in a phospholipase pseudogene from the Mojave rattlesnake, there is high conservation of the intronic regions but much less in the protein-coding regions (John et al., 1996). On the contrary, in the green alga *Chlamydomonas reinhardtii,* a cystein-rich protein pseudogene and the functional gene are similar within coding regions, but the similarity drops dramatically in the intron and in the

5'- and 3'-noncoding regions (Matters and Goodenough, 1992). The first and second exon sequences show a 91% and 74% similarity, respectively, while the 5'- and 3'-flanking regions are only 51% and 41% similar. Similarity between the intron sequences is even lower, 40%, and is clustered towards the exon-intron boundaries (Matters and Goodenough, 1992).

We have studied a pseudogene within the *β-esterase* gene cluster of *D. melanogaster*. The *β-esterase* gene cluster is on the left arm of chromosome 3 of *D. melanogaster*, at 68F7-69A1 in the cytogenetic map. The cluster comprises two tandemly duplicated genes, first described as *Est-6* and *Est-P* (Collet et al., 1990). The coding regions are 1686 and 1691 bp long, respectively, and consist of two exons (1387 bp and 248 bp) and a small (51 bp in *Est-6* and 56 bp in *Est-P*) intron (Oakeshott et al., 1987). The *Est-6* gene is well characterized (reviews in Oakeshott et al., 1993; Richmond et al., 1990). The gene encodes the major *β*-carboxylesterase (EST-6) that is transferred by *D. melanogaster* males to females in the seminal fluid during copulation (Richmond et al., 1980) and affects the female's consequent behavior and mating proclivity (Gromko et al., 1984). Less information is available for *Est-P*. It was first described as a functional gene, based on several lines of evidence: transcriptional activity, intact splicing sites, no premature termination codons, and presence of initiation and termination codons (Collet et al., 1990). Dumancic et al. (1997) showed that some alleles of *Est-P* produce a catalytically active esterase, corresponding to the previously identified EST-7 isozyme (Healy et al., 1991) and renamed the gene *Est-7*. However, there are indications that *Est-P* might be a pseudogene (and has, accordingly, been designated as ψ*Est-6*; Balakirev and Ayala, 1996).

There are 11 premature stop codons among 28 studied sequences; the number of amino acid replacements is 2.9 times higher in ψ*Est-6* than in *Est-6*, and some of them are drastic; nucleotide polymorphism is 2.1 times higher in ψ*Est-6;* structural entropy analysis reveals significantly lower structural regularity and higher structural divergence for ψ*Est-6*, in accordance with the expectations if it is a pseudogene or nonfunctional gene (Balakirev et al., 2003). However, as noted, the gene can be expressed (Collet et al., 1990) and some alleles of ψ*Est-6* produce a catalytically active esterase (Dumancic et al., 1997), although this is detected in late larvae and adults of *both* sexes, whereas the functional *Est-6* genes transcripts are found in all life stages but predominantly in adult males (Collet et al., 1990; Dumancic et al., 1997), consistent with the significant role of EST-6 in male mating (Richmond et al., 1980; Gromko et al., 1984). Moreover, the rate of synonymous substitutions is higher than the rate of nonsynonymous substitutions and neutrality tests are significant (Balakirev et al., 2003).

3.   PSEUDOGENE TRANSCRIPTION AND EXPRESSION

The original definition of pseudogenes implied that they should be transcriptionally and translationally silent and hence not subject to selection (Li et al., 1981). However, nonfunctional as well as functional transcripts of pseudogenes have been described in many organisms.

We have just noted the case of ψEst-6. The human examples are numerous and include interferon pseudogene $\psi LeIFNE$ (Goeddel et al., 1981; *IFNAP22*, Díaz et al., 1994); glyceraldehyde-3-phosphate dehydrogenase pseudogene $\psi Gapdh$ (Arcari et al., 1984; Tso et al., 1985); glucocerebrosidase pseudogene *psGBA* (Sorge et al., 1990); dopamine D5 pseudogene $\psi DRD5$-1 (Nguyen et al., 1991); dopamine $D_{1\beta}$ pseudogene (Weinshank et al., 1991); complement factor H-related protein 1 pseudogene *H36-2* (Skerka et al., 1991); DNA topoisomerase 1 pseudogene *TOP1* (Zhou et al., 1992); serotonin 5-hydroxytryptamine receptor $\psi 5HT_{1D\alpha}$ (Bard et al., 1995); glutamine synthetase ΨGS (Chakrabarti et al., 1995); neuropeptide Y1-like receptor ψNPY *Y1-like* (Rose et al., 1997); steroid 21-hydrolase pseudogene *CYP21P* or *CYP21A* (Endoh et al., 1998); type I hair keratin $\Psi hHaA$, which is transcribed in humans (Rogers et al., 1998), while its orthologs are also expressed in chimpanzee and gorilla (Winter et al., 2001); tumor repressor $\Psi PTEN$ (Fujii et al., 1990); 5-HT7 receptor $\Psi 5$-*HT7* (Olsen and Schechter, 1999); translationally-controlled tumor protein pseudogene *TPT1*, transcribed in all tissues investigated, but with widely variable expression in different human tissues, while its orthologs are fully expressed in the rabbit (Thiele et al., 2000); and myosin pseudogene *MYO15BP* (Boger et al., 2001). Human pseudogene translation activity has been shown *in vivo* (McCarrey and Thomas, 1987; Bristow et al., 1993) as well as *in vitro* (Misra-Press et al., 1994). The human glucocerebrosidase pseudogene *psGBA* is transcriptionally active *in vivo* and the amount of pseudogene-derived mRNA is sometimes quite comparable to the amount of mRNA derived from the active gene, indicating that pseudogene transcription occurs at a high level (Sorge et al., 1990).

Pseudogene transcripts have been detected in other organisms including the mouse, cow, sheep, and silk moth (Balakirev and Ayala, 2003). Plant examples are the *rps19* pseudogene in *Oenothera berteriana* (Schuster and Brennicke, 1991); the *rps14* pseudogene (Aubert et al., 1992) and myrosinase pseudogene *TGG3* (Zhang et al., 2002) in *Arabidopsis thaliana*; potato ribosomal protein $\psi rps14$ (Quiñones et al., 1996); and liverwort NADH dehydrogenase pseudogene (Takemura et al., 1995). Other examples are erythrocyte binding protein $\Psi EBA165$ in *Plasmodium falciparum* (Triglia et al., 2001) and yeast adenine phosphoribosyltransferase pseudogene *APT2* (Alfonzo et al., 1999).

4. REGULATION OF GENE EXPRESSION

A possible role for pseudogenes in development has been hypothesized (McCarrey and Riggs, 1986). Pseudogenes could be the source of the antisense RNA that hybridizes with the sense RNA from the determinator genes, thereby blocking their expression. Pseudogenes, or portions thereof, may be transcribed from the opposite strand relative to their functional counterparts, which would make them a source of antisense RNA. If the inhibitors act at the protein level, then they should be identifiable, for example, as proteins that modify the electrophoretic migration of determinator proteins. Korneev et al. (1999) have shown that a *nitric oxide synthase* pseudogene (*pseudo-NOS*) and its paralogous functional gene (*nNOS*) are co-expressed in identifiable neurons of the mollusk *Lymnaea stagnalis*. The *pseudo-NOS* transcript includes a region with significant antisense homology to the *nNOS* mRNA. The antisense region of the *pseudo-NOS* RNA specifically suppresses the synthesis of the *nNOS* protein (Korneev et al., 1999). Thus the *pseudo-NOS* transcript acts as an antisense regulator of *nNOS* protein synthesis.

There is additional evidence that pseudogenes may have regulatory roles, particularly for the genes from which they have been derived. Healy et al. (1996) have shown that 3' sequences that lie within the ψ*Est-6* pseudogene transcription unit of *D. melanogaster* contain elements that modulate the expression of *Est-6*, which obviously implies some regulatory function for ψ*Est-6*. The *D. melanogaster Suppressor of Stellate* locus [*Su(Ste)*] is located on the long arm of the *Y* chromosome (Hardy et al., 1984) and contains tandemly repeated pseudogene sequences with premature stop codons and frameshift mutations (Balakireva et al., 1992). Nevertheless, the pseudogenes are transcribed (Kalmykova et al., 1998). The *Ste* gene is homologous to a moderately repeated sequence that maps to the *Su(Ste)* region (Livak, 1984). The *Y* chromosome-located *Su(Ste)* sequences are important in regulating the activity of the *X* chromosome *Ste* locus (Hardy et al., 1981; Livak, 1984, 1990). Particularly, they repress transcription and alter the splicing of transcripts from the X-linked *Stellate* (*Ste*) locus (Livak, 1990; Balakireva et al., 1992). A number of models for suppression have been suggested (Livak, 1990; Danilevskaya et al., 1991; Kalmykova et al., 1998) including the competitive interaction between *Su(Ste)* and *Ste* for the positively acting transcription factor (s) (Livak, 1990; Kalmykova et al., 1998), the *Su(Ste)* antisense transcription (Danilevskaya et al., 1991), and the DNA-protein or protein-protein interaction affecting *Ste* transcription (Kalmykova et al., 1998).

An interesting example of gene/pseudogene cooperation in the human involves *cytokeratin 17* expression. A detailed examination of *cytokeratin* transcription regulation using gene/pseudogene chimeric constructs has identified specific promoter/enhancer elements that are inactive by themselves, but can interact to induce strong transcriptional activity of reporter genes. The process includes the interaction between the proximal region of the inactive *cytokeratin* pseudogene promoter and the distal upstream region of the actively transcribed *cytokeratin* gene. The conclusion is that *cis* elements in the proximal 5'-upstream region of the pseudogene promoter can cooperate with distal enhancer elements of the functional gene to induce strong transcriptional activity in transfected HeLa cells (Troyanovsky and Leube, 1994).

5.   GENERATION OF ANTIBODY DIVERSITY

Pseudogenes may be a source of new genetic variation in the immune response of humans, chicken, rabbit, and other vertebrates (review in Balakirev and Ayala, 2003). Immunoglobulin gene diversity is generated by somatic gene conversion events in which sequences derived from alleles or paralogous genes generate new gene sequences. Gene conversion may also occur between genes and pseudogenes, so that selection may contribute to conserving the sequence and other functional characteristics of pseudogenes. In chicken, the process has been shown to occur in the germ line as well as somatically during antibody diversification (Benatar and Ratcliffe, 1993).

Gene conversion events have been reported in several bacterial and other pathogens as a mechanism for generating sequence diversity in expressed antigenic genes. In the bacterium *Borrelia hermsii* a *vmp* pseudogene serves as a source of antigenic variation during relapsing fever (Restrepo et al., 1994). Bacteria of the genus *Borrelia* generate antigenic diversity of the *vmp/vls* coat proteins through recombination from a tandem array of silent partial pseudogene cassettes into a telomeric expression site on a linear plasmid (Zhang et al., 1997). Antigenic variation is generated in *Anaplasma marginale* (a member of ehrlichial genogroup II) by recombination of pseudogenes into the functional expression site (Brayton et al., 2001, 2002). Gonococci generate pilin variants that allow evasion of the host immune system: a pseudogene partial pilin gene sequence is introduced into the expression locus to form a complete pilin gene (Haas et al., 1992). Pilin phase variation can also occur by deletions between direct repeats internal to the expressed pilin gene or between a pilin gene and an adjacent silent pseudogene (Hill et al., 1990). In *Mycoplasma,* antigenic variation of surface proteins is achieved by recombination between multiple pseudogenes and the expressed *vlhA* gene; the multiple pseudogenes serve as a repertoire for the creation of *vlhA* sequence diversity, probably by site-specific recombination, with recombination between the single complete *vlhA* gene and one of the multiple partial copies, creating new *vlhA* gene variants. Subsequent gene conversion, drawing on the shorter pseudogenes, may introduce further variability into this new variant (Noormohammadi et al., 2000).

Antigenic diversity is generated in *Trypanosoma cruzi* by segmental gene conversion (Allen and Kelly, 2001), as well as by recombination between surface protein pseudogenes (Takle et al., 1992; Taylor et al., 1999). In *Trypanosoma equiperdum* antigenic diversity may be generated by recombination between *vsg* genes and pseudogenes (Thon et al., 1989).

6. OTHER FUNCTIONS

Intensive intergenic gene conversion in the human olfactory receptor (OR) multigene family leads to segment shuffling in the odorant binding site, an evolutionary process reminiscent of somatic combinatorial diversification in the immune system (Sharon et al., 1999). OR pseudogenes may importantly contribute to the generation and maintenance of receptor diversity: OR pseudogenes adopt noncoding functions as CpG islands (Glusman et al., 2000), enhancers (Buettner et al., 1998), and matrix attachment regions (Gimelbrant and McClintock, 1997).

In some instances, the pseudogene function has been restored *in vitro* by mutagenesis (Zhang et al., 1992), transfection (Seiser et al., 1990), or *in vivo* by site-specific (or intermolecular) recombination (Restrepo et al., 1994; Benevolenskaya et al., 1997). Possible functional restoring (by gene conversion mechanism) has been suggested for the ribonuclease pseudogene in artiodactyl evolution (Trabesinger-Ruef et al., 1996). In chlorella virus PBCV-1 a nonfunctional, nontranscribed cytosine DNA methyltransferase pseudogene differs by eight amino acids from a functional counterpart; a single amino acid change or deletion by site-directed point mutagenesis restores the pseudogene activity (Zhang et al., 1992). A similar restoration has been observed in the yeast *Schizosaccharomyces pombe* (Pinarbasi et al., 1996). A strain of *E. coli* deleted for the *lacZ* gene has been shown to give rise to spontaneous lactose-utilizing mutants as a result of the activation of the *ebg* pseudogene (for a review, see Hall, 1990).

7. CONCLUDING OBSERVATIONS

Pseudogenes are defined as nonfunctional, but the examples given challenge the generality of such assumptions. Pseudogenes often are not without purpose: sometimes they are expressed; often they represent sources of regulatory sequences and reservoirs of genetic diversity. These features are in discrepancy with the common opinion that pseudogenes are nonfunctional sequences of genomic DNA ("junk" DNA). An extensive and fast-increasing literature does not justify a sharp division between functional genes and pseudogenes. Some pseudogenes approach the functionality of typical genes, while others seem to have completely degenerated toward nonfunctionality. But pseudogenes often exhibit characteristics that are intermediate between those of fully functional genes and those of virtually degenerated DNA sequences. A good example of a pseudogene with intermediate character-ristics is ψ*Est-6* in the β-*esterase* gene cluster of *D. melanogaster.*

It seems appropriate to conclude that pseudogenes are often important components of genomes as a repertoire of sequences for the evolution of new or different functions. Some pseudogenes interact with their ancestral genes, forming functional complexes ("intergenes") in which each component part cannot separately accomplish the final functional role. Examples of such intergenes that we have cited are the complexes of *Est-6* and ψ*Est-6* (Balakirev and Ayala, 1996; Balakirev et al., 2003) and *Ste* and [*Su(Ste)*] (Hardy et al., 1981; Livak, 1984, 1990) in *Drosophila melanogaster; nNOS* and *pseudo-NOS* in the mollusk *Lymnaea stagnalis* (Korneev et al., 1999); *cytokeratin 17* gene and *cytokeratin 17* pseudogene (Troyanovsky and Leube, 1994) in humans. Pseudogenes often serve as reservoirs for generating genetic diversity.

It is unwarranted to assume that "all mutations occurring in pseudogenes are selectively neutral and become fixed in the population with equal probability" (Graur and Li, 2000, p. 124). Many pseudogenes have been identified in all sorts of organisms on the grounds that they are duplicated genes that exhibit stop codons or other disabling mutations in their DNA sequences, so that they cannot have the full function of the original genes from which they derived. In many of these cases, however, it remains unknown because it has not been investigated, whether the pseudogenes, described only on the basis of DNA sequence, may have acquired regulatory or other functions, or plays a role in generating genetic variability. There seems to be the case that some functionality has been discovered in all or nearly all cases whenever this possibility has been pursued with suitable investigations.

Once a pseudogene appears in a population, presumably in a single genome, by duplication or retroposition, the pseudogene may at first follow neutral population dynamics and become extinct or not. A pseudogene may rapidly become fixed if it is closely linked to a gene experiencing a selective sweep, or slowly increase in frequency by neutral drift. However, a mutant that provides a functional role for a pseudogene may be favored by natural selection and, thus, has a higher probability of becoming fixed than its nonfunctional alleles.

8. ACKNOWLEDGMENTS

We are grateful to V. V. Lobzin, V. R. Chechetkin, G. McVean, D. A. Filatov, J. K. Kelly, J. H. McDonald, J. D. Wall, J. M. Comeron, F. Depaulis, and J. Rozas for useful advice on analyses and for providing computer programs. We thank Elena Balakireva, Andrei Tatarenkov, Victor DeFilippis, Martina Zurovkova, and Carlos Márquez for encouragement and help; and W. M. Fitch, B. Gaut, R. R. Hudson, A. Long, and two anonymous reviewers for detailed and valuable comments. This work is supported by NIH grant GM42397 to F. J. Ayala.

9. REFERENCES

Allen C.L. and Kelly J.M. 2001. *Trypanosoma cruzi*: mucin pseudogenes organized in a tandem array. Exp Parasitol. 97, 173-177.

Alfonzo J.D., Crother T.R., Guetsova M.L., Daignan-Fornier B. and Taylor M.W. 1999. *APT1*, but not *APT2*, codes for a functional adenine phosphoribosyltransferase in *Saccharomyces cerevisiae*. J Bacteriol. 181, 347-352.

Andersson J.O. and Andersson S.G.E. 2001. Pseudogenes, junk DNA, and the dynamics of rickettsia genomes. Mol Biol Evol. 18, 829-839.

Arcari P., Martinelli R. and Salvatore F. 1984. The complete sequence of a full length cDNA for human liver glyceraldehyde-3-phosphate dehydrogenase: evidence for multiple mRNA species. Nucl Acids Res. 12, 9179-9189.

Aubert D., Bisanz-Seyer C. and Herzog M. 1992. Mitochondrial *rps14* is a transcribed and edited pseudogene in *Arabidopsis thaliana*. Plant Mol Biol. 20, 1169-1174.

Balakirev E.S. and Ayala F.J. 1996. Is esterase-P encoded by a cryptic pseudogene in *Drosophila melanogaster*? Genetics 144, 1511-1518.

Balakirev E.S. and Ayala F.J. 2003. Pseudogenes: Are They "Junk" or Functional DNA? Ann Rev Genet. 37, 123-151.

Balakirev E.S., Chechetkin V.R., Lobzin V.V. and Ayala F.J. 2003. DNA polymorphism in the *β-esterase* gene cluster of *Drosophila melanogaster*. Genet. 164, 533-544.

Balakireva M.D., Shevelyov Y.Y., Nurminsky D.I., Livak K.J. and Gvozdev V.A. 1992. Structural organization and diversification of Y-linked sequences comprising *Su(Ste)* genes in *Drosophila melanogaster*. Nucl Acids Res. 20, 3731-3736.

Bard J.A., Nawoschik S.P., O'Dowd B.F., George S.R., Branchek T.A. and Weinshank R.L 1995. The human serotonin 5-hydroxytryptamine$_{1D}$ receptor pseudogene is transcribed. Genetics 153, 295-296.

Benatar T. and Ratcliffe M.J.H. 1993. Polymorphism of the functional immunoglobulin variable region genes in the chicken by exchange of sequence with donor pseudogenes. Eur J Immunol. 23, 2448-2453.

Benevolenskaya E.V., Kogan G.L., Tulin A.V., Philipp D. and Gvozdev V.A. 1997. Segmented gene conversion as a mechanism of correction of 18S rRNA pseudogene located outside of rDNA cluster in *D. melanogaster*. J Mol Evol. 44, 646-651.

Blankenstein T., Bonhomme F. and Krawinkel U. 1987. Evolution of pseudogenes in the immunoglobulin VH-gene family of the mouse. Immunogene. 26, 237-248.

Boger E.T., Sellers J.R. and Friedman T.B. 2001. Human myosin XVBP is a transcribed pseudogene. J Muscle Res Cell Motility 22, 477-483.

Brayton K.A., Knowles D.P., McGuire T.C. and Palmer G.H. 2001. Efficient use of a small genome to generate antigenic diversity in tick-borne ehrlichial pathogens. Proc Natl Acad Sci, USA 98, 4130-4135.

Brayton K.A., Palmer G.H., Lundgren A., Yi J. and Barbet A.F. 2002. Antigenic variation of *Anaplasma marginale msp2* occurs by combinatorial gene conversion. Mol Microbiol. 43, 1151-1159.

Bristow J., Gitelman S.E., Tee M.K., Staels B. and Miller W.L. 1993. Abundant adrenal-specific transcription of the human P450c21A "pseudogene". J Biol Chem. 268, 12919-12924.

Brosius J. and Gould S.J. 1992. On "genomenclature": A comprehensive (and respectful) taxonomy for pseudogenes and other "junk DNA". Proc Natl Acad Sci, USA 89, 10706-10710.

Buettner J.A., Glusman G., Ben-Arie N., Ramos P., Lancet D. and Evans G.A. 1998. Organization and evolution of olfactory receptor genes on human chromosome 11. Genom. 53, 56-68.

Chakrabarti R., McCracken J.B., Chakrabarti D. and Souba W.W. 1995. Detection of a functional promoter/enhancer in an intronless human gene encoding a glutamine synthase-like enzyme. Genetics 153, 163-199.

Collet C., Nielsen K.M., Russell R.J., Karl M., Oakeshott J.G. and Richmond R.C. 1990. Molecular analysis of duplicated esterase genes in *Drosophila melanogaster*. Mol Biol Evol. 7, 9-28.

Currie P.D. and Sullivan D.T. 1994. Structure, expression and duplication of genes which encode phosphoglyceromutase of *Drosophila melanogaster*. Genetics 138, 352-363.

Danilevskaya O.N., Kurenova E.V., Pavlova M.N., Bebehov D.V., Link A.J., Koga A., Vellek A. and Hartl, D.L. 1991. He-T family DNA sequences in the *Y* chromosome of *Drosophila melanogaster* share homology with the *X*-linked *Stellate* genes. Chromosome 100, 118-124.

De Wind N., Dekker M., Berns A., Radman M. and Riele H.T. 1995. Inactivation of the mouse Msh2 gene results in mismatch repair deficiency, methylation tolerance, hyperrecombination and predisposition to cancer. Cell 82, 321-330.

Díaz M., Pomykala H.M., Bohlander S.K., Maltepe E., Malik K., et al. 1994. Structure of the human type-I interferon gene cluster determined from a YAC clone contig. Genom. 22, 540-552.

Dumancic M.M., Oakeshott J.G., Russell R.J. and Healy, M.J. 1997. Characterization of the *EstP* protein in *Drosophila melanogaster* and its conservation in Drosophilids. Biochem Gene. 35, 251-271.

Dvořák J., Luo M.C. and Yang Z.L. 1998. Restriction fragment length polymorphism and divergence in the genomic regions of high and low recombination in self-fertilizing and cross-fertilizing Aegilops species. Genetics 148, 423-434.

Elliott B., Richardson C., Winderbaum J., Nickoloff J.A. and Jasin M. 1998. Gene conversion tracts in mammalian cells from double-strand break repair. Mol Cell Biol. 18, 93-101.

Endoh A., Yang L. and Hornsby P.J. 1998. CYP21 pseudogene transcripts are much less abundant than those from the active gene in normal human adrenocortical cells under various conditions in culture. Mol Cell Endocrin. 137, 13-19.

Ferguson S.E., Rudikoff S. and Osborne B.A. 1988. Interaction and sequence diversity among T15 V_H genes in CBA/J mice. J Exp Med. 168, 1339-1349.

Fujii G.H., Morimoto A.M., Berson A.E. and Bolen J.B. 1990. Transcriptional analysis of the PTEN/MMAC1 pseudogene, ΨPTEN. Oncogene. 18, 1765-1769.

Gimelbrant A.A. and McClintock T.S. 1997. A nuclear matrix attachment region is highly homologous to a conserved domain of olfactory receptors. J Mol Neurosci. 9, 61-63.

Givol D., Zakut R., Effron K., Rechavi G., Ram D. and Cohen J.B. 1981. Diversity of germ-line immunoglobulin VH genes. Nature 292, 426-430.

Glusman G., Sosinsky A., Ben-Asher E., Avidan N., Sonkin D., et al. 2000. Sequence, structure, and evolution of a complete human olfactory receptor gene cluster. Genom. 63, 227-245.

Goeddel D.V., Leung D.W., Dull T.J., Gross M., Lawn R.M., McCandliss R., Seeburg P.H., Ulrich A., Velverton E. and Gray P.W. 1981. The structure of eight distinct cloned human leukocyte interferon cDNAs. Nature (Lond.) 290, 20-26.

Gojobori T., Li W.H. and Graur D. 1982. Patterns of nucleotide substitution in pseudogenes and functional genes. J Mol Evol. 18, 360-369.

Gottlieb L.D. and Ford V.S. 1997. A recently silenced, duplicate *PgiC* locus in *Clarkia*. Mol Biol Evol. 14, 125-132.

Graur D. and Li W.H. 2000. Fundamentals of Molecular Evolution. 2nd (Ed.), Sinauer, Sunderland, Mass.

Gromko M.H., Gilbert D.F. and Richmond R.C. 1984. Sperm transfer and use in the multiple mating system of *Drosophila*. In: Smith R.L. (Ed.), Sperm Competition and the Evolution of Animal Mating Systems, pp. 371-426 Academic Press, New York.

Haas R., Veit S. and Meyer T.F. 1992. Silent pilin genes of *Neisseria gonorrhoeae* MS11 and the occurrence of related hypervariant sequences among other gonococcal isolates. Mol Microbiol. 6, 197-208.

Hall B.G. 1990. Directed evolution of a bacterial operon. BioEssays 12, 551-558.

Hardy R.W., Tokuyasu K.T. and Lindsley D.L. 1981. Analysis of spermatogenesis in *Drosophila melanogaster* bearing deletions for *Y* chromosome fertility genes. Chromosoma 83, 593-617.

Hardy R.W., Lindsley D.L., Livak K.J., Lewis B., Siversten A.L., Joslyn G.L., Edwards J. and Bonaccorsi S. 1984. Cytogenetic analysis of a segment of the *Y* chromosome of *Drosophila melanogaster*. Genetics 107, 591-610.

Harrison P.M. and Gerstein M. 2002 Studying genomes through the aeons: protein families, pseudogenes and proteome evolution. J Mol Biol. 318, 1155-1174.

Harrison P.M., Echols N. and Gerstein M.B. 2001. Digging for dead genes: an analysis of the characteristics of the pseudogene population in the *Caenorhabditis elegans* genome. Nucl Acids Res. 29, 818-830.

Harrison P.M., Kumar A., Lan N., Echols N., Snyder M. and Gerstein M. 2002. A small reservoir of disabled ORFs in the yeast genome and its implications for the dynamics of proteome evolution. J Mol Biol. 316, 409-419.

Harrison P.M., Milburn D., Zhang Z., Bertone P. and Gerstein M. 2003, Identification of pseudogenes in the *Drosophila melanogaster* genome. Nucl Acids Res. 31, 1033-1037.

Healy M.J., Dumancic M.M. and Oakeshott J.G. 1991. Biochemical and physiological studies of soluble esterases from *Drosophila melanogaster*. Biochem Gene. 29, 365-388.

Healy M.J., Dumancic M.M., Cao A. and Oakeshott J.G. 1996. Localization of sequences regulating ancestral and acquired sites of esterase 6 activity in *Drosophila melanogaster*. Mol Biol Evol. 13, 784-797.

Hill S.A., Morrison S.G. and Swanson J. 1990. The role of direct oligonucleotide repeats in gonococcal pilin gene variation. Mol Microbiol. 4, 1341-1352.

Jacq C., Miller J.R. and Brownlee G.G. 1977. A pseudogene structure in 5S DNA of *Xenopus laevis*. Cell 12, 109-120.

Jeffs P.S. and Ashburner M. 1991. Processed pseudogenes in *Drosophila*. Proc R Soc Lond B. 244, 151-159.

Jeffs P.S., Holmes E.C. and Ashburner M. 1994. The molecular evolution of the Alcohol dehydrogenase and alcohol dehydrogenase-related genes in the *Drosophila melanogaster* species subgroup. Mol Biol Evol. 11, 287-304.

John T.R., Smith J.J. and Kaiser I.I. 1996. A phospholipase A$_2$-like pseudogene retaining the highly conserved introns of Mojave toxin and other snake venom group II PLA$_2$s, but having different exons. DNA and Cell Biol. 15, 661-668.

Kalmykova A.I., Dobritsa A.A. and Gvozdev V.A. 1998. *Su(Ste)* diverged tandem repeats in a *Y* chromosome of *Drosophila melanogaster* are transcribed and variously processed. Genetics 148, 243-249.

Kavsan V.M., Koval A.P. and Palamarchuk A.J. 1994. A growth hormone pseudogene in the salmon genome. Genetics 141, 301-302.

Kimura M. 1980. A simple method for estimating evolutionary rate of base substitutions through comparative studies of nucleotide sequences. J Mol Evol. 16, 111-120.

Korneev S.A., Park J.H. and O'Shea M. 1999. Neuronal expression of neural nitric oxide synthase (nNOS) protein is suppressed by an antisense RNA transcribed from an NOS pseudogene. J Neurosci. 19, 7711-7720.

Larhammar D. and Risinger C. 1994. Why so few pseudogenes in tetraploid species? Trends Gene. 10, 418-419.

Lawrence J.G., Hendrix R.W. and Casjens S. 2001. Where are the pseudogenes in bacterial genomes? Trends Microbiol. 9, 535-540.

Li W.H. 1983. Evolution of duplicate genes and pseudogenes. In Nei M and Koehn RK (eds) Evolution of Genes and Proteins, pp. 14-37, Sinauer Associates, Sunderland, MA.

Li W.H., Gojobori T. and Nei M. 1981. Pseudogenes as a paradigm of neutral evolution. Nature 292, 237-239.

Little P.F. 1982. Globin pseudogenes. Cell 28, 683-684.

Livak K.J. 1984. Organization and mapping of a sequence on the *Drosophila melanogaster X* and *Y* chromosomes that is transcribed during spermatogenesis. Genetics 107, 611-634.

Livak K.J. 1990. Detailed structure of the *Drosophila melanogaster* Stellate genes and their transcripts. Genetics 124, 303-316.

Matters G.L. and Goodenough U.W. 1992. A gene/pseudogene tandem duplication encodes a cysteine-rich protein expressed during zygote development in *Chlamydomonas reinhardtii*. Mol Gen Gene. 232, 81-88.

McCarrey J.R. and Riggs A.D. 1986. Determinator-inhibitor pairs as a mechanism for threshold setting in development: a possible function for pseudogenes. Proc Natl Acad Sci, USA 83, 679-683.9+

McCarrey J.R. and Thomas K. 1987. Human testis-specific PGK gene lacks introns and possesses characteristics of a processed gene. Nature (Lond.) 326, 501-505.

McCormack W.H. and Thompson C.B. 1990. *IgL* variable gene conversion display pseudogene donor preference and 5' to 3' polarity. Genes Dev. 4, 548-558.

Mighell A.J., Smith N.R., Robinson P.A. and Markham A.F. 2000. Vertebrate pseudogenes. FEBS Letters 468, 109-114.

Misra-Press A., Cooke N.E. and Liebhaber S.A. 1994. Complex alternative splicing partially inactivates the human chorionic somatomammotropin-like (*hCS-L*) gene. J Biol Chem. 269, 23220-23229.

Nguyen T., Sunahara R., Marchese A., Van Tol H.H.M., Seeman P. and O'Dowd B.F. 1991. Transcription of a human dopamine pseudogene. Biochem Biophys Res Commun. 181, 16-21.

Noormohammadi A.H., Markham P.F., Kanci A., Whithear K.G. and Browning G.F. 2000. A novel mechanism for control of antigenic variation in the haemagglutinin gene family of *Mycoplasma synoviae*. Mol Microbiol. 35, 911-923.

Oakeshott J.G., Collet C., Phillis R., Nielsen K.M., Russell R.J., Chambers G.K., Ross V. and Richmond R.C. 1987. Molecular cloning and characterization of esterase 6, a serine hydrolase from *Drosophila*. Proc Natl Acad Sci, USA 84, 3359-3363.

Oakeshott J.G., van Papenrecht E.A., Boyce T.M., Healy M.J. and Russell R.J. 1993. Evolutionary genetics of *Drosophila* esterases. Genetics 90, 239-268.

Olsen M.A. and Schechter L.E. 1999. Cloning, mRNA localization and evolutionary conservation of a human 5-HT7 receptor pseudogene. Genetics 227, 63-69.

Phillips J.P., Tainer J.A., Getzoff E.D., Boulianne G.L., Kirby K. and Hilliker A.J. 1995. Subunit-destabilizing mutations in *Drosophila* copper/zinc superoxide dismutase: Neuropathology and a model of dimmer disequilibrium. Proc Natl Acad Sci USA 92, 8574-8578.

Pinarbasi E., Elliott J. and Hornby D.P. 1996. Activation of a yeast pseudo DNA methyltransferase by deletion of a single amino acid. J Mol Biol. 257, 804-813.

Powell J.R. 1997. Progress and Prospects in Evolutionary Biology. The *Drosophila* Model. Oxford University Press, Oxford and New York.

Pritchard J.K. and Schaeffer S.W. 1997. Polymorphism and divergence at a *Drosophila* pseudogene locus. Genetics 147, 199-208.

Proudfoot N. 1980. Pseudogenes. Nature 286, 840-841.

Quiñones V., Zanlungo S., Moenne A. and Gómez I. 1996. The *rpl5-rps14-cob* gene arrangement in *Solanum tuberosum*: *rps14* is a transcribed and unedited pseudogene. Plant Mol Biol. 31, 937-943.

Ramos-Onsins S. and Aguadé M. 1998. Molecular evolution of the *Cecropin* multigene family in *Drosophila*: Functional genes *vs.* pseudogenes. Genetics 150, 157-171.

Restrepo B.I., Carter C.J. and Barbour A.G. 1994. Activation of a *vmp* pseudogene in *Borrelia hermsii*: an alternate mechanism of antigenic variation during relapsing fever. Mol Microbiol. 13, 287-299.

Richmond R.C., Gilbert D.G., Sheehan K.B., Gromko M.H. and Butterworth F.M. 1980. Esterase 6 and reproduction in *Drosophila melanogaster*. Science 207, 1483-1485.

Richmond R.C., Nielsen K.M., Brady J.P. and Snella E.M. 1990. Physiology, biochemistry and molecular biology of the *Est-6* locus in *Drosophila melanogaster*. In: Barker JSF, Starmer W.T. and MacIntyre R.J. (Eds.), Ecological and Evolutionary Genetics of *Drosophila* pp. 273-292, Plenum Press, New York.

Rogers M.A., Winter H., Wolf C., Heck M. and Schweizer J. 1998. Characterization of a 190-kilobase pair domain of human type I hair keratin genes. J Biol Chem. 273, 26683-26691.

Rose P.M., Lynch J.S., Frazier S.T., Fisher S.M., Chung W., et al. 1997. Molecular genetic analysis of a human neuropeptide Y receptor. The human homolog of the murine "Y5" receptor may be a pseudogene. J Biol Chem. 272, 3622-3627.

Rothenfluh H.S., Blanden R.V. and Steele E.J. 1995. Evolution of V genes: DNA sequence structure of functional germline genes and pseudogenes. Immunogene. 42, 159-171.

Schiff C., Milili M. and Fougereau M. 1985. Functional and pseudogenes are similarly organized and may equally contribute to the extensive antibody diversity of the *IgVHII* family. EMBO J. 4, 1225-1230.

Schuster W. and Brennicke A. 1991. RNA editing makes mistakes in plant mitochondria: editing loses sense in transcripts of a *rps19* pseudogene and in creating stop codons in *coxI* and *rps3* mRNAs of *Oenothera*. Nucl Acids Res. 19, 6923-6928.

Seiser C., Beck G. and Wintersberger E. 1990. The processed pseudogene of mouse thymidine kinase is active after transfection. FEBS Letters 270, 123-126.

Sharon D., Glusman G., Pilpel Y., Khen M., Gruetzner F., Haaf T., et al. 1999. Primate evolution of an olfactory receptor cluster: Diversification by gene conversion and recent emergence of pseudogenes. Genom. 61, 24-36.

Skerka C., Horstmann R.D. and Zipfel P.F. 1991. Molecular cloning of a human serum protein structurally related to complement factor H. J Biol Chem. 266, 12015-12020.

Sorge J., Gross E., West C. and Beutler E. 1990. High level transcription of the glucocerebrosidase pseudogene in normal subjects and patients with Gaucher disease. J Clin Invest. 86, 1137-1141.

Sudo K., Maekawa M., Luedemann M.M., Deaven L.L., Li S.S.L. 1990. Human lactate dehydrogenase-B processed pseudogene: nucleotide sequence analysis and assignment to the X-chromosome. Biochem Biophys Res Com. 171, 67-74.

Sullivan D.T., Starmer W.T., Curtiss S.W., Menotti-Raymond M. and Yum J. 1994. Unusual molecular evolution of an *Adh* pseudogene in *Drosophila*. Mol Biol Evol. 11, 443-458.

Takemura M., Nozato N., Oda K., Kobayashi Y., Fukuzawa H. and Ohyama K. 1995. Active transcription of the pseudogene for subunit 7 of the NADH dehydrogenase in *Marchantia polymorpha* mitochondria. Mol Gen Gene. 247, 565-570.

Takle G.B., O'Connor J., Young A.J. and Cross G.A. 1992. Sequence homology and absence of mRNA defines a possible pseudogene member of the *Trypanosoma cruzi* gp85/sialidase multigene family. Mol Biochem Parasitol. 56, 117-128.

Taylor M.C., Muhia D.K., Baker D.A., Mondragon A., Schaap P. and Kelly J.M. 1999. *Trypanosoma cruzi* adenylyl cyclase is encoded by a complex multigene family. Mol Biochem Parasitol. 104, 205-217.

Thiele H., Berger M., Skalweit A. and Thiele B.J. 2000. Expression of the gene and processed pseudogenes encoding the human and rabbit translationally controlled tumour protein (TCTP). Eur J Biochem. 267, 5473-5481.

Thon G., Baltz T. and Eisen H. 1989. Antigenic diversity by the recombination of pseudogenes. Genes Develop. 3, 1247-1254.

Trabesinger-Ruef N., Jermann T., Zankel T., Durrant B., Frank G. and Brenner S.A. 1996. Pseudogenes in ribonuclease evolution: a source of new biomacromolecular function? FEBS Letters 382, 319-322.

Triglia T., Thompson J.K. and Cowman A.F. 2001. An EBA175 homologue which is transcribed but not translated in erythrocytic stages of *Plasmodium falciparum*. Mol Biochem Parasitol. 116, 55-63.

Troyanovsky S.M. and Leube R.E. 1994. Activation of the silent human cytokeratin 17 pseudogene-promoter region by cryptic enhancer elements of the cytokeratin 17 gene. Eur J Biochem. 223, 61-69.

Tso J., Sun X.H., Kao T.H., Reece K. and Wu R. 1985. Isolation and characterization of rat and human glyceraldehyde-3-phosphate dehydrogenase cDNAs: genomic complexity and molecular evolution of the gene. Nucl Acids Res. 13, 2485-2502.

Vanin E. 1985. Processed pseudogenes: characteristics and evolution. Ann Rev Gene. 19, 253-272.

Weiner A.M., Deininger P.L. and Efstratiadis A. 1986. Nonviral retroposons: gene, pseudogenes, and transposable elements generated by the reverse flow of genetic information. Ann Rev Biochem. 55, 631-661.

Weinshank R.L., Adham N., Macchi M., Olsen M.A., Branchek T.A. and Hartig P.R. 1991. Molecular cloning and characterization of a high affinity dopamine receptor (D1$_\beta$) and its pseudogene. J Biol Chem. 266, 22427-22435.

Wilde C.D. 1986. Pseudogenes. CRC Crit Rev Biochem. 19, 323-352.

Whitcombe D.M., Albertson D.G. and Cox T.M. 1994. Molecular analysis of functional and nonfunctional genes for human ferrochelatase: isolation and characterization of a FECH pseudogene and its sublocalization on chromosome 3. Genom. 20, 482-486.

Winter H., Langbein L., Krawczak M., Cooper D.N., Jave-Suarez L.F., Rogers M.A., Praetzel S., Heidt P.J. and Schweizer J. 2001. Human type I hair keratin pseudogene *ψhHaA* has functional orthologs in the chimpanzee and gorilla: evidence for recent inactivation of the human gene after the *Pan-Homo* divergence. Hum Genet. 108, 37-42.

Zhang J.R., Hardham J.M., Barbour A.G. and Norris S.J. 1997. Antigenic variation in lyme disease Borrelia by promiscuous recombination of VMP-like sequence cassettes. Cell 89, 275-285.

Zhang J., Pontoppidan B., Xue J., Rask L. and Meijer J. 2002. The third myrosinase gene TGG3 in *Arabidopsis thaliana* is a pseudogene specifically expressed in stamen and petal. Physiol Plantarum. 115, 25-34.

Zhang X.M., Cathala G., Soua Z., Lefranc M.P. and Huck S. 1996. The human T-cell receptor gamma variable pseudogene *V10* is a distinctive marker of human speciation. Immunogene. 43, 196-203.

Zhang Y., Nelson M. and Van Etten J.L. 1992. A single amino acid change restores DNA cytosine methyltransferase activity in a cloned chlorella virus pseudogene. Nucl Acids Res. 20, 1637-1642.

Zhou B.S., Beidler D.R. and Cheng Y.C. 1992. Identification of antisense RNA transcripts from a human DNA topoisomerase I pseudogene. Cancer Res. 52, 4280-4285.

GENOME SIGNATURE COMPARISONS OF THE PROTEOBACTERIA

Andrew J. Gentles and Samuel Karlin
Department of Mathematics, Stanford University, Stanford, CA 94305, USA

Abstract: We present a comparative analysis of proteobacteria based on genome signature assessments of oligonucleotide relative abundances. Although α- and γ-proteobacteria form largely coherent groups in terms of dinucleotide relative abundance distance, there are notable exceptions. The obligate intracellular parasites *Rickettsia prowazekii* and *R. conorii* are extremely distant from all other α-proteobacteria. *Brucella melitensis* and *B. suis* are weakly or distantly similar to most other α-proteobacteria, but moderately similar to the nitrogen-fixing Rhizobiales *Sinorhizobium meliloti* and *Mesorhizobium loti*. Among the γ-proteobacteria, *Pseudomonas* and *Buchnera* sp. stand out, being significantly distant from the rest of the class. The β-proteobacteria *Neisseria meningitidis* and *Ralstonia solanacearum* also are closer to other proteobacteria than to each other. ε-proteobacteria, represented by *Helicobacter pylori* and *Campylobacter jejuni* are moderately similar to each other, but distant from other proteobacteria. We also examined individual dinucleotide and tetranucleotide relative abundances. GpC is overrepresented in many species, except *Pseudomonas* and some α-proteobacteria, while ApT is broadly over-represented in the α-proteobacteria. TpA dinucleotides are widely suppressed. Among the tetranucleotides, CTAG is prominently suppressed in many species, while ATAG/CTAT (a one nucleotide change from CTAG) is elevated in most α-proteobacteria and in *R. solanacearum*. We discuss possible interpretations of these observations in terms of genome-wide processes.

S.P. Wasser (ed.), Evolutionary Theory and Processes: Modern Horizons,
Papers in Honour of Eviatar Nevo
A.J. Gentles and S. Karlin. Genome Signature Comparisons of the Proteobacteria, 195-206.
© 2004 *Kluwer Academic Publishers.*

1.   INTRODUCTION

Dinucleotide relative abundances in DNA sequences can be evaluated using the dinucleotide odds ratio $\rho_{XY} = f_{XY}/f_X f_Y$ where f_X, f_Y, and f_{XY} are the frequencies of the nucleotides X and Y, and the dinucleotide XY, respectively. A symmetrized version ρ^*_{XY} is calculated for double-stranded DNA, from the frequencies of X, Y, and XY computed over the sequence concatenated with its inverted complement. Dinucleotide relative abundances (ρ^*-1) assess the contrast between the observed dinucleotide frequencies and the expectation from the component mononucleotide frequencies. We have previously demonstrated that the vector of independent dinucleotide biases $\{\rho^*_{XY}\}$, evaluated for disjoint 50 kb DNA sequence windows from the same species, are essentially constant across each genome and are more similar to each other than to 50 kb windows from other species (Gentles and Karlin, 2001). On this basis, the profile $\{\rho^*_{XY}\}$ is referred to as a genome signature, and it is an intrinsic characteristic of an organism's DNA pervading both coding and noncoding DNA (Karlin and Mrázek, 1996, 1997). The remarkable stability of dinucleotide relative abundances across a given genome had previously been noted in early biochemical experiments measuring nearest-neighbor frequencies, and termed "general designs" (Russell et al., 1973, 1976; Russell and Subak-Sharpe, 1977).

Corresponding expressions can be constructed for higher-order oligonucleotide measures. Specifically, for tetranucleotides the relative abundance measure factoring out lower-order biases is (Karlin and Cardon, 1994)

$$\tau^*_{XYZW} = \frac{f^*_{XYZW} f^*_{XY} f^*_{XNZ} f^*_{XN_1N_2W} f^*_{YZ} f^*_{YNW} f^*_{ZW}}{f^*_{XYZ} f^*_{XYNW} f^*_{XNZW} f^*_{YZW} f^*_X f^*_Y f^*_Z f^*_W},$$

where X, Y, Z, W each take values from $\{A,T,C,G\}$, and f^*_{XNZ} and $f^*_{XN_1N_2W}$ are respectively the frequencies of all trinucleotides XNZ and tetranucleotides XN1N2W.

A useful measure of the difference between two sequences p and q is the dinucleotide absolute relative abundance distance $\delta^*(p,q) = (1/16)\sum_{XY}|\rho^*_{XY}(p) - \rho^*_{XY}(q)|$, with the summations extending over all dinucleotides XY. Usually, δ^* is multiplied by 1000 and averaged over all pairwise comparisons of 50 kb contigs in the sequences p and q. With this definition, the following levels of δ^* similarity can be distinguished: $\delta^* \leq 50$ is "close" (e.g., generally within genomes, human vs. cow); $55 \leq \delta^* \leq 85$ is "moderately similar" (e.g., human vs. mouse, *Escherichia coli* vs. *Haemophilus influenzae*); $90 \leq \delta^* \leq 110$ is "weakly similar" (e.g., human vs. sea urchin); $120 \leq \delta^* \leq 150$ is "distant"(e.g., *Arabidopsis thaliana* vs. mouse, *Methanococcus janaschii* vs. *Methanobacterium thermoautotrophicum*); $160 \leq \delta^* \leq 195$ is "very distant" (e.g., human vs. *Caenorhabditis elegans, E. coli* vs. *Helicobacter pylori*); and $\delta^* > 200$ is "extremely distant" (e.g., human vs. *Leishmania major*, human vs. *E. coli*). Genome signature anomalies have been used in the identification of pathogenicity islands and laterally transferred segments in *H. pylori, Vibrio cholerae,* and *M. tuberculosis* (Karlin, 2001).

The proteobacteria (purple bacteria and relatives) are the second largest class of bacteria, comprising around 32% of known bacterial species. All are gram-negative and are generally subclassified on the basis of 16S rRNA comparisons into the subgroups α-, β-, γ-, δ-, and ε-proteobacteria. The diversity among proteobacteria is high, ranging from sulphur- and iron-oxidizing bacteria to nitrogen-fixing plant symbionts, and bioluminescent bacteria. Understandably, sequencing efforts have typically focused on significant microbes, which impact directly on human health, or have economic consequences.

The α-proteobacteria are highly variable in terms of their lifestyles and habitats. One classification scheme subdivides them into three subgroups on the basis of comparisons between RecA sequences: *A1 (Caulobacter, Rhizobium, Brucella, Agrobacterium), A2 (Rhodobacter),* and *A3 (Rickettsia, Ehrlichia)* (Karlin et al., 1995). *C. crescentus* is a non-pathogenic aquatic species which is of great interest in bioremediation, because of its capability to convert toxic metals including mercury, copper, cadmium, and cobalt into less soluble forms (Nierman et al., 2001). *B. melitensis* is a facultative intracellular pathogen of mammals, which causes spontaneous abortion in goats and sheep, and *B. suis* raises concerns as a threat in bioterrorism; while *A. tumefaciens* is a plant pathogen causing gall tumours and has been used extensively in the production of transgenic plants (Goodner et al., 2001; DelVecchio et al., 2002; Paulsen et al., 2002). *S. meliloti* and *M. loti* are both nitrogen-fixing plant symbionts which, however, display considerable differences in their gene content and organization, e.g., many genes found on the pSymA and pSymB megaplasmids of *Sinorhizobium* are distributed around the *Mesorhizobium* genome (Kaneko et al., 2000; Galibert et al., 2001).

Among the β-proteobacteria, complete genome sequences are available for *Neisseria meningitidis* MC58, responsible for meningitis and septicemia in humans and for the virulent plant pathogen *Ralstonia solanacearum* (Tettelin et al., 2000; Salanoubat et al., 2002).

Numerous γ-proteobacterial genomes have been sequenced including the human pathogens *Escherichia coli* O157, *Salmonella typhimurium, Haemophilus influenzae, Vibrio cholerae,* the plague bacterium *Yersinia pestis,* and the opportunistic *Pseudomonas aeruginosa* and *P. putida* (Fleischmann et al., 1995; Heidelberg et al., 2000; Stover et al., 2000; Hayashi et al., 2001; May et al., 2001; Parkhill et al., 2001; Nelson et al., 2002). Other γ-proteobacterial species we discuss are the endocellular aphid symbiont *Buchnera aphidicola*, and *Shewanella oneidensis* (Shigenobu et al., 2000; Heidelberg et al., 2002). *S. oneidensis* is another bacterium with potential use in bioremediation because of its ability to reduce nitrates and soluble chromium and uranium to insoluble oxides (Heidelberg et al., 2002).

Finally, ε-proteobacteria are represented in the following comparisons by the gastric ulcer-producing *Helicobacter pylori*, and the related *Campylobacter jejuni*, which is the major causative agent of food-related diarrhoea worldwide (Tomb et al., 1997; Parkhill et al., 2000). There are currently no complete δ-proteobacterial genomes available; however, there is substantial sequence data for the soil bacterium *Myxococcus xanthus*, which has one of the largest bacterial genomes known (9.6 Mb). *Myxococcus* is a bacterial predator which ingests prey bacteria through swarming behavior and secretion of digestive enzymes.

2.   DINUCLEOTIDE RELATIVE ABUNDANCE DISTANCES

δ^* distances between the proteobacterial genomes are shown in Figure 1. Within-species distances are appropriately small, in the range δ^* ~23-43. Comparing across different species, the α-proteobacteria *C. crescentus, S. meliloti, A. tumefaciens*, and *M. loti* form a coherent group, with δ^* of 23-31 within species (closely similar), and δ^* around 50-90 between species (moderately similar). *B. melitensis* and *B. suis* are closely similar to each other (δ^*=28), and moderately similar to *A. tumefaciens* and to *M. loti*, with which they share considerable synteny (Paulsen et al., 2002); but only qualify as weakly or distantly similar to the other α-proteobacteria with δ^* distances of 81-124. Both *Rickettsia*, while moderately similar to each other in signature, are very distant or extremely distant from any other α-proteobacteria, with most δ^* values exceeding 200. Intriguingly, the *Rickettsia* genomes are much closer in signature to the aphid endocellular symbiont *Buchnera* sp. (δ^*~97-104), and to the metal-reducing bacterium *S. oneidensis* (δ^*=100), than they are to other α-proteobacteria. Could the nature of the host interactions of *Rickettsia* (which is transmitted to humans from lice causing typhus fever), and *Buchnera* (aphid symbiont) contribute to the similarity in their genome signature? The anoxygenic photosynthetic nitrogen-fixing bacterium *Rhodobacter capsulatus* is moderately similar to other α-proteobacteria, except *Brucella* and *Rickettsia*. The obligate intracellular *Ehrlichia chaffeensis* is distant or very distant from most other proteobacteria. However, it is weakly similar to *R. prowazekii* (δ^*=95) and *Buchnera* sp., with δ^*=118.

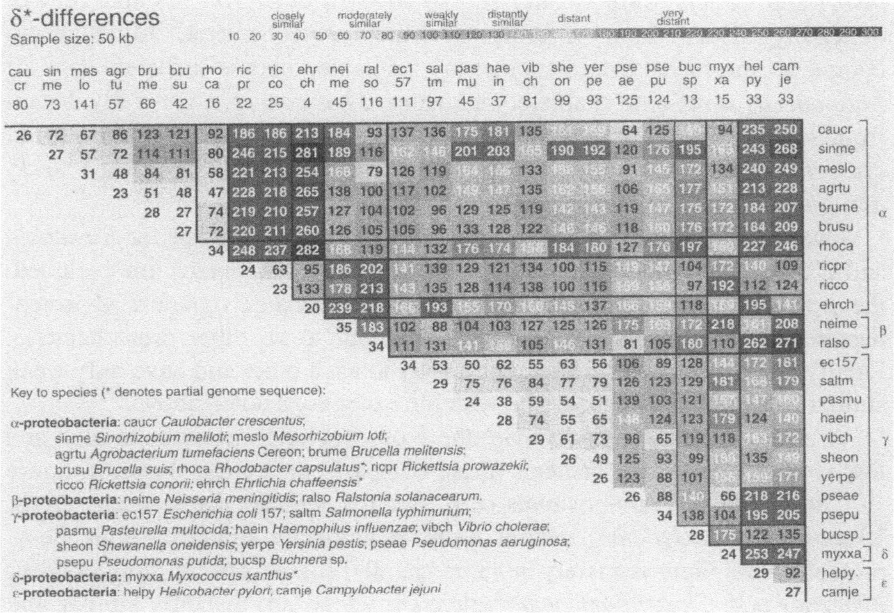
δ*-differences
Sample size: 50 kb
closely similar
moderately similar
weakly similar
distantly similar
distant
very distant
cau cr 80
sin me 73
mes lo 141
agr tu 57
bru me 66
bru su 42
rho ca 16
ric pr 22
ric co 25
ehr ch 4
nei me 45
ral so 116
ec1 57 111
sal tm 97
pas mu 45
hae in 37
vib ch 81
she on 99
yer pe 93
pse ae 125
pse pu 124
buc sp 13
myx xa 15
hel py 33
cam je 33
caucr sinme meslo agrtu brume brusu rhoca ricpr ricco ehrch α
neime ralso β
ec157 saltm pasmu haein vibch sheon yerpe pseae psepu bucsp γ
myxxa δ
helpy camje ε
Key to species (* denotes partial genome sequence):
α-proteobacteria: caucr Caulobacter crescentus;
sinme Sinorhizobium meliloti; meslo Mesorhizobium loti;
agrtu Agrobacterium tumefaciens Cereon; brume Brucella melitensis;
brusu Brucella suis; rhoca Rhodobacter capsulatus*; ricpr Rickettsia prowazekii;
ricco Rickettsia conorii; ehrch Ehrlichia chaffeensis*
β-proteobacteria: neime Neisseria meningitidis; ralso Ralstonia solanacearum.
γ-proteobacteria: ec157 Escherichia coli 157; saltm Salmonella typhimurium;
pasmu Pasteurella multocida; haein Haemophilus influenzae; vibch Vibrio cholerae;
sheon Shewanella oneidensis; yerpe Yersinia pestis; pseae Pseudomonas aeruginosa;
psepu Pseudomonas putida; bucsp Buchnera sp.
δ-proteobacteria: myxxa Myxococcus xanthus*
ε-proteobacteria: helpy Helicobacter pylori; camje Campylobacter jejuni

Figure 1. δ^*-distances between complete proteobacterial genomes.

The β-proteobacteria *N. meningitidis* and *R. solanacearum* are very distant from each other (δ^*~182) and from the ε-proteobacteria, and distant from most of the γ-proteobacteria (but both are closer to the γ-proteobacteria than they are to each other). *N. meningitidis* is also very distant from all α-proteobacteria; however, *R. solanacearum* is moderately similar (δ^*~79) to *M. loti* and weakly similar to the other α-proteobacteria. It further qualifies as moderately similar to the γ-proteobacterium *P. aeruginosa*, but is only weakly similar to *P. putida*. *P. aeruginosa* and *P. putida* are only weakly similar to each other.

With the exceptions of *Pseudomonas* sp. and *Buchnera* sp., the γ-proteobacteria form a group which are closely or moderately similar to each other, with between-species δ^* distances of 38-79. *Pseudomonas* and *Buchnera* are distant both from each other and from the other γ-proteobacteria. *P. aeruginosa* is moderately similar to the α-proteobacteria *C. crescentus* and *S. meliloti*, and is also moderately similar to *R. solanacearum*. In contrast, *Buchnera* is distant from all other γ-proteobacteria except *S. oneidensis* and *Y. pestis* (with which it shows weak similarity), and very distant from all α- and β-proteobacteria, except for a weak similarity to the two *Rickettsias*. The remaining γ-proteobacteria are at most weakly similar to the α-proteobacteria, and are generally distant or very distant from them. The two complete ε-proteobacterial genomes, *H. pylori* and *C. jejuni* are weakly similar to each other (δ^*=92) and very or extremely distant from virtually all other proteobacteria. Although no complete δ-proteobacterial genomes are available, there is sufficient data for *Myxococcus xanthus* to enable comparisons. *Myxococcus* is generally distant from α-, β-, and γ-proteobacteria (δ^*≥118), but is weakly similar to *C. crescentus*, and *R. solanacearum*, and moderately similar (δ^*=66) to *P. aeruginosa*.

Overall, the α-proteobacteria are grouped together, except for *Rickettsia* and *Ehrlichia*. Similarly, if *Pseudomonas* spp. and *Buchnera* are excluded, the γ-proteobacteria form a coherent group in terms of signature. ε-proteobacteria are weakly similar to each other, but to no other proteobacteria. Finally, the β-proteobacteria are dissimilar to each other and have only weak similarity to other proteobacteria, with the exceptions noted above.

A useful contrast is to compare the proteobacterial genomes to high and low G+C gram-positive bacteria (these comparisons are not shown in figure 1). The low G+C gram-positives (e.g., *Bacillus subtilis, Lactococcus lactis, Ureaplasma urealyticum*) exhibit weak to moderate similarity with the γ-proteobacteria (approximately in the range 60-120), except for *Streptococcus pyogenes* and *Clostridium acetobutylicum*, which are distantly similar and distant, respectively. *C. acetobutylicum* is distant from most proteobacteria with the exceptions of the two *Rickettsia, Buchnera*, and *Ehrlichia* (δ^*=92, 112, 86, 68, respectively). It is interesting that *Buchnera* is weakly similar to all complete low G+C gram-positive genomes. α-proteobacteria are distant from low G+C gram-positive bacteria, except for some weak similarities with *Brucella, Rickettsia*, and *Ehrlichia*. There are no close relations between any proteobacteria and the high G+C gram-positive bacteria, except that *M. xanthus* is weakly similar to *Streptomyces coelicolor*.

3.   DINUCLEOTIDE AND TETRANUCLEOTIDE RELATIVE ABUNDANCE EXTREMES

Tables 1 and 2 display symmetrized dinucleotide (ρ^*) and tetranucleotide (τ^*) relative abundances for cases which show significant overrepresentation (>1.23) or underrepresentation (<0.78). Among the dinucleotides, GC is strongly over represented ($1.24 \leq \rho^*_{GC} \leq 1.75$) in all β-, γ-, and ε-proteobacteria except *Pseudomonas* (ρ^*=1.17, 1.21 for *P. aeruginosa, P. putida*); but is normal in the α-proteobacteria except for *Brucella* and *Rickettsia* (ρ^*_{GC}=1.30-1.53), which, as noted previously, are also separated from the other α-proteobacteria by their δ^*-distances. TA is strongly suppressed ($\rho^*_{TA} \leq 0.53$), and AT is overrepresented in α-proteobacteria other than the *Rickettsia*, where both are in the normal range. TA is strongly suppressed also in the β-proteobacteria ($\rho^*_{TA} \leq 0.64$), although here AT is over-represented in *Ralstonia* ($\rho^*_{TA} = 1.38$), but not *Neisseria* (1.04). The pattern among the γ-proteobacteria is not so clear: AT is normal in all cases, but TA is normal in some (*Shewanella, Salmonella, Yersinia, Buchnera*) and marginally suppressed in the remainder, although the underrepresentation is not as pronounced as among the α-proteobacteria.

Table 1. Overrepresented and underrepresented dinucleotide and tetranucleotide relative abundances in the proteobacteria.

	CG	GC	TA	AT	CC GG	AA TT	AC GT	AG CT	GA TC
caucr	1.16	1.11	<u>0.45</u>	**1.29**	0.85	1.09	0.86	0.96	1.22
sinme	**1.29**	1.15	<u>0.47</u>	**1.39**	0.82	1.18	<u>0.77</u>	0.89	**1.28**
meslo	1.23	1.21	<u>0.44</u>	**1.41**	0.81	1.17	0.79	0.87	1.18
agrtu	**1.24**	1.21	<u>0.47</u>	**1.37**	0.86	**1.26**	<u>0.75</u>	0.82	1.14
brume	1.20	**1.30**	<u>0.52</u>	**1.36**	0.88	**1.30**	<u>0.70</u>	0.81	1.06
brusu	<u>1.21</u>	**1.30**	<u>0.52</u>	**1.36**	0.88	1.30	<u>0.70</u>	0.81	1.06
ricpr	<u>0.77</u>	**1.53**	0.98	0.98	1.03	1.05	0.86	1.06	0.91
ricco	1.03	**1.51**	0.99	0.97	1.09	1.11	0.81	1.03	0.90
neime	**1.31**	**1.28**	<u>0.64</u>	1.04	0.97	**1.45**	0.84	<u>0.70</u>	0.90
ralso	1.20	**1.24**	<u>0.46</u>	**1.38**	<u>0.77</u>	1.07	0.90	0.88	1.06
ec157	1.14	**1.27**	<u>0.76</u>	1.10	0.92	1.20	0.88	0.82	0.93
saltm	**1.24**	**1.31**	0.84	1.16	0.91	1.22	0.84	0.81	0.92
pasmu	1.07	**1.30**	<u>0.76</u>	0.97	1.01	1.23	0.90	0.81	0.89
haein	1.09	**1.43**	<u>0.75</u>	0.95	1.01	**1.25**	0.85	0.82	0.87
vibch	1.04	**1.30**	<u>0.69</u>	0.99	0.88	1.20	0.89	0.90	0.95
shepu	1.00	**1.36**	0.84	1.03	0.97	1.18	0.85	0.91	0.86
yerpe	0.99	**1.24**	0.83	1.12	1.03	1.15	0.88	0.84	0.88
pseae	1.10	1.17	<u>0.54</u>	1.17	0.84	1.08	0.86	1.02	1.10
psepu	0.98	1.21	<u>0.56</u>	1.09	0.89	1.13	0.91	0.96	0.93
bucsp	0.87	**1.25**	0.85	0.95	1.22	1.14	0.81	0.94	1.02
helpy	0.93	**1.56**	<u>0.73</u>	0.86	1.17	**1.37**	<u>0.67</u>	0.97	0.87
camje	<u>0.62</u>	**1.75**	<u>0.77</u>	0.84	1.11	**1.25**	<u>0.71</u>	1.09	0.92

Overrepresentation corresponds to $\rho^* > 1.23$ (bold-face) and underrepresentation $\rho^* < 0.78$ (underlined).

Table 2. Selected tetranucleotide (τ^*) relative abundances showing under- and over-representation in proteobacteria.

	CCGG	GCGC	TCGA	ACGT	GGCC	CGCG	TTAA	ATAT	GTAC	TATA	GATC	CATG	CTAG	ATAG CTAT
caucr	0.90	1.01	1.06	0.90	0.99	0.96	**1.44**	0.66	0.91	0.89	0.99	1.01	0.35	**1.65**
sinme	1.02	1.00	0.95	0.95	0.99	0.97	1.04	0.81	0.88	0.99	1.03	1.00	0.28	**1.43**
meslo	1.02	1.02	0.97	0.99	1.06	0.98	1.20	0.79	0.82	0.94	1.02	0.97	0.34	**1.43**
agrtu	1.02	0.98	0.96	0.98	1.04	0.96	0.92	0.87	0.83	1.01	1.05	0.98	0.39	**1.35**
brume	0.98	0.95	0.96	0.92	1.01	0.98	0.83	0.86	1.00	1.02	1.03	0.96	0.35	**1.31**
brusu	0.98	0.95	0.96	0.92	1.01	0.98	0.83	0.85	1.00	1.02	1.03	0.96	0.34	**1.32**
ricpr	0.91	0.83	0.90	1.03	0.70	0.94	0.97	1.00	0.98	0.95	0.94	0.76	0.86	1.02
ricco	0.95	0.31	0.80	0.99	0.48	0.79	0.97	1.01	1.00	0.95	0.84	0.71	0.92	1.01
neime	0.81	0.81	0.84	0.80	0.50	0.89	0.99	0.94	0.88	0.92	0.43	0.61	0.84	1.12
ralso	0.98	1.01	0.95	0.97	1.05	0.97	**1.33**	0.88	0.89	0.96	1.02	0.92	0.27	**1.42**
ec157	0.87	0.91	1.07	0.95	0.86	1.00	0.94	1.00	1.03	0.85	0.93	1.00	0.28	1.12
saltm	0.88	0.92	0.97	0.95	0.95	1.02	0.93	0.98	1.06	0.84	0.94	0.98	0.21	1.17
pasmu	0.93	0.85	0.90	1.03	0.28	0.90	0.87	1.00	0.94	0.72	1.02	0.89	0.71	0.89
haein	0.37	0.62	0.78	0.91	0.50	0.70	0.85	1.00	0.78	0.71	0.87	0.43	0.63	0.92
vibch	0.94	0.94	1.00	0.98	1.01	1.00	0.90	0.98	1.08	0.72	0.91	0.95	0.85	0.99
shepu	0.70	0.95	1.04	0.98	0.99	0.89	0.91	1.02	1.05	0.69	0.86	0.92	0.77	1.01
yerpe	0.90	0.96	1.05	1.00	1.03	0.94	0.89	0.95	1.01	0.80	0.95	0.96	0.61	1.10
pseae	0.91	1.01	0.91	1.01	1.04	0.99	**1.38**	0.89	0.84	0.94	1.02	1.00	0.23	**1.31**
psepu	0.89	1.02	0.93	0.96	1.06	0.98	0.98	0.94	0.92	0.88	1.00	0.98	0.44	**1.30**
bucsp	1.18	1.00	0.94	0.99	1.04	0.92	1.01	1.07	0.93	0.96	0.90	0.89	0.82	0.95
helpy	**1.43**	0.68	0.12	0.16	0.67	0.88	0.92	0.93	0.16	0.84	0.89	0.93	0.88	0.96
camje	0.42	0.53	0.80	0.33	0.59	0.92	0.98	0.92	0.51	0.92	0.78	0.62	0.82	1.08

CTAG is also strongly suppressed, with corresponding overrepresentation of ATAG, in the α-proteobacterium R. capsulatus ($\tau^*_{CTAG}=0.22$, $\tau^*_{ATAG}=1.59$) and the δ-proteobacterium M. xanthus (0.33, 1.46).

Among symmetric tetranucleotides (palindromes), CTAG is extremely suppressed ($0.28 \leq \tau^*_{CTAG} \leq 0.54$) in all α-proteobacteria (including the nearly complete *R. capsulatus* genome), except *Rickettsia* ($\tau^*=0.86$ in *R. prowazekii*, $\tau^*=0.92$ in *R. conorii*) and *Ehrlichia* ($\tau^*=0.89$), and in all γ-proteobacteria, except *Vibrio* and *Buchnera* ($\tau^*=0.85$ and 0.82, respectively). On the other hand, CTAG is normal in both ε-proteobacteria; while in the β-proteobacteria, CTAG is very underrepresented in *Ralstonia*, with $\tau^*=0.27$, but normal in *Neisseria*, with $\tau^*=0.84$. ATAG/CTAT, which differs from CTAG by a single substitution, is overrepresented in all the α-proteobacteria ($1.24 \leq \tau^* \leq 1.65$), again with the exception of the *Rickettsias* ($\tau^*=1.02$, 1.01). In the δ-proteobacterium *M. xanthus*, CTAG is strongly suppressed with $\tau^*=0.33$, with ATAG correspondingly high at $\tau^*=1.46$. It is also high in

Ralstonia and *Pseudomonas*, which have τ^*_{ATAG}=1.42 and 1.31, but is in the normal range in the other proteobacteria. There is a strong negative correlation between CTAG and ATAG relative abundances, with $r(\tau^*_{CTAG}, \tau^*_{ATAG/CTAT}) = -0.77$, suggesting that ATAG may be high specifically due to avoidance of CTAG. In comparison, tetranucleotide relative abundances are in the normal range in most high and low G+C gram-positive bacteria. The sole exceptions are *M. tuberculosis* and *S. coelicolor*, which have (τ^*_{CTAG}=0.62, τ^*_{ATAG} =1.20) and (τ^*_{CTAG}=0.25, τ^*_{ATAG} =1.62), respectively.

The γ-proteobacterium *H. influenzae* and the ε-proteobacteria *H. pylori* and *C. jejuni* are distinguished by the number of tetranucleotides for which they show under- or overrepresentation (7 in *Haemophilus*, 6 in *Campylobacter*, and 5 in *Helicobacter*). CCGG is suppressed in *Haemophilus* and *Campylobacter*, but overrepresented in *Helicobacter*. GCGC, GGCC, and GTAC are suppressed in all three genomes, with *Helicobacter* having an extremely low τ^*_{GTAC}=0.16. *H. pylori* also shows the most highly suppressed tetranucleotide, TCGA, with τ^*=0.12; TCGA is normal in all other proteobacterial genomes. Overall, tetranucleotides are more commonly underrepresented than elevated, probably due to restriction system avoidance. The only exceptions are: ATAG/CTAT as indicated in connection with CTAG; TTAA in *C. crescentus, R. solanacearum,* and *P. aeruginosa* (but not *P. putida*); and CCGG in *H. pylori*.

4. DISCUSSION

Both *R. prowazekii* and *R. conorii* are obligate intracellular parasites, which are classified as α-proteobacteria on the basis of 16S RNA comparisons (Andersson et al., 1998). Yet, it is clear from Figure 1 that the *Rickettsia* species deviate strongly from the other α-proteobacteria in terms of genome signature, with the δ^* distance identifying them as "very" to "extremely distant". The obligate parasitic nature of *Rickettsia* may be significant in their distance from other α-proteobacteria; the isolation of such organisms in their hosts has led to the suggestion that genetic exchange may be especially low among obligate intracellular parasites (Read et al., 2000). *Brucella* is a facultative intracellular parasite, and it appears to spend little time outside its host, even though it is able to survive independently (Tsolis, 2002).

The wide host range plant pathogen *Ralstonia* is unexpectedly close in signature (δ^* distance) to *Pseudomonas* and to *Mesorhizobium*. All share the common feature that they are capable of living in soil. *Pseudomonas* is ubiquitous in aquatic environments and in soil, while the Rhizobiales are found in close association with plants. While *Caulobacter* is most often found in dilute aquatic, often nutrient poor habitats, it is also found in soil.

Many factors may influence DNA dinucleotide relative abundances including dinucleotide stacking energies, DNA packaging, and DNA repair/replication processes. Many DNA repair enzymes function by recognizing the shape of lesions in DNA, rather than specific primary sequences, and the efficiency of repair processes is impacted by the context provided by neighbouring base-pairs (Echols and Goodman, 1991). Stacking energies, charge interactions, and conformational tendencies also influence local DNA structure affecting DNA curvature.

One explanation for the suppression of CTAG tetranucleotides has been proposed on the basis of a model of biased DNA repair (Bhagwat and McClelland, 1992; Burge et al., 1992). In this model, biases in the activity of the VSP (Very Short Patch) DNA repair system in *E. coli* lead to depletion of CTAG tetranucleotides. On the other hand, the model also predicts an enrichment of CCAG tetranucleotides, which we do not observe in any of the genomes; and in fact, CCAG is depressed in *R. conorii* and *C. jejuni*. Another possible explanation for CTAG underrepresentation is that it fulfils a specific functional role in many bacterial genomes, and, therefore, is avoided in contexts where it is not performing this function. In the context of DNA coiling/supercoiling, it has been observed in crystallographic studies of complexes between TrpR-DNA and MetJ-DNA that kinks are formed in the *E. coli* DNA at CTAG sites, multiple copies of which are involved in *trp* and *metJ* repressor control binding sites. These tetranucleotides might, therefore, be undesirable from the point of structural stability at other locations in the genome. Furthermore, it has been demonstrated experimentally that (CTAG). (CTAG) promotes cross-linking between complementary DNA strands by UV irradiation, at a much greater rate than any other tetra-nucleotides (Nejedlây et al., 2001). Since such cross-linking is considered generally to have deleterious effects, avoidance of CTAG motifs would be a natural consequence. CTAG is also recognized as a restriction site by at least one type II endonuclease, *XspI* in *Xanthomonas* sp., which cleaves between C and T (Song et al., 1998). There is therefore a variety of reasons why CTAG might be avoided in bacterial genomes. ATAG is strongly over represented in most genomes in which CTAG is suppressed; in fact, the correlation coefficient over all proteobacterial genomes between τ^{*}_{CTAG} and τ^{*}_{ATAG} is -0.77. In certain contexts, ATAG is recognized as being relatively unstable and prone to forming non-B type DNA, which would suggest it should be under represented, not elevated (Slebos et al., 2002). Avoidance of restriction systems is a likely reason for many tetranucleotide under representations. For example, *H. pylori* has the restriction enzymes *HpaII* (CCGG), *Hfp2* (CATG), *HaeIII* (GGCC), *HhaI/Hind*p1 (GCGC), and *Hin*1056I (CGCG).

5. REFERENCES

Andersson S.G., Zomorodipour A., Andersson J.O., Sicheritz-Pontâen T., Alsmark U.C., Podowski R.M., et al. 1998. The genome sequence of *Rickettsia prowazekii* and the origin of mitochondria. Nature 396(6707), 133-140.

Bhagwat A.S. and McClelland M. 1992. DNA mismatch correction by Very Short Patch repair may have altered the abundance of oligonucleotides in the *E. coli* genome. Nucl Acids Res. 20(7), 1663-1668.

Burge C., Campbell A.M. and Karlin S. 1992. Over- and underrepresentation of short oligonucleotides in DNA sequences. Proc Natl Acad Sci USA 89(4), 1358-1362.

DelVecchio V.G., Kapatral V., Redkar R.J., Patra G., Mujer C., Los T., et al. 2002. The genome sequence of the facultative intracellular pathogen *Brucella melitensis*. Proc Natl Acad Sci USA 99(1), 443-448.

Echols H. and Goodman M.F. 1991. Fidelity mechanisms in DNA replication. Ann Rev Biochem. 60.

Fleischmann R.D., Adams M.D., White O., Clayton R.A., Kirkness E.F., Kerlavage A.R., et al. 1995. Whole-genome random sequencing and assembly of *Haemophilus influenzae* Rd. Science 269(5223), 496-512.

Galibert F., Finan T.M., Long S.R., Puhler A., Abola P., Ampe F., et al. 2001. The composite genome of the legume symbiont *Sinorhizobium meliloti*. Science 293(5530), 668-672.

Gentles A.J. and Karlin S. 2001. Genome-scale compositional comparisons in eukaryotes. Genome Res. 11(4), 540-546.

Goodner B., Hinkle G., Gattung S., Miller N., Blanchard M., Qurollo B., et al. 2001. Genome sequence of the plant pathogen and biotechnology agent *Agrobacterium tumefaciens* C58. Science 294(5550), 2323-2328.

Hayashi T., Makino K., Ohnishi M., Kurokawa K., Ishii K., Yokoyama K., et al. 2001. Complete genome sequence of enterohemorrhagic *Escherichia coli* O157:H7 and genomic comparison with a laboratory strain K-12. DNA Res. 8(1), 11-22.

Heidelberg J.F., Eisen J.A., Nelson W.C., Clayton R.A., Gwinn M.L., Dodson R.J., et al. 2000. DNA sequence of both chromosomes of the cholera pathogen *Vibrio cholerae*. Nature 406(6795), 477-483.

Heidelberg J.F., Paulsen I.T., Nelson K.E., Gaidos E.J., Nelson W.C., Read T.D., et al. 2002. Genome sequence of the dissimilatory metal ion-reducing bacterium *Shewanella oneidensis*. Nat Biotechnol. 20(11), 1118-1123.

Kaneko T., Nakamura Y., Sato S., Asamizu E., Kato T., Sasamoto S., et al. 2000. Complete genome structure of the nitrogen-fixing symbiotic bacterium *Mesorhizobium loti*. DNA Res. 7(6), 331-338.

Karlin S. 2001. Detecting anomalous gene clusters and pathogenicity islands in diverse bacterial genomes. Trends Microbiol. 9(7), 335-343.

Karlin S. and Cardon L.R. 1994. Computational DNA sequence analysis. Ann Rev Microbiol. 48.

Karlin S. and Mrázek J. 1996. What drives codon choices in human genes? J Mol Biol. 262(4), 459-472.

Karlin S. and Mrázek J. 1997. Compositional differences within and between eukaryotic genomes. Proc Natl Acad Sci, USA 94(19), 10227-10232.

Karlin S., Weinstock G.M. and Brendel V. 1995. Bacterial classifications derived from recA protein sequence comparisons. J Bacteriol. 177(23), 6881-6893.

May B.J., Zhang Q., Li L.L., Paustian M.L., Whittam T.S. and Kapur V. 2001. Complete genomic sequence of *Pasteurella multocida*, Pm70. Proc Natl Acad Sci, USA 98(6), 3460-3465.

Nejedlây K., Kittner R., Pospâisilovâa S. and Kypr J. 2001. Cross-linking of the complementary strands of DNA by UV light: dependence on the oligonucleotide composition of the UV irradiated DNA. Biochim Biophys Acta. 1517(3), 365-375.

Nelson K.E., Weinel C., Paulsen I.T., Dodson R.J., Hilbert H., Martins dos Santos V.A., et al. 2002. Complete genome sequence and comparative analysis of the metabolically versatile *Pseudomonas putida* KT2440. Environ Microbiol. 4(12), 799-808.

Nierman W.C., Feldblyum T.V., Laub M.T., Paulsen I.T., Nelson K.E., Eisen J.A., et al. 2001. Complete genome sequence of *Caulobacter crescentus*. Proc Natl Acad Sci, USA 98(7), 4136-4141.

Parkhill J., Wren B.W., Mungall K., Ketley J.M., Churcher C., Basham D., et al. 2000. The genome sequence of the food-borne pathogen *Campylobacter jejuni* reveals hypervariable sequences. Nature 403(6770), 665-668.

Parkhill J., Wren B.W., Thomson N.R., Titball R.W., Holden M.T., Prentice M.B., et al. 2001. Genome sequence of *Yersinia pestis*, the causative agent of plague. Nature 413(6855), 523-527.

Paulsen I.T., Seshadri R., Nelson K.E., Eisen J.A., Heidelberg J.F., Read T.D., et al. 2002. The *Brucella* suis genome reveals fundamental similarities between animal and plant pathogens and symbionts. Proc Natl Acad Sci, USA 99(20), 13148-13153.

Read T.D., Brunham R.C., Shen C., Gill S.R., Heidelberg J.F., White O., et al. 2000. Genome sequences of *Chlamydia trachomatis* MoPn and *Chlamydia pneumoniae* AR39. Nucl Acids Res. 28(6), 1397-1406.

Russell G.J., McGeoch D.J., Elton R.A. and Subak-Sharpe J.H. 1973. Doublet frequency analysis of bacterial DNAs. J Mol Evol. 2(4), 277-292.

Russell G.J. and Subak-Sharpe J.H. 1977. Similarity of the general designs of protochordates and invertebrates. Nature 266(5602), 533-536.

Russell G.J., Walker P.M., Elton R.A. and Subak-Sharpe J.H. 1976. Doublet frequency analysis of fractionated vertebrate nuclear DNA. J Mol Biol. 108(1), 1-23.

Salanoubat M., Genin S., Artiguenave F., Gouzy J., Mangenot S., Arlat M., et al. 2002. Genome sequence of the plant pathogen *Ralstonia solanacearum*. Nature 415(6871), 497-502.

Shigenobu S., Watanabe H., Hattori M., Sakaki Y. and Ishikawa H. 2000. Genome sequence of the endocellular bacterial symbiont of aphids *Buchnera* sp. APS. Nature 407(6800), 81-86.

Slebos R.J., Oh D.S., Umbach D.M. and Taylor J.A. 2002. Mutations in tetranucleotide repeats following DNA damage depend on repeat sequence and carcinogenic agent. Cancer Res. 62(21), 6052-6060.

Song T., Sik Kang B. and Min Kim Y. 1998. XspI, a new type II restriction endonuclease from a *Xanthomonas* species. Mol Cells 8(3), 370-373.

Stover C.K., Pham X.Q., Erwin A.L., Mizoguchi S.D., Warrener P., Hickey M.J., et al. 2000. Complete genome sequence of *Pseudomonas aeruginosa* PAO1, an opportunistic pathogen. Nature 406(6799), 959-964.

Tettelin H., Saunders N.J., Heidelberg J., Jeffries A.C., Nelson K.E., Eisen J.A., et al. 2000. Complete genome sequence of *Neisseria meningitidis* serogroup B strain MC58. Science 287(5459), 1809-1815.

Tomb J.F., White O., Kerlavage A.R., Clayton R.A., Sutton G.G., Fleischmann R.D., et al. 1997. The complete genome sequence of the gastric pathogen *Helicobacter pylori*. Nature 388(6642), 539-547.

Tsolis R.M. 2002. Comparative genome analysis of the alpha-proteobacteria: Relationships between plant and animal pathogens and host specificity. Proc Natl Acad Sci USA 99(20), 12503-12505.

Part Three

Phylogeography and Phylogeny

A MAXIMUM LIKELIHOOD FRAMEWORK FOR CROSS VALIDATION OF PHYLOGEOGRAPHIC HYPOTHESES

Alan R. Templeton

Department of Biology, Washington University, St. Louis, Missouri 63130-4899, USA

Abstract: Phylogeographic inferences are frequently drawn from the analysis of an evolutionary haplotype tree of a single DNA region or type of DNA, such as mitochondrial DNA. However, a phylogeographic event is only detectable in principle if an appropriate mutation or mutations occurred in the right time and place in the evolutionary history of the genetic variation being screened. In addition, like all forms of statistical inference, phylogeographic inference is subject to both false positives and false negatives. One way of both increasing the resolution of phylogeographic analyses and reducing the error rates is to survey multiple DNA regions and cross validate phylogeographic inferences across these DNA regions. A formal hypothesis testing framework based upon likelihood ratio tests is developed for performing such cross validation. This framework is used to test hypotheses about recent human evolution based upon ten different DNA regions. This same framework can also be used to test interspecific phylogeographic hypotheses of a single historical event influencing more than one species simultaneously. This interspecific use is illustrated by testing the hypothesis that both humans and malaria spread out of Africa together. The added power and reduced error of cross-validated analyses indicate that phylogeographic studies should be based upon multiple loci or DNA regions.

S.P. Wasser (ed.), Evolutionary Theory and Processes: Modern Horizons,
            Papers in Honour of Eviatar Nevo
A.R. Templeton. A Maximum Likelihood Framework for Cross Validation of Phylogeographic Hypotheses, 209-230.
© 2004 *Kluwer Academic Publishers.*

1.   INTRODUCTION

Intraspecific phylogeography refers to a species' evolutionary history over both space and time, including past fragmentation events that subdivided the species into isolates, range expansions, or colonization events that increased the geographical range of the species and/or brought formerly isolated populations into contact once again and patterns of genetic interchange over space and time. The era of modern phylogeographic studies based on evolutionary trees of DNA regions began with the pioneering work of Avise et al. (1979) on mitochondrial DNA (mtDNA) variation in the mouse genus *Peromyscus*. In this and subsequent studies, genetic variation in mtDNA was scored, and the resulting haplotypes were used to estimate an evolutionary tree that shows how mutations accumulated in the DNA lineages over time to produce the current array of sampled haplotypes. The haplotype trees in turn were overlaid upon the geographical locations of the samples to make phylogeographic inferences.

Although visual overlays of haplotype trees upon geography can be suggestive of phylogeographic events or processes, such overlays do not constitute a formal estimation or hypothesis testing framework. There is no determination of whether or not enough individuals and geographical sites have been sampled to ensure that any observed phylogeographic patterns could not have arisen by chance alone. Even when one accepts the phylogeographic patterns as real and not as artifacts of inadequate sampling, there is no formal, explicit interpretative framework for making biological conclusions from the observed phylogeographic patterns. Templeton et al. (1995) addressed these inadequacies through the development of nested clade phylogeographic analysis (NCPA).

NCPA first defines a series of hierarchically nested clades (branches within branches) from the haplotype tree using a set of explicit nesting rules (Templeton et al., 1987; Templeton and Sing, 1993). Haplotypes themselves are the lowest level of nesting, and these in turn are nested together into 1-step clades, which in turn are nested into 2-step clades, and so on until a nesting level is reached such that the next higher-nesting level would result in only a single clade spanning the entire original haplotype network. Nested haplotype trees contain much temporal information. When rooted, the oldest clade is known in any given nested category. Even if the haplotype tree were unrooted, coalescent theory predicts that clades on the tips of the tree are highly likely to be younger than interior clades to which the tips are connected (Castelloe and Templeton, 1994). In this manner, turning a haplotype tree into a series of nested clades captures much information about relative temporal orderings that is used to analyze the spread of haplotypes and clades through space and time. NCPA next quantifies the spatial distribution of haplotypes and clades. The geographical data are quantified in two main fashions (Templeton et al., 1995). The first is the clade distance, D_c, which measures how widespread the clade is spatially. The clade distance is determined by calculating the average latitude and longitude for all observations of the clade in the sample, weighted by the local frequencies of the clade at each location. This estimates the geographical center for the clade. Next, the great circle distance from a location containing one or more members of the clade to the geographical center is calculated, and these distances are averaged over all locations containing the clade of interest, once again weighted by the frequency of the clade in the local sample. Sometimes geographical distance is not the most appropriate measure of space. For example, suppose a sample is taken of a riparian fish species. Because rivers do not flow in straight lines and because the fish are confined in their movements to the river, the geographical distance between two sample sites in the river is not relevant to the fish; rather, the important distance in this case is the distance between the two points going only along the river. In cases such as these, the investigator should define the distances between any two sample points in the most biologically relevant fashion, and the clade distance is now calculated as the average pairwise distance between all observations of the clade, once again weighted by local frequencies.

The second measure is the nested clade distance, D_n, that quantifies how far away a haplotype or clade is located from the center of its nesting clade. For geographical distance, the geographical center is calculated for all individuals bearing members not only of the clade of interest, but also bearing any other clades that are nested with the clade of interest at the next higher level of nesting. This is the geographical center of the nesting clade. The nested clade distance is then calculated as the average distance that an individual bearing a haplotype from the clade of interest lies from the geographical center of the nesting clade. Once again, all averages are weighted by local frequencies. When the investigator defines the distances between sample locations, the nested clade distance is the average pairwise distance between an individual bearing a haplotype from the clade of interest to individuals bearing any haplotype from the nesting clade that contains the clade of interest.

Because of sampling artifacts, it is dangerous to make biological inferences from a visual overlay of geography upon a haplotype tree or from just the observed values of quantitative distance measurements. To adjust for sampling, the nested clade analysis quantifies the degree of confidence in the distance measures by testing the null hypothesis that the haplotypes or clades nested within a higher-level nesting clade show no geographical associations. This null hypothesis is tested by randomly permuting the observations within a nesting clade across geographical locations in a manner that preserves the overall clade frequencies and sample sizes per locality (Templeton et al., 1995). After each random permutation, the clade and nested clade distances can be recalculated. Repeating these a thousand or more times simulates the distribution of these distances under the null hypothesis of no geographical associations for a fixed frequency. The observed clade and nested clade distances are then contrasted to this null distribution to determine which distances are statistically significant. Statistical power can be enhanced by averaging the clade and nested clade distances for all tips pooled together and subtracting the tip average from the corresponding average for the older interiors. The average interior-tip difference captures the temporal contrast of old versus young within a nesting clade.

Statistical significance is not the same as biological significance. Statistical significance tells us that we can be confident that geographical associations exist with the haplotype tree. However, statistical significance alone does not tell us how to interpret those geographical associations. Indeed, no single test statistic discriminates between recurrent gene flow, past fragmentation, and past range expansion in the nested clade analysis. Rather, it is a pattern formed from several statistics that allows discrimination. Also, many different patterns can sometimes lead to the same biological conclusion, and sometimes a statistically significant pattern has ambiguous biological interpretations because of inadequate geographical sampling. Finally, NCPA searches out multiple, overlaying patterns within the same data set. In light of these complexities (which reflect the reality of evolutionary possibilities and sampling constraints), an inference key was provided as an appendix to Templeton et al. (1995), with the latest version being available at http://bioag.byu.edu/zoology/crandall_lab/geodis.htm along with the program GEODIS for implementing the nested clade analysis.

Although the nested clade approach to phylogeographic inference has many strengths, inference is limited by 1) sample size and sample sites, 2) insufficient genetic resolution to detect an event or process that actually occurred, and 3) false inferences arising from the evolutionary stochasticisty of the coalescent process itself or by the haplotype tree being skewed or otherwise altered by natural selection. Another limitation, the failure to make use of recombination, is addressed elsewhere (Templeton, 2003a).

Because biological interpretation is limited to those distance statistics that result in a significant rejection of the null hypothesis of no geographical associations, the ability to make inference in NCPA is obviously limited by sample size. A sample based upon only a few individuals has little chance of yielding meaningful inference, no matter how dramatic the resulting genetic patterns may appear. Even when significant geographical associations are detected, the inference key may lead to the conclusion that there has been an inadequate geographical sampling for unambiguous biological interpretation. When these sampling limitations are encountered, an investigator can only circumvent them by additional sampling. NCPA, therefore, ensures that biological inferences are based upon adequate sampling of individuals and geographical sites. Moreover, when the inference key results in the conclusion of inadequate sampling, the regions that need to be sampled are identified. Hence, the nested clade analysis provides specific guidance for future sampling efforts.

The last two limitations are the failure to detect events or processes because of insufficient genetic resolution (as opposed to inadequate sampling) and the danger of false inferences. Both of these are real possibilities. For example, Templeton (1998, 2003b) validated the original 1995 inference criteria by examining biological examples for which strong prior evidence exists of their phylogeographic history. Overall, NCPA did well with these examples with *a priori* expectations, but these same worked examples reveal that NCPA sometimes fails to detect known phylogeographic events and more rarely leads to false inferences. The failure to detect known events was not always due to inadequate sampling, but sometimes was due to the failure of an appropriate mutation to occur in the right place and time to mark the event. This shows that no one locus or DNA region can capture the totality of a species' population structure and recent evolutionary history. Some false inferences were due to inadequate sampling or resolution and were easily corrected by modifications of the original inference key (hence the version of the inference key on the website should be used rather than previously published versions). However, inadequate sampling and resolution does not explain all false positives. The processes of mutation and genetic drift, which shape the haplotype tree upon which the nested analysis is based, are both random processes. Therefore, the expected pattern for a particular event or process can sometimes arise just by chance alone, leading to a false biological inference. Moreover, natural selection can skew both the shape of the haplotype tree and the geographical distribution of certain haplotypes, thereby creating patterns that do not necessarily reflect phylogeographic processes. This can also yield false biological inferences.

These last two types of limitations can be circumvented by performing NCPA simultaneously upon many loci or gene regions rather than just one (Templeton, 2002). By studying multiple DNA regions, an investigator increases the chances that an appropriately placed mutation occurred in time and space in one or more DNA regions to detect a real event or process. Moreover, all phylogeographic information is lost once all the haplotype variation at a particular DNA region coalesces to a common ancestral molecule, and different DNA regions coalesce at different times. Therefore, by studying several regions, a greater time period is generally sampled than with just one DNA region. Finally, the chances of making a false inference can be reduced with multiple loci by cross-validating inferences across DNA regions.

This multilocus approach with cross validation was used in a nested clade analysis of recent human evolutionary history using ten different DNA regions; the human mitochondrial genome and nine nuclear genome regions, including Y-linked, X-linked, and autosomal regions (Templeton, 2002). No one DNA region detected all events or processes, illustrating the incompleteness of inference based upon a single DNA region or type. When inferences were made, they were generally cross validated by two or more of the ten DNA regions, indicating that the problem of false inferences may not be very common as the failure to make an inference. Cross validation in this case was based upon temporal concordance across DNA regions of NCPA inferences of the same type and geographical location. However, Templeton (2002) did not provide a complete, formal framework of statistical hypothesis testing for accessing such concordance. The purpose of this chapter is to present a formal statistical framework of temporal concordance based upon log-likelihood ratio tests. The use of this maximum likelihood framework will be illustrated by an analysis of the human data given in Templeton (2002). Moreover, this same framework can be used to test the hypothesis that two or more species were affected by a common historical event or process. This use of the framework will be illustrated by testing the hypothesis that both humans and their malarial parasites expanded out of Africa at the same time.

2.   A LIKELIHOOD FRAMEWORK FOR TEMPORAL CONCORDANCE

When NCPA infers an event or process, the time of that event or process can be estimated as the age of the youngest monophyletic clade that contributed in a statistically significant fashion to the inference (Templeton, 2002). This represents the time in the population's history during which all haplotype lineages affected by the event were present, but more derived haplotype lineages not affected by the event had not yet evolved. Point estimates of these ages can be obtained by the method of Takahata et al. (2001); henceforth called the TLS estimator. Because the TLS estimator does not take into account the potentially large error associated with the coalescent process, Templeton (2002) fit a gamma distribution with a mean equal to the date from the TLS estimator and with a variance given by a neutral coalescent process conditioned upon the amount of pairwise diversity observed within the monophyletic clade being dated. The rationale for this approach is given in Templeton (1993), but the derivation was not given in Templeton (2002) due to space limitations. Therefore, this derivation is now presented.

Tajima (1983) showed that under neutrality, the time to coalescence to a common ancestral molecule of a monophyletic clade of DNA sequences is a random variable with mean

$$T = \frac{\theta(1+k)}{2\mu(1+n\theta)} \tag{1}$$

and variance

$$\sigma^2 = \frac{\theta^2(1+k)}{4\mu^2(1+n\theta)^2} \tag{2}$$

where k is the pairwise divergence among present-day haplotypes measured by the number of nucleotide differences, n is the number of nucleotides that were sampled, μ is the mutation rate, and θ is the expected nucleotide heterozygosity (this is a different but equivalent parameterization of that given in Tajima). The TLS estimator gives a direct estimate of T based on the ratio of average pairwise divergences within the clade to be aged relative to the interspecific differences with an outgroup species. If the age of the split between the outgroup and ingroup species is known, the age of the ingroup clade can be estimated from this ratio without the need to know all of the underlying parameters in equation (1). Equation (2) can be expressed as

$$\sigma^2 = \frac{T^2}{(1+k)} \tag{3}$$

so given T, we need only know k to obtain the variance.

The distribution of time to coalescence is well approximated by a gamma distribution (Kimura, 1970). One standard form of the gamma distribution is

$$f(t \mid \alpha, \beta) = \frac{1}{\Gamma(\alpha)\beta^\alpha} t^{\alpha-1} e^{-x/\beta} \tag{4}$$

where t is the random variable (time to the most recent common ancestral molecule for the clade), α and β are parameters of the distribution, and $\Gamma()$ is the gamma function. With this parameterization, the mean and variance of the gamma distribution are

$$Mean = \alpha\beta$$

$$\sigma^2 = \alpha\beta^2 \tag{5}$$

Substituting equations (1) and (3) into (5) yields

$$\alpha = 1 + k$$

$$\beta = \frac{T}{1+k} \tag{6}$$

Let t_i be the time of a phylogeographic event or process inferred from DNA region i. The distribution of t_i is obtained by substituting equations (6) into equation (4) and adding subscripts to specify DNA region i to be

$$f(t_i \mid T_i, k_i) = \frac{t_i^{k_i} e^{-t_i(1+k_i)/T_i}}{\left(\dfrac{T_i}{1+k_i}\right)^{1+k_i} \Gamma(1+k_i)} \tag{7}$$

where k_i is the average pairwise nucleotide diversity among the haplotypes in DNA region i in the youngest monophyletic clade that contributed in a statistically significant fashion to the NCPA inference of interest, and T_i is the age obtained by the TLS estimator (or perhaps some other method) for this inference from DNA region i. In this manner, Templeton (2002) used the TLS estimator not as a point estimator but rather to define a probability distribution that explicitly incorporates coalescence error and the fact that phylogenetic dating procedures such as that used by Takahata et al. (2001) have increasing accuracy (decreasing variance) with increasing mutational resolution (as measured by k_i) (Rannala and Bertorelle, 2001).

Equation (7) provides the basis for a formal maximum likelihood framework. Consider first the problem of inference from only a single DNA region. There is only one observation of the age of the event/process, t_i. The log-likelihood of T_i associated with equation (7) given t_i and k_i is

$$\ln L_i(T_i \mid t_i, k_i) = k_i \ln t_i - \frac{t_i(1+k_i)}{T_i} - (1+k_i)[\ln T_i - \ln(1+k_i)] - \ln \Gamma(1+k_i) \tag{8}$$

The maximum likelihood estimator of T_i is obtained by taking the derivative of equation (8) with respect to T_i, setting the derivative equal to zero, and solving for T_i as follows:

$$\frac{d \ln L_i(T_i \mid t_i, k_i)}{dT_i} = \frac{t_i(1+k_i)}{T_i^2} - \frac{1+k_i}{T_i} = 0$$

$$\Rightarrow \frac{t_i(1+k_i)}{T_i} = 1 + k_i \tag{9}$$

$$\Rightarrow \hat{T}_i = t_i$$

Since the one observation of age in this case is the TLS estimator, equation (9) shows that the TLS estimator is treated as both the *a posteriori* maximum likelihood estimator and the *a priori* random variable in this situation.

Now consider the problem in which j DNA regions all lead to a geographically concordant NCPA inference. The use of equation (7) to combine inferences across loci depends upon whether the inference is an event or a recurrent process such as gene flow restricted by isolation by distance. For a recurrent process, there is no biological restriction that the estimated ages at which the process is detected should be concordant across all j DNA regions because by definition the process is continuing throughout an interval of time rather than just occurring at a particular point in time. For populations in a geographical region interconnected by gene flow, different DNA regions may detect statistically significant signals of that gene flow at different times depending upon the mutation rates (and hence genetic resolutions) at each region and evolutionary stochasticity in the appearance of appropriate mutations in the right time and place to mark the process in a manner detectable by NCPA. For recurrent processes, the time of the process is not biologically meaningful because the process involves an interval of time rather than a point in time. However, it is meaningful to determine the time by which we can be confident that the process had started. For DNA region i that yields an inference of gene flow at t_i, the probability of gene flow in the interval from the present to T in the past is, from equation (7),

$$\int_0^T \frac{t_i^{k_i} e^{-t_i(1+k_i)/T_i}}{\left(\dfrac{T_i}{1+k_i}\right)^{1+k_i} \Gamma(1+k_i)} dt_i \tag{10}$$

The probability that all j regions indicate gene flow between the present and T is

$$\prod_{i=1}^{j} \int_0^{T_i} \frac{t_i^{k_i} e^{-t_i(1+k_i)/T_i}}{\left(\dfrac{T_i}{1+k_i}\right)^{1+k_i} \Gamma(1+k_i)} dt_i \tag{11}$$

Hence, the probability of one or more DNA regions indicating gene flow older than T is

$$\text{Pr(gene flow by } T) = 1 - \prod_{i=1}^{j} \int_0^{T_i} \frac{t_i^{k_i} e^{-t_i(1+k_i)/T_i}}{\left(\dfrac{T_i}{1+k_i}\right)^{1+k_i} \Gamma(1+k_i)} dt_i \tag{12}$$

The time at which we can be 95% confident that gene flow was established in a geographical area based upon all DNA regions inferring gene flow in that area is obtained by setting equation (12) equal to 0.95 and solving for T and using the maximum likelihood estimators of the T_i.

Now consider the case when NCPA infers an event (*e.g.*, fragmentation or range expansion) from two or more DNA regions in a geographically concordant fashion. One hypothesis, say hypothesis A, is that each of these inferred events is marking a separate historical event, in which case there are no constraints on the T_i's. The log-likelihood under hypothesis A is

$$\ln L_A(T_i, i = 1,...j \mid t_i, k_i, i = 1,...j) = \sum_{i=1}^{j} \ln L_i(T_i \mid t_i, k_i) \tag{13}$$

The estimators that maximize this likelihood are equations (9) for every i. Now consider the alternative hypothesis, say hypothesis B, that all inferred events are the same event. This means that $T_i = T$ for every i and

$$\ln L_B(T \mid t_i, k_i, i = 1,...j) = \sum_{i=1}^{j} \ln L_i(T \mid t_i, k_i) \tag{14}$$

Differentiating equation (14) with respect to T and setting it equal to 0 yields

$$\sum_{i=1}^{j} \left[\frac{t_i(1+k_i)}{T^2} - \frac{1+k_i}{T} \right] = 0 \tag{15}$$

which yields the maximum likelihood estimator of the age of the common event to be

$$\hat{T} = \frac{\sum_{i=1}^{j} t_i(1+k_i)}{\sum_{i=1}^{j}(1+k_i)} \tag{16}$$

The variance of the above estimator is obtained from the variances of the t_i's (equations 5 and 6) as

$$Var(\hat{T}) = \frac{\sum_{i=1}^{j}(1+k_i)^2 Var(t_i)}{\left(\sum_{i=1}^{j}(1+k_i)\right)^2} = \frac{\sum_{i=1}^{j}(1+k_i)t_i^2}{\left(\sum_{i=1}^{j}(1+k_i)\right)^2} \tag{17}$$

The standard log-likelihood ratio test of hypotheses A versus B is

$$G = -2\left(\ln L_B\left(\hat{T} \mid t_i, k_i, i = 1,\dots j\right) - \ln L_A\left(\hat{T}_i, i = 1,\dots j \mid t_i, k_i, i = 1,\dots j\right)\right)$$
$$= -2\sum_{i=1}^{j}(1+k_i)\left(1 - t_i \Big/ \hat{T} + \ln t_i - \ln \hat{T}\right) \tag{18}$$

where G is asymptotically distributed as a chi-square with j-1 degrees of freedom. Small values of G favor hypothesis B of a single event, whereas large values favor hypothesis A of many distinct events.

When the hypothesis of a single event is accepted, the pooled data should be approximately distributed as a gamma distribution with mean given by equation (16) and variance by equation (17). Substituting equations (16) and (17) into equations (5) yields the standard gamma parameters of the pooled distribution to be

$$\alpha = \frac{\hat{T}^2}{Var(\hat{T})}$$
$$\beta = \frac{\hat{T}}{\alpha} \tag{19}$$

Once the parameters have been estimated by (19), the upper and lower 2.5% tails of the resulting gamma distribution define a 95% confidence interval for the pooled estimate of the age of the event given by equation (16).

3.   TESTING HYPOTHESES ABOUT RECENT HUMAN EVOLUTION

Much of the debate about the origin of anatomically modern humans has centered around two hypotheses. Both hypotheses accept the fossil evidence that the human lineage evolved in Africa and spread out into the southern parts of Eurasia as *Homo erectus* by at least 1.7 million years ago (Gabunia et al., 2000). Under the multiregional trellis model (Wolpoff et al., 2000), populations in Africa, Europe, and Asia maintained genetic interchange after this initial out-of-Africa expansion, although the genetic contact was restricted by isolation by distance. Isolation by distance allowed regional genetic differentiation, but gene flow caused all of humanity to evolve as a single evolutionary lineage over long periods of time. In contrast, the out-of-Africa replacement theory (Stoneking and Soodyall, 1996) posits that anatomically modern humans only evolved in Africa (thus assuming insufficient genetic interchange for Africans and Eurasians to evolve as a single lineage), and that a second out-of-Africa expansion occurred around 100,000 years ago involving these anatomically modern humans. Moreover, these modern humans did not interbreed with the Eurasian populations they made contact with, but rather drove them to genetic extinction and completely replaced them.

Templeton (2002) presented the results of NCPA upon ten different gene regions: mitochondrial DNA (mtDNA), Y-chromosome DNA, two X-linked regions, and six autosomal regions. The results of these analyses were not in complete concordance with either the trellis or replacement models, but rather had elements of both plus some novel inferences. As postulated by the replacement model, NCPA detected an out-of-Africa expansion event around 100,000 years ago with both mtDNA and Y-DNA. However, NCPA detected events or processes involving Eurasian populations that were older than 100,000 years ago for all autosomal and X-linked DNA regions. If complete replacement had occurred, these genetic signals from early Eurasian populations should have been erased, so their detection indicates that complete replacement did not occur. Moreover, evidence for recurrent gene flow involving African and Eurasian populations extends back at least 610,000 years ago with 95% confidence as ascertained by equation (12); a feature in accordance with the trellis model. Finally, NCPA of three autosomal regions indicated an intermediate out-of-Africa expansion event around 700,000 years ago; an expansion event predicted by neither the replacement nor trellis models. All of these conclusions were cross-validated in the sense that NCPA detected the same type of event in a geographically concordant manner in two or more DNA regions at about the same estimated times. The temporal depth of the recurrent gene flow was tested with equation (12), but temporal concordance of events was only tested heuristically. The results of the previous section will now be used to formally test these hypotheses about recent human evolution.

Consider first the hypothesis that there were two, not one, major expansions of human populations out-of-Africa after the first expansion by *Homo erectus*. NCPA detected population expansion events involving African/Eurasian populations at five different DNA regions: mtDNA, Y-DNA, the β-hemoglobin locus (*β-Hb*), the melanocortin 1 receptor locus (*MC1R*), and the *MS205* hypervariable minisatellite region on chromosome 16 (Templeton, 2002). Table 1 presents the TLS estimators of the ages of these expansion events using a 6 million year calibration for the split between humans and chimps (Templeton, 2002) as well as the average pairwise nucleotide differences within the haplotype clade used to age the event. In addition, Table 1 presents the pooled estimator of the expansion event using equation (16).

Table 1. Estimated Ages of Out-of-Africa Expansion Events

DNA Region	Estimated Age (t_i) in millions of years	*Pairwise Nucleotide Differences (k_i)*
mtDNA	0.1308	38.5
Y-DNA	0.0916	3.25
β-Hb	0.8212	1.87
MC1R	0.6390	1.00
MS205	0.6306	1.87
Pooled ($\hat{T}$)	0.2136	

A likelihood ratio test of the hypothesis of a single out-of-Africa expansion event is performed by evaluating equation (18) with the data in Table 1, yielding $G = 27.63$ with 4 degrees of freedom. This yields a p-value of 0.000015. Therefore, the hypothesis of a single out-of-Africa expansion event is strongly rejected.

As Table 1 reveals, the uniparental inherited elements indicate an expansion event around 100,000 years ago, whereas the autosomal loci appear to detect an older out-of-Africa expansion event around 700,000 years ago (Templeton, 2002) – a time period outside the informative range of mtDNA and Y-DNA because of their shallow coalescent times. Therefore, equations (16) and (18) were applied separately to the unisexually inherited DNA regions and to the autosomal regions. For the unisexual DNA, the pooled estimate for the age of the out-of-Africa expansion is 127,000 years ago, and $G = 0.44$ with 1 degree of freedom, yielding a p-value of 0.51. Hence, the hypothesis that the mtDNA and Y-DNA are detecting a single expansion event is accepted and fits the data well. This justifies combining the mtDNA and Y-DNA distributions, which jointly yield a 95% confidence interval for this out-of-Africa expansion of 92,000 to 167,000 years ago, using equations (19) to fit a gamma distribution to the pooled data. For the autosomal DNA, the pooled estimate for the age of the expansion is 703,400 years ago, and $G=0.1233$ with 2 degrees of freedom, yielding a p-value of 0.94. This result strongly indicates that the autosomal regions are detecting a single expansion event. This justifies combining the three autosomal distributions to yield a joint 95% confidence interval of 296,000 to 1,284,000 years ago for this out-of-Africa expansion event. Hence, this log-likelihood ratio analysis strongly supports the idea of two separate out-of-Africa expansion events after the initial expansion of *Homo erectus*.

Perhaps the most contentious hypothesis about recent human evolution is that the most recent out-of-Africa expansion event was also a replacement event that resulted in the total genetic extinction of the previous inhabitants of Eurasia. NCPA can only detect events in the history of populations that have living descendents, so if total replacement had occurred, there should be no detectable events or processes involving Eurasian populations prior to the most recent out-of-Africa expansion event. As seen above, this most recent out-of-Africa expansion is detected and cross-validated by mtDNA and Y-DNA. NCPA also detected significant events or processes (gene flow restricted by isolation by distance) involving Eurasian populations (such as the out-of-Africa expansion events shown in Table 1) for *all* the autosomal and X-linked regions studied (Templeton, 2002). Table 2 presents the estimated ages for the oldest events or processes detected by each DNA region involving Eurasian populations along with the pairwise nucleotide differences of the haplotype clade used to age the event or process.

Table 2. Estimated Ages of the Oldest Events or Processes

DNA Region	Estimated Age (t_i) in millions of years	*Pairwise Nucleotide Differences (k_i)*
mtDNA	0.1308	38.5
Y-DNA	0.0916	3.25
β-Hb	0.8212	1.87
MC1R	0.6390	1.00
MS205	0.6306	1.87
EDN	0.6912	2.00
ECP	0.5824	1.33
PDHA1	0.8597	3.05
Xq13.3	0.4800	2.00
MX1	3.408	1.67
Pooled ($\hat{T}$)	0.4258	

If the out-of-Africa replacement hypothesis were true, none of the ages in Table 2 should be older than the ages for the out-of-Africa expansion event detected by mtDNA and Y-DNA. Because all the ages for the oldest events detected by bisexually inherited DNA regions are older than the out-of-

Africa expansion detected by the uniparental DNA regions, a conservative test of the replacement hypothesis is achieved by testing the hypothesis that the ages of all the events or processes detected by the bisexually inherited DNA regions are the same age as the out-of-Africa expansion events detected by mtDNA and Y-DNA. This conservative hypothesis is tested by using the data in Table 2 in equation (18) to yield G=77.30 with 9 degrees of freedom. The p-value is 6×10^{-13}, indicating an extremely strong rejection of the replacement hypothesis. However, Table 2 indicates that the autosomal *MX1* locus is an outlier, and indeed it has many unusual features (Templeton, 2002). Therefore, the analysis was repeated excluding *MX1* because of its outlier status. With this exclusion, the pooled time estimate is 0.303 million years, and the G statistic is 44.41 with 8 degrees of freedom, yielding a p-value of 5×10^{-7}. Thus, regardless of whether or not *MX1* is included, it is impossible to reconcile the genetic data with the replacement hypothesis. Note also that every DNA region that is informative about replacement (that is, those DNA regions with older coalescent times than mtDNA and Y-DNA) cross-validate this rejection of the replacement hypothesis. Thus, the genetic evidence strongly supports the idea that the most recent expansion of humans out of Africa was not a total replacement event.

4. TESTING PHYLOGEOGRAPHIC HYPOTHESES ACROSS SPECIES

The area of interspecific phylogeography has long acknowledged the fact that the geographical distributions of many species can be influenced by the same historical events (Futuyma, 1986). The intraspecific phylogeography of many different species can also be influenced by common events (Strange and Burr, 1997). In particular, interspecific interactions can cause correlations in the phylogeography of two different species. For example, some parasites are strongly associated with their hosts, so events or processes affecting the movement of the hosts through space could also have a direct effect on the movement of the parasites. One parasite of humans is *Plasmodium falciparum*, the malarial parasite. There has been speculation that this parasite could have moved out of Africa along with the most recent out-of-Africa expansion event of its human hosts, or alternatively it spread much more recently due to changes associated with the emergence of agriculture (Joy et al., 2003). To address these issues, mtDNA sequence variation was recently surveyed in 100 worldwide isolates of the malarial parasite (Joy et al., 2003), but no formal phylogeographic analyses were performed on these data. These data on malarial parasites will therefore be analyzed using

NCPA, and then the statistics developed in this paper will be used to test the hypothesis that malaria spread out of Africa with the human expansion out of Africa marked by human mtDNA and Y-DNA.

Figure 1 presents the haplotype tree estimated for the mtDNA haplotypes found in the survey of 100 *P. falciparum* isolates. Haplotypes are designated by the numbers given in Joy et al. (2003). Two loops found in the original haplotype tree (Joy et al., 2003) were resolved by the use of frequency criteria based on coalescent theory (Crandall and Templeton, 1993), and Figure 1 shows this completely resolved tree. The tree was then used to define nested clades, as also shown in Figure 1. Haplotypes are boxed together by thin lines into 1-step clades, indicated by "1-#", which in turn are boxed together by thick lines into 2-step clades, indicated by "2-#."

The nested design shown in Figure 1 was used for an NCPA. There are only two clades at the highest level of nesting, and clade 2-1 was regarded as the interior clade because the outgroup probabilities (Castelloe and Templeton, 1994) are strongly concentrated in that clade (Joy et al., 2003). Because the sample sizes per locality were often small, nearby localities were pooled and assigned a geographically central location. The results of the NCPA are given in Table 3.

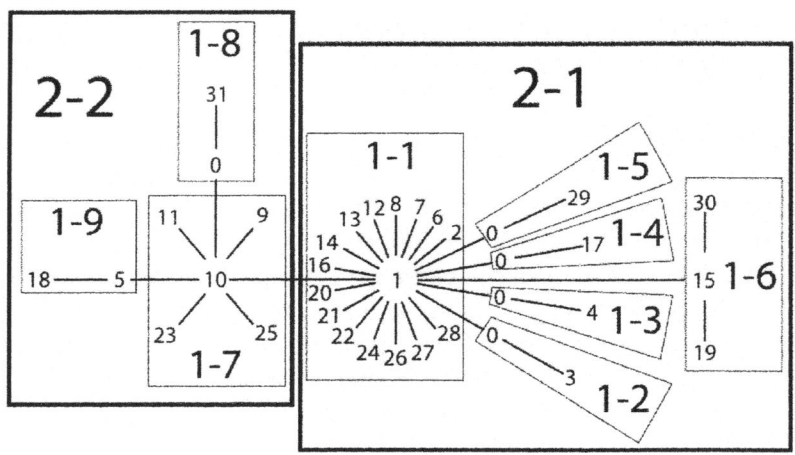

Figure 1. The mtDNA haplotype tree of *Plasmodium falciparum*

Several statistically significant results are shown in Table 3, and these were interpreted biologically with the inference key at the GEODIS website. The haplotypes nested within clade 1-1 reveal several statistically significant distances that yield the biological interpretation of recurrent gene flow restricted.by isolation resulting from distance. Isolation by distance therefore dominants the most recent evolution of *P. falciparum* since clade 1–1 contains most of the diversity and covers most of the geographical range of modern *P. falciparum*. There are also significant distances within clades 1–7, 2–1, and 2–2, but none of these has an unambiguous interpretation because of inadequate sampling of Asia in the study of Joy et al. (2003). Although several Asian populations were sampled, they were all tightly clustered in Southeast Asia and were therefore pooled in the NCPA. However, the pooling was not the problem; rather, the problem was the lack of any other Asian samples, particularly in the Indian subcontinent and the Middle East. This creates a large sampling gap between Africa and Asia/New Guinea, and it is this sampling gap that creates difficulties in biological interpretation.

The final statistically significant distances shown in Table 3 are those found for the two clades, 2–1 and 2–2, nested within the total haplotype tree. Here, the inference key does lead to an unambiguous inference: contiguous range expansion, most likely out of Africa, but out of Asia is a possibility (once again, this would be clarified by better geographical sampling). This is the inference that is germane here: both humans and their malarial parasite have experienced a range expansion that is concordant with being out of Africa. Given that there is concordance in event type and geographical location, the hypothesis is raised that these are the same event for both the host and its parasite. If so, there should also be temporal concordance, and this hypothesis can be tested with the log-likelihood ratio tests developed here.

The age of the malarial range expansion must first be calculated. As can be seen from Table 3, clade 2-2 has a significantly large nested clade distance, and this contributed to the biological inference of range expansion. Given that the outgroup probabilities are high within clade 2-1, this means that clade 2-2 is the youngest clade contributing to the inference, and therefore the age of this clade best approximates the age of the event (Templeton, 2002). The outgroup in this case is *P. reichenowi*, whose host is the chimpanzee (Joy et al., 2003). Using synonymous and noncoding sites with Jukes-Cantor correction, Joy et al. (2003) estimated the age of the *P. falciparum* mtDNA haplotype tree as the ratio of the average nucleotide difference within *P. falciparum* (0.0012) to the average difference between *P. faciparum* and *P. reichenowi* (0.0856) multiplied by a calibration constant, say *C*. This is the age of clade 2–1, the oldest clade in the tree. To

Table 3. Nested Clade Analysis of Malarial mtDNA. Interior clades are indicated by shaded boxes. Superscripts above the clade distances, Dc, and nested clade distances, Dn, indicate significantly large (L) or small (S) values.

Haplotypes			1–Step Clades			2-Step Clades		
Clade	D_c	D_n	Clade	D_c	D_n	Clade	D_c	D_n
1	8156^L	8055^L						
2	0	7156						
6	0	7548						
7-14	0	7156						
16	0^S	3095^S						
20-22	0	7156						
24	0	3095						
26-27	0	7548						
28	0	7156						
I-T	8156^L	1948^L	1-1	7278	7310^L			
3	-	-	1-2	0	6806			
4	-	-	1-3	0	6806			
17	-	-	1-4	0	15748^L			
29	-	-	1-5	0	3258			
15	2303	2183	1-6	2062^S	3998^S			
19	0	1091	I-T	5850^L	1825^L	2-1	6771^S	7565^S
30	0	1091						
I-T	2303	1093						
10	0	6425						
9,11,23	0	6425						
25	0^S	8282^L						
I-T	0	-743	1-7	7239	9629	2-2	7049	10481^L
						I-T	-278	-2916^S
31	-	-	1-8	0	3334			
5	0	0	1-9	0^S	5723			
18	0	0	I-T	7239^L	4020			
I-T	0	0						

obtain the age of clade 2–2, the average absolute number of synonymous and noncoding nucleotide differences between clades 2–1 and 2–2 is calculated as 2.31, and the average within clade 2–2 is 1.22. Hence, the estimator for the age of the 2-2 clade is:

$$t = \frac{0.0012}{0.0856} \times \frac{1.22}{2.31} \times C = C \times 0.007 \qquad (20)$$

Assuming the two malarial species split at the same time as humans and chimpanzees, C is equal to 6 million years, yielding an age of the expansion event of 42,000 years. However, it is possible that the malarial parasites diverged after the human-chimpanzee split (Joy et al., 2003), and this possibility is made more likely by recent evidence for continued contact between humans and chimpanzees after their initial divergence (Navarro and Barton, 2003). Therefore, a calibration of 5 million years was also used (Joy et al., 2003), yielding an estimate of 35,000 years.

These estimates, along with the $k=1.22$ for clade 2–2, can now be used to test the hypothesis that this malarial expansion event occurred at the same time as the most recent human out-of-Africa expansion event found in human mtDNA and Y-DNA. The log-likelihood ratio test (equation 18) of the hypothesis that the expansion events found in human mtDNA, human Y-DNA, and malarial mtDNA have a common time is 2.33 with 2 degrees of freedom for the 6 million year calibration, and 2.89 with 2 degrees of freedom for the 5 million year calibration. Neither is significant at the 5% level, so the hypothesis cannot be rejected that malaria expanded out-of-Africa along with its human hosts. An alternative hypothesis is that malaria spread out-of-Africa around 6,000 years ago with the spread of agriculture (Joy et al., 2003). The gamma distribution (equation 7) can also be used to define a 95% confidence interval for t, the time of the malarial expansion out of Africa. This interval is 113,000 to 5,900 years ago for the 6 million year calibration, and is 93,700 to 4,900 years ago for the 5 million year calibration. Thus, the malarial mtDNA is compatible with either hypothesis.

5.   DISCUSSION

The results of the log-likelihood ratio tests on the human data confirm and strengthen the conclusions made in Templeton (2002) concerning recent human evolution. In particular, the implausibility of a total replacement model can now be quantified within the context of a formal hypothesis testing framework and cross validation. This is an important result in the area of human evolution as the replacement hypothesis has been a particularly contentious issue. The work presented here is the first formal statistical test of complete replacement versus the alternative of some interbreeding. Most studies only examine the fit of the data to the replacement hypothesis without any formal evaluation of how the data discriminate between the alternatives (Templeton, 1994). Hypothesis compatibility is quite different from hypothesis testing, and much of the contention in this area has arisen from the false premise that data compatibility with replacement automatically means that the hypothesis of some interbreeding has been rejected (Templeton, 1994). For example, mtDNA and Y-DNA only sample the time period back to around 100,000 years ago with any degree of statistical power, as the out-of-Africa expansion event is the oldest and deepest detectable signal in either type of DNA (Templeton, 2002). But tests of replacement require an examination of the time *before* the out-of-Africa expansion; that is, one must examine the time periods and populations supposedly affected by replacement. Because mtDNA and Y-DNA contain no information about events or processes before the out-of-Africa expansion, mtDNA and Y-DNA are compatible with replacement for the simple reason that they contain no information whatsoever about replacement. Hypothesis *testing* of two alternatives requires an explicit comparison of how well the data fit the two alternatives. Such a test is provided here by a likelihood ratio of the two alternatives, and the data strongly reject the complete replacement hypothesis. Indeed, this rejection could not be any stronger; out of eight loci that sample the time period before 100,000 years ago and hence are potentially informative about replacement, *all* eight cross validate each other in rejecting complete replacement.

More generally, the work presented in this chapter adds an important feature to NCPA; a formal statistical framework for cross-validation when data from multiple DNA regions exists. Because any one DNA region samples only a limited period of historical time and because both false positives and false negatives can arise in NCPA, using data from multiple loci in an integrated statistical analysis increases the temporal breadth of phylogeographic analyses, augments the genetic resolution, and provides a method for validating biological conclusions. These advantages indicate that intraspecific phylogeographic studies would be much stronger and more informative if based upon multiple DNA regions rather than upon a single DNA region or singly inherited unit of DNA such as mtDNA.

This same statistical framework for cross validation within a species can also be used for testing hypotheses about shared phylogeographic constraints across species, as illustrated by the human/malaria example. The inability of the malarial mtDNA to resolve between the ancient versus the agriculturally driven hypotheses of expansion out-of-Africa could be due to the inadequate geographical sampling mentioned earlier. Clade 1–7 has one of the hallmarks of a range expansion: a tip clade with a significantly small clade distance but a significantly large nested clade distance (Templeton et al., 1995). However, the lack of samples between Africa and Asia/New Guinea undermine the ability to make a clear biological inference. If such samples had been obtained and the resulting NCPA had yielded clade 1-7 as the youngest clade marking the malarial expansion event, the time estimates would be substantially lower and would indeed reject the hypothesis that malaria spread out of Africa at the same time as the most recent human out-of-Africa expansion. Indeed, it is also possible that with better sampling, there would be evidence for two expansions out-of-Africa by the malarial parasite; an ancient one associated with the human population expansion, and a more recent one associated with the spread of agriculture. All of these are testable hypotheses with the statistical framework given in this chapter, but their testing must wait for additional samples of malarial mtDNA from southern central and southwest Asia. This illustrates the requirement for adequate geographical sampling when testing phylogeographic hypotheses. The ability of the nested clade approach to identify areas where additional sampling is needed is one of the strengths of NCPA because it protects against over interpreting the data and provides guidance for future studies.

6. ACKNOWLEDGMENTS

Support from a Burroughs-Wellcome Fund Innovation Award is gratefully acknowledged. Also, the opportunity to know and converse with Eibi Nevo has helped shape my thinking on many matters related to evolutionary genetics, and I thank him for the intellectual stimulation that arose from these conversations as well as his boundless energy, enthusiasm, and encouragement.

7. REFERENCES

Avise J.C., Lansman R.A. and Shade R.O. 1979. The use of restriction endonucleases to measure mitochondrial DNA sequence relatedness in natural populations. I. Population structure and evolution in the genus *Peromyscus*. Genetics 92(1), 279-295.

Castelloe J. and Templeton A.R. 1994. Root probabilities for intraspecific gene trees under neutral coalescent theory. Mol Phylogen Evol. 3, 102-113.

Crandall K.A. and Templeton A.R. 1993. Empirical tests of some predictions from coalescent theory with applications to intraspecific phylogeny reconstruction. Genetics 134, 959-969.

Futuyma D.J. 1986. Evolutionary Biology (2nd ed.). Sinauer Associates, Inc, Sunderland, Massachusetts.

Gabunia L., Vekua A., Lordkipanidze D., Swisher C.C., Ferring R., Justus A., et al. 2000. Earliest Pleistocene hominid cranial remains from Dmanisi, Republic of Georgia: Taxonomy, geological setting, and age. Science 288(5468), 1019-1025.

Joy D.A., Feng X.R., Mu J.B., Furuya T., Chotivanich K., Krettli A.U., et al. 2003. Early origin and recent expansion of *Plasmodium falciparum*. Science 300(5617), 318-321.

Kimura M. 1970. The length of time required for a selectively neutral mutant to reach fixation through random frequency drift in a finite population. Gene Res. 15, 131-133.

Navarro A. and Barton N.H. 2003. Chromosomal speciation and molecular divergence - Accelerated evolution in rearranged chromosomes. Science 300(5617), 321-324.

Rannala B. and Bertorelle G. 2001. Using linked markers to infer the age of a mutation. Hum Mut. 18(2), 87-100.

Stoneking M. and Soodyall H. 1996. Human evolution and the mitochondrial genome. Cur Opin Gene Develop. 6(6), 731-736.

Strange R.M. and Burr B.M. 1997. Intraspecific phylogeography of North American highland fishes - a test of the pleistocene vicariance hypothesis. Evolution 51(3), 885-897.

Tajima F. 1983. Evolutionary relationship of DNA sequences in finite populations. Genetics 105, 437-460.

Takahata N., Lee S.H. and Satta Y. 2001. Testing multiregionality of modern human origins. Mol Biol Evol. 18(2), 172-183.

Templeton A.R. (1993). The "Eve" hypothesis: a genetic critique and reanalysis. Amer Anthropol. 95, 51-72.

Templeton A.R. 1994. "Eve": hypothesis compatibility versus hypothesis testing. Amer Anthropol. 96, 141-147.

Templeton A.R. 1998. Nested clade analyses of phylogeographic data: testing hypotheses about gene flow and population history. Molec Ecol. 7, 381-397.

Templeton A.R. 2002. Out of Africa again and again. Nature 416, 45-51.

Templeton A.R. 2003a. Using Haplotype Trees for Phylogeographic and Species Inference in Fish Populations. In A. J. Gharrett (Ed.), Genetics of Subpolar Fish and Invertebrates.

Templeton A.R. 2003b. Statistical phylogeography: methods of evaluating and minimizing inference errors. Molecular Ecology, (in press).

Templeton A.R., Boerwinkle E. and Sing C.F. 1987. A cladistic analysis of phenotypic associations with haplotypes inferred from restriction endonuclease mapping. I. Basic theory and an analysis of Alcohol Dehydrogenase activity in *Drosophila*. Genetics 117, 343-351.

Templeton A.R., Routman E. and Phillips C. 1995. Separating population structure from population history: a cladistic analysis of the geographical distribution of mitochondrial DNA haplotypes in the Tiger Salamander, *Ambystoma tigrinum*. Genetics 140, 767-782.

Templeton A.R. and Sing C.F. 1993. A cladistic analysis of phenotypic associations with haplotypes inferred from restriction endonuclease mapping. IV. Nested analyses with cladogram uncertainty and recombination. Genetics 134, 659-669.

Wolpoff M.H., Hawks J. and Caspari R. 2000. Multiregional, not multiple origins. Amer J Physic Anthropol. 112(1), 129-136.

MODERN VIEW ON THE ORIGIN AND PHYLOGENETIC RECONSTRUCTION OF HOMOBASIDIOMYCETES FUNGI

Ivan V. Zmitrovich[1] and Solomon P. Wasser[2,3]

[1]*V.L. Komarov Botanical Institute, Russian Academy of Sciences, 2, Popova St., St. Petersburg, 197376, Russia;* [2]*Institute of Evolution, University of Haifa, Mt. Carmel, 31905, Haifa, Israel;* [3]*N.G.Kholodny Institute of Botany, National Academy of Sciences of Ukraine, 2,Tereshchenkovskaya St., Kiev, 01601, Ukraine*

Abstract: The problem of the origin and phylogeny of Homobasidiomycetes remains controversial. However, various data accumulated in recent decades provide an opportunity to reach some definite conclusions. These conclusions are listed below. 1) Final establishment of the circumscription of the class Homobasidiomycetes (Eumycetes with dolipore septa without nanopores, with perforated, or rarely unperforated, parenthosomata; fungi that have lost their ability to produce secondary spores and microconidia), and the proof of their evolutionary advancement compared with Heterobasidiomycetes. 2) The loss of taxonomic status of such groups as Hymenomycetes/Gasteromycetes and Aphyllophorales. 3) The appearance, in the course of studies of rDNA, of some "molecular concepts", the most important of which are the rooting of Homobasidiomycetes within the Dacryomycetales-Auriculariales and the basal position of the order Cantharellales. 4) Proof of evolutionary advancement of brown-rot fungi; considerable devaluation of the type of rot as a character in taxonomy Basidiomycetes. 5) Proof of secondary origins of corticioid and gasteroid forms. Any phylogenetic reconstruction of Homobasidiomycetes phylogeny should be based on the concepts listed above. Some difficulties and contradictions still remaining in this field of taxonomy are analyzed.

S.P. Wasser (ed.), Evolutionary Theory and Processes: Modern Horizons,
Papers in Honour of Eviatar Nevo)
I.V. Zmitrovich and S.P. Wasser. *Modern View on the Origin and Phylogenetic Reconstruction of Homobasidiomycetes Fungi, 231-263.*
© 2004 *Kluwer Academic Publishers.*

1.   INTRODUCTION

Despite the progress achieved during recent decades, the taxonomy of Homobasidiomycetes remains rather unstable. The "good times", when this group was "easily" subdivided into three large subgroups - Aphyllophorales, Agaricales, and Gasteromycetes ended by the 1980s when this traditional taxonomic practice was at least questioned and doubted (Zerova and Wasser, 1974; Zerova et al., 1979; Kühner, 1980; Nuss, 1980; Wasser, 1980, 1982, 1985; Locquin, 1981; Jülich, 1981; Moser, 1983). The main shift of the classification paradigm, which is associated chiefly with the work by Jülich (1981), turned the phylogenetic system of Homobasidiomycetes into chaos; unfortunately, this disorder still prevails today. Multiple convergence (terminology of Donk, 1971) accompanying the huge evolutionary radiation of the group, which covers not only macromorphology (so-called 'life forms'; Bondartseva, 1993), but also microscopic structures of basidiomata, is the main reason why, when trying within the chaos of Homobasidiomycetes to guess any order, we are skating on very thin ice.

It is hardly possible to agree that the current application of molecular data to the taxonomy of this group (e.g., Hibbett and Donoghue 1995; Hibbett et al., 1997; Boidin et al., 1998; Hibbett and Thorn, 2001) has solved all the problems. Certainly, these methods have helped to prove the monophyly of many compact groups (although the monophyletic status of the majority of these was already well tested by traditional morphological methods). Nevertheless, the reliability of any separation of large groups in various topologies based on single genes remains problematic and inconsistent, whereas the resulting consensus trees leave many open possibilities for future interpretations.

An approach using **a set of characters** in phylogenetic reconstruction, which would allow allocating and comparing the **types of organization** of existing groups of organisms, seems more promising. Thus, some differenttiation of the morphological and molecular fields and data would be more satisfactory methodologically, taking into account that "molecular taxa" are only partly justified, and the traditionally understood taxa are mainly the objects of a "morphological" semantic field (Vasilyeva, 1999; Karatygin, 2000).

2.   A "MORPHOLOGICAL" TREE

2.1   The problem of the phylogenetic roots of Homobasidiomycetes

> Thus, the consideration of the evolutionary patterns within the Basidio-
> mycetes depends on answers to two questions; on the one hand, are they
> monophyletic or polyphyletic and, on the other hand, which type of
> basidium is primary, and thus, which of the existing groups of Basidio-
> mycetes stands close to the ancestral forms of the class?
>
> D.K. Zerov (1972)

The idea of polyphyletic origins of basidial fungi from various groups of Ascomycetes was expressed in its day by Gäumann (1964) and supported by Petersen (1971).

The reflection of this idea on classification schemes can be found in the systems of Arx (1967) and Kreisel (1969), where ustilaginaceous (as well as exobasidial) fungi are united with Taphrinales and Hemiascomycetes as a separate class, Endomycetes.

In later studies, adherents of the idea of an isolated position of usti-laginaceous and exobasidial fungi were forced to reject the concept of their relationship with Endomycetes (representatives of that group have no dikaryophase and no chitin in the cell walls, whereas their GC-pairs content < 50 %). Instead they created for this group a separate class, Ustomycetes (Moore, 1971; Arx, 1979).

The splitting of basidial fungi into three classes - Ustomycetes (Ustila-ginomycetes), Teliomycetes (Uredinomycetes), and 'true' Basidiomycetes - is maintained in modern manuals (e.g., Nordic Macromycetes, 1997; Ainsworth and Bisby's Dictionary of the Fungi, 2001), although it is impossible to understand from those taxonomic schemes whether an inde-pendent origin of these three groups of basidial fungi was postulated, or the elevation of their taxonomic status to the class rank was a result of some general trend of devaluation of taxonomic ranks.

Finally, we should note the system proposed by Wells (1994), in which the division (phylum) Basidiomycota is split into two subdivisions (sub-phyla), Teliomycotina (Uredinales s. l. + Septobasidiales + Atractiellales (etc.) + Ustilaginales s. l. + Exobasidiales) and Basidiomycotina (classes Heterobasidiomycetes and Homobasidiomycetes). We see that this split was made on rather different principles from those of other classifications.

Cytological data provide arguments for removing the ustilaginaceous and exobasidial fungi (as well as Uredinales, Septobasidiales, etc.) from Basidio-mycetes: these groups are characterized by discoid spindle pole bodies

(SPB) and simple (or, in some Ustilaginales, rather specific in their structure) septal pores, whereas Heterobasidiomycetes s.str. and Homobasidiomycetes are characterized by biglobular SPB and dolipore septa (Jones, 1973; Oberwinkler, 1977; Littlefield and Heath, 1979; Khan et al., 1981; Wells, 1994; Moore, 1996; Bauer et al., 1997, etc.).

Yet these cytological data obviously testify against the concept of the possible polyphyly of Basidiomycetes: Ustomycetes (the group most frequently separated from other Basidiomycetes) cytologically are closer to Uredinales. The linkage of the latter with doliporous Heterobasidiomycetes through Septobasidiales-Platygloeales is now obvious.

However, the question arises, as to which characters of the type of organization of Basidiomycetes can be possibly considered primitive in this group? We shall try to solve this problem with reference to two essential features of basidial fungi, their hyphal ultrastructure, and the type of basidium.

Note, however, that simple (ascomycetous) septal pores and discoid SPB should be considered primitive characters, which makes this group more similar to ascomycetes, since the doliporous state is an obvious apomorphy. The basal position of taxa with simple septal pores was also confirmed by molecular data (Swann and Taylor, 1993, 1995).

Considering the issue of an initial type of basidium, we should reject the thesis about the morphological simplicity of the holobasidium as its supposedly primitive feature (Raitviir, 1964; Oberwinkler, 1965, etc.). First, it is impossible to ignore the case of the obviously secondary origin of holobasidium in Tilletiaceae (this example was discussed in detail by Linder, 1940). Second, the greatest "primitiveness" of ultrastructural features is characteristic for fungi with a septate basidium (more precisely, a transversely septate phragmobasidium). The latter is connected, in turn, with a longitudinally septate tremelloid basidium (some intermediates are observed in several representatives of Auriculariales and Exidiales).

The origin of **holobasidium** seems to be polyphyletic. The holobasidium of Ustomycetes (including Exobasidiales) represents a structurally "rationalized" promycelium adapted to synchronous maturation of sporidia.

The holobasidium of Dacryomycetales is comparable with **phragmobasidium** of some rust fungi, like *Coleosporium*, where the tetrakaryotic basidial body is formed inside the teleutospore, whence it sprouts by long sterigmata.

The holobasidia of *Tulasnella* or *Ceratobasidium* types in many respects are comparable with longitudinally septate cruciate basidia of the *Sebacina* type. Recently, it was assumed that the formation of holobasidium in *Oliveonia* progressed independently of similar processes in Tulasnellales-Ceratobasidiales (Roberts, 1998).

The holobasidium of Filobasidiales is also of independent origin: ultra-structure as well as life cycle features in this group testify to its close relationships with tremellaceous (Tremellales s. str.) dimorphic fungi having cruciate phragmobasidia (see Oberwinkler et al., 1984).

All these types of holobasidia are at the same time **heterobasidia**, since they are subdivided into the hypo- and epibasidial segments, whereas during their ontogenesis the stages of probasidium and metabasidium can be easily observed. This character is also clearly correlated with the mode of germination of basidiospores: they are either divided into several segments, prolife-rating by hyphae or microconidia (Exobasidiales, Dacryomycetales), or they germinate as a series of secondary spores (Ceratobasidiales, Tulasnellales), or else begin to bud (Filobasidiales). Such features are not characteristic of the **homobasidium** in the majority of representatives of the class: the stage of probasidium is greatly reduced or not expressed at all, and the dividing of a mature basidium into hypo- and epibasidial segments is poorly expressed.

The problem of ancestral forms of Homobasidiomycetes is still now far from being positively solved (as well as a century ago). Many mycologists (in the 20th century and now) specify the group of corticioid fungi as a pos-sible link between Heterobasidiomycetes and Homobasidiomycetes (or vice versa).

Indeed, there are remarkable parallels in the formation of basidiome microstructures in some corticioid Homobasidiomycetes and resupinate Heterobasidiomycetes (Oberwinkler, 1972).

Moreover, the presence of long-lasting probasidia is characteristic of some corticioids (*Corticium, Coniophora*, etc.), whereas in other groups mature basidia are obviously differentiated into hypo- and epibasidial segments, although the probasidial stage in ontogenesis is not so clearly expressed. These are, first of all, **urniform (urnlike) basidia** (*Sistotrema*) characterized by ovoid egg-like (oval) hypobasidium and narrowed epibasidium; secondly, **utriform (utricular) basidia** (*Hyphoderma, Punctu-laria*) having hypo- and epibasidial segments approximately equal in their width; thirdly, **stalked basidia (podobasidia** - *Radulomyces, Sphaero-basidium*) with narrow, twisting hypobasidial and extending epibasidial segments. All these types are interconnected by transitions. Such "aberrant" morphology of homobasidia is usually observed in wood-inhabiting fungi with poorly differentiated hymenophores. These types of basidia are most typical for Corticiaceae s.str., having a loose hymenial layer or a hyphidial hymenium (catahymenium, according to Lemke, 1964; the same type of hymenium is also characteristic for jelly fungi). A catahymenium is a good protection for developing basidia; in such a case, however, for effective sporulation they should protrude through a dense hyphidial layer. This is probably a secondary phenomenon connected with development of various strategies of effective sporulation. The same is most likely true for **pleuro-**

basidia of Xenasmataceae adhering to the substrate, or **repetobasidia** unprotected by sterile layers (in some Xenasmataceae), which are comparable with microcyclic conidiophores of some aspergills. Stalked basidia are also characteristic for some wood-inhabiting Heterobasidiomycetes (*Protodontia, Myxarium,* etc.). Interpretations of such original morphology and possible origin of "holo-hetero-basidia" are discussed by many authors (Rogers, 1934; Martin, 1949; Donk, 1954a,b, 1956; Oberwinkler, 1965, 1977, 1978, 1982; Parmasto, 1969; Chadefaud, 1975, etc.).

Better success might result from a search for a "relict" homobasidium, considering data on the orientation of the microtubular spindle during the cell division process before the basidiospore formation. This character was first considered by Juel (1898), and later its importance was noted by Linder (1940) and Donk (1964). Unfortunately, nowadays it is hardly ever used in taxonomy; moreover, because of an unfortunate typo in Ainsworth and Bisby's Dictionary of the Fungi (2001, p. 62: the figures 9C and 9D are misnamed) the concept of modern researchers regarding this character could become even more confused. According to Juel (1898), basidia (he discussed mainly holobasidia) are either stichic (the spindle is arranged longitudinally, nuclei are divided in a plane parallel to the longitudinal axis of the basidium) or chiastic (the nuclei are divided in a plane perpendicular to the basidial axis). **Stichobasidia** are characteristic only for a few representatives of Homobasidiomycetes (Clavulinaceae, Cantharellaceae, *Sistotrema, Hydnum* s. str.) as well as for some Exobasidiales, whereas **chiastobasidia** are characteristic for the majority of homobasidial fungi (see Donk, 1964, pp. 221-222). In fact, Juel's terminology can be also easily extended to heterobasidia. Thus, cruciate tremelloid basidia (and related holobasidia of the *Oliveonia* or *Ceratobasidium* types) may be interpreted as chiastobasidia, while the transversely septate phragmobasidia, characteristic for Uredinales, Ustilaginales s. l., Septobasidiales, and some Auriculariales, can be interpreted as stichobasidia. Most importantly, the two-spored (bisporous) basidium of Dacryomycetales is also stichic, and thus represents a kind of link between the phragmobasidium of Uredinales (Coleosporieae) and the two-spored homobasidium of Clavulinaceae. In our opinion, stichobasidial Clavulinaceae and Sistotremataceae fungi are close to the archetype of Homobasidiomycetes, and a "link" uniting this group with ancestral rust fungi is heterobasidiomycetous fungi of the order Dacryomycetales.

2.2   Evolutionary trends in Homobasidiomycetes

As we see, the formation of homobasidium was a major evolutionary acquisition of the group, connected with its transition to terrestrial life and adaptation to conditions of terrestrial and independent (non-endobiotic) existence, which triggered its active speciation and wide taxonomic diversification. At present there are about 9000 species of heterobasidio-mycetous fungi (in the broad sense, including all basidiomycetes having the probasidial stage and pleomorphic haplophase; average estimations) and more than 14,000 species of Homobasidiomycetes.

A wide diversification of the taxa was accompanied in the evolution of Homobasidiomycetes by a most impressive large-scale convergence pheno-mena, probably comparable only to the patterns observed in red algae. No wonder: in both these groups the main evolving structures are apically grow-ing filaments with centripetal septa, so the entire range of morphogenetic processes is determined by the "work" of apical cells.

Such a situation has created a rather complex problem for taxonomists. The problem of phylogeny of Holobasidiomycetes is presently among the most debated issues of phylogenetic taxonomy in general, at least judging from the extensive available literature.

In delimiting large groups, the most essential role is probably played by characters of hyphal ultrastructure, since these characters are most conser-vative. With reference to the Homobasidiomycetes group, we shall consider first the peculiarities of parenthosomata: continuous (unperforated) parentho-somata in Cantharellaceae, Botryobasidiaceae, Sistotremataceae, and Hyme-nochaetaceae (which are similar in that respect to Heterobasidiomycetes) and perforated parenthosomata in other Homobasidiomycetes.

Mutual relationships between hypo- and epibasidial segments are also essential for taxonomy; they are among the leading characters in disting-uishing the orders (see Table 1). The ratio between the width of the epiba-sidial segment and subhymenial hyphae is also important. This ratio remains stable in representatives of different orders (Zmitrovich et al., 2003). Other features require a somewhat differential approach depending on the set of correlated characters.

As noted earlier, biochemical and molecular data require further accumulation and reassessment in light of new (not yet generated) para-digms. However, one biochemical character, the type of rotting produced by fungi, has to be considered since many researchers, beginning with Nobles (1958), actively use this character in taxonomy although, nowadays, its taxonomic value has been somewhat revised.

After the discovery of the laccase gene-specific sequences in the genome of brown-rotters (D'Souza et al., 1996) and the oxidase-dependent (not hydrolytic!) mechanism of cellulose degradation (Hyde and Wood, 1997), it has become obvious that the brown-rot fungi represent a heterogeneous assemblage of the most advanced wood-rotters (Ander and Marzullo, 1997). Phylogenetically, this means that brown-rot fungi are probably independent terminal branches of the phyleme of Homobasidiomycetes (Gilbertson, 1980), but certainly not domain of its basal radiation and not its basal clades (Nobles, 1971).

We conclude that detecting evolutionary trends in Homobasidiomycetes is a very complicated task with no straightforward solution, at least for now. Any further advance in this area is possible only by means of the whole sets of characters reflecting the types of organization. Below we consider some possible trends in transformation of basic elements of basidiome structures in Homobasidiomycetes.

Basidiome Morphology. There are four basic versions of the transformation of basidiomata during the evolution of the group: 1) from corticioid crustothecia to sessile and stipitate, and from these to gasteroid forms (Bessey, 1950; Raitviir, 1964; Parmasto, 1968, 1969; Zerov, 1972, among many others); 2) from gasteroid to pileate cantharelloid forms, then to sessile and resupinate aphyllophoroid forms (Singer, 1962, 1971, 1975, 1986; Kreisel, 1969; Wasser, 1974); 3) from poorly differentiated "mycelial clusters" to terrestrial stipitate (+ gasteroid) on the one hand, and to wood-inhabiting sessile or resupinate on the other (Parmasto, 1986; Bondartseva, 1997; Zmitrovich, 1997, with some reservations); 4) from clavarioid to pileate forms, and from them to resupinate and gasteroid ones (Corner, 1954, 1964, 1966, 1968, 1970).

Version 3 is the most popular at present. However, as rightly noted, it does not take into account the possibility of heterochronous wood settling by various groups of terrestrial fungi (Mukhin, 1993).

We will demonstrate the validity of this possibility using one example. Practically all **merulioid fungi**, including those having basidiomata completely adhering to the substrate, demonstrate some anatomic features approaching those in agaricoid forms. These fungi have three-layered basidiomata consisting of a more or less gelatinized hymenophore, medullar tissue, and a superficial abhymenial layer constructed mainly according to the trichodermis type; the superficial layer also occurs in wholly prostrate forms. Merulioid fungi have as a rule lateral or dorsal attachment, and if basidiomata is adhering, they usually have free, inrolling margins. The presence of a characteristic abhymenial layer both in pileate and effused (prostrate) forms of merulioid fungi allows them to be considered not primary effused (prostrate), but **resupinate** (upturned), representing in fact dorsally attached pilei with their margins adhering to the substrate (Corner, 1971; Zmitrovich, 2001). The microscopic features of the hymenial layer make it possible to trace, without any reasonable doubt, the origin of merulioid fungi from agaricoids; and here two basic evolutionary lines can be clearly distinguished: Panelleae-Merulieae and Paxillaceae-*Serpula*-Amylocorticieae. Most of resupinate fungi with smooth (or almost smooth) hymenophores are easily traced evolutionarily from such transitional forms, assuming the most efficient adaptation to the positive geotropism as the basic line of evolution in that group (Corner, 1954). The remaining groups of corticioid fungi also can be connected either to reduced agarics (Myceneae-Aleurodisceeae-

Stereeae; this line in such a case will be parallel to several "cyphelloid" lines - see Singer, 1945) or to less differentiated clavarioid and cantharelloid forms (Clavulinaceae-Sistotremataceae; Ramariaceae-Xenasmatceae, etc.).

Examples of evidently secondary development of resupinate forms in Thelephoraceae are also well known (Corner, 1968; Kõljalg and Renvall, 2000).

Thus we arrive at the *"Clavaria* hypothesis" (version 4) proposed by Corner, which seems to be capable of explaining most successfully many inconsistent facts in the field of comparative morphology of Homobasidiomycetes (Petersen, 1971; Thiers, 1971; Watling and Largent, 1977; Jülich, 1981; Pegler and Young, 1981; Kovalenko, 1984; Gorovoy, 1989). It is most important that, in conformity with Corner's hypothesis, we have to place clavarioid and cantharelloid taxa, having a set of really primitive characters (hyphal ultrastructure, basidial karyology), at the base of the evolutionary tree.

The *"Clavaria* hypothesis" concurs well with the idea of stichobasidial Clavulinaceae as the most primitive group of Homobasidiomycetes.

If we accept Corner's hypothesis, the evolutionary tree of Homobasidiomycetes will look like a fountain, with "jets" formed by cantharelloid and agaricoid clades, and "splashes and drops" composed of various groups of aphyllophoroid resupinates. The *"Clavaria* hypothesis" also postulates that Gasteromycetes are descendants of various groups of agaricoid fungi.

The hypothesis of the origin of secotiaceous and hymenogastraceous Gasteromycetes from various groups of Agaricales s.l. is not new; from the 1950s it was actively developed by Gäumann (1964), Heim (1971), Zerov (1972), Wasser (1974, 1980) and other authors, as opposed to the group of researchers who considered Gasteromycetes the ancestors of basidial fungi (Buchgoltz, 1902 with some reservations; Singer and Smith, 1960; Singer, 1962, 1975; Demoulin, 1974). However, Heim (1971) and his followers tried not to emphasize the question of the origin of Endogasteromycetes, considering this group is rather primitive. Corner's hypothesis in this respect is more radical, since it assumes the origin of Endogasteromycetes from secotioid fungi through their further loss of the initial agaricoid habit and transformation of sporogeneous tissues (Corner, 1964).

A brief summary of this **gasteromycetization** process was provided by Thiers in his remarkable article "The secotioid syndrome" (1984). According to the author, among agaricoid fungi adapting to open habitats, there was a selection of the forms in which the hymenium was protected against the loss of moisture. This protection proved most effective at fixing one of the juvenile stages of sporocarp development, namely, the stage with a closed pileus. At this stage, the function of active spore discharging was lost. This process developed independently in various groups of Agaricales s.l. affected by the "secotioid syndrome". When this active spore discharge ability was lost, the stipe also lost its selective advantages; it was preserved in some groups only as an enclosed columella. The loss of the stipe gradually led many fungi to the formation of underground carpophores. The completely angiocarpous development of the hymenophore also became possible due to stipe loss. Thus a typical gasteroid basidiocarp was formed. The angiocarpous hymenophore development under the peridium resulted in its reduction to a system of lacunas.

The latest data show that in the morphologically fairly simple white-spored agaricoids [e.g., *Lentinus tigrinus* (Bull.:Fr.) Fr.], the mutations can be induced that lead to the formation of secotioid basidiocarps (Hibbett et al., 1997). Sometimes such mutations occur in nature; however, under the current conditions they are not preserved by natural selection.

The basic trends of basidiome transformations in the evolution of Homobasidiomycetes according to the proposed concept can be reflected in the following scheme (Figure 1).

Shape and Structure of Hymenophore. Opinions of various authors on the patterns of evolution of hymenophore are also inconsistent. Undoubtedly, in the first Homobasidiomycetes, the hymenophore was formed by a smooth palisade layer. This type is observed, on the one hand, in clavarioids with negatively geotropic, vertically growing basidiomata, and on the other hand in positively geotropic resupinates, which, beginning from Herter (1910), were placed in the large family Corticiaceae (Bondarzew and Singer, 1941; Eriksson, 1958; Donk, 1964, 1971; Parmasto, 1968, 1986; Jülich, 1972, 1974; Hallenberg and Parmasto, 1998, etc.). It is remarkable that the latter group was most frequently considered initial in Homobasidiomycetes evolution, whereas the possibility of a secondary shift to a smooth hymenophoral surface during their evolution was mostly overlooked.

However, gradually other facts were collected, which allowed a different look at the seemingly stable evolutionary concepts. First, systematists following "classical" opinions began to show some concern over the case of cyphelloid fungi, since the reduced nature of their hymenophore could hardly be denied. On the other hand, in many "hemiagaricoid" fungi (representatives of Tricholomataceae, Lentinaceae, Paxillaceae, Gomphidiaceae, and other groups), as well as in some cantharelloids, the trend to formation of the gill anastomoses is also rather obvious. This tendency is especially clearly expressed in wood-inhabiting fungi. When basidiomata transform from laterally attached to resupinate (turned out, with dorsal attachment), the radial orientation of the gills (folds) becomes completely lost, and the anastomoses become perfectly "equal" elements of the resulting merulioid (irregularly folded) hymenophore. This explanation is readily supported by the case of hymenophore variability in *Serpula panuoides* (Fr.:Fr.) Zmitrovich and *S. olivacea* (Schwein) Zmitrovich, depending on the basidiome attachment to the substrate (Zmitrovich, 2001). A merulioid hymenophore under certain conditions can be transformed into a smooth or nearly smooth one. In some species [*Serpulomyces borealis* (Romell) Zmitrovich, *Amylocorticiellum molle* (Fr.:Fr.) Spirin et Zmitrovich] the initially merulioid hymenophore becomes smoothed on drying (Zmitrovich and Spirin, 2002).

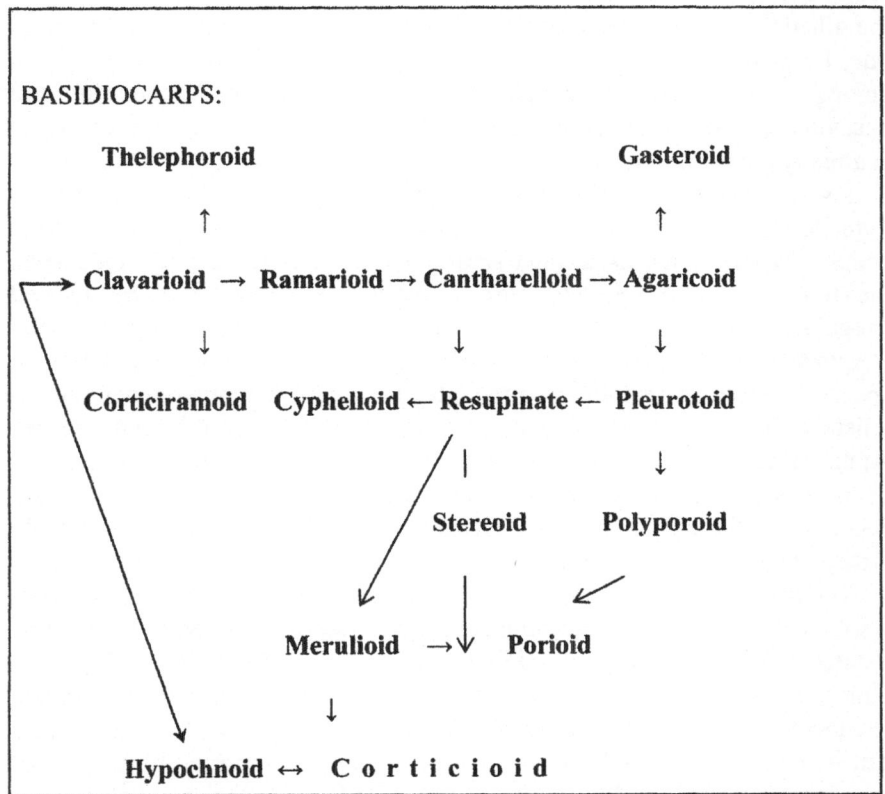

Figure 1. The basic trends of basidiome transformation in the evolution of Homobasidiomycetes.

Thus, the view of corticioid fungi as a homogeneous primitive family of Hymenomycetes should be abandoned. Note that attempts to subdivide this aggregate into more natural and smaller units have been made by Kreisel (1969); he was the first mycologist who stressed the need to use the priority name Peniophoraceae Lotsy, 1907 for a redefined group of Corticiaceae (Parmasto, 1968, 1986, 1995; Jülich, 1981). Despite these efforts, some difficulties still remain.

Let us consider other types of hymenophores, which are derivatives of the merulioid type. The polyporoid type can be considered a specific "update" of the merulioid type: the main difference is that the folds (or tubes) become deeper, and their edges become sterile. Transitions from the merulioid to polyporoid hymenophore are evident in various species: *Gloeoporus panno-cinctus* (Romell) J. Erikss., *Ceriporia tarda* (Berk.) Ginns, *Oxyporus late-marginatus* (Dur. et Mont.) Donk, *Castanoporus castaneus* (Lloyd) Ryvarden, etc. However, in some lines of development the gilled hymenophore can become directly transformed into the polyporoid one, by passing the merulioid stage. A number of examples can be found among Lentinaceae, in which there are species with plentiful gill anastomoses; for other representatives the cellular (favoloid) hymenophore is already characteristic.

As polyporoid aberrations, one interprets the labyrinthiform (daedaleoid) hymenophore, which is connected by intermediate forms with radially-folia-ceous (*Daedaleopsis, Lenzites, Cerrena, Gloeophyllum*) and irpicoid, when the tubes are split up, and with formation of teeth or plates located irregularly (*Irpex*), radially (*Antrodiella, Trichaptum* spp.), or concentrically (*Cyclomyces*). The polyporoid nature of these deviations is obvious to specialists (Bondartsev, 1953), although with regard to morphogenetic paral-lelism with Lentinaceae, this fact, as far as we know, has not been discussed. At the same time, the secondary formation of plates by polyporoid fungi may be regarded as an atavistic phenomenon, evidence of the existence of a segment of "hereditary memory" responsible for the formation of radially arranged elements of the hymenophore.

The hymenophore of *Schizophyllum* presents a problem: it is regarded as a gilled derivative by some authors (of the Lentinaceae type; Pegler and Young, 1971), as polycyphelloid by others (Cooke, 1961; Donk, 1962, 1964; Ginns, 1974; Singer, 1975), or as a modified merulioid one by still others (Stalpers, 1988; Nordic Macromycetes, 1997). Probably all these three ideas can be reasonably united. Since both the cyphelloid and merulioid types of hymenophore originated from the radially-gilled (-folded) one, it is logical to assume that the hymenophore of *Schizophyllum* represents a specialized, originally agaricoid or merulioid hymenophore adapted to sporulating under dry conditions. The morphogenesis of *Sch. commune* Fr.:Fr. was actively studied; however, interpretations of experimental results by various authors are controversial. For example, Gorovoy (1989) assumed that the polycyphelloid structure of its fruit bodies is formed initially. Stalpers (1988), by contrast, specifies that the germinating basidiomata of *Sch. com-mune* are in fact single cupulate basidiomata, resembling those of *Auricula-riopsis ampla* (Lév.) Maire, whereas a more complex structure of hymeno-phore appears at subsequent stages, when its surface becomes ridged, and invaginations arise on ridge edges; it starts splitting further on its external and internal borders. In our opinion, *Schizophyllum* is one of the preserved links between hemiagaricoid (like *Panellus*) and merulioid (*Auriculariopsis,*

Plicaturopsis, Merulius) fungi; it was stabilized in its secondary, xeromorphic state, which has developed from some disturbance at the initial stages of basidiome morphogenesis. Other polycyphelloids (*Fistulina, Stromatoscypha*, etc.) probably also represent some secondary results of the process of development of polyporoid and merulioid types of hymenophore.

We should also briefly consider the issue of origin of the toothed hymenophore. It seems incorrect to treat its origin as a result of some complication of the smooth hymenium: under the initial "fountain" hyphal arrangement in negatively geotropic basidiomata, a more economical way would be expansion of the hymenial surface through forming radial folds. Studies of variation of the hymenophore and its tramal structure in various representatives of *Phlebia, Byssomerulius, Phanerochaete, Hyphoderma*, and *Hyphodontia* oblige us to conclude that the toothed hymenophore is more logically connected with the merulioid type via phlebioid and raduloid (scraper-like) types. The latter has evolved in two directions: through tuberculate to smooth and to odontoid (consisting of regular conical aculei) types. The odontoid hymenophore, in turn, had the potential for further progressive complicated development (formation of the pseudocystidial core, elongation of aculei), and for simplification to the grandinioid (spot-like), farinaceous, or smooth types. Essentially the same processes probably occurred in precursors of stipitate hydnums. The irpicoid hymenophore, as stated above, originated from the polyporoid type (Trameteae - *Irpex* - *Steccherinum*; Trameteae - *Schizopora* - *Hyphodontia*).

Below we propose a scheme reflecting possible trends of transformation of the hymenophore during the process of evolution in Homobasidiomycetes (Figure 2).

Summarizing our discussion of the evolutionary pathways of macrostructure in Homobasidiomycetes, we emphasize the convergent nature of its evolution. The ability to form both negative and positive geotropic (resupinate) sporocarps is observed practically in all phyla of basidial fungi and it is obviously adaptive; here again we see some parallels with red algae, where closely related coralloid and crustaceous forms occur. The above conclusions are also true in respect to the problem of evolution of the hymenophore.

Hyphal Structure. To clarify phylogenetic relationships between taxa, some information on the features of the hyphal structure of carpophores can be useful. Following the publications of Corner (1932a,b, 1953), the concept of a hyphal system (monomitic, dimitic, and trimitic) was introduced in basidiomycete morphology. The type of hyphal system began to be considered as a taxonomic character; however, soon it became clear that its successful use was possible only in combination with other data and characters. The loss of the plasmatic contents by hyphae and formation of sclerified elements is characteristic practically for all groups of homobasidial fungi: clavarioid, gasteroid, agaricoid (especially wood-inhabiting), polyporoid, and stereoid ones; however, various groups are characterized by the specificity of their hyphal differentiation. The pigment content of hyphal walls is also of some taxonomic importance (see Wasser, 1980, 1982, 1985, 2002; Besl et al., 1986; Singer, 1986). A case of a very successful use of hyphal structure characters in taxonomy and phylogeny of a group can be exemplified by comparative studies of Lentinaceae and Polyporaceae.

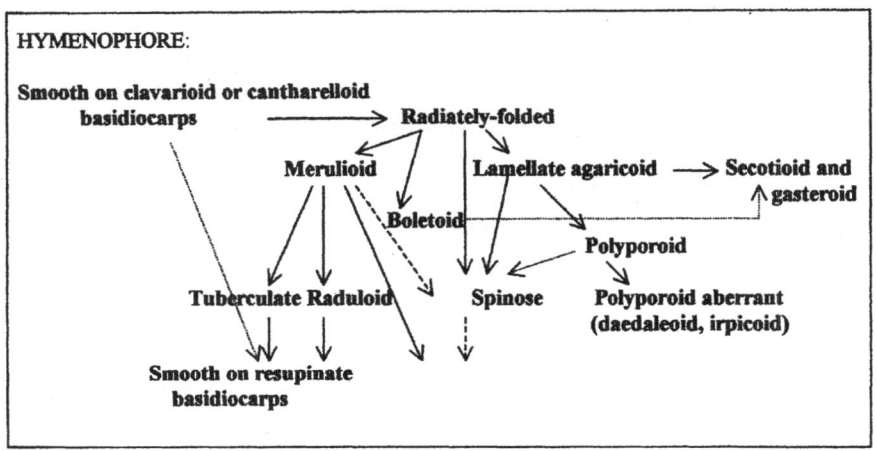

Figure 2. Possible trends of transformation of the hymenophore during the process of evolution in Homobasidiomycetes.

In fact, it is impossible to reveal any anatomic differences between the gilled genera *Lentinus*, *Pleurotus*, *Phyllotopsis* and the tubular genera *Polyporus* or *Favolus*. This had already become obvious to taxonomists by the middle of the last century (in many respects due to work of Corner), which was reflected in taxonomic interpretations of the position of Lentinaceae (we omit the extensive literature on this issue as it has been summarized by Stankovicova, 1973). However, numerous parallels with Lentinaceae in the microscopic structures are revealed not only in the stipitate polypores (Polyporaceae s. str.), but also in trametoid and fomitoid ones. The concept of Polyporaceae without trametoid and fomitoid elements is thus rather vulnerable because of the presence of some groups representing in their microstructure characters a continuous series of transitions: (Polyporeae - [*Microporus, Xerotus, Dichomitus*] - Trameteae - [*Hexagonia, Daedaleopsis*] - Fomiteae).

The peculiarities of internal hyphal structures are usually not used in taxonomy, though based on these characters (the presence of secondary septation), for example, the genus *Pseudocraterellus* has been described (Corner, 1966; Petersen, 1971). The value of the presence/absence of clamp-connections in various taxa was discussed above. In some cases the pattern of arrangement of these structures is useful for taxonomists. For example, *Coniophora* and some representatives of *Phanerochaete* are characterized by multiple clamp connections on one septum, whereas in the genera *Serpula*, *Amylocorticium*, *Ceraceomyces*, or *Parmastomyces* the clamps have a central aperture (observed as a small "eye").

Some researchers consider the presence of differentiated gloeoplerous elements (gloeocystidia, pseudocystidia, macrocystidia, gloeoplerous hyphae, vesicles) in the basidiocarp as an important taxonomic character. In particular, based on the presence of gloeocystidia and gloeoplerous hyphae in the basidiocarp, in combination with the character of amyloidity of spore walls, Stalpers (1996) united a number of diverse agaricoid, clavarioid, and corticioid taxa in the large order Hericiales. However, in our opinion, the taxonomic value of gloeopleroid elements, which actually represents analogues of secretor formations of algae and higher plants (Corner, 1964, 1970), is greatly exaggerated. Besides "Hericiales", these elements can be expressed in various taxa that are not connected by direct phylogenetic relationships (*Clavulicium, Gomphus, Russula, Mycena, Albatilus, Laricifomes, Fomitopsis, Inonotus, Antrodia, Oxyporus, Cylindrobasidium, Punctularia*), whereas in the majority of Homobasidiomycetes the gloeoplerous system is completely reduced. In fact, this character is not correlated in any way with other characters, in particular with amyloidity of the spore wall (Table 1). In this respect it is enough to analyze the pattern of distribution of these two characters in the following pairs of closest relatives: *Stereum* (amyloid) - *Peniophora* (inamyloid), *Aleurodiscus* (amyloid) - *Corticium* (inamyloid), *Dichostereum* (amyloid) - *Scytinostroma* (reaction varying). The reaction of the gloeocystidia content with sulfoaldehyde, which many taxonomists originally regarded as a kind of "panacea", proved to be rather variable, sometimes even within one species, depending on the age stage and humidity of the specimen (Larsen and Burdsall, 1976).

In a phylogenetic context, we may believe that in ancestral Homobasidiomycetes (*"Proto-Clavulinaceae"*) the gloeoplerous hyphal system was well advanced, and during its evolution in different phyla, it was to various degrees, preserved or modified, or disappeared. However, sterile elements of the hymenophore (except hyphidia and pseudocystidia) could be considered as its derivatives. Studies of peculiarities of hymenial elements (and their transitional stages) in *Stereum, Peniophora, Metulodontia, Oxyporus, Hyphoderma*, and *Subulicystidium* clearly show that lamprocystidia and various grades of leptocystidia (guttating and those that have lost this ability), as well as various "cysts", are connected by intermediate stages with gloeocystidia and vesicles, whereas tubular cystidia, metuloids, and setae demonstrate a certain similarity with pseudocystidia.

The peculiarities of basidia as basic hymenial elements indeed play an important part in taxonomy. Note, however, that their phylogenetic value greatly varies depending on the "background" determined by the specificity of evolutionary tendencies of any concrete phylum. The shape of the homobasidium is a rather deceptive character. Thus, the karyologically similar basidia of Clavulinaceae, Cantharellaceae, or Sistotremataceae are very different in their habit. Basidia of Cantharellaceae are surprisingly similar to long chiastobasidia of some stereoid fungi (like *Veluticeps*); basidia of *Sistotrema* remind suburniform basidia of *Hyphodontia*, and basidia of *Clavulicium* (Clavulinaceae) are similar to podobasidia of *Radulomyces* (Cyphellaceae). On the other hand, in many cases the particular shape of basidia correlates with other groups of characters, and in such cases, we probably observe well-distinguished natural groups of Homobasidiomycetes.

In trying to distinguish any basic trends in the morphological transformations of homobasidia, we must proceed from the obvious fact that elongate cylindrical basidia are characteristic for negatively geotropic fungi with smooth or venose hymenophores. Shorter clavate basidia are developed on gilled and tubular hymenophores of negative or positive geotropic basidiomata, and in many groups not connected by close relationships; they have a common tendency to widening and truncation. In positively geotropic fungi with smooth hymenophores, we see some regularity in development of basidia (as well as other elements of the hymenium): all of them, to a greater or lesser degree, are inclined to adhere to the substrate. This tendency is maximally realized in *Xenasma*-like fungi with pleurobasidia and multiradicate cystidia. In this respect, we cannot agree with the theory developed by Oberwinkler (1977, 1978, 1982) on the primary, initial position of pleurobasidia in the evolution of Homobasidiomycetes. The medial constriction, which is evident in many types of homobasidia, is thus better interpreted as a borderline between the pro- and metabasidial segments, but not as a truncated hyphal end, as supposed by the author.

The spore characters (shape, size, color, ornamentation) are also extremely important in the taxonomy of the group; however, with the use of these characters it is even more hazardous to disregard the peculiarities of the "body plan" of the classified organisms. Let us explain this with positive and negative examples.

It is widely known that Polyporaceae and Lentinaceae are currently regarded as related based on their hyphal systems, the shape and dimensions of basidia, the presence of characteristic metuloids and hyphal pegs, an irregular trama, and also the tendency to spore elongation (from ellipsoid to fusoid-cylindrical). In this example, the spore features are well correlated with a complex of other characters determining the specificity of this group. As a negative example, we mention the proposal by Locquin (1981) to recognize the order of agaricoid fungi, Mycenales, which is in fact based on a single character - thin-walled non-ornamented spores.

The thickening, melanization, and ornamentation of the spore wall is quite an ordinary phenomenon in the evolution of the plant and fungal kingdoms. Some degree of convergence is possible here even between the representatives of various phyla, for example, Zygomycota (Mucorales) and Eumycota (Boletogastreae). In that respect it is probably inappropriate in phylogenetic reconstructions to segregate as the primary (basal) clades the dark-spored and light-spored lines of development of agaricoid fungi (Miller and Watling, 1987), especially since the phylogenetic connections between representatives of dark-spored and light-spored agarics are evident (Wasser, 1982, 2002). It is also hardly justified to unite various groups of amyloid-spored Aphyllophorales (Ryvarden, 1991; Stalpers, 1996); the so-called "sporocentrical" approach to solving difficult problems of gasteromycete phylogeny (Locquin, 1981) is also questionable.

Concerning the evolutionary transformation of the form (shape) of basidiospores, even in onthogenesis, its obvious tendency to elongation is well expressed (Corner, 1947). In different groups of ballistosporous (as well as statismosporous) Homobasidiomycetes; certain stages of lengthening were registered depending upon ecological conditions. In gasteroid statismo-sporous fungi, the basidiospores are usually spherical or regularly ellipsoid. Regularly spherical spores are not characteristic for ballistosporous Homo-basidiomycetes. Their spores are commonly more or less extended along their longitudinal axis. Most typical for all these groups are ellipsoid spores with a suprahilar depression. After a longitudinal excess of some critical length (usually Q > 2.5-3), tendencies are evident of some incurving of spores (so-called allantoid or sigmoid shape), or of concentration of mass around the equatorial plane (fusoid), less often on one of the poles (pyriform or amygdaloid shape). Regularity is the existence of a common correlation between the spore size and the length of basidia. In many evolutionary lines, a trend toward some reduction of basidia and diminishing of spores is observed (*Merulius - Gloeoporus - Skeletocutis*; Trameteae – *Diplomito-porus*; *Serpula - Parmastomyces*, etc.).

2.3   Phylogenetic lines of Homobasidiomycetes

On the basis of the above discussion, we shall try to draw some conclusions on the possible phylogenetic lines in Homobasidiomycetes in general (Figure 3).

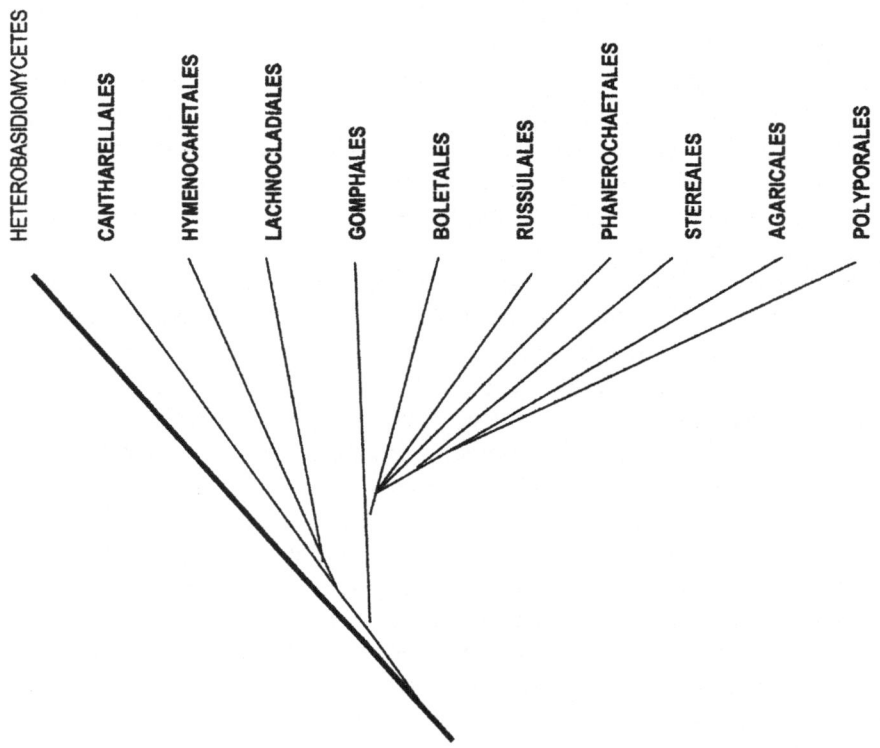

Figure 3. Cladistic interpretation of possible relationships between body plans of Homobasidiomycetes (data sets are presented in Table 1).

Figure 3 presents a hypothetical phylogenetic tree of Homobasidiomycetes, which provides data on the origin and deviations from the common
archetype of large groups of that class; circumscriptions and diagnostic
characters are provided in Table 1.

As we see from Table 1 above, most archetypal features are preserved in
Cantharellales. This small group is characterized by basidia of a primitive
structure (stichobasidia), having an unstable number of basidiospores (2-8)
and **continuous parenthosomata**. Practically all representatives of the
group demonstrate the tendency to form negatively geotropic basidiomata
(including even some representatives of *Sistotrema* and *Trechispora*,
normally having effused prostrate basidiocarps).

Table 1. Taxonomic subdivision of Homobasidiomycetes based on generalized morphological (including ultrastructural) data sets

Orders	Diagnostic characters	General families
1. Cantharellales	Parenthosomata continuous; basidia stichic, utriform to urniform or ovate, 2-6-spored; hyphae/epibasidium ratio ≥ 2; spores thin-walled or with distinct walls, smooth; basidiomata cantharelloid (clavarioid) to resupinate, monomitic; clamp connections more or less inflated.	Clavulinaceae Cantharellaceae Sistotremataceae Botryobasidiaceae
2. Hymenochaetales	Parenthosomata continuous; basidia chiastic, suburniform to clavate, 2-4 (6)-spored; hyphae/ epibasidium ratio ≥ 2; spores thin- to thick-walled, smooth, inamyloid; basidiomata clavarioid to sessile or resupinate, mono- to pseudodimitic or dimitic; clamp connections, if present, of normal appearance.	Clavariachaetaceae Hymenochaetaceae Inonotaceae Phellinaceae Rigidoporaceae (incl. Schizoporaceae)
3. Lachnocladiales	Parenthosomata perforated; basidia chiastic, suburniform, utriform, or clavate, 2-4-spored; hyphae/ epibasidium ratio ≥ 2; spores thin-walled, amyloid or dextrinoid; basidiomata clavarioid, sessile, or resupinate, dimitic or trimitic with dextrinoid, often strikingly ramified skeletals and often also gloeoplerous hyphae; clamp connections, if present, of normal appearance.	Lachnocladiaceae Dichostereaceae Perenniporiaceae Ganodermataceae
4. Gomphales	Parenthosomata perforated; basidia chiastic, clearly utriform, 2-4 (6)-spored; hyphae/epibasidium ratio ≥ 2; spores with ± distinct wall, commonly thick-walled, Melzer's-negative; basidiomata clavarioid, gasteroid, or resupinate, monomitic or pseudodimitic, clamp connections opened, inflated, in some cases of normal appearance.	Gomphaceae Ramariaceae Xenasmataceae Clavariaceae Atheliaceae Cyphellaceae Hyphodermataceae Thelephoraceae Clathraceae Phallaceae Hysterangiaceae Singeromycetaceae
5. Boletales	Parenthosomata perforated; basidia chiastic, utriform to clavate, (2) 4-6 spored; hyphae/epibasidium ratio ≥ 2; spores Melzer's positive, with distinct to thick wall, smooth or ornamented, commonly pigmented; basidiomata agaricoid to resupinate or gasteroid, monomitic to sarcotrimitic; gloeoplerous hyphae present or not; clamp connections, when present, slightly to massively inflated.	Gomphidiaceae (incl. Rhizopogonaceae) Paxillaceae (incl. Meiorganaceae, Gyrodontaceae, Amylocorticiaceae) Coniophoraceae Boletaceae (incl. Xerocomaceae) Strobilomycetaceae Sclerodermataceae Astraeaceae
6. Russulales	Parenthosomata perforated; basidia chiastic, clavate, 4-spored; hyphae/epibasidium ratio ≥ 2; spores ± amyloid, commonly ornamented, mostly with distinct wall; basidiomata clavarioid, sessile polyporoid, agaricoid,	Leucopaxillaceae Albatrellaceae (incl. Amylosporaceae) Hericiaceae

	gasteroid, or resupinate, mono- to sarco(pseudo)dimitic; gloeoplerous hyphae commonly present; clamp connections, when present, slightly inflated.	(incl.Gloeocystidiellaceae) Auriscalpiaceae Bondarzewiaceae Russulaceae (incl. Elasmomycetaceae, Hybogasteraceae)
7. Stereales	Parenthosomata perforated; basidia chiastic, utriform to cylindrical, (2) 4-spored; hyphae/epibasidium ratio 1.5-3; spores amyloid or not, thin-walled, smooth, or minutely rugulose, without pigmentation; basidiomata agaricoid to resupinate, monomitic to pseudodimitic; gloeoplerous hyphae present or not; clamp connections, if present, of normal appearance.	Mycenaceae (incl. Aleurodiscaceae) Peniophoraceae (incl. Corticiaceae, Punctulariaceae) Stereaceae Schizophyllaceae Cylindrobasidiaceae
8. Phanerochaetales	Parenthosomata perforated; basidia chiastic, clavate-cylindrical, (2) 4-spored; hyphae/epibasidium ratio ≤ 2; spores inamyloid, smooth, mostly with distinct wall; pleurotoid to resupinate or sessile, monomitic to sarcotrimitic, hyphae with rectangular branching pattern; gloeoplerous hyphae present or not; commonly efibulate or with inflated (multiple) clamp connections.	Hohenbueheliaceae Fistulinaceae (incl. Grifolaceae, Meripilaceae, Sparassidaceae) Laetiporaceae (incl. Phaeolaceae) Phanerochaetaceae
9. Agaricales	Parenthosomata perforated; basidia chiastic, utriform to clavate, (2) 4-spored; hyphae/epibasidium ratio ≥ 2; spores Melzer's negative or positive, thin-walled or with distinct wall, pigmented or not; basidiomata agaricoid to resupinate or gasteroid, monomitic to trimitic; gloeoplerous hyphae present or not; clamp connections, when present, of normal appearance. Cf. Boletales and Gomphales.	Tricholomataceae (incl. Xerulaceae, Podoscyphaceae) Laccariaceae (incl. Hydnangiaceae) Hygrophoraceae Catathelasmataceae Amanitaceae Pluteaceae Entolomataceae Agaricaceae (incl. Secotiaceae, Podaxaceae) Battareaceae Tulostomataceae Lycoperdaceae Nidulariaceae Coprinaceae Bolbitiaceae Strophariaceae Cortinariaceae (incl. Crepidotaceae) Hymenogastraceae Gastrosporiaceae
10. Polyporales	Parenthosomata perforated; basidia chiastic, subutriform to clavate, 4-spored; hyphae/epibasidium ratio ≥ 2; spores inamyloid, mostly thin-walled and smooth, without	Polyporaceae (incl. Lentinaceae, Coriolaceae,

pigmentation; basidiomata pleurotoid to sessile or resupinate, mainly di-, trimitic; clamp connections commonly present, of normal appearance.	Fomitaceae) Irpicaceae Fomitopsidaceae

Continuous parenthosomata are characteristic for another primitive group, Hymenochaetales, which according to recent studies, includes setae-bearing taxa, and also the genera *Hyphodontia, Oxyporus,* and *Trichaptum* (Langer, 1994, 1998; Keller, 1997). Basidia in that group are already chiastic, but in many representatives they preserve their suburniform shape, reminiscent of *Sistotrema*. A hypothetical archetype of the group was probably similar to the tropical genus *Clavarichaete* characterized by clava-rioid basidiomata, 6-spored basidia, and tramatic setae. A sister group to Hymenochaetales is probably Lachnocladiales. However, the latter are characterized by the presence of dextrinoid skeletals with more or less regular dichotomous (arboriform) branching and Melzer's-positive (some-times appreciably ornamented) basidiospores.

The order Gomphales, characterized by **perforated parenthosomata**, utriform chiastobasidia, and more or less rough/thick-walled, often ornamen-ted spores, is also phylogenetically connected with Cantharellales, which can be seen in multiple parallels between such taxa as *Clavulina* (Cantharel-lales) - *Clavaria* (Gomphales), *Craterellus* (Cantharellales) - *Gomphus*, and some others.

Results of a comparative morphological analysis testify to a high probability that the basic phylogenetic radiation of Homobasidiomycetes (Boletales-Russulales-Agaricales) is connected with this group. All these groups, like Gomphales, have utriform to clavate chiastobasidia with more or less expressed medial constrictions, in many agaricoid and boletoid taxa with an inflated epibasidial part (see Oberwinkler, 1977, 1982; Stalpers, 1992). This phylum is the foremost in its species diversity; its development proceeded mostly in the direction of colonizing mainly terrestrial soil habitats, although in various groups fairly successful attempts of adaptation to the wood substrate are observed.

Representatives of Tricholomataceae have in their basic features preserved the *Cantharellus*-like type of organization, though in many evolu-tionary lines of the family we see tendencies to transformations toward com-plication and reduction.

The problem of monophyly of Agaricales is still unresolved. If we assume the monophyly concept, this group may be successfully traced phylogenetically from Leucopaxillaceae (Russulales). In this case, Hygro-phoraceae, as well as various dark-spored taxa, will necessarily be phylo-genetically deduced from Tricholomataceae. Extremely simplifying this issue, we may propose such lines as *Omphaliaster*-Hygrophoraceae, *Phaeo-lepiota*-Cortinariaceae, and *Catathelasma*-Agaricaceae. In the case of polyphyly, the Tricholomataceae seem to be rooted in the domain of Boletales /Russulales divergence (Leucopaxillaceae), whereas the Hygrophoraceae(ales) and Cortinariaceae(ales) took root near to the Gomphales (in this case, it would be worthwhile examining parallels in the pairs Hygrophoraceae-Clavariaceae, Entolomataceae-Clavulinopsis, and Cortinariaceae-Ramariaceae). In this case, the dark-spored taxa can be also rooted phylogenetically "lower" than Tricholomataceae (several possibilities exist to connect these lines with Cortinariaceae or Paxillaceae).

The various groups of Agaricales were involved in the process of gasteromycetization, and among these groups we shall look for the roots of such a large group of Gasteromycetes as Lycoperdaceae.

The order Polyporales is definitely evolutionarily connected with clitocyboid Tricholomataceae (*Clitocybe-Pleurotus-Lentinus-Polyporus*-Trametoideae), whereas another group of wood-inhabiting fungi, Stereales, is connected with Russulales (*Cantharellula* from the family Leucopaxillaceae is probably the taxon closest to the archetype). Probably two lines were separated at the earliest stages of evolution of Stereales: (*i*) *Cantharellula*-Mycenaceae-Peniophoraceae (Stereaceae), and (*ii*) *Cantharellula-Sarcomyxa*-Cylindrobasidiaceae and Schizophyllaceae (Meruliaceae). The orders Polyporales and Stereales are joined by the related small group Phanerochaetales, which is also rooted in the phylogenetic zone of Leucopaxillaceae - clitocyboid Tricholomataceae. In that group there are many parallels with Albatrellaceae, but the spores of Phanerochaetales are Melzer's-negative, the hyphae/epibasidium ratio does not exceed 2, and striking rectangularily branched hyphae are present. Several sarcotrimitic polypores (such as *Abortiporus, Grifola, Osteina,* and *Pilatoporus*) were recently transferred to this group (Zmitrovich et al., 2003).

3.   MOLECULAR TREE (MOLECULAR PHYLOGENY)

Several evolutionary trees of Homobasidiomycetes have been reconstructed based on data derived from sequences of 18S nuclear and 12S mitochondrial (n/mSSU) rDNA genes (Hibbett et al., 1997; Hibbett and Thorn, 2001; Redhead et al., 2002; Figure 4) or from the comparative studies of sequences of nuclear large subunit (nLSU) rDNA genes (Larsson et al., 2002; Redhead et al., 2002; Moncalvo et al., 2003). An attempt has also been made to build a global tree based on the ITS region (Boidin et al., 1998); however, this region is too polymorphic to address higher-level phylogeny issues.

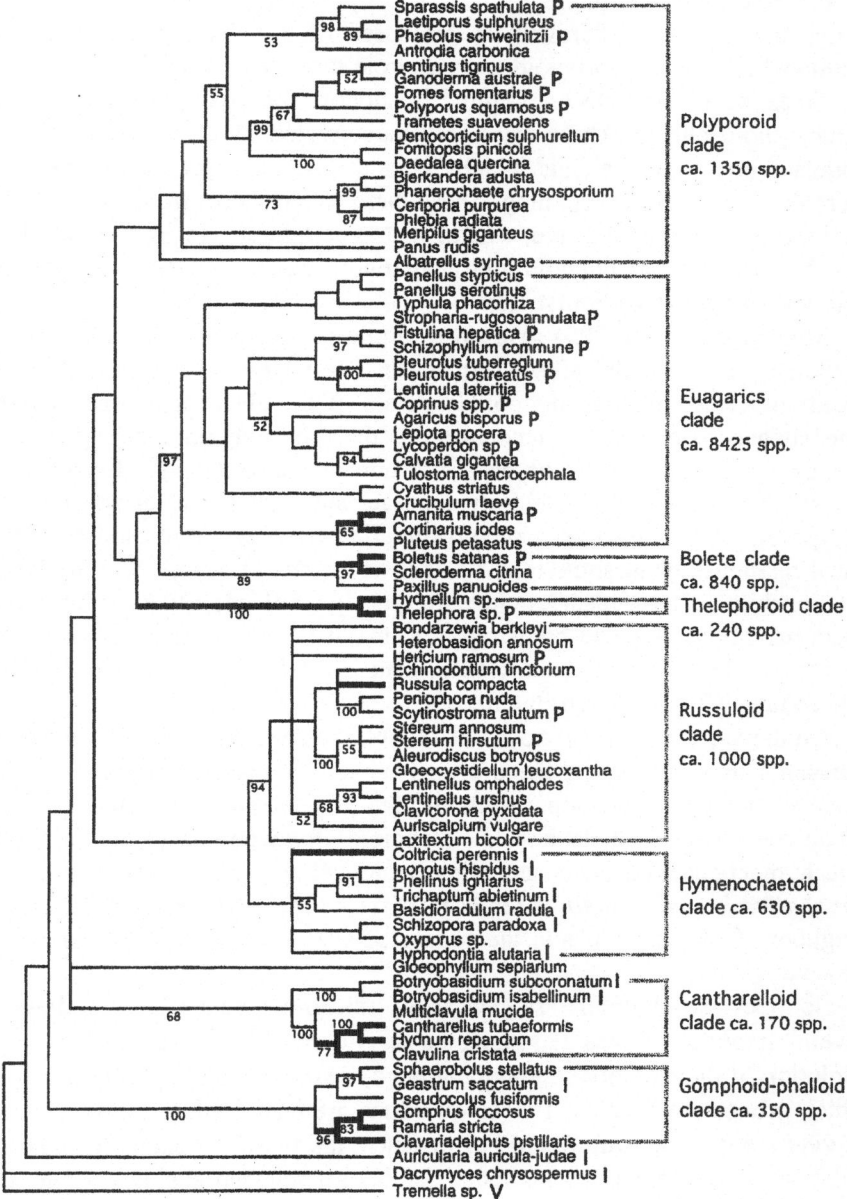

Figure 4. Large-scale Homobasidiomycetes phylogeny based on analysis of nuc-SSU and mt-SSU rDNA sequences (Hibbett and Thorn, 2001).

The common traits of the SSU and nLSU trees are their "roots" (sequences of representatives of Auriculariales-Dacryomycetales) and their topology: the longest (hence, the oldest) branches are formed by cantharelloid and gomphoid-phalloid clusters; a hymenochaetoid cluster develops above them, and the "crown" is composed of initially agaricoid groups, within which various corticioid and gasteroid taxa are dispersed.

Trees of nLSU rDNA (Figure 5) support the basal position of the cantharelloid cluster. These trees are characterized by a larger quantity of compact groups as corticioid, phlebioid, and athelioid clades are singled out from the "polyporoid" grouping. These three clusters shift from the "crown" to the central part of the tree, where, with a certain degree of rotation, they prove to be close to boletoid and euagaricoid clades. The latter forms the "crown" of Homobasidiomycetes.

Moncalvo et al. (2003) give an example of elegant work on euagarics phylogeny using nLSU rDNA sequences. Their large-scale tree of Homobasidiomycetes supports monophyly of six of the above-mentioned clades (the euagarics, bolete, hymenochaetoid, thelephoroid, gomphoid-phalloid, and cantharelloid clades), the basal position of the cantharelliod clade, and the reciprocal monophyly of Polyporaceae and corticoid clades within the polyporoid clade. However, some difficulties arose in the resolution of the basal relationships in some major clades of the tree. For instance, the basal relationships of the euagarics clade are poorly resolved, although the maximum parsimony tree and bootstrap 50% majority rule consensus tree represent more than 100 distinct clades with considerable bootstrap support (exceeding 70% for most of them).

On the whole, the basal nodes of Monclavo's tree correspond to those of Larsson's tree. Moreover, the order of separating clades starting from the base of the tree to the top maintains, the general pattern: cantharelloid - gomphoid-phalloid - thelephoroid - corticioid - hymeno-chaetoid - russuloid. The hymenochaetoid clade including a number of agaricoid taxa (such as *Omphalina brevibasidiata* or representatives of the genus *Rickenella*) is the neighbor of the bolete clade, the suggested sister group of the euagarics in Moncalvo's tree.

So trees based on the sequences of ribosomal genes have addressed several important issues regarding basic patterns of Homobasidiomycetes evolution: if there are monophyletic groups; if so, what these groups are, and which groups are basal. All these questions are answered, but only in part, leaving many issues still open. For instance, the russuloid clade, the agaric-bolete-russuloid "triangle", and others, although included in studies of genes, vary in respect of rates of evolution (18S evolves slower than the nLSU), and, in general, should allow a better resolution of higher-level phylogenetic groupings (terminal as well as deeper levels of branching).

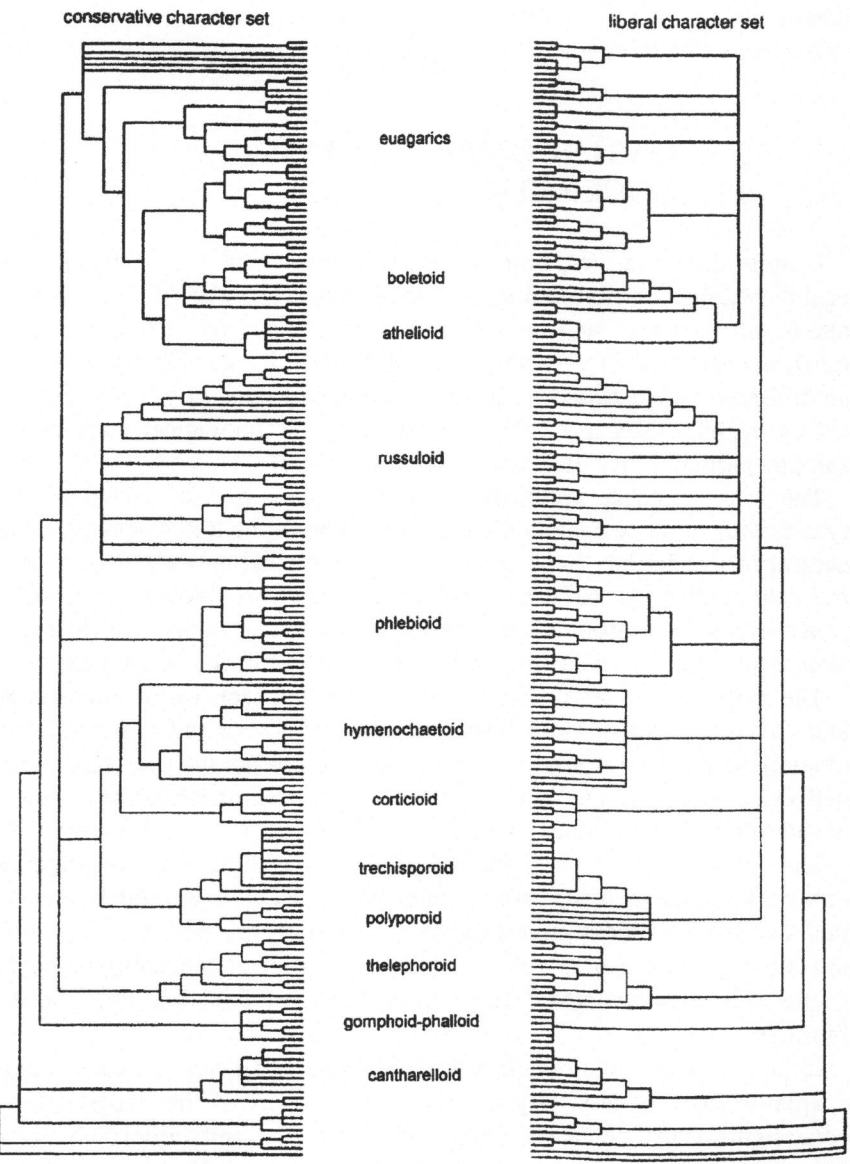

Figure 5. Large-scale Homobasidiomycetes phylogeny based on analysis of nuc-LSU rDNA sequences (Larsson et al., 2002).

However, apart from these question marks, the phylogenies confirm many groups previously outlined by classical methods (based on morphology, microstructures, and hyphal ultrastructure). It would be correct to address the following question: to what extent do both types of the trees correspond to the morphological tree?

4.   RECONSTRUCTING THE TRUE PHYLOGENETIC TREE

Despite dissimilarities, both groups of trees accord well with the most general features of morphological topologies. This accord is evident in patterns of placement and mutual distribution of groups of long and short branches, conditioned by the significant difference in age between them. The age difference is expressed in different degrees of rDNA polymorphisms as well as in the existence of the corresponding morphological markers (for example, particularities of the hyphal ultrastructure).

The long branches appearing from the crown part of Heterobasidio-mycetes thus produce cantharelloid and gomphoid-phalloid clades. It is also necessary to underline that traditional morphological as well as ultrastruc-tural data confirm the basal position of Cantharellales (including Sisitostre-mataceae and Botryobasidiaceae) as these fungi have nonperforated paren-tosomes and sticho basidia with an indeterminate number of sterigmata.

The most general reason for the differences between various trees is the nature of divergence of Homobasidiomycetes, characterized as evolutionary radiation several times above. The major Homobasidiomycetes taxa sepa-rated and diverged during an insignificant period of time (from the middle to the end of the Carboniferous), which later developed in parallel lineages.

Thus, the interpretation of the tree of Homobasidiomycetes evolution as deriving from the existing trees cannot be clear-cut. It should be kept in mind that in reality these groups did not "overbuild" one on top of another in the course of evolution as a result of an even order of branching of phylo-genetic trees; they diverged during a very short period of time from the area of primary radiation.

If the gomphoid-phalloid clade is understood under this radiation region, it explains more or less why the bootstrap support of the nodes situated above significantly decreases (from 100% for the cantharelloid cluster to 57% of the node of large radiation in Larsson's tree); rDNA does not have enough resolution for reliable clusterization of the top polytomy.

All the foregoing confirms the suggestion that the basic trees of Homobasidiomycetes are associated with the gomphoid area, and their further development was parallel. This conforms to Corner's phylogeny, postulating a cantharelloid ancestor of the group (1966) and further "gomphoid-agaricoid" radiation (1972) accompanied by gasteromycetization and corticoid simplification, although it leaves open a number of other questions. In this respect, claims such as "gilled mushrooms evolved at least six times, from morphologically diverse precursors" (Hibbett et al., 1997) probably need some amendment: more likely is the existence of several (let us say, six) lines, within which the initial cantharelloid or agaricoid ancestors gave rise to the formation of non-gilled forms.

* *
*

Reconstruction of the phylogenesis of Homobasidiomycetes is far from complete. No serious factual and methodological assumptions for advancing the process toward its end are seen. Molecular systematics, factors of which captured the minds of taxonomists several years ago, enter into a new and more constructive phase (in light of previous experience). The appearance of numerous noncongruent molecular topologies has assisted the process to a great extent.

Nowadays, despite the long-lasting debate on molecular versus morphological systematics, it is very important to realize that any reconstruction of phylogenies should be based on some analysis of molecular characters together with morphological, anatomical, and ultrastructural traits. However, incorporation of such different data sets into a joint analysis entails a number of difficulties already widely discussed. The issues considered are the assigning of weight to molecular and nonmolecular characters, violation of weighting methods for various kinds of evolutionary changes on the incorporation of characters from different sources into one data set, etc. (Kluge, 1983; Miyamoto, 1983; Lavin and Doyle, 1991). Possible solutions are to analyze morphological and molecular characters separately and to create one hypothesis later on using agreed methods (Adams, 1972); to conduct combined analysis (Hillis, 1987; Donoghue and Sanderson, 1991); to record recoding of terminal taxa of the molecular tree as states of one trait followed by its inclusion into a general (molecular-morphological) matrix (Schantser, 2003). Each of the solutions has its own drawbacks, as well as advantages: combined analysis maximizes parsimony for all putative homologies, while independent analysis could reveal disagreements between the data sets, which might not be apparent in a combined analysis (Hibbett, 1992).

Either way, progress in this field should be attributed to the accumulation of new data (both molecular and morphological) as well as the maintenance of one of the most important principles of systematics: the totality of traits and combinations of characters is always more important than one trait or one character alone - no matter how important this sole character seems to be.

5.   ACKNOWLEDGMENTS

We are very grateful to Professor S.L. Mosyakin (Ukraine), Mr. M. Rosovsky (Israel), Dr. W.A. Spirin (Russia), and Ph.D. student M.Ya. Didukh (Ukraine) for skillful comments on the contents, language, and style of the manuscript.

6.   REFERENCES

Adams E. 1972. Consensus techniques and the comparison of taxonomic trees. Syst Zool. 21, 390-397.

Ainsworth and Bisby's Dictionary of the Fungi. 2001. 9th ed. Kirk P.M., Cannon P. F., David J.C., and Stalpers J. A. (Eds.), CABI, Wallingford.

Ander P. and Marzullo L. 1997. Sugar oxidoreductases and vetatril alcohol oxidase as related to lignin degradation. J Biotechnol. 53, 115-131.

Arx J.A. 1967. Pilzkunde. Lehre. 356 S.

Arx J.A. 1979. Propagation in the yeasts and yeast-like fungi. In The Whole Fungus, v. 2. pp. 555-566. Ottawa.

Bauer R., Oberwinkler F. and Vánky K. 1997. Ultrastructural markers and systematics in smut fungi and allied taxa. Can J Bot. 75, 1273-1314.

Besl H., Bresinsky A. und Kammerer A. 1986. Chemosystematic der Coniophoraceae. Z Mykol. 52, 277-286.

Bessey E.A. 1950. Morphology and Taxonomy of Fungi. Philadelphia/Toronto, 791 pp.

Boidin J., Mugnier J. and Canales R. 1998. Taxonomie moleculaire des Aphyllophorales. Mycotaxon 66, 445-491.

Bondartseva M.A. 1993. Life forms of higher fungi in European ecosystems. In Fungi in Europe: Investigation, recording and conservation, Pegler D.N., Boddy L., Kirk P.M., (Eds.), Roy Bot Gard., Kew, pp. 157-170.

Bondartseva M.A. 1997. Evolutionary relationships and taxonomic position of tinder fungi (Polyporaceae s. lato). Mikol Fitopatol. 31, 76-83 (in Russian).

Bondartsev A.S. 1953. The Polyporaceae of the European part of the USSR and the Caucasia. Acad. Sci. USSR, Moscow-Leningrad (in Russian).

Bondarzew A. und Singer R. 1941. Zur Systematik der Polyporaceae. Ann Mycol. 39, 43-65.

Buchgoltz F.V. 1902. Materials to the Morphology and Systematics of Hypogean Fungi (Tuberaceae and Gastromycetes pr. p.). Riga (in Russian).

Chadefaud M. 1960. Les Végétaux non Vasculaires (Cryptogamie). In Traite de Botanique Systématique, Chadefaud M. et Emberger L., (Eds.), Masson et Cie Editeurs, Paris, t. 1, 1-1018.

Chadefaud M. 1975. Sur le "basidies a répétition" des *Repetobasidium*. Rev Mycol. 39, 173-179.

Cooke W.B. 1961. The genus *Schizophyllum*. Mycologia. 53, 575-599.

Corner E.J.H. 1932a. The fruitbody of *Polystictus xanthopus* Fr. Ann Bot. 46, 71-101.

Corner E.J.H. 1932b. A *Fomes* with two systems of hyphae. Trans Brit Myc Soc. 17, 51-81.

Corner E.J.H. 1947. Variation in the size and shape of spores, basidia, and cystidia in Basidiomycetes. New Phytol. 46, 195-228.

Corner E.J.H. 1953. The construction of polypores. 1. Introduction. Phytomorph. 3, 152-167.

Corner E.J.H. 1954. The classification of the higher fungi. Proc Linn Soc Lond. 165, 4-6.

Corner E.J.H. 1964. The Life of Plants. Clowes and Sons, London.

Comer E.J.H. 1966. A Monograph of Cantharelloid Fungi. Oxford Univ. Press, Oxford.

Comer E.J.H. 1968. A monograph of *Thelephora*. Beih Nova Hedwigia 27, 1-110.

Comer E.J.H. 1970. Supplement to "A monograph of *Clavaria* and allied genera". Beih Nova Hedwigia 33, 1-299.

Comer E.J.H. 1971. Merulioid fungi in Malaysia. Gdns'. Bull Singapore. 25, 355-381.

Comer E.J.H. 1972. *Boletus* in Malaysia. Gov. Printing Office, Singapore.

Demoulin V. 1974. The origin of Ascomycetes and Basidiomycetes. The case for a red algal ancestry. Bot Rev. 40, 315-345.

Donk M.A. 1948. Notes on Malesian fungi. Bull Bot Gdns Buitenzorg. 17, 473-483.

Donk M.A. 1954a. A note on sterigmata in general. Bothalia. 6, 301-302.

Donk M.A. 1954b. Notes on resupinate Hymenomycetes. I. On *Pellicularia* Cooke. Reinwardtia 2, 425-434.

Donk M.A. 1956. Notes on resupinate Hymenomycetes. II. The tulasnelloid fungi. Reinwardtia 3, 363-379.

Donk M.A. 1962. Notes on Cyphellaceae - 2. Persoonia 2, 331-348.

Donk M.A. 1964. A conspectus of the families of Aphyllophorales. Persoonia 3, 199-324.

Donk M.A. 1971. Multiple convergence in the polyporaceous fungi. In Evolution in the Higher Basidiomycetes, Tennessee Press, Knoxville. pp. 393-422.

Donoghue M. and Sanderson M. 1991. The suitability of molecular and morphological evidence in reconstructing plant phylogeny. In Molecular Systematics in Plants, Soltis D., Soltis P., and Doyle J. (Eds.).

D'Souza T.M., Boominathan K. and Reddy C.A. 1996. Isolation of laccase gene-specific sequences from white rot and brown rot fungi by PCR. Appl Environ Microbiol. 62, 3739-3744.

Eriksson J. 1958. Studies in the Heterobasidiomycetes and Homobasidiomycetes-Aphyllophorales of Muddus National Park in North Sweden. Uppsala.

Gäumann E. 1926. Vergleichende Morphologie der Pilze. Gustav Fischer, Jena.

Gäumann E. 1964. Die Pilze. Birkäuser Verl., Basel/Stuttgart.

Gilbertson R.L. 1980. Wood-rotting fungi of North America. Mycologia 72, 1-49.

Ginns J. 1974. *Schizophyllum commune*. Fungi Canadensis, 42.

Gorovoy L.F. 1989. Morphogenesis in the Gilled Fungi. Naukova Dumka Press, Kiev, 167 pp. (in Russian).

Hallenberg N. and Parmasto E. 1998. Phylogenetic studies in species of Corticiaceae growing on branches. Mycologia 90, 640-654.

Heim R. 1971. The interrelationships between the Agaricales and Gasteromycetes. In Evolution in the Higher Basidiomycetes, Tennessee Press, Knoxville. pp. 505-534.

Herter W. 1910. Pilze. 2. Autobasidiomycetes. In Kryptogamenflora der Mark Brandenburg, bd 6. S. 1-192.

Hibbet D.S. 1992. Ribosomal RNA and fungal systematics. Trans Myc Japan 33, 533-556.

Hibbett D.S. and Donoghue M.J. 1995. Progress toward a phylogenetic classification of the Polyporaceae through parsimony analysis of mitochondrial ribosomal DNA sequences. Can J Bot. 73, suppl. 1: sect. E-H, 853-861.

Hibbett D.S., Pine E.M., Langer E., Langer G. and Donoghue M.J. 1997. Evolution of gilled mushrooms and puffballs inferred from ribosomal DNA sequences. Proc Natl Acad Sci USA. 94, 12002-12006.

Hibbett D.S. and Thorn R.G. 2001. Basidiomycota: Homobasidiomycetes. The Mycota VII Part B. McLaughlin D.J., McLaughlin E.G., and Lemke P.A., (Eds.), pp. 121-168. Springer Verlag, Berlin, Heidelberg, NY.

Hillis D. 1987. Molecular versus morphological approaches in systematics. Ann Rev Ecol Syst. 18, 23-42.

Hyde S.M. and Wood P.M. 1997. A mechanism for production of hydroxyl radicals by brown rot fungus *Coniophora puteana*: Fe (III) reduction by cellobiose dehydrogenase and Fe (II) oxidation at a distance from the hyphae. Microbiol. 143, 259-266.

Jones D.R. 1973. Ultrastructure of septal pore in *Uromyces dianthi*. Trans Brit Myc Soc. 60, 227-235.

Juel H.O. 1898. Die Kernteilung in den Basidien und die Phylogenie der Basidiomyceten. Jahrb Wiss Bot. 32, 361-388.

Jülich W. 1972. Monographe der Athelieae (Corticiaceae, Basidiomycetes). Willdenowia Bh. 7, 1-283.

Jülich W. 1974. The genera of the Hyphodermoideae (Corticiaceae). Persoonia 8, 59-97.

Jülich W. 1981. Higher Taxa of Basidiomycetes. J. Cramer, Vaduz.

Karatygin I.V. 2000. Fungal macrosystematics in the recent times. Bot Zurn. 85, 19-34 (in Russian).

Keller J. 1997. Atlas des Basidiomycetes vus aux microscopes électroniques. Union des Societies Suisses de Mycologie, Neuchâtel.

Khan S.R., Kimbrough J.W. and Mims C.W. 1981. Septal ultrastructure and the taxonomy of Exobasidium. Can J Bot. 59, 2450-2457.

Kluge A. 1983. Cladistics and the classification of the great apes. In New Interpretations of Ape and Human Ancestry, Ciochon R. and Corrucini R., (Eds.), pp. 151-177. Plenum Press, New York.

Kõljalg U. and Renvall P. 2000. *Hydnellum gracilipes* - a link between stipitate and resupinate Hymenomycetes. Karstenia 40, 71-77.

Kovalenko A.E. 1984. The modern views on phylogenetic relationships and systematics of agaricoid fungi. In Evolution and Systematics of the Fungi. Leningrad. pp. 118-136 (in Russian).

Kovalenko A.E. 1989. Definitorium Fungorum URSS. Ordo Hygrophorales. Nauka, Leningrad (in Russian).

Kreisel H. 1969. Grundzüge eines Natürlichen Systems der Pilze. G. Fischer Verlag, Jena.

Kühner R. 1980. Hymenomycetes agaricoïdes, étude générale et classification. Bull Soc Linn Lyon 49 (No spec.), 1-1025.

Langer E. 1998. Evolution of *Hyphodontia* (Corticiaceae, Basidiomycetes) and related Aphyllophorales inferred from ribosomal DNA sequences. Folia Cryptog Estonica 33, 57-62.

Larsen M.J. and Burdsall H.H. 1976. A consideration of the term gloeocystidium. Mem NY Bot Gard. 28, 123-130.

Larsson K.-H., Larsson E. and Kõljalg U. 2002. Evolution of crust-like homobasidiomycetes inferred from nuclear ribosomal DNA sequences (a manuscript).

Lavin M. and Doyle J. 1991. Tribal relationships of Sphinctospermum (Leguminosae): integration of traditional and chloroplast DNA data. Syst Bot. 16, 162-172.

Lemke P.A. 1964. The genus *Aleurodiscus* (sensu lato) in North America. Can J Bot. 42, 723-768.

Linder D.H. 1940. Evolution of the Basidiomycetes and its relation to the terminology of the basidium. Mycologia 32, 419-447.

Littlefield L.J. and Heath M.C. 1979. Ultrastructure of Rust Fungi. Acad Press, New York.

Locquin M.V. 1981. Entaxie, taxotropie, néguentropie, valeur et qualité en taxonomie géneralisée. Ark'All Com. 2, 45-86.

Locquin M. 1984. Mycologie Générale et Structurale. Masson, Paris etc.

Martin G.W. 1949. The genus *Ceracea* Cragin. Mycologia 41, 77-86.

Miller O.K. (Jr.) and Watling R. 1987. Whence cometh the agarics? A reappraisal. In Evolutionary Biology of the Fungi. Cambridge Univ. Press, Cambridge. pp. 435-448.

Miyamoto M. 1983. Frogs of the *Eleutherodactylus rugulosus* group: a clasdistic study of allozyme, morphological, and caryological data. Syst Zool. 32, 109-124.

Moncalvo J.-M., Vilgalys R., Redhead S.A., Johnson J.E., James T.Y., Aime M.C., Hofstetter V., Verduin S.J.W., Baroni T.J., Thorn R.G., Jacobsson S., Clémençon H. and Miller O.K. Jr. 2003. One hundred and seventeen clades of euagarics (a manuscript).

Moore R.T. 1971. An alternative concept of fungi based on their ultrastructure. Recent Adv Microbiol. 10, 49-64.

Moore R.T. 1996. An inventory of the phylum Ustomycota. Mycotaxon 59, 1-31.

Moser M. 1983. Die Röhrlinge and Blätterpilze (Polyporales, Boletales, Agaricales, Russulales). 5. Aufl. Gustav Fischer, Stuttgart; New York. 533 S. (Bd. Ii, b/2, Kleine Kryptogamenflora. Begründet von H. Gams).

Mukhin V.A. 1993. A biote of xylotrophic Basidiomycetes of the West Siberian Plain. Nauka, Ekaterinburg (in Russian).

Nobles M.K. 1958. Cultural characters as a guide to the taxonomy and phylogeny of the Polyporaceae. Can J Bot. 36, 883-926.

Nobles M.K. 1971. Cultural characters as a guide to the taxonomy of the Polyporaceae. In Evolution in the Higher Basidiomycetes. Tennessee Press, Knoxville. pp. 169-195.

Nordic Macromycetes. Vol. 2: Polyporales, Boletales, Agaricales, Russulales. 1992. Nordsvamp, Copenhagen.

Nordic Macromycetes. Vol. 3: heterobasidioid, aphyllophoroid and gastromycetoid Basidiomycetes. 1997. Nordsvamp, Copenhagen.

Nuss I. 1980. Untersuchungen zur systematischen Stellung der Gattung *Polyporus*. Hoppea. 39, 127-198.

Oberwinkler F. 1965. Primitive Basidiomyceten. Ann Myc Ser. II. 19, 17-72.

Oberwinkler F. 1972. The relationships between the Tremellales and Aphyllophorales. Persoonia 7, 1-16.

Oberwinkler F. 1977. Das neue System der Basidiomyceten. In Beiträge zur Biologie der niederen Pflanzen, Frey W. et al., (Eds.), Fischer Verlag, Stuttgart, New York. pp. 59-105.

Oberwinkler F. 1978. Was ist ein Basidiomycet? Z Mykol. 44, 13-29.

Oberwinkler F. 1982. The significance of the morphology of the basidium in the phylogeny of Basidiomycetes. In Basidium and Basidiocarp, Wells K. and Wells E.K., (Eds.), pp. 9-35.

Oberwinkler F., Bandoni R. J., Bauer R., Deml G. and Kisimova-Horovitz L. 1984. The life-history of *Christiansenia pallida*, a dimorphic, mycoparasitic heterobasidiomycete. Mycologia 76, 9-22.

Parmasto E. 1968. Conspectus Ssystematis Corticiacearum. Inst Zool Bot, Tartu.

Parmasto E.H. 1969. The main problems of systematics of Aphyllophorales. Mikol Fitopatol. 3, 322-330 (in Russian).

Parmasto E. 1969. *Paullicorticium curiosum* Parm. et Žukov sp. nov. and the phylogenetical development of the basidium of the corticiaceous fungi. Česká Mykol. 23, 73-78.

Parmasto E. 1986. On the origin of the Hymenomycetes (What are corticioid fungi?) Windahlia 16, 3-19.

Parmasto E. 1995. Corticioid fungi: a cladistic study of a paraphyletic group. Can J Bot. 73, suppl. 1: sect. E-H, 843-852.

Pegler D.N. and Young T.W.K. 1971. Basidiospore Morphology in the Agaricales. Beih Nova Hedwigia 35, 1-210.

Pegler D.N. and Young T.W.K. 1981. A natural arrangement of the Boletales, with reference to spore morphology. Trans Brit Myc Soc. 76, 103-146.

Petersen R.H. 1971. Interfamilial relationships in the clavarioid and cantharelloid fungi. In Evolution in the Higher Basidiomycetes, Tennessee Press, Knoxville. pp. 345-371.

Petersen R.H. 1974. The rust fungus life cycle. Bot Rev. 40, 453-513.

Raitviir A.G. 1964. Review of heterobasidial fungi of USSR in connection with some problems of phylogeny and mycogeography: Ref. of Ph.D. thesis. Tartu (in Russian).

Redhead S.A., Lutzoni F., Moucalvo J.M. and Vilgalis R. 2002. Phylogeny of Agarics: Partial systematics solutions for core omphalinoid genera in the Agaricales (euagarics). Mycotaxon 83, 19-57.

Roberts P. 1998. *Oliveonia* and the origin of the Holobasidiomycetes. Folia Crypt Estonica. 33, 127-132.

Rogers D.P. 1934. The basidium. Univ Iowa Studies Nat Hist. 16, 160-183.

Ryvarden L. 1991. Genera of polypores. Nomenclature and taxonomy. Fungiflora, Oslo.

Schantser I.A. 2003. About discrepancy between "phylogeny" of molecules and true phylogeny. Proceedings of the XIth International Symposium on Plant Phylogeny, Moscow, Russia, pp. 109-111.

Singer R. 1945. The *Laschia*-complex (Basidiomycetes). Lloydia 8(3), 170-230.

Singer R. 1962. The Agaricales in Modern Taxonomy. 2nd ed. Weinheim.

Singer R. 1971. A revision of the genus *Melanomphalia* as a basis of the phylogeny of the Crepidotaceae. In Evolution in the Higher Basidiomycetes, Tennessee Press, Knoxville. pp. 475-480.

Singer R. 1975. The Agaricales in Modern Taxonomy. 3rd ed. J. Cramer, Vaduz.

Singer R. 1986. The Agaricales in Modern Taxonomy. 4th ed. Koeltz Scientific Books, Koenigstein.

Singer R. and Smith A.H. 1960. Studies on secotiaceous fungi VII. *Secotium* and *Neosecotium*. Madroño 15, 152-158.

Stalpers J.A. 1988. *Auriculariopsis* and the Schizophyllales. Persoonia 13, 495-504.

Stalpers J.A. 1992. *Albatrellus* and Hericiaceae. Persoonia 14, 537-541.

Stalpers J.A. 1996. The aphyllophoraceous fungi - II. Keys to the species of the Hericiales. Stud Mycol. 40, 1-185.

Stankovicova L. 1973. Hyphal structure in some pleurotoid species of Agaricales. Nova Hedwigia 24, 61-85.

Swann E.C. and Taylor J.W. 1993. Higher taxa of Basidiomycetes: An 18S rRNA gene perspective. Mycologia 85, 923-936.

Swann E.C. and Taylor J.W. 1995. Phylogenetic perspectives on Basidiomycetes systematics: Evidence from the 18S rRNA gene. Can J Bot. 73 Supplement, S862-S868.

Thiers H.D. 1971. Some ideas concerning the phylogeny and evolution of the boletes. In Evolution in the Higher Basidiomycetes, Tennessee Press, Knoxville. pp. 423-436.

Thiers H.D. 1984. The secotioid syndrome. Mycologia 76, 1-8.

Vasilyeva L. 1999. Systematics in Mycology. Bibliotheca Mycologica, Bd. 178. J. Cramer in der Gebr. Borntraeger Verlagsbuchhandlung, Berlin; Stuttgart, 253 pp.

Wasser S.P. 1974. Problems of systematics of secotiaceous mushrooms. In Progress of mycology and lichenology in the Soviet Baltic republics. Estonian Acad Sci Press, Tartu, pp. 62-65.

Wasser S.P. 1980. Flora Fungorum RSS Ucrainicae: Agaricaceae Cohn. Naukova Dumka Press, Kiew. 328 pp. (in Russian).

Wasser S.P. 1982. Family Agaricaceae (Fr.) Cohn of the Soviet Union (morphology, anatomy, ecology, geography, origin, evolution, and practical application). Dr. Sci. (Biol.) Thesis, Ukr Acad Sci, Kiev, 965 pp.

Wasser S.P. 1985. Family Agaricaceae of the USSR. Naukova Dumka, Press Kiev (in Russian).

Wasser S.P. 2002. Family Agaricaceae (Fr.) Cohn (Basidiomycetes) of Israel mycobiota. I. Tribe Agariceae Pat. Nevo E. and Volz A., (Eds.), A.R.A. Gantner Verlag K.-G., Ruggell/Liechtenstein 211 pp.

Watling R. and Largent D. 1977. Macro- and microscopic analysis of the cortical zones of basidiocarps of selected agaric families. Nova Hedwigia. 28, 569-617.

Wells K. 1994. Jelly fungi, then and now! Mycologia 86, 18-48.

Zerov D.K. 1972. Outlines of Non-Vascular Plant Phylogeny. Naukova Dumka Press, Kiev (in Russian).

Zerova M.Ya., Sosin P.E. and Rozhenko G.I. 1979. Identification key of the fungi of Ukraine. Vol. 5: Basidiomycetes. Book 2. Naukova Dumka Press, Kiev (in Ukranian).

Zerova M.Ya. and Wasser S.P. 1974. Current problems of taxonomy of Agricales s.l. In Progress of Mycology and Lichenology in the Soviet Baltic Republics. Estonian Acad. Sci. Press, Tartu, 22-24.

Zmitrovich I.V. 1997. Corticioid fungi: modern systematics and problems of phylogenetics. Mikol Fitopatol. 31, 79-91 (in Russian).

Zmitrovich I.V. 2001. On the systematics of genus *Serpula* s. lato. Nov Syst Pl non Vasc. 35, 70-89.

Zmitrovich I.V. and Spirin W.A. 2002. On the taxonomy of corticioid fungi. II. The genera *Serpula* (Pers.) Gray, *Serpulomyces* gen. nov., *Amylocorticiellum* gen. nov. Mikol Fitopatol. 36, 11-26.

Zmitrovich I.V., Malysheva V.F. and Spirin W.A. 2003. A new arrangement of Phanero-chaetales (a manuscript).

MACROEVOLUTIONARY EVENTS AND THE ORIGIN OF HIGHER TAXA

Valentin A. Krassilov
Institute of Evolution, University of Haifa , Mount Carmel, Haifa 31905, Israel

Abstract: The role of macromutation as a source of evolutionary novelty is discussed in relation to the repeatedly appearing reticulate pollen grain morphology; the origin of stomata by pedomorphic transformation of gametangial conceptacles in the alga-like precursors of vascular plants; the fixation of environmentally induced anomalous stomatography as a genus-specific character; the origin of specialized seedling morphology in mangroves through changes of developmental rates; the cyclic evolution (retroconvergence) of reproductive and foliar characters; the possibility of pathogenic cecidogenous origin of plant organs; and the diversification burst (anastrophes) on the basis of macropolymorphic post-crisis populations. Macromutation can be involved in both micro- and macroevolutionary processes, yet the origin of higher taxa is not reducible to a single macromutation event. Rather, it is an accumulation of macromutational novelties arising in different lineages, as well as a retrieval of ancestral characters lost in the immediate ancestors, but retained in the genomic memory and their integration in the new adaptive context.

1. INTRODUCTION

The term macroevolution is commonly applied to evolutionary events at or above the species level, while the modern evolutionary studies are focused, mostly or exclusively, on the processes that are not, or at least not in the time-scale of the experiment, consummated in speciation and are, therefore, considered as microevolutionary. A well-trod way from micro-

S.P. Wasser (ed.), Evolutionary Theory and Processes: Modern Horizons,
        Papers in Honour of Eviatar Nevo
V. A. Krassilov. Macroevolutionary Events and the Origin of Higher Taxa, 265-289.
© 2004 *Kluwer Academic Publishers.*

evolutionary observation to macroevolutionary deduction lies in the assumption that macroevolution is a gradual accumulation of micro-evolutionary events through geological time.

A different line of thought ascribes macroevolution to saltational changes, or macromutations, that are rare and abnormal for the gradual microevolutionary developments. The latter theory was forcefully advanced by Goldschmidt (1958 and earlier work), who defined macromutation (systemic mutation) as "a shift of developmental processes by a single mutation event to such an extent that a structural departure appears which is of the order of magnitude of macroevolutionary differences". Arguing with Dobzhansky and Stebbins, who dubbed his theory "cataclismic", Goldschmidt (1958, p. 491) bitterly remarked: "If cataclisms enter this theorizing at all, it is the cataclism of the orthodox and extreme Neo-Darwinism".

It was later discovered that at least some macromutations are the developmental effects of genome – plasmon interactions, potentially of a great evolutionary significance (Jinks, 1964; Løvtrup, 1974). Developmental studies lend support to macromutational effects of rate changes, or hetero-chronies (Arthur, 1975; Gould, 1977), in particular those occurring early in the ontogeny (archallaxes of Severtsov, 1939).

Long before the advent of macromutation theory, the synthesis of developmental and paleontological evidence gave rise to the theory of orthogenesis, which held that morphological evolution results from either acceleration or retardation of developmental rates, accompanied by addition or deletion of developmental stages (e.g., in the evolutionary series of ammonites: Neumayr, 1879). The orthogenetic views shared by such eminent 19th century paleontologists as Cope, Hyatt, Neumayr, Grabau, and others (reviewed in Krassilov, 1977) were expelled from the modern synthesis and all the voluminous supportive evidence came to be viewed with suspicion. Yet there are more recent paleontological examples of speciation through halted development (Gould, 1970; Soulé, 1973; Niklas et al., 1976), and even the modern studies of interpopulation polymorphism occasionally reveal morphological and/or genetic distinctions that correspond to consecutive developmental stages or retention of successional isozymes (Nair et al., 1977).

Paleontological models of episodic high rate macroevolution go back to the XIX century catastrophism (mass extinctions) and anastrophism (bursts of diversification at higher taxonomic levels), later furnished with the macromutation mechanism (Schindewolf, 1958). Their milder versions are represented by the quantum evolution concept (Simpson, 1944) and its derived model of punctuated equilibrium (Eldredge and Gould, 1972).

The Goldschmidtian macromutations were rediscovered as "genetic revolutions" in founder populations (Mayr, 1954) or saltational reorganizations of supergenes (Carson, 1975). Yet, it is commonly held (e.g., Stanley, 1979) that although quantum evolution is a possibility, the extreme views of Goldschmidt, Schindewolf, and their followers only discredited the idea.

Thus, the old dilemma of gradualism vs. saltationism is still with us, but the problem of macroevolution is far more comprehensive than that.

2.   MATERIALS AND METHODS

This study is based on paleobotanical materials from the Devonian, Permian, Mesozoic and Paleogene deposits of the Far East, Central Asia, Eastern Europe, and the Middle East preserved as impressions and compressions of macroscopic plant remains, pollen grains, gut compressions of fossil insects, fossil galls, and other traces of plant – animal interactions. Stratigraphic, taphonomic, and paleoecological studies were conducted in the major localities. The material was subjected to macro- and micromorphological studies, the latter assisted with the scanning electron microscopy. Fossils selected for the analysis of macroevolutionary events came from the following localities:

Cutinized thalloid plant *Shuguria ornata* Tschirkova-Zalesskaya: The Pavlosk Quarry, Voronezh Region, central European Russia, Middle Devonian (Givetian).

Cutinized thalloid plant *Orestovia* devonica Ergolskaya: The Barzas River, Kuznetsk Basin, Western Siberia, Middle Devonian.

Conifer *Ullmannia* cf. *bronnii* Goeppert: The Kitchmenga River near Nedubrovo Village, Vologda Region, northern European Russia, the latest Permian to lowermost Triassic.

Dispersed pollen grains *Reticulatina* sp. (courtesy of Natalia Zavialova): The Tschekarda locality, Cis-Urals, Lower Permian (Kungurian).

Pollen grains *Lunatisporites* sp., from the gut compression of Permian insect *Idelopsocus diradiatus* Rasnitsyn: The Tschekarda locality, Cis-Urals, Lower Permian (Kungurian).

Peltasperm *Scytophyllum vulgare* (Prynada) Dobruskina (courtesy of Vera Vladimirovitch): The Bogoslovskoye Coal Mine, the Eastern Urals, Middle Triassic.

Cycadophyte *Baruligyna disticha* Krassilov et Doludenko: The Barula River, Georgia, Middle Jurassic (Callovian).

Pollen grains from the angiosperm flower *Freyantha sibirica* Krassilov et Golovneva: The Kem' River, Tchulymo-Yeniseysk Basin, West Siberia, mid-Cretaceous (Cenomanian).

Aquatic fern megaspores: The Kem' River, Tchulymo-Yeniseysk Basin, West Siberia, mid-Cretaceous (Cenomanian).

Fossil seedlings: The Gerofit locality, southern Negev, Israel, mid-Cretaceous (Turonian).

Fossil galls: The Gerofit locality, southern Negev, Israel, mid-Cretaceous (Turonian).

The collections are deposited in the Institute of Biology and Pedology, Vladivostok (Russia), Institute of Paleontology, Moscow (Russia), and the Institute of Evolution, University of Haifa (Israel).

3.   PROPHETIC CHARACTERS: RETICULATE POLLEN GRAINS

Characters may appear earlier, sometimes much earlier, than the taxon for which they are diagnostic. For instance, practically all the diagnostic characters of angiosperms, including double fertilization, occur in phyllo-genetically older seed plant lineages (Krassilov, 1997). The mystery of angiosperm origins, although not resolved yet, is becoming less abominable with accumulation of paleobotanical evidence. Transitional forms combining angiospermous and gymnospermous characters are presently known from the Late Jurassic to Early Cretaceous deposits (Krassilov, 1997; Krassilov and Bugdaeva, 1999, 2000).

It is commonly believed that the easily dispersed pollen grains, rather than macrofossils, provide the most reliable records of angiosperm entries into the Mesozoic plant communities. In particular, the records of semi-tectate reticulate pollen grains are widely used for dating the angiosperm origins (Hughes, 1976). A steady increase in both frequencies and diversity of such pollen morphotypes over the Early Cretaceous is commonly held as evidence of high rate macroevolution in early angiosperms. Actually, such pollen grains have been extracted from staminate organs of more than one group of mid-Cretaceous angiosperms (Friis et al., 1986; Krassilov and Shilin, 1986; Krassilov and Golovneva, 2001), thus, suggesting a widespread parallelism of pollen morphology at this level.

Yet the angiosperm-like pollen grains are occasionally found in much older deposits. Fig. 1A depicts a reticulate pollen grain at least 230 million years older than the earliest Cretaceous reticuliform records. Such pollen morphotypes (*Reticulatina* Koloda) are known from the Early and Middle Permian (Kungurian to Kazanian) of northern Russia and the stratotypic Cis-Uralian Region (Koloda, 1996). They are presently studied with SEM and TEM from the Kungurian of Tschekarda, the Cis-Urals (Zavialova, in preparation) where *Reticulatina* spp. constitute up to 4% of palynological assemblage dominated by the widespread Permian pollen types *Lunati-sporites*, *Protohaploxypinus*, *Cladaitina*, *Platisaccus*, *Vittatina*, and *Cor-*

daitina. These latter have either smooth or taeniate (with the surface layer divided into parallel stripes, or taeniae) exine. In *Reticulatina,* the surface reticulum has thick walls (muri) enclosing the irregularly polygonal, triangular, elliptical or slit-like lumina (Fig. 1B). Joints of the muri are sometimes perforated with a minute circular pore. Occasional lumina are traversed by a thin wall. The reticulum is coarser in the polar region fining out to the periphery. All these characters occur in the reticulate angiosperm pollen found *in situ* in the anthers of *Freyantha sibirica* Krassilov et Golovneva, the mid-Cretaceous angiosperm (Fig. 1C, D).

One can suspect that *Reticulatina* was produced by some primordial angiosperms hiding in the uplands of the proto-Uralian ranges. Yet this is scarcely the case. Not only is the time gap (150 million years) between the Permian and Cretaceous reticulate forms too large for their phylogenetic continuity, but also the Permian reticulatinas were both morphologically and ultrastructurally similar to the common types of non-reticulate Permian pollen grains found both dispersed and *in situ* in the pollen organs of Permian pteridosperms. Almost all types of taeniate pollen grains occur in the pollen load of Permian insects (Krassilov and Rasnitsyn, 1997, 1998) providing evidence of widespread pollen-feeding. In the pollen clumps extracted from the gut compressions, some grains are fully intact (Fig. 1C), whereas in the others the surface layer is partly digested exposing the infrastructure, which is reticulate and practically identical to the surface reticulum in *Reticulatina* (Fig. 1F). Reticulate infrastructure was observed also in smooth pollen grains of Cretaceous proangiosperms such as *Baisia* (Krassilov and Bugdaeva, 1982).

These findings indicate that sexinal reticulum first developed as infrastructure covered with a non perforate tectum. Dispensing of the latter as a late developmental stage resulted in surfacing of the reticulum in few Permian species, and the same repeated on a larger scale in the Cretaceous.

The surface reticulum in early angiosperms is interpreted (Zavada, 1984) as a component of self-incompatibility syndrome related to the pollen – stigma interaction. In the Permian seed plants, its functional meaning might have been entirely different. Our data show that most Permian pollinivores were visually attracted by taeniate surfaces promoting their parallel development in a number of insect-pollinated gymnosperm lineages (Krassilov and Rasnitsyn, 1997, 1998). With the build-up of ecological and morphological diversity through the Early Permian, the reticulate surface structures might have appeared as an alternative of taeniate structures breaking the monotony of contemporaneous pollen grain morphology.

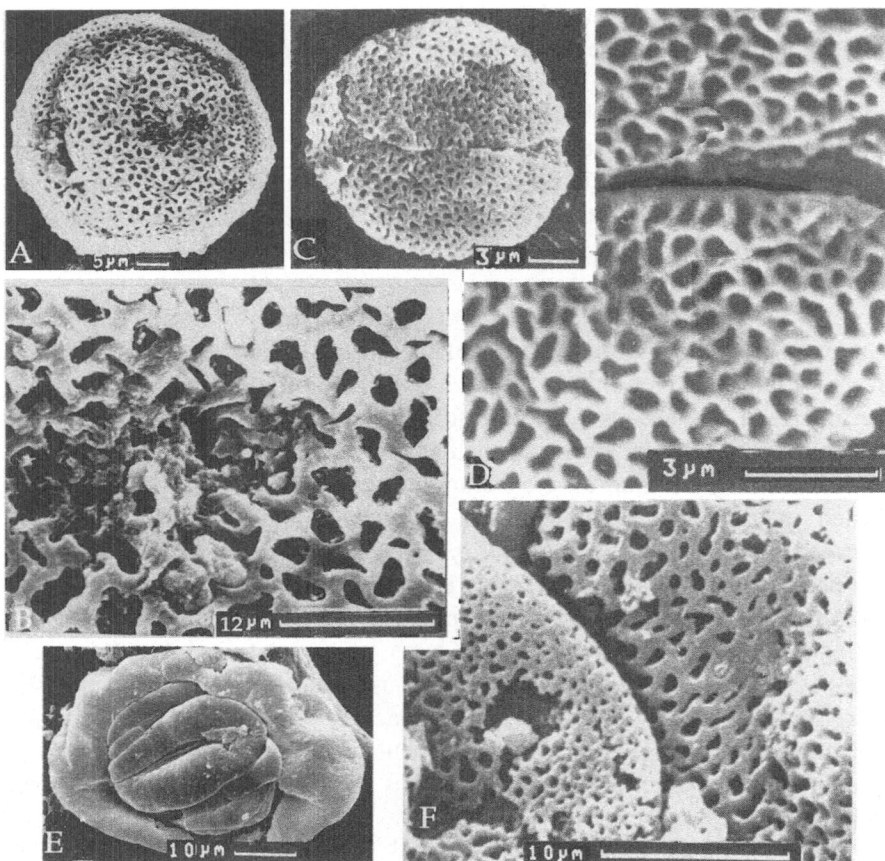

Figure 1. Reticulate pollen grains as an example of prophetic character repeatedly appearing due to the surfacing of infratectal structures: A. *Reticulatina* sp., dispersed pollen grain from the Lower Permian (Kungurian) of the Cis-Urals (courtesy of Natalia Zavialova); B. Reticulate surface structure of the same grain, enlarged; C. *In situ* pollen grain of *Freyantha sibirica* Krassilov et Golovneva, angiosperm flower from the mid-Cretaceous (Cenomanian) of West Siberia (Krassilov and Golovneva, 2001); D. Surface structure of the same grain, enlarged; E. *Lunatisporites* sp., taeniate pollen grain from gut compression of fossil insect *Idelopsocus diradiatus* Rasn. from the Lower Permian (Kungurian) of the Cis-Urals (Krassilov and Rasnitsyn, 1997); F. Reticulate infratectal structure of another *Lunatisporites*-type pollen grain from the same pollen load, with the tectum partly eroded by intestinal exudates.

Irrespective of functional interpretation, the loss of continuous tectum - surfacing of infratectal structure is a striking example of macromutation through abbreviated development resulting in a distinctive new morphology. This example also shows that a single character, however typical, does not unequivocally testify to the origin of the respective higher taxon. The character may develop prophetically in a different adaptive context to acquire a new function much later in evolutionary history.

4. GENOCOPY: FIXATION OF ANOMALOUS STOMATOGRAPHY

Epidermal structures are a promising object of developmental studies. In particular, the types and arrangements of stomata are thought to reflect a morphogene concentration that (1) induces differentiation of epidermis into stomatal and stomata-free areas, (2) inhibits stomatal initials (meristemoids) in the vicinity of the developing stoma, and (3) modifies the adjacent epidermal cells (Bloch, 1965; Ursprung, 1966; Lewis et al., 1977). These features are of a certain taxonomic significance. At the same time, the ratio of stomata to ordinary epidermal cells is under strong environmental (atmospheric CO_2 concentration) control (Kürschner, 1996) and the other stomatographic characters are also liable to environmental induction.

Uneven distribution of stomata over leaf surface is sometimes ascribed to standing waves of morphogene concentration (Ursprung, 1966). Yet stomata typically tend to develop between veins, in parallel bands or polygonal areas depending on the type of venation. Such distribution has a functional meaning in respect to concentration of vapor and CO_2 in the boundary layer separating leaf surface from the ambient air. A decoupling efficiency of the boundary layer depends on the leaf surface micromorphology (Meinzer, 1993). It is typically higher between the vein ridges, hence, the intercostal concentration of stomata. My study of bennettitalean stomata (Krassilov, 1978a) has shown that in the microshadow of prominent lateral veins, not only is the distribution of stomata less regular, but also the frequency of contiguous stomata (only one cell apart) is considerably higher than in the stomatal mid-zones.

The configuration of stomatal bands in respect to vascular bundles is diagnostic for species or even genera in many angiosperm families and in the vast majority of gymnosperms. It can persist even after the loss of the veins. Thus, in conifers, several longitudinal stomatal bands are sometimes retained in single-veined needle-leaves derived from broad leaves with several veins (e.g., in the single-veined podocarp leaves with three stomatal bands: Florin, 1931; Krassilov, 1967). Yet the differentiation of stomatal initials in respect to veins is often obliterated or nearly so in enrolled leaves or those with deeply embedded vascular bundles. Epidermal topography thus falls under an intricately combined control of genetically determined morphogenic gradients and environmental induction.

The taxonomically significant configurations of subsidiary cells are related to the meristemoid developmental rates. Ontogenetically, subsidiary cells are either perigenous, derived from adjacent epidermal cells, or mesogenous, produced by the rapid successive divisions of meristemoid itself. The latter can be seen as an atavistic mutation betraying the evolutionary origin of stomata from multicellular structures (part 5). Yet the mesogenous subsidiaries are distinctive for a number of angiosperm families and for bennettites among gymnosperms (Florin, 1933; Krassilov, 1978a). At the same time, in the bennettitalean cataphylls, the configuration of subsidiaries is different from that in the foliage leaf and is more typically perigenous. Thus, in this group at least, the stomatotype depends on the developmental program of organogenesis.

An exceptionally high frequency of epidermal anomalies such as irregular epidermal zones, interrupted stomatal rows, underdeveloped and/or contiguous stomata, disorganized arrangements of subsidiary cells, etc., are recorded in the transboundary Permian - Triassic fossil plant assemblage of Nedubrovo, central European Russia (Krassilov et al., 1999). As this stratigraphic level is marked by huge volcanic eruptions, the anomalous epidermal developments might have been related either to the massive ash falls or to the mutagenic effect of UV radiation (increased with the discharge of ozone-destroying volcanic aerosols) or both.

A conspicuous anomaly is an occasional increase of subsidiary cells from 4 – 6, typical for peltasperms and conifers, to 8 – 12, exceptional for these plant groups. Remarkably, in the monotypic conifer genus *Ullmannia* Goeppert, appearing during the large-scale environmental disturbance in the latest Permian, a ring of 10 – 12 subsidiary cells is a diagnostic character (Fig. 2A). The subsidiary cells are considerably smaller than the ordinary epidermal cells and they are radially disposed encircling the guard cells. Occasional aborted stomata appear as a group of minute cells resulting from proliferation of meristemoid, in which the guard cells failed to differentiate. These are evidence of mesogenous origin for all or most subsidiaries (the intruding larger cells, as in Fig. 2B, are probably perigenous).

Since the typical stomatal structure in *Ullmannia* corresponds to the anomalous environmentally induced condition in other contemporaneous conifers, a fixation through genocopy is the most probable mechanism of macromutational origin for this genus-specific epidermal character.

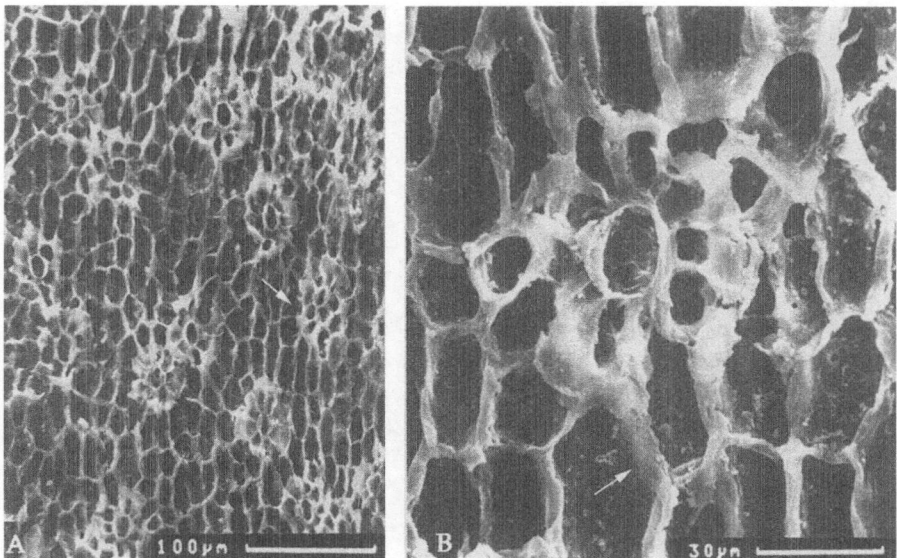

Figure 2. Teratological origin of stomatal structure in *Ullmannia* cf. *bronnii* Goeppert., a conifer from the transitional Permian – Triassic deposits of Nedubrovo, European Russia: A. Stomata with numerous (10 – 12) small subsidiary cells and a similar configuration of small cells produced by prolific divisions of a meristomoid that failed to develop into stoma (arrow); B. Stoma with a ring of small mesogenous cells intruded by a larger perigenous cell (arrow).

5.   PEDOMORPHIC ORGANOGENESIS: ORIGIN OF STOMATA

Stomata are obligatory for photosynthetic tissues of terrestrial plants, therefore, conceivably appearing at a very early stage of their evolution. Genuine transitional stages between early land plants and their algal ancestors are not known, yet the conservative morphology of thickly cutinized (hence terrestrial) alga-like Devonian plants give some idea of how the basic structures might look. The most primitive among the cutinized thalloids is *Shuguria* Tchirkova-Zalesskaya emend Krassilov from the mid-Devonian of European Russia having tubular shoots with two-layered cortex. The cortical layers probably correspond to the mesoderm and subcortex of brown algae. The medullar tissues comprise the subcortical parenchymatous layer and the central bundle of pitted conducting cells resembling the conducting tubes of extant laminarialean algae and *Ascoseira* (Moe and Henry, 1982).

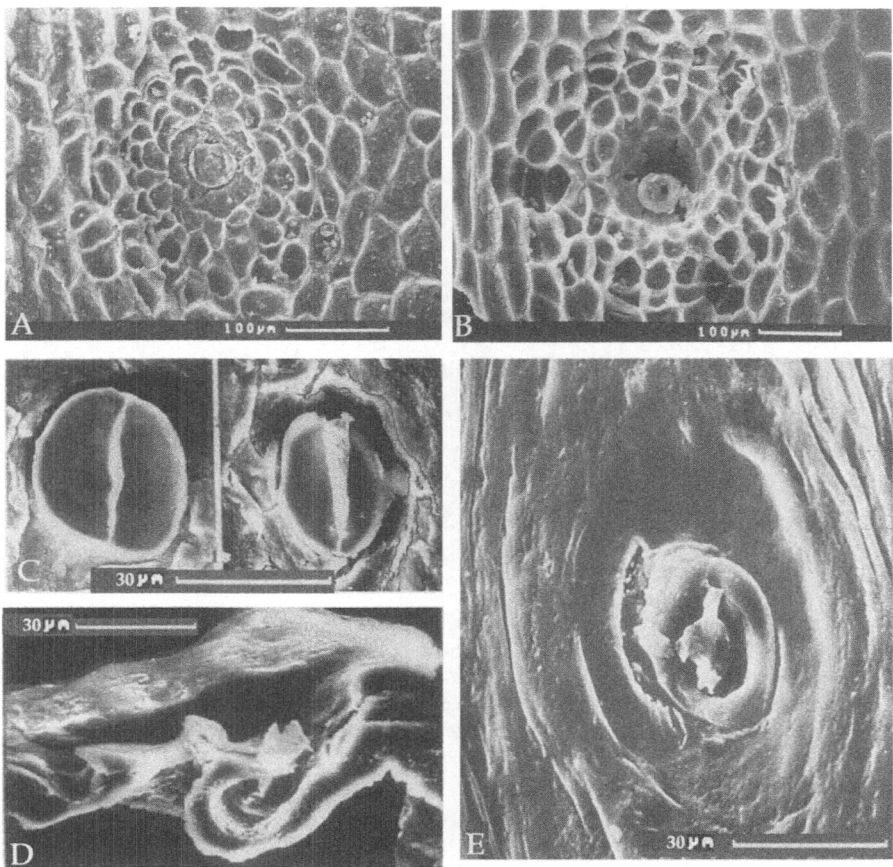

Figure 3. The supposed homology of sex organs and stomata in the Devonian alga-like plants:
A. Gametangial conceptacle of *Shuguria ornata* Tchirkova-Zalesskaya B. Conceptacle of the
same plant with the neck cell detached; C – E. Stomata of *Orestovia devonica* Ergolskaya,
interior view and transverse section showing cuticle of guard cells extending to substomatal
chamber (Krassilov, 1981a).

The most conspicuous structures of *Shuguria*-type thalloids are spherical
bodies developing subcortically and opening by a pore that is covered with a
detachable neck cell (Fig. 3A). These structures are closely comparable with
gametangial conceptacles of brown algae, in particular the Fucales and
Ascoseirales, in which they develop from the initial cells, or prospores, that
differentiate in the meristoderm and are developmentally shifted to the
subcortex. Consecutive developmental stages preserved in *Shuguria* are
closely comparable with those in extant *Ascoseira* (Moe and Henry, 1982).
In both, the conceptacle initial appears as an enlarged thick-walled cell of the
inner cortical layer producing by successive anticlinal divisions the basal
chamber with gametangial or sporangial sori and with a neck reaching to the

shoot surface. The surrounding cortical cells undergo mitotic divisions forming concentric rings.

In *Shuguria*, the terminal neck cell is puck-shaped, about 50 μm in the larger diameter, divided by thin median septa, with the parts gaping in the middle forming the central canal. The neck cell appears as a terminal periclinal derivative of the stem cell. After maturation, the neck cell is detached as a lid and drops out or sticks near the circular orifice (Fig. 3B). With continuing stretching of the cortex, the conceptacles are destroyed, their orifices developing into irregular pores scattered over the shoot surface.

New conceptacles are often initiated in close proximity to, occasionally even contiguous with, the older ones. The conceptacles of the same developmental stage, however, are nearly equidistant from each other. Such arrangements suggest an inhibitory interaction between simultaneously developing structures and the lack of such between successive generations.

Shuguria has been compared and even synonymized (Istchenko and Istchenko, 1981) with *Orestovia* Ergolskaya from the Devonian of Siberia, to which it is similar in the general habit and the cell pattern of cutinized cortical tissues. In *Orestovia*, however, the conducting cylinder is much thicker, consisting of tracheid-like cells with helical or reticulate thickenings (Krassilov, 1981a). Moreover, *Orestovia* shows the regularly disposed stomata or stomata-like structures lacking in *Shuguria*, the tubular shoots of which might have been ventilated through the pores left of the shed conceptacles. *Orestovia* clearly represents a more advanced stage of terrestrial plant life requiring a more specialized organ of gas exchange. Here, the stomata-like structures consist of two cells with a cutinized canal between them extending to the cavity beneath (Fig. 3C - E). They are similar to the detachable neck cells of gametangial conceptacles in *Shuguria*, thus confirming the hypothetical derivation of stomata from sex organs (Pant, 1960). In their distinctive shape and morphogenic control over development of surrounding cells, the conceptacle initials are exactly like stomatal initials.

The morphologically feasible steps in derivation of *Orestovia*-type stomata from *Shuguria*-type conceptacles would have been an abbreviation of proliferative activity in the stem cells after a few divisions, a reduction of the gametangial chamber leaving the residual substomatal chamber, and a direct transformation of stem cell into the neck complex bypassing the intermediate stages. Instead of being detached at maturation, the neck complex would remain intact providing a more sophisticated mechanism of gas exchange than the residual pores of the *Shugura*-type conceptacles.

6.   RESTRUCTURING THROUGH DEVELOPMENTAL RATES: MANGROVE SEEDLINGS

All morphological changes involve developmental rates. Yet examples of a rapid and conspicuous restructuring solely resulting from a definable shift of developmental rates are not so common. Perhaps the best of these are provided by the viviparous seedlings of mangroves, the woody plants of intertidal mangal communities. In the typically viviparous *Rhizophora* L., the proximal seedling axis, or hypocotyl, emerges from the fruit while it is still attached to the parent tree. As a matter of fact, neither fruit nor seed take any part in propagation that is totally assigned to the hypocotyl, which detaches itself from the cotyledons (left with the fruit), falls to the muddy ground, and develops into a new plant. This peculiar form of propagule results from the early shift of high rate growth from cotyledons to the intercallary meristematic zone below the cotyledonary node (Juncosa, 1984). As a result, the cotyledons never develop as distinct foliar structures, but instead are completely fused into the gaustorial cotyledonary body. Prior to detachment, growth is once more shifted to the base of cotyledons forming the cotyledonary collar that protrudes from the fruit enclosing the epicotyl. Even the radicle meristem is suppressed in favor of lateral roots arising from the hypocotyl. A high rate growth of hypocotyl is shared by the other rhyzophoracean genera, although the base of cotyledons may not be so excessively developed, and the abscission line may remain inside the fruit.

Due to their relative robustness, mangrove seedlings are often preserved as fossils. Various types of seedling development are recorded in the mid-Cretaceous mangrove assemblage (recognized as such on the basis of root morphology primarily) of Gerofit, southern Israel. One of them was cryptoviviparous, germinating from the seed, but enclosed in the fruit that developed from epigynous flower (Fig. 4A). The hypocotyl apex emerged from the fruit after shedding (Fig. 4B). Another type of propagule is represented by a detached seedling consisting of a relatively short and warty hypocotyl with a well-developed pair of lateral roots near the apex and with shorter lateral roots above (Fig. 4C). It shows no evidence of radicle, the development of which might have been totally suppressed. Here the developmental pattern should have been much as in *Rhizophora*, yet with the hypocotyl less excessively elongated before shedding.

Fig. 5A – C shows an uprooted seedling with fruit remains attached near the base and with a well-developed cotyledonary collar emerging from it. The epicotyledonary axis and leaves are well developed. This seedling morphology appears less specialized than in the *Rhizophoraceae*, with epicotyl overgrowing the cotyledons and the fruit abscised with the seedling.

However, the cotyledonary collar indicates an excessive basal growth of cotyledonary body as an early appearing specialized feature (playing a role in regulation of salt content) of mangrove seedlings. In this respect, the seedling morphology in the Cretaceous mangrove plant is more advanced than in the extant *Bruguiera* Lamark, in which the intercalary meristematic zone between the cotyledons and hypocotyl remains inside the fruit (Jancosa, 1982, 1984). Fossil seedlings thus add new variants of heterochronous development unknown in the extant mangroves.

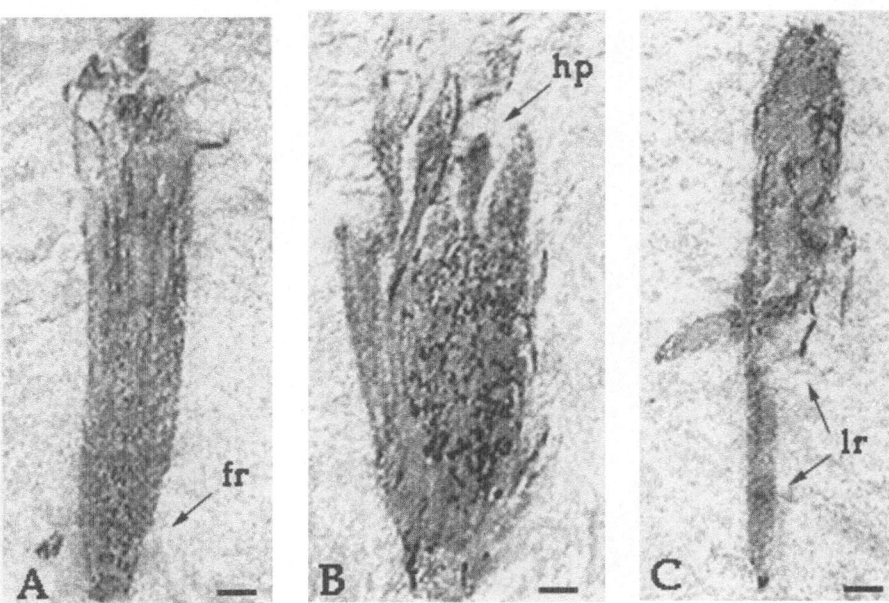

Figure 4. Seedling morphologies in the early mangroves (mid-Cretaceous of Gerofit, Israel) arising through developmental shifts: A. Early stage of fruit (fr) development from the epigynous flower; B. Flower in fruit with the hypocotyl (hp) emerging from the fruit; C. Detached hypocotyl with lateral roots (lr). Scale bar 0.5 mm.

The diversity of specialized seedling morphologies in the mid-Cretaceous assemblage is remarkable considering that this is the earliest record of just appearing angiosperm mangroves, which attests to a high rate macromorphological evolution through the shifting growth rates in the early development.

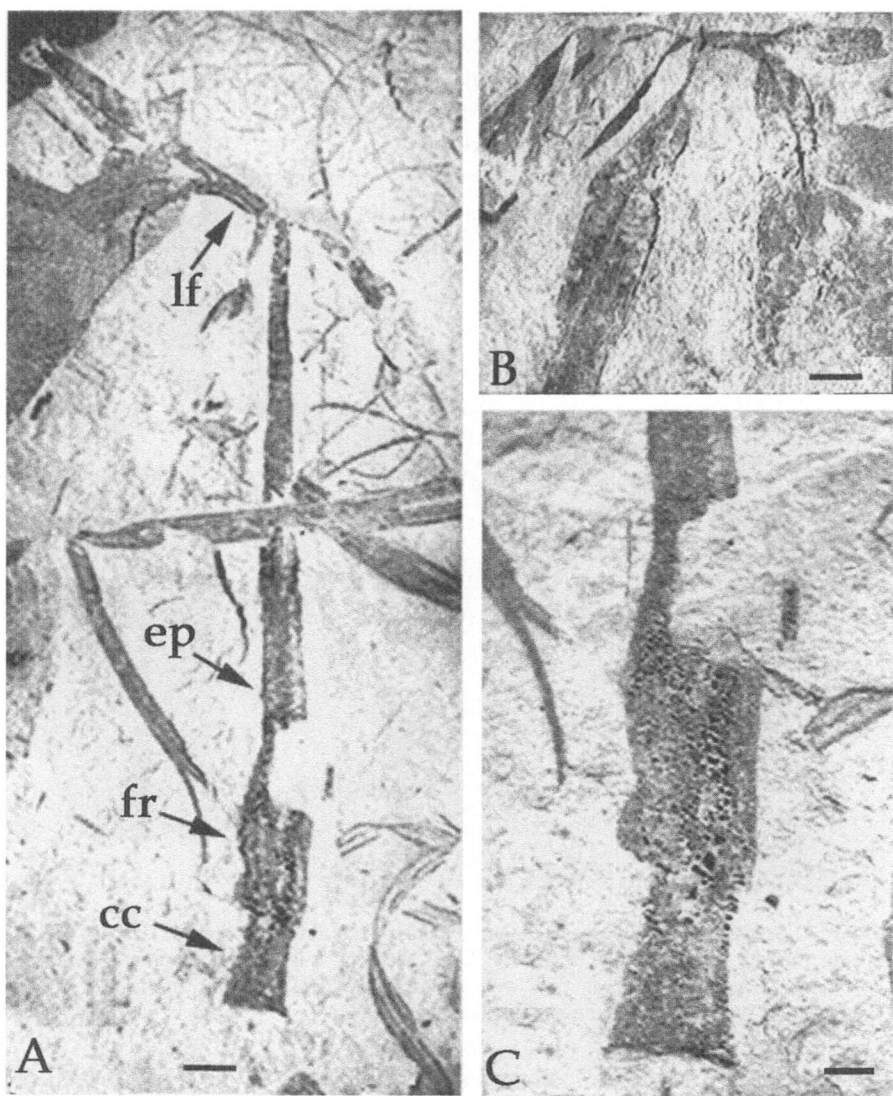

Figure 5. Seedling morphology in the early mangroves (mid-Cretaceous of Gerofit, Negev, Israel) arising through developmental shifts: A. Seedling uprooted below the cotyledonary collar (cc – cotyledonary collar emerging from the fruit, ep – epicotyl, fr – fruit coat, lf - leaf); B. Proximal part enlarged to show cotyledonary collar; C. Seedling leaf from counterpart of the same specimen. Scale bar 0.3 mm.

7.   RETROCONVERGENT MORPHOLOGY

This category comprises macroevolutionary events, which bring back character states that had once appeared and were lost in the course of historic development. The well-known examples from plant morphology are the leafy "megasporophylls" of *Cycas* L., as well as the anomalous ovules on leaves in *Ginkgo biloba* L. As both these structures resemble leafy ovulate organs of Paleozoic pteridosperms, the putative forerunners of all modern gymnosperms, they were interpreted as a retention of ancestral character state, a peculiar case of evolutionary conservatism. Plant morphologists even considered *Ginkgo* (and, by implication, *Cycas*) as "living pteridosperms" (Meyen, 1987).

Leafy ovuliphores, however, never occurred in the Mesozoic ginkgophytes and cycadophytes that intervened between the Paleozoic and extant forms. Typical of the Mesozoic ginkgophytes were ovulate strobili (Krassilov, 1970) that showed no evidence of either leafy prototype or secondary planation.

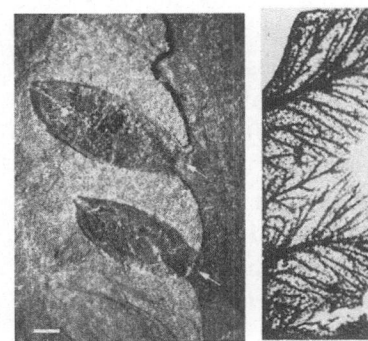   

Figure 6. Retroconvergence of *Cycas* like "megasporophyll" in Jurassic cycad *Baruligyna* Krassilov and Doludenko, with scars of subtending bracts (arrows) betraying its origin from a planated strobilate structure.

Figure 7. Retroconvergence of pinnule shape and venation in Triassic pteridosperm *Scytophyllum vulgare* (Prynada) Dobruskina: A. Proximal fusion of pinnules with open dichotomous venation. Scale bar 1.5 mm; B. Fused portion of the pinna with areolate venation around interstitial pinnule incorporated in the webb. Scale bar 0.75 mm; C. Later stage appearing as a simple pinnule with dichotomous lateral veins, except the occasional anastomoses and irregular branching of basal veins. Scale bar 1.5 mm (Krassilov, 1995).

The reproductive structures of Mesozoic cycadophytes were likewise strobilate, yet showing a planation tendency in at least two forms of ovulate organs, *Semionogyna* from the Early Cretaceous of Transbaikalia (Krassilov and Bugdaeva, 1988) and *Baruligyna* from the Jurassic of Georgia. Both were flat axes bearing ovules in two lateral rows. In *Semionogyna*, the ovules developed in the axils of subtending bracts that were shed at maturity. *Baruligyna*, although chronologically an earlier form, was further advanced in the direction of *Cycas*-type "megasporophyll" morphology. Here the subtending bracts were lost early in development leaving inconspicuous scars at the base of the ovules that were decurrent at the base, webbing the axis (Fig. 6). These forms preceded the first appearance of *Cycas* in the Late Cretaceous (Krassilov, 1978b) definitely suggesting a derivation of leaf-like ovulate organs from a strobilate precursor. The pteridospermous aspect of "megsporophylls" in *Cycas* and the anomalous ovuliferous leaves in *Ginkgo* are, thus, of secondary nature, falling in the category of retroconvergence (Krassilov, 1995).

The phenomenon of retroconvergence is most spectacular in the case of complex structures homomorphous to their constituent units, as in the compound strobili (of conifers) converging upon their progenitorial simple cones, inflorescences imitating solitary flowers, fruits shaped as seeds, permanent pollen tetrads resembling solitary pollen grains, etc. In the Mesozoic peltasperm *Scytophyllum vulgare* (Prynada) Dobruskina, segments of bi- to tripinnate leaves fuse marginally webbing the axes and, at a more advanced stage, forming the entire leaf blade. Although the pinnules are lost in fusion, their venation remains distinct over the coalescent blade. Areolate venation develops by arching and anastomosing of veins around an interstitial pinnule imbedded in the web (Fig. 7A, B). Yet at the later stages, all traces of fusion are obliterated, the venation is reduced to the conventional dichotomous pattern, and the coalescent blade appears as a segment of the initial pinnate leaf (Fig. 7C). An adaptive meaning of such cyclic transformations is scarcely conceivable. Rather, the meaning is related to the nature of genomic memory retaining images of lost basic structures, which can be conferred upon their derivative organs.

8.   PATHOGENIC ORGANOGENESIS: ORGANOID GALLS

Symbionts render not only physiological but also morphological effects (reviewed in Buchner, 1965), most conspicuous in the case of organoid galls, which are induced by various organisms, from bacteria to arthropods. Gall formation, or cecidogenesis, often simulates normal organogenesis, even to the extent of pathogenic "organocopies". Familiar examples are the "green petal" disease induced by mycoplasms, the polyclady (witches' brooms) caused by rust fungi that inflict also hypertrophy of leaf tissues and fusion of leaf lobes, the inhibition of apical growth bringing about a sympodial branching or aggregation of lateral shoots or pinnae (e.g., in the familiar fern species *Athyrium filix-femina* (L.) Roth galled by dipteran larvae), the swelling of petioles and pathogenic leaf dehiscence, the production of epiphyllous buds and dwarf shoots, the transformation of paniculate inflorescences into heads by inhibition of internodes (by cecidogenous larvae of a homopteran species *Philaenus spumarius* L.), the enlargement and closure of corolla in flower galls or, conversely, the inhibition of corolla (e.g., by a heteropteran galler *Copium cornatum* Thumb.), the development of pouch galls resembling ascidiform carpels, inflicted by various insect gallers such as aphids, gall midges, etc. (reviewed in Meyer, 1987).

In the above examples, the cecidogenesis involves both hypertrophy and inhibition of growth. The parallelism of evolutionary derived morphogenetic processes and pathogenic organogenesis indicates that developmental pathways may be the same in both. If the galls are organocopies of evolutionary developments, can the reverse be true for non pathogenic characters that look like galls? In other words, can certain plant morphologies, such as the fusion of needles, cecidogenic in *Pinus silvestris* L., but a generic character in *Sciadopitys verticillata* Sieb et Zucc. or the ascidiform gynoecia of archaic angiosperms that look like pouch galls, be acquired as genocopies of respective pathogenic transformations?

Since galling is mediated by the gene transducing plasmids (Manulis, 1992), it seems fairly possible that the same transposons are involved in normal morphogenetic processes and their cecidogenous organocopies. A genomic study addressing this problem would have been rewarding. From the morphological point of view, the possibility is confirmed by the existence of pseudogalls, or plant structures of nonpathogenic origin that appear like galls and are colonized by cecidicolous insects (Monod and Schmitt, 1968). Pseudogalls obviously result from a genetic fixation of pathogenic induction. If so, then certain plant organ morphologies might, in principle, originate in the same way.

A paleobotanical observation perhaps relevant to the problem of cecidogenous organocopies is that the diversity of both organogenesis and gall formation had immensely and simultaneously increased with the advent of flowering plants, the coincidence that was scarcely accidental. Although galls are recorded since the Carboniferous at least (Van Amerom, 1973), their fossil record is meager until the mid-Cretaceous and the causative agents remained unknown. In contrast, in the Turonian fossil plant assemblage of Gerofit, southern Israel, almost every leaf is parasitized by gallers, sometimes of two or more kinds. The most common are the pouch galls (Fig. 8A) induced by cecidomiids (their larvae are occasionally preserved). As their extant progeny, the Cretaceous cecidomiids were able to break loose their galls, which are found detached, abandoned by imago, with the pupal horns emerging from the exit (Fig. 8B). Such remains can easily be confused with detached fruits from the same locality, and it is still to be learned whether this morphology first appeared in fruits or in galls. Anyway, the abundance and diversity of galls on early angiosperms testifies to an explosive cecidogenesis for which the appropriate mechanism has to be sought.

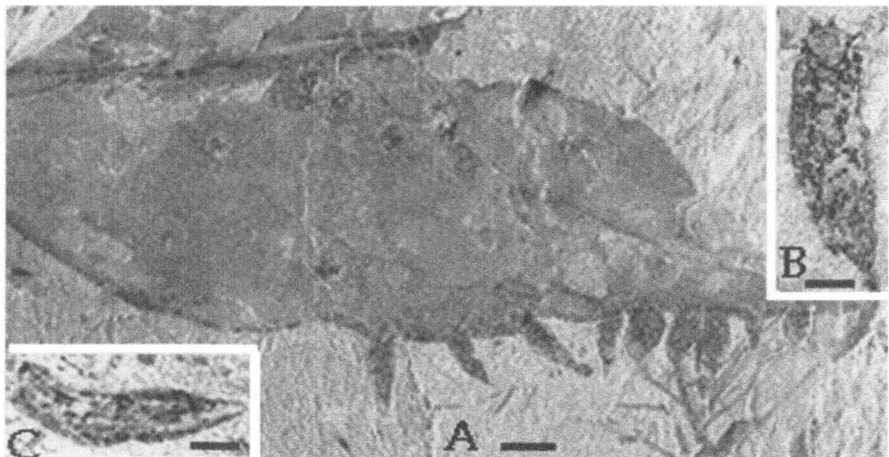

Figure 8. Gall and fruit remains from the mid-Cretaceous of Gerofit, southern Negev, Israel: A. Poach galls on leaf margin, scale bar 2 mm; B. Poach gall detached, with cephalic horns of the pupa protruding from the exit, scale bar 0.3 mm. C. Detached fruit of *Gerofitia* Krassilov et Dobruskina for comparison with the gall, scale bar 0.3 mm.

9.   ANASTROPHES

No matter how loyal to the *natura non facit saltum* slogan and how suspicious of the fossil record one may be, one has to accept the non-uniform evolution rates as an undeniable fact of nature.

New orders of plants and animals appear in dozens within rather short intervals of geological time followed by the much longer periods of evolutionary quiescence (examples in Simpson, 1944, 1949; House, 1963; Newell, 1967; Boucot, 1975; Krassilov, 1977; Vermeij, 1987 and elsewhere). Seed plant phylogeny (Fig. 9) clearly shows the clustered appearance of orders in gymnosperms and proangiosperms, whereas angiosperms accomplished their basic radiation in the mid-Cretaceous (Krassilov, 1997).

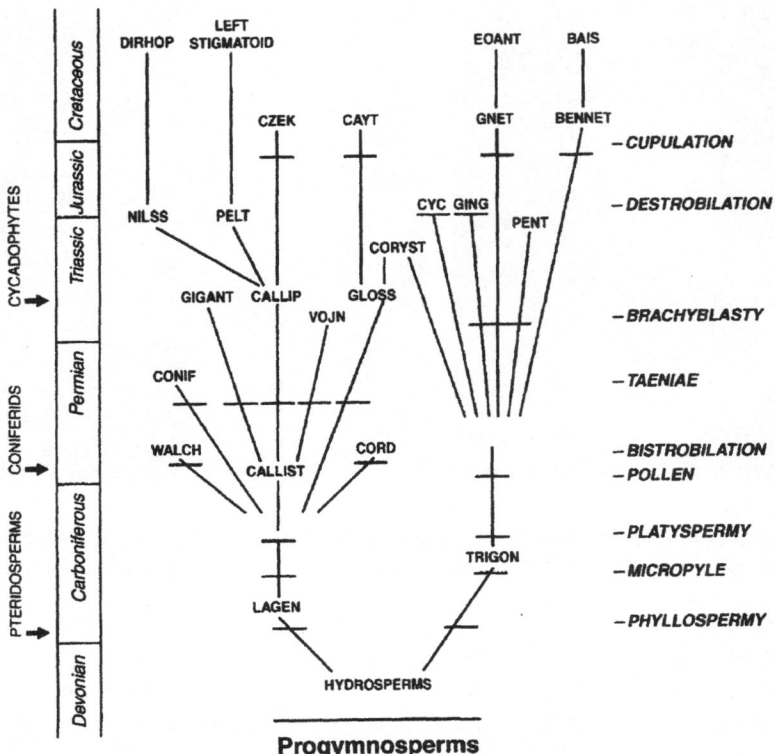

Figure 9. Phylogeny of gymnosperms and proangiosperms (Krassilov, 1997) showing radiation bursts (anastrophes) and parallel developments of morphological traits (horizontal dashes). Abbreviations: Bais, *Baisia*; Bennet, bennettites; Callist, callistophytes; Callipt, callipterids; Conif, conifers; Cord, cordaites; Cayt, catonias; Cyc, cycads; Czek, czekanowskians; Dirhop, *Dirhopalostachys;* Eoant, *Eoantha;* Gigant, gigantopterids; Ging, ginkgoaleans, Gnet, gnetophytes; Gloss, glossopterids; Lagen, lagenopstoms; Lept, *Leptostrobus*; Nilss, nilssonias; Pelt, peltasperms; Trigon, trigonocarps; Walch, walchians.

The anastrophes are commonly thought to be consequential to colonization of new biotopes, which explains why diversification bursts closely followed the first appearances, e.g., of land plants in the Devonian. A striking example is the rapid diversification of aquatic ferns and angiosperms at the initial stage of their entry into lacustrine ecosystems in the mid-Cretaceous including the bizarre forms producing amphisporions, the megaspores bearing microspores (Fig. 10) retroconvergent upon the Devonian *Kryshtofovichia* Nikitin (Krassilov and Golovneva, 1999). Likewise, the diversity of specialized seedling morphologies in the mid-Cretaceous mangrove assemblage of southern Israel (part 6) is evidence of explosive macroevolution at the first appearance of a new adaptive type.

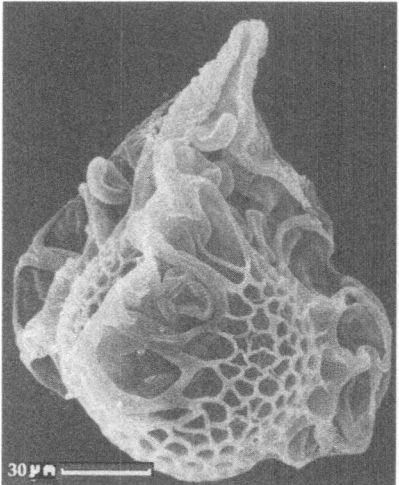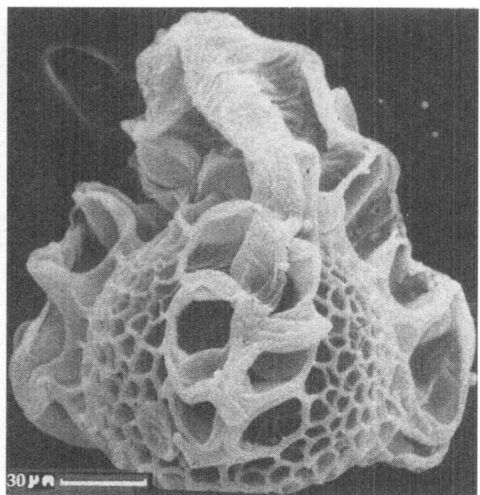

Figure 10. Innovation at anastrophe in aquatic ferns: the water-borne megaspores bearing microspores in their laesural pockets as a single dispersal unit (Krassilov and Golovneva, 1999).

Yet anastrophes are sometimes delayed for several periods, as in the case of mammals (the first appearance in the mid-Triassic, anastrophe in the Paleogene), or they happen more than once, as in ammonoids (in the Devonian and Jurassic) or dinosaurs (in the Jurassic and mid-Cretaceous). In such cases, the bursts of diversification were preceded by near-extinctions, which corroborate a causal link between catastrophes and anastrophes.

The coincidence of anastrophes in dinosaurs and angiosperms can be explained by their links through trophic cascades (Krassilov, 1981b). However, such trophically unrelated groups as planktonic foraminifers, marine bivalves (aucellinas, rudistids), craniid brachiopods, terrestrial insects, etc., also experienced anastrophes at the same time, which requires a more general explanation. According to the model of coherent vs. non-coherent evolution (Krassilov, 1969, 1977, 2003, and elsewhere), anastrophes are consequential to disruption of co-evolved biotic communities (simultaneous in marine and terrestrial realms under global environmental change). With stabilizing selection relaxed, high premium is placed on accelerate development and early reproduction promoting heterochronic shifts of growth rates. All kinds of macromutation described above may appear jointly because they are interrelated. In effect, the morphological disparity is greatly enhanced bringing about macropolymorphic populations typical of post-crisis situations.

Examples of post-crisis macropolymorphism could have been concealed by the routine paleontological treatment of any deviating morphology as a separate taxon. This, in fact, has been revealed in my studies (Krassilov,

1976, 1989) of morphological variation in the Early Paleocene species *Trochdendroides arctica* (Heer) Berry and *Platanus raynoldsii* Newberry, previously described under a dozen generic names each (Kryshtofovich and Baikovskaya, 1966). At this pre-anastrophic stage, macromutation does not result in the origin of new taxon and must be considered as a micro-evolutionary event. Yet the macropolymorphic population might serve as a basis for diversification at the species and higher levels with evolution recessing to the coherent coevolution mode.

10.   CONCLUSIONS

There is ample evidence of taxonomically important characters arising from developmental changes that can hardly be anything but saltational. Yet the appearance of such characters does not mean that the higher taxa, for which they are typical, came into being in the same saltational way. Characters appear, disappear, and reappear each time in a different adaptive context. They become typical for a higher taxon as a part of its adaptive syndrome through interaction with other typical characters having a separate history of macroevolutionary developments.

Higher taxa are commonly conceived of as constant combinations of characters indicating a monophyletic origin. However, although the latter remains an unverifiable assumption, a more realistic cause for morphological distinctness is its functional adaptive significance. Conifers or angiosperms are essentially the distinct adaptive types in the same way that amphibians or mammals are. As such they make use of structures that appeared in a different adaptive context and for different functions. Such prophetic structures (e.g., the reticulate pollen grains, part 3) might have evolved in a remote rather than immediate ancestor of the group to be retrieved as retroconvergent characters (e.g., the "pteridospermous" megasporophyll morphology in cycads, part 6) from a silent part of genomic memory. In particular, the angiosperms show typical characters of cycads, ferns, horsetails, charophytes even. Each of these groups has been claimed to be ancestral at one time or another and, in a sense, they all contributed to the origin of angiosperms (Krassilov, 2002). An essential feature of the latter is the accelerate development opening the way to a retrieval of the seemingly long-forgotten traits. The phenomenon of retroconvergence suggests that, once appearing, a morphological trait would seldom or maybe never escape from the genomic memory.

Ample opportunities for macromutational developmental shifts arise with accelerated growth and precocious sexual maturation. Moreover, pedomorphic forms are especially vulnerable to environmental induction including the pathogenic organogenesis. Therefore, one can expect a high rate macroevolution under circumstances that encourage precocious development. Just such circumstances arise with global environmental crises and the collapse of stable biotic communities. The post-crisis populations tend to be macropolymorphic incorporating morphological effects of macromutations. In such populations, macromutations are microevolutionary, rather than macroevolutionary, events enhancing morphological disparity that is not accompanied by an adequate increase in taxonomic diversity. The diversification bursts start with adaptive segregation of macropolymorphic phenotypes.

The traditional speciation theory holds that separation by a physical barrier is necessary for interruption of gene flow. It has been shown, however, that ecological differences may be even more important than geographic separation (Ehrlich and Raven, 1969). Rapid accumulation of evidence for sympatric genetic divergence in respect to differentiation of habitats, greatly enhanced by Eviatar Nevo and his collaborators (Nevo et al., 1997a,b, and elsewhere), makes clear that the traditional theory of isolation first – adaptive divergence next, just started at the wrong end. Effective interruption of gene flow is achievable, without geographic barrier, through a phenotypic modification (such as dwarfing or a shift of flowering time) that interferes with crossing.

Vast evidence collected under the orthogenetic paradigm (part 1) and a few more recent examples attest to developmental shifts as an immediate response to environmental induction. The effects typically fall in the category of phenocopies (Goldschmidt, 1958), which means that the environmentally induced phenotype is reproducible by mutation and, therefore, can be fixed as genocopy.

Thus, recognition of the evolutionary role of macromutation does not necessarily imply that mechanisms of micro- and macroevolution are different. It implies only that our understanding of both is impeded by the tenacious, yet hopefully surmountable, dogmas of traditional evolutionism.

11. REFERENCES

Arthur W. 1975. Mechanisms of Morphological Evolution. Wiley, Chichester.

Bloch R. 1965. Historical foundation of differentiation and development in plants. In Encyclopedia of Plant Physiology (15), Ruhland W., (Ed.), Springer, Berlin, pp. 140-188.

Boucot A.J. 1975. Evolution and Extinction Rate Controls. Elsevier, Amsterdam.

Buchner P. 1965. Endosymbiosis of Animals with Plant Microorganisms. Interscience, New York.

Carson H. L. 1975. The genetics of speciation at the diploid level. Amer Natur. 109, 83-92.

Ehrlich P.R. and Raven P.H. 1969. Differentiation of populations. Science 165, 1228-1231.

Eldredge N. and Gould S.J. 1972. Punctuated equilibria: an alternative to phyletic gradualism. In Models in Paleobiology, Schopf T.M. (Ed.), Freeman, San Francisco, pp. 82-115.

Florin R. 1931. Die Epidermisstruktur der recenten Koniferen und ihre Verwendbarkeit für die Systematik, Kgl. Svensk. Vetenskapsakad. Handl., Tredje Ser. 10, 45-496.

Florin R. 1933. Studien über Cycadales des Mesozoikums nebst Erörterungen über die Spaltöffnunggsapparate der Bennettitales, Kgl. Svensk. Vetenskapsakad. Handl., Tredje Ser. 12, 1-134.

Friis E.M., Crane P.R. and Pedersen K.R. 1986. Floral evidence for Cretaceous chloranthoid angiosperms. Nature 320, 163-164.

Goldschmidt, R.B. 1958. Theoretical Genetics. California UP, Berkley.

Gould, S.J. 1969. Land snail communities and Pleistocene climates in Bermuda: a multivariate analysis of microgastropod diversity. Proc North Amer Paleont Conf., Pt. E., 486-521.

Gould S.J. 1977. Ontogeny and phylogeny. Harvard UP, Cambridge, Mass.

House M.R. 1963. Bursts in evolution. Adv Sci. 19, 499-507.

Hughes N.F. 1976. Paleobiology of angiosperm origins.Cambridge UP, Cambridge.

Istchenko T.A. and Istchenko A.A. 1981. Middle Devonian Flora of Voronezh Anticlise. Naukova Dumka, Kiev (in Russian).

Juncosa A.M. 1982. Embryo and seedling development in the Rhizophoraceae .Am JBot. 69, 1599-1611.

Juncosa A.M. 1984. Embryogenesis and seedling developmental morphology in Bruguiera exaristana Ding How (Rhizophoraceae). Amer J Bot. 71, 180-191.

Jinks J.L. 1964. Extrachromosomal Inheritance. Prentice-Hall, Englewood Cliff, New Jersey.

Koloda, N.A. 1996. New data on systematics of quasimonosaccate pollen. Proc Komi Geol Inst. RAS. 89, 49-59 (in Russian).

Krassilov V.A. 1967. Early Cretaceous Flora of South Primorye and its Stratigraphic Significance. Nauka, Moscow (in Russian).

Krassilov V.A. 1969. Phylogeny and systematics. In Problems in Phylogeny and Systematics, Far East Branch Russsian Acad Sci. Gramm M.N. and Krassilov V.A. (Eds.), Nauka USSR, Vladivostok (in Russian).

Krassilov V.A. 1970. Approach to the classification of Mesozoic "ginkgoalean" plants from Siberia. Palaeobotanist 18, 12-19.

Krassilov V.A. 1976. Tsagajan Flora of Amur Province. Nauka, Moscow (in Russian).

Krassilov V.A. 1977. Evolution and Biostratigraphy. Nauka, Moscow (in Russian).

Krassilov V.A. 1978a. Bennettitalean stomata. Palaeobotanist 25, 179-184.

Krassilov V.A. 1978b. Late Cretaceous gymnosperms from Sakhalin and the terminal Cretaceous event. Palaeontol. 21, 893-905.

Krassilov V.A. 1981a. Orestovia and the origin of vascular plants. Lethaia 14, 235-250.

Krassilov V.A. 1981b. Changes of Mesozoic vegetation and the extinction of dinosaurs. Palaeogeogr Palaeoclimatol Palaeoecol. 34, 207-224.

Krassilov V.A. 1989. Vavilov's species concept and evolution of variation. Evol Th. 9, 37-44.

Krassilov V.A. 1995. *Scytophyllum* and the origin of angiosperm leaf characters. Paleontol J. 29(A), 63-75.

Krassilov V.A. 1997. Angiosperm Origins: Morphological and Ecological Aspects. Pensoft, Sophia.

Krassilov V.A. 2002. Character parallelism and reticulation in the origin of angiosperms. In Horizontal Gene Transfer. 2nd ed. Syvanen M. and Kedo C.I. (Eds.), Acad. Press, San Diego e.a., pp. 373-382.

Krassilov V.A. 2003. Terrestrial Palaeoecology and Global Change. Pensoft, Sofia.

Krassilov V.A., Afonin S.A. and Lozovsky V.R. 1999a. Floristic evidence of transitional Permian-Triassic deposits of the Volga - Dvina Region. Permophiles 34, 12-14.

Krassilov V.A. and Bugdaeva E.V. 1982. Achene-like fossils from the Lower Cretaceous of the Lake Baikal area. Rev Palaeobot Palynol. 36, 279-295.

Krassilov V.A. and Bugdaeva E.V. 1988. Protocycadopsid pteridosperms from the Lower Cretaceous of Transbaikalia and the origin of cycads. Palaeontographica Abt B. 208, 27-32.

Krassilov V.A. and Bugdaeva E.V. 1999. An angiosperm cradle community and new proangiosperm taxa. Acta Palaeobot Suppl. 2, 111-127.

Krassilov V.A. and Bugdaeva E.V. 2000. Gnetophyte assemblage from the Early Cretaceous of Transbaikalia. Palaeontographica Abt B. 253, 139-151.

Krassilov V.A. and Dobruskina I.A. 1998. Gramminoid plant from the Cretaceous of Middle East. Palaeontol J. (Moscow) 4, 106-110 (in Russian with English translation).

Krassilov V.A. and Golovneva L.B. 1999. A new heterosporous plant from the Cretaceous of West Siberia. Rev Palaeobot Palynol. 105, 75-84.

Krassilov V.A. and Golovneva L.B. 2001. Inflorescence with tricolpate pollen grains from the Cenomanian of Chulymo-Yeniey Basin, West Siberia. Rev Palaeobot Palynol. 115, 99-106.

Krassilov V.A. and Rasnitsyn A.P. 1997. Pollen in the guts of Permian insects: first evidence of pollinivory and its evolutionary significance. Lethaia 29, 369-372.

Krassilov V.A. and Rasnitsyn A.P. 1998. Plant remains from the guts of fossil insects: evolutionary and palaeoecological inferences. Proc 1st Intern. Palaeoentomol Conf., Moscow, AMBA Projects AM/PFICM-98/1. 99, 65-72.

Krassilov V.A. and Shilin P.V. 1995. New platanoid heads from the mid-Cretaceous of Kazakhstan. Rev Palaeobot Palynol. 85, 207–211.

Kryshtofovich A.N. and Baikovskaya T.N. 1966. Late Cretaceous flora of Tsagajan in Amur Province. In: Selected Work, 3, Kryshtofovich A.N., Nauka, Moscow, pp.184-320 (in Russian).

Kürschner W.M. 1996. Leaf Stomata as Biosensors of Palaeoatmospheric CO_2. LPP Contributions series 5, Utrecht: LPP Foundation, Utrecht.

Lewis J., Slack J.M.W. and Wolpert L. 1977. Thresholds in development, J Theor Biol. 65, 579-590.

Løvtrup S. 1974. Epigenetics. A Treatise on Theoretical Biology. Wiley, London.

Schindewolf O.H. 1950. Grundfragen der Paläontologie. Schweizerbart'sche Verlag, Stuttgart.

Manulis S. 1992. Evaluation of DNA probe for detecting Ervinia herbicola strains pathogenic on *Gypsophila paniculata*. Plant Pathol. 41,342-347.

Mayr E. 1954. Change of genetic environment and evolution. In: Evolution as a Process,J Huxley A.C. Hardy and Ford E.B. (Eds.), Allen and Unvin, London, pp.157-180.

Meinzer F.C. 1993. Stomatal control of transpiration. Trends in Ecol Evol. (TREE) 8, 289-294.

Meyen S.V. 1984. Basic features of gymnosperm systematics and phylogeny as evidenced by fossil record. Bot Rev. 50, 1-111.

Meyer J. 1987. Plant Galls and Gall Inducers. Gebrüder Bortentraeger, Berlin, Stuttgart.

Moe R.L. and HenryE.C. 1982. Reproduction and early development of *Ascoseira mirabilis* Skottsberg (Phaeophyta), with notes on Ascoseirales Petrov. Phycologia 21, 55-66.

Monod T. and Schmitt C. 1968. Contribution à létude des pseudo-galles formicaires chez quelques Acacias africains, Bull Inst Franç Afrique Noir 30, 953-1012.

Nair P.S., Carson H.L. and Sene E.M. 1977. Isozyme polymorphism due to regulatory influence. Amer. Natur. 77, 789-791.

Neumayr M. 1879. Zur Kentniss der Fauna der untesten Lias in der Nordalpen, Abh. Geol. Reichsanst. 7, 5.

Nevo E., Kirzhner V., Beiles A. and Korol A. 1997a. Selection versus random drift: long-term polymorphism persistence in small populations (evidence and modeling), Phil Trans Roy Soc Lond. B 352, 381-389.

Nevo E., Rashkovetsky E., Pavlicek T. and Korol A. 1997b. A complex adaptive syndrome in *Drosophila* (fruit fly) caused by microclimatic contrasts. Heredity 79, 1-8.

Newell N.D. 1967. Revolutions in the history of life. Geol Soc Amer Spec. 89, 63-91.

Niklas K.J., Phillips T.L. and Carozzi A.V. 1976. Morphology and paleoecology of Protosalvinia from the Upper Devonian (Famenian) of Middle Amazon basin of Brazil. Paleontographica, Abt. B155, 1-30.

Pant D.D. 1960. The gametophyte of the Psilophytales. Proc Summer School Bot Darjeeling, 226-301.

Severtsov A.N. 1939. Morphological Regularities of Evolution. Acad Sci USSR, Moscow, Leningrad.

Simpson G.G. 1944. Tempo and Mode in Evolution. Columbia UP, New York.

Simpson G.G. 1949. The Meaning of Evolution. Yale UP, New Haven, Conn.

Soulé.M. 1973. The epistasis cycle: a theory of marginal population. Ann Rev Ecol Syst. 4, 165-187.

Stanley S.M. 1979. Macroevolution. Pattern and Process, Freeman, San Francisco.

Ursprung H. 1966. The formation of patterns in development.In: Major Problems in Developmental Biology, Locke M. (Ed.), Academic Press, New York, London, pp. 177-216.

Van Amerom H.W.J. 1973. Gibt es Cecidien im Karbon bei Calamiten und Asterophylliten? Septiéme Congrés International de Stratigraphie et de Géologie du Caronifére, Compte Rendu 2, 63-76.

Vermeij G.L. 1987. Evolution and Escalation, An Ecological History of Life. Princeton UP, Princeton.

Zavada M.S. 1984. Angiosperm origins and evolution based on dispersed pollen ultrastructure. Ann Mo Bot Gard. 71, 444-463.

Part Four

Human Evolution and Ecology

HUMANKIND'S PLACE IN A PHYLOGENETIC CLASSIFICATION OF LIVING PRIMATES

Derek E. Wildman[1,2] and Morris Goodman[1]

[1]*Department of Anatomy and Cell Biology, Wayne State University School of Medicine, 540 E. Canfield Ave, Detroit, MI 48201, USA* [2]*Center for Molecular Medicine and Genetics, Wayne State University School of Medicine, 540 E. Canfield Ave, Detroit, MI 48201, USA*

Abstract: In order to accurately place humankind in a phylogenetic classification of Primates it is necessary to know the phylogenetic relationships among all members of the order. Toward this end, we present a consensus view of primate phylogeny based on DNA nucleotide sequence evidence and local molecular clock calculations constrained by fossil evidence for the majority of primate genera. These data place the two chimpanzee species as the sister group of humans, and the three species together form a clade that is the sister group of gorillas. Humans and chimpanzees shared a most recent common ancestor (mrca) between five and six million years ago. Using the objective yardstick of time of origin as the means for assigning taxonomic rank, we find that humans and chimpanzees should be recognized as members of different subgenera within the genus *Homo*. Under this arrangement, humankind is called *Homo (Homo) sapiens* while the common chimpanzee and bonobo chimpanzee are called *Homo (Pan) troglodytes* and *Homo (Pan) paniscus*, respectively. All extant apes are placed in the family Hominidae. Additionally, the age based phylogenetic classification of the order Primates suggests that among cercopithecid genera Theropithecus and Lophocebus should be reduced to subgeneric rank within the genus *Papio*, *Mandrillus* should be reduced to subgeneric rank within the genus *Cercocebus*, and *Erythrocebus* should be reduced to subgeneric rank within the genus *Chlorocebus*. Furthermore, the platyrrhine genus *Cacajao* should be reduced to subgeneric rank within the genus *Chiropotes*. The average age of generic crown groups of mammals was estimated across the class Mammalia and was found to be >8 Ma, much earlier

S.P. Wasser (ed.), Evolutionary Theory and Processes: Modern Horizons,
Papers in Honour of Eviatar Nevo
D.E. Wildman and M. Goodman. *Humankind's Place in a Phylogenetic Classification of Living Primates, 293-311.*
© 2004 *Kluwer Academic Publishers.*

than the mrca of chimpanzees and humans. The age of sampled mammalian generic crown groups ranges from 2 to 21 million years ago. Recognizing chimpanzees as members of the human genus and all the apes as members of the human family, highlights the urgent need to protect these close relatives from extinction and to conserve the natural habitats of primates.

1. INTRODUCTION

As in the previous Festschrift paper honoring Eviatar Nevo (Goodman et al., 1999), we again delineate the place of humans in primate phylogeny. The cumulative DNA evidence on primate phylogeny combined with fossil evidence shows that genetically humans are quite similar to other primates and highly similar to their closest living relatives, the common and bonobo chimpanzees, that all apes should be placed in the human family Hominidae, and that the two species of chimpanzees should be placed in the human genus *Homo* (Goodman et al., 1998, 1999, 2001, 2002; Goodman, 1999; Wildman et al., 2003). According to this arrangement, the two chimpanzees constitute the subgenus *Homo* (*Pan*), and humans along with archaic human-like apes, constitute the subgenus *Homo* (*Homo*). Rejecting the view that functionally important genetic characters will show chimpanzees to be closer to gorillas than to humans, our recent study used coding nucleotide sequences of different primates and, within these sequences, examined the nonsynonymous (amino acid changing) substitutions at those DNA sites that are scrutinized and shaped by natural selection. The results of our examination of these functionally important genetic characters show that humans and chimpanzees are closest to each other and equally different from gorillas (Wildman et al., 2003). It may be considered fitting that we dedicate our present paper to Professor Nevo who continues to elucidate the central role of natural selection in shaping the course of evolution.

In order to update the previous Festschrift paper (Goodman et al., 1999), we first briefly recapitulate the taxonomic philosophy we employ to place humans in a phylogenetic classification of living primates. We have taxa that represent clades and, where known, the ages of the clades determine the ranks of the taxa. We then present our phylogenetic classification for all living primate genera and some subgenera, e.g., *Homo* (*Homo*) and *Homo* (*Pan*). Next, to test the validity of congenerically grouping humans and chimpanzees, we compare the ages of a variety of mammalian genera. In agreement with the classification having congeneric humans and chimpanzees, the age of the enlarged genus *Homo* is well within the range for other mammalian genera. Finally, we note that chimpanzees, gorillas, and orangutans are facing extinction because of human assaults and disappearing habitats. Recognizing chimpanzees as members of the human genus and all the apes as members of the human family, should highlight the urgency of protecting both these close relatives of humans as well as humankind itself from extinction.

2. TAXONOMIC PHILOSOPHY

The philosophy we follow is essentially that of Charles Darwin and Willi Hennig. Darwin proposed that a "Natural System" of classification should be simply genealogical (Darwin, 1859, 1871; Ghiselin and Jaffe, 1973). Elaborating on what Darwin envisioned, Hennig (1966) formulated three principles to use in constructing a phylogenetic classification. The first principle is that each taxon should represent a monophyletic group or clade; that is, it should represent all species descended from a common ancestor. The second and paramount principle is that the hierarchical groupings of lower-ranked taxa into higher-ranked taxa should describe the phylogenetic relationships of the clades. For example, the different species within any genus nested within a family should have a most recent common ancestor (mrca) that is more recent in time than the mrca of the genera within that family. Although the hierarchical ranking of taxa in a nested series captures the relative ages of the clades represented by the series, it does not ensure that taxa assigned the same rank are evolutionarily equivalent by an objective measure. To have an objective measure, Hennig's third principle is that the absolute age of a clade should determine the clade's taxonomic rank. By this principle, taxa at the same hierarchical level or rank would represent clades that are equally old, i.e., are at an equivalent evolutionary age.

The fossil record by itself allows estimates of the ages of only a scattering of branch-points in primate phylogeny. However, the model of local molecular clocks applied to the branch lengths of phylogenetic trees constructed from DNA data provided estimates of the ages of all branch-points in these trees. (Goodman, 1986; Bailey et al., 1991, 1992; Barroso et al., 1997; Porter et al., 1997a,b; Goodman et al., 1998; Chaves et al., 1999; Meireles et al., 1999; Page et al., 1999; Page and Goodman, 2001; Wildman et al., 2003). The model of local molecular clocks differs from that of a global molecular clock by not assuming that all lineages accumulate nucleotide substitutions at the same rate; local molecular clock calculations are much more constrained by fossil evidence on branch-times than are global molecular clock calculations. Local molecular clock estimates of branch-times adjust for lineage variation in rates by having each base substitution occur over a longer period of time in a more slowly evolving lineage than in a more rapidly evolving lineage.

Also, different kinds of DNA and different parts of genomes evolve at different rates. For example, because of purifying natural selection, nonsynonymous substitutions generally occur at much slower rates than the nucleotide substitutions that do not cause amino acid changes. Moreover, because during descent the positive form of natural selection might act more on certain genes than on others, the rate of nonsynonymous substitutions will vary from one gene to another. Thus, simply using degrees of interclade DNA divergence by itself would not prove to be an accurate objective measure for assigning taxonomic ranks among phylogenetic groups. However, an objective yardstick is attainable by utilizing the fossil record to convert the different molecular phylogenetic distances into the common currency of time.

3.   AN AGE BASED PHYLOGENETIC CLASSIFICATION OF LIVING PRIMATES

The time scale for the phylogenetic tree shown in Fig. 1, and the divergence dates in the age-based phylogenetic classification shown in Table 1, portray a series of phylogenetic branchings during the course of primate evolution. Table 2 relates this time scale to the geological epochs in which the branchings occurred and to the ranks of the taxa produced by the branchings. The division of a higher-ranked taxon into subordinate lower-ranked taxa denotes a phylogenetic branching. The age (in Ma) placed after the name of a taxon (Table 1) is the estimated age of that taxon treated as a *crown group* but also of that taxon's closest (at a step below in rank) subordinate taxa treated as *total groups*. A crown group includes both the mrca of the extant species in a clade and all descendant species (extinct and extant) of the mrca, but does not include the stem of the mrca (Jeffries, 1979). The total group includes, in addition to all members of the crown group, the stem of the mrca and all extinct offshoots of the stem. Thus the age of 63 Ma for the mrca of all living primates – that is, the age for Primates as a crown group – is the age for both Strepsirrhini and Haplorhini as total groups. In turn, the ages of 50 Ma and 58 Ma listed alongside of Strepsirrhini and Haplorhini, respectively, are the ages for these two taxa treated as crown groups.

After this first major branching, in the early Paleocene epoch, into semi-ordinal clades, subordinal clades emerged. The late Paleocene haplorhines divided into Tarsiiformes and Anthropoidea. The anthropoideans of the middle Eocene epoch (at ~40 Ma) divided into the infra-orders Platyrrhini and Catarrhini. As total groups, families originated from superfamilial clades within infra-orders in the middle to late Oligocene epoch (~28-25 Ma), subfamilies in the early Miocene epoch (~23-22 Ma), tribes in the early to middle Miocene (~20-15 Ma), subtribes in the middle to late Miocene (~14-10 Ma), genera in the late Miocene (~11-7 Ma), and subgenera in the late Miocene to early Pliocene epoch (~7-4 Ma).

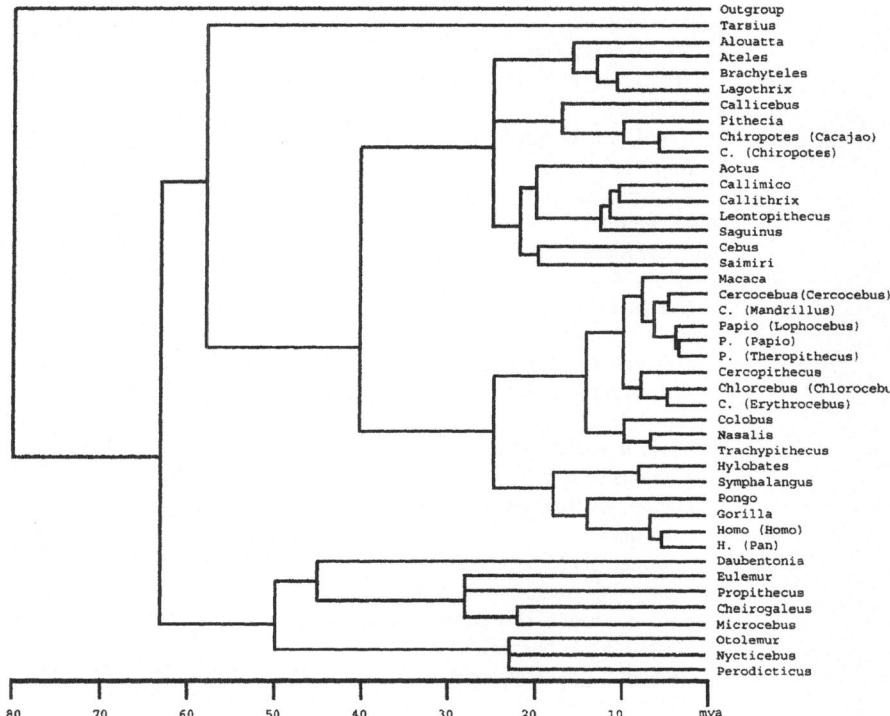

Figure 1. Generic Phylogeny of the Primates. This phylogenetic tree shows the best estimate of generic relationships among primates based on DNA evidence complemented by fossil evidence (Goodman et al., 1998, 1999, 2001, 2002). Shown are the 40 genera whose phylogenetic relationships were elucidated with DNA sequence data. The time scale is based on fossil evidence and molecular clock calculations as reviewed in Goodman et al. (1998). All sister-group relationships depicted in this figure are also depicted in the classification shown in Table 1.

Table 1. Provisional classification of extant Primates at the genus level[a].
Order Primates (63 Ma)
Semiorder Strepsirrhini (50 Ma)
Suborder Lemuriformes (45 Ma)
Infra-order Chiromyiformes
Family Daubentoniidae
Daubentonia: aye-aye
Infra-order Eulemurides
Superfamily Lemuroidea (28 Ma)
Family Cheirogaleidae (22 Ma)*

   Phaner: fork-marked lemur
  Unnamed Rank
   Cheirogaleus: dwarf lemurs
  Unnamed Rank
  Unnamed Rank
   Allocebus: hairy eared dwarf lemur
  Unnamed Rank
   Microcebus: mouse lemurs
   Mirza: coquerel's dwarf lemur
  Family Indridae
   Avahi: woolly lemurs
   Indri: indri
   Propithecus: sifakas
  Family Lemuridae*
  Unnamed Rank
  Unnamed Rank
   Eulemur: brown lemurs
  Unnamed Rank
   Hapalemur: bamboo lemurs
   Lemur: ring-tailed lemur
  Unnamed Rank
   Varecia: ruffed lemur
  Family Megaladipae
   Lepilemur: sportive lemurs
Suborder Loriformes
 Family Loridae (23 Ma)
  Subfamily Galagoninae
   Euoticus: needle-clawed bushbabies
   Galago: lesser bushbabies
   Otolemur: greater galagos
  Subfamily Perodictinae
   Perodicticus: potto
   Arctocebus: angwantibo
  Subfamily Lorinae
   Loris: slender lorises
   Nycticebus: slow lorises
Semiorder Haplorhini (58 Ma)
Suborder Tarsiiformes
 Family Tarsiidae
   Tarsius: tarsiers (6 Ma)
Suborder Anthropoidea (40 Ma)
Infra-order Platyrrhini
 Superfamily Ceboidea (26 Ma)
 Family Cebidae (22 Ma)
  Subfamily Cebinae (20 Ma)
   Cebus: capuchin monkeys (5 Ma)
   Saimiri: squirrel monkeys (2 Ma)
  Subfamily Aotinae
   Aotus: owl monkeys (3 Ma)
  Subfamily Callitrichinae
  Tribe Callitrichini (14 Ma)

Subtribe Saguinina
 Saguinus: tamarins (6 Ma)
Subtribe Leontopithecina
 Leontipithecus: lion tamarins
 Subribe Callitrichina
  Callithrix: marmosets (5 Ma)
  Callimico: goeldi's marmoset
Family Pitheciidae
 Subfamily Pitheciinae (19 Ma)
 Tribe Callicebini
  Callicebus: titis (7 Ma)
 Tribe Pitheciini
  Subtribe Pitheciina (11 Ma)
  Pithecia: saki monkeys
  Chiropotes: (6 Ma)
   C. (*Chiropotes*): bearded sakis
   C. (*Cacajao*): uakari
Family Atelidae
 Subfamily Atelinae (17 Ma)
 Tribe Alouattini
  Alouatta: howler monkeys (7 Ma)*
 Tribe Atelini (12 Ma)
  Subtribe Atelina
  Ateles: spider monkeys (4 Ma)
  Subtribe Brachytelina (10 Ma)
   Brachyteles: woolly spider monkeys (muriqui)
  Lagothrix: woolly monkeys
Infra-order Catarrhini
Superfamily Cercopithecoidea (25 Ma)
 Family Cercopithecidae
 Subfamily Cercopithecinae (15 Ma)
  Tribe Colobini (10 Ma)
   Subtribe Colobina
   Colobus: black-and-white colobus (5 Ma)
   Piliocolobus: red colobus
   Procolobus: olive colobus
   Subtribe Presbytina (7 Ma)
   Nasalis: proboscis monkey
   Presbytis: surilis
   Pygathrix: doucs
   Rhinopithecus: snub-nosed monkeys
   Semnopithecus: sacred langurs (gray langurs)
   Simias: pig-tailed langur (simakobu)
   Trachypithecus: lutungs
  Tribe Cercopithicini (10 Ma)
   Subtribe Cercopithecina (9 Ma)*
   Allenopithecus: Allen's swamp monkey
   Miopithecus: talapoins
   Cercopithecus: arboreal guenons
   Chlorocebus: (5 Ma)
    C. (*Chlorocebus*): vervets (green monkeys, grivets)

      C. (***Erythrocebus***): patas monkey
      C. (***Allochrocebus***): l'hoesti guenons, terrestrial guenons*
    Subtribe Papionina (9 Ma)
     Macaca: macaques (5.25 Ma)*
     Cercocebus: (5 Ma)
      C. (***Cercocebus***): white-eyelid mangabeys
      C. (***Mandrillus***): mandrill, drill (4 Ma)
     Papio (4 Ma)
      P. (***Papio***): papio baboons
      P. (***Theropithecus***): gelada baboons
      P. (***Lophocebus***): crested mangabeys
  Family Hominidae
    Subfamily Homininae (18 Ma)
    Tribe Hylobatini
     Subtribe Hylobatina (8 Ma)
     Hylobates: lar-group gibbons
     Symphalangus: siamangs
    Tribe Hominini (14 Ma)
     Subtribe Pongina
     Pongo: orangutans
     Subtribe Hominina (6-7 Ma)
     Gorilla: gorilla
     Homo: (5-6 Ma)
      H. (***Homo***): human
      H. (***Pan***): chimpanzees (3 Ma)

[a]Taxa in bold have DNA sequence data available in the public databases (e.g., GenBank). When available, the age (in parenthesis) placed after the name of a taxon in this hierarchical classification represents both the age of that taxon treated as a crown group and the age of the next lower taxon treated as a total group. *In addition to DNA studies reviewed in Goodman et al. (1998, 1999, 2001, 2002), we provisionally use those of Pastorini et al. (2001, 2002), Yoder (1994), and Yoder et al. (1996) for Lemuriformes, Meireles et al. (2003) for *Tarsius*, Cortés-Ortiz et al. (2003) for *Alouatta*, and (Tosi et al., 2003) for select cercopithecins. We did not assign ranks to the clades within the Cheirogaleidae and Lemuridae because molecular clock calculations were not included in those analyses. With regard to Cercopithecina, we support the suggestion of Tosi et al. (2003) that *Erythrocebus* be reduced to subgeneric rank within *Chlorocebus* because the two taxa diverged at about 5 Ma (Page et al., 1999; Page and Goodman, 2001).

The classification shown in Table 1 includes all recognized extant primate genera. As indicated in the footnote to the table, some of the genera are not yet represented by DNA data, and among those that are, some crown group ages have yet to be calculated. Future work on primate phylogeny should fill in these gaps. We provisionally recognize the family Megaladipae because *Lepilemur* appears to have originated as a total group earlier than 24 Ma (Yoder et al., 2003). Although Groves (2001) recognizes two loriform families (Galagonidae and Loridae), we recognize only one loriform family

(Loridae) consistent with local molecular clock calculations that place the mrca for extant loriform genera at 23 Ma (Goodman et al., 1998, 1999, 2001). Recent fossil finds (Seiffert et al., 2003) of Eocene loriforms are unclear as to whether stem loriforms more closely resemble galagonines, perodictines, or lorines. Thus, the new finds may have merely dated the loriform clade as a total group. In our opinion, the new finds do not seriously challenge the date employed for keeping all loriforms within a single family Loridae.

Table 2. Geologic time frame[b]

Epoch	Begin (Ma)	End (Ma)	Crown groups (Ma)	Total groups (Ma)
Early Paleocene	65.0	60.9	Orders (63)	Semi-orders (63)
Late Paleocene	60.9	54.8	Semi-orders (58-50)	Suborders (58-50)
Early Eocene	54.8	49.0	Semi-orders	Suborders
Middle Eocene	49.0	37.0	Suborders (45-40)-Infra-orders (39-29)	Infra-orders (45-40)-Superfamilies (39-29)
Late Eocene	37.0	33.7	Infra-orders	Superfamilies
Early Oligocene	33.7	28.5	Infra-orders	Superfamilies
Late Oligocene	28.5	23.8	Superfamilies (28-25) Families (23-22)-Subfamilies	Families (28-25) Subfamilies (23-22)-Tribes (20-
Early Miocene	23.8	16.4	(20-15)	15)
Middle Miocene	16.4	11.2	Subfamilies-Tribes (14-10) Tribes-Subtribes (11-7)-	Tribes-Subtribes (14-10)
Late Miocene	11.2	5.3	Genera (6-4)	Subtribes-Genera (11-7)
Early Pliocene	5.3	3.6	Genera -Subgenera (<4)	Subgenera (6-4)
Late Pliocene	3.6	1.8	Subgenera	Species (<4)
Pleistocene	1.8	0.01		
Recent	0.01	0.0		

[b]Ages of geological epochs are from McKenna and Bell (1997). Approximate ages for taxonomic ranks as both crown and total groups are shown in parentheses.

The family Tarsiidae has only one extant genus, *Tarsius*. Meireles et al. (2003) have shown two of the extant members of the genus, *Tarsius bancanus* and *Tarsius syrichta* had a most recent common ancestor between 5 and 6 Ma. There are at least seven recognized species in the genus *Tarsius*, and since only two have been studied, the time of origin of the crown group is unknown. The earliest fossil assigned to the genus *Tarsius* is from the middle Eocene of China (Beard et al., 1994; Gunnell and Rose, 2002), not only should this early tarsier have a different genus name, but it is also possible that the extant species may constitute more than one genus.

Among extant neotropical primates we recognize three families (Atelidae, Cebidae, and Pitheciidae). In contrast, Groves (2001) raises the taxonomic status of the New World owl monkeys from the cebid subfamily Aotinae to the family Nyctipithecidae, for a total of four extant platyrrhine families. Because the mrca of *Aotus*, the cebines (*Cebus* and *Saimiri*), and the callitrichines appears to be no older than 22 Ma, the age expected for a subfamily total group, there is no justification for raising *Aotus* to family status. However, the earliest member of the genus *Aotus* from the fossil record is *Aotus dindensis* from the middle Miocene at about 12 Ma (Hartwig and Meldrum, 2002). This suggests that owl monkeys may indeed comprise more than one genus using the time-based phylogenetic approach.

The earliest known fossil platyrrhine is *Branisella* from Bolivia in the Oligocene approximately 25-26 Ma (Fleagle and Tejedor, 2002). This taxon appears to be a stem platyrrhine, and therefore represents the maximum age of the crown group of New World monkeys. The earliest stem cebids include *Dolichocebus* and *Chilecebus* from the Miocene about 20-21 Ma. There is a great degree of taxonomic uncertainty regarding fossil pitheciids and atelids. For example, Rosenberger (2002) recognizes eleven fossil pitheciid genera, while Kay et al. (1998) recognize only three. Moreover, some authors disband the family Pitheciidae and place the subfamily Pitheciinae into the family Atelidae (Fleagle and Tejedor, 2002; Hartwig and Meldrum, 2002). Regardless of this uncertainty, the earliest stem pitheciid or ptheciine appear to be either *Soriacebus* or *Carlocebus* with a Miocene age of earlier than 16 Ma. (Fleagle and Tejedor, 2002). The earliest atelid fossils belong to the Miocene genus *Stirtonia* from the Colombian locality of La Venta, and are from between 12.6 and 13.7 Ma (Hartwig and Meldrum, 2002).

We recognize only one extant Old World monkey family, the Cercopithecidae. This family has an origin of between 20 and 25 Ma, and is the most speciose extant primate family with Groves (2001) recognizing 21 genera and 131 species. The cercopithecids represent nearly one third of all extant primate diversity (Groves, 2001). The earliest stem Old World monkey in the fossil record is *Prohylobates* from Uganda dated between 18 and 20 Ma, and the earliest crown cercopithecid is *Microcolobus* from about 11-10 Ma in Kenya (Benefit and McCrossin, 2002; Jablonski, 2002).

Our age-related classification places all living Old World monkeys in the subfamily, Cercopithecinae, and divides the subfamily into two tribes, the Colobini (leaf-eating monkeys) and the Ceropithecini (cheek-pouched monkeys). These two tribes are equivalent to the traditional two subfamilies, Colobinae and Cercopithecinae. The reason for reducing the rank from subfamily to tribe is that the crown group age of each of these two clades is that of a tribe. In turn, there is reasonable support for dividing the Colobini into two subtribes, one for the African colobins (three genera) and the other for the Asian colobins (seven genera).

The Cercopithecini are the best-studied Old World monkeys from a molecular perspective. The relationships within the Papionina have been elucidated starting with Sarich and Cronin (1976), and followed by Disotell et al. (1992), Disotell (1994), Harris and Disotell (1998), Harris (2000), Page et al. (1999), and Page and Goodman (2001). The dates of divergence among the papionan genera suggest that there should be fewer genera than are commonly recognized. The clade that includes in most classifications *Papio*, *Theropithecus*, and *Lophocebus* should be reduced to a single genus *Papio* with three subgenera *Papio* (*Papio*) for the dozen or so "kinds" of baboons, *Papio* (*Theropithecus*) for the gelada baboon, and *Papio* (*Lophocebus*) for the gray-cheeked and black mangabeys.

Tosi et al. (2002, 2003) have resolved the relationships among five genera in the subtribe Cercopithecina. Based on Y chromosome DNA sequence data, they show that *Chlorocebus* and *Erythrocebus* are sister taxa to the clade made up of *Allenopithecus*, *Miopithecus*, and *Cercopithecus*. Page et al. (1999), and Page and Goodman (2001) dated the mrca of *Chlorocebus* and *Erythrocebus* at approximately 5 Ma. The Page studies did not have other cercopithecins represented; however, the recent divergence date suggests that the taxonomic rank of these taxa be reduced to the subgeneric level. Because *Chlorocebus* Gray, 1870 has taxonomic priority over *Erythrocebus* Trouessart 1897, the genus encompassing patas monkeys, green monkeys, vervets, tantalus monkeys and grivets should be referred to as *Chlorocebus* with two subgenera *Chlorocebus* (*Erythrocebus*) and *Chlorocebus* (*Chlorocebus*). *Chlorocebus* (*Erythrocebus*) has only one extant species, the patas monkey, while *Chlorocebus* (*Chlorocebus*) has six extant species (Groves, 2001). More recently, Tosi et al. (2003) have asserted that the genus *Cercopithecus* is not monophyletic, and that the semi-terrestrial guenon *C. lhoesti* be included in the genus *Chlorocebus*.

We recognize one ape family Hominidae with only one extant subfamily (Homininae). Within the Homininae, there are at least five and possibly seven extant genera. These are the three hominin genera *Homo* (humans and chimpanzees), *Gorilla* (gorillas), *Pongo* (orangutans), and the two hylobatin genera, *Hylobates* (gibbons) and *Symphalangus* (siamangs). Two other potential hylobatin genera (*Bunopithecus* and *Nomascus*) may also warrant generic rank (Zehr, 1999; Roos and Geissmann, 2001); however, this remains to be validated because there is considerable incongruence among the available nuclear and mitochondrial gene trees. The earliest agreed upon stem hominid from the fossil record is *Morotopithecus* from the Miocene of Africa with an age of between 15 and 21 Ma (Gebo et al., 1997; Harrison, 2002). There are no fossils assigned to the hylobatin lineage, and the earliest unambiguous and uncontroversial member of the crown group that includes orangutans, gorillas, chimpanzees, and humans is the pongan taxon *Sivapithecus* from about 12-13 Ma in Pakistan (Kelley, 2002).

4.   ENLARGING THE GENUS *HOMO*

Concerning our own human place in primate phylogeny, to recapitulate, the age-equals-rank system (Table 1) places all living apes and humans in the subfamily Homininae. A phylogenetic branching (at ~18 Ma) divided Homininae into tribes Hylobatini and Hominini. Within Hylobatini, the phylogenetic branching (at ~8 Ma) in the subtribe Hylobatina separated *Symphalangus* (siamangs) from *Hylobates* (gibbons). Within Hominini, a phylogenetic branching (at ~14 Ma) separated the monogeneric subtribe Pongina for *Pongo* (orangutans) from Hominina. Within Hominina, a phylogenetic branching (at ~7 Ma) separated *Gorilla* from *Homo*. Within *Homo*, a phylogenetic branching (at ~5-6 Ma) separated the subgenus for common chimpanzees and bonobos – that is, *H. (Pan)* – from the subgenus for humans – that is, *H. (Homo)*. Thus, the principle of rank equivalence with other primate clades of the same age requires grouping the chimpanzee clade with the human clade within the same genus.

Could our objective yardstick for assigning ranks, in particular the generic rank, hold not only for primates but also for the class Mammalia as a whole? On the basis of results presented in Fig. 2 our answer is yes. Figure 2 shows the range of calculated crown group times of origin for a representative sample of mammalian genera. We conducted a literature search in order to assess whether the proposal of placing humans and chimpanzees together in the genus *Homo* based on a divergence of between five and six million years ago was unreasonable. We were able to find crown group times of origin for about thirty mammalian genera from ten orders (see Table 3 for a list of these taxa). In order to estimate the crown group time of origin, a genus must have two extant species and have been dated using molecular data calibrated to the fossil record. Our sample size is small (n=33), but the sample did pass the test for normality. The age of these genera treated as crown groups ranged from 2 (*Bos*) to 21 (*Elephantulus*) million years ago. The mean crown group time of origin is 8.16 million years ago, and the 95% confidence interval falls between 6.61 and 9.71 Ma. The elephant shrew genus *Elephantulus* is not monophyletic according to Douady et al. (2003). Even so, if the largest monophyletic *Elephantulus* clade were considered as the group constituting the genus it would still have a time of origin as a crown group of 18 Ma. If *Bos* and *Elephantulus* are excluded from the analysis, the mean date of generic origin for crown groups is 7.94 Ma.

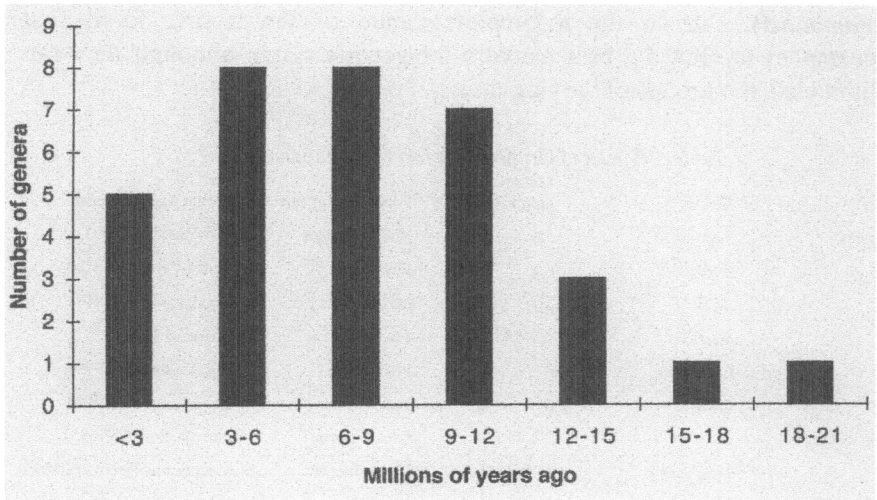

Figure 2. Time of origin for generic crown groups. The time of origin for mammalian generic crown groups is plotted in a frequency histogram with bin sizes of three million years. The times of origin were taken from the literature sources as shown in Table 3. All crown origin times were estimated from molecular data. While the sample size is relatively small (n=33), it is distributed normally. The mean time of origin for the mammalian genera sampled is 8.16 million years ago.

These dates are on average much less recent (p=0.0014, one sample t test) than the mrca of humans and chimpanzees. Thus, if one uses an age based yardstick to define genera as crown groups, then it is not unreasonable to place those groups that are more recent than the lower limit of the 95% confidence interval (6.61 Ma) in the same genus. Conversely, if a group of congeneric species has an origin as a crown group earlier (less recent) than the upper 95% confidence interval of 9.71 Ma, then it would be reasonable to place those taxa in separate genera. For example, the genus *Elephantulus* would be split into multiple genera using this scheme. Because the time of the mrca of *Homo* and *Gorilla* falls within the 95% confidence interval with age estimates ranging from 6-8 Ma (Goodman et al., 1998, 1999; Chen and Li, 2001; Wildman et al., 2003), there is at this point, no compelling evidence to place *Gorilla* in the genus *Homo*. Doing so would obscure the key finding that chimpanzees and humans are each other's sister group.

The average crown group age for genera is also much earlier than the first appearance at approximately 2 Ma of those fossil species that are commonly cited as the first members of the genus (*Homo ergaster* and *Homo habilis*) (Wood and Richmond, 2000). Table 3 also shows the age of the first fossil appearance of genera examined. For the most part, genera originate in the late Pliocene or early Miocene; however, some genera are quite ancient (e.g., *Tarsius*) while others have no fossil representatives (e.g.,

Funambulus). Due to the incomplete nature of the record, fossil first appearances are not the best markers for generic origin although they can delimit cladistic groups of varying rank.

Table 3. Times of Origin for Select Mammalian Genera[c]

Order	Genus	Minimum crown group age (MA)	Fossil 1st. appearance	Crown Age Reference
Primates	*Tarsius*	5.6	Middle Eocene	Meireles et al. (2003)
	Alouatta	6.8	Pleistocene	Cortés-Ortiz et al. (2003)
	Aotus	3	Middle Miocene	Goodman et al. (1999)
	Macaca	5.6-4.9	Late Miocene	Tosi et al. (2003)
	Callicebus	6	Pleistocene	Goodman et al. (1999)
	Colobus	6	Late Miocene	Goodman et al. (1999)
Cetartiodactyla	*Bos*	2	Late Pliocene	Janacek et al. (1996)
	Tragelaphus	11.2±0.52	Late Miocene	Matthee and Robinson (1999)
	Mus	5	Late Miocene	Juénet and Bonhomme (2003)
	Tamias	5.7	Early Miocene	Mercer and Roth (2003)
	Funambulus	10.5-7.4	Recent	Mercer and Roth (2003)
	Sciurus	8.6	Middle Miocene	Mercer and Roth (2003)
	Sciurillus	10.5-7.4	Recent	Mercer and Roth (2003)
	Ratufa	7.4-4.8	Early Miocene	Mercer and Roth (2003)
	Exilisciurus	7.4-4.8	Recent	Mercer and Roth (2003)
Lagomorpha	*Ochotona*	2.4	Late Miocene	Yu et al. (2000)
	Lepus	5-4	Early Pliocene	Yamada et al. (2002)
Perrisodactyla	*Equus*	3	Early Pliocene	Norman and Ashley (2000)
	Rhinocerus	11.7±1.9	Late Miocene	Tougard et al. (2001)
	Tapirus	3.3-22.7	Late Miocene	Norman and Ashley (2000)
	Ursus	3.5-2	Early Pliocene	Waits et al. (1999)
Eulipotyphla	*Sorex*	11.5	Late Miocene	Fumagalli et al. (1999)
Proboscidea	*Loxodonta*	2.63	Late Miocene	Roca et al. (2001), Eggert et al. (2002)
Macroscelidea	*Elephantulus*	21±3	Late Pliocene	Douady et al. (2003)
Dasyuromorphia	*Dasyurus*	11.3	Late Pliocene	Krajewski et al. (2000)
	Myoictis	6.6	Recent	Krajewski et al. (2000)
	Pseudantechinus	9.3	Pleistocene	Krajewski et al. (2000)
	Antechinus	8.6	Pleistocene	Krajewski et al. (2000)
	Murexia	11.5	Recent	Krajewski et al. (2000)
	Phascogale	14.0	Pleistocene	Krajewski et al. (2000)
	Ningaui	12.8	Recent	Krajewski et al. (2000)
	Sminthopsis	15.7	Pleistocene	Krajewski et al. (2000)
	Planigale	10.7	Early Pliocene	Krajewski et al. (2000)

[c]Shown are sampled mammalian genera, their estimated time of origin as crown groups based on molecular data, their time of fossil first appearance (from McKenna and Bell, 1997), and citations for the molecular studies.

5.   REASONS FOR RECLASSIFYING HUMANS

In proposing that the ages of clades serve as an objective yardstick for determining the ranks of taxa, we are confident that the emerging wealth of genomic DNA data will allow this approach to be applied to all groups of mammals. In our opinion, having an objective nonanthropocentric view of the place of humans in evolution is correct not only scientifically, but also ethically. There is now overwhelming evidence that chimpanzees, gorillas, and orangutans are close to extinction, and the major cause for this crisis is human predation and destruction of the habitats of these close relatives of ours (Walsh et al., 2003). It should not be denied that the destruction of natural habitats places our own species at risk. Recognizing that chimpanzees are members of the human genus, and that all apes are members of the human family, highlights the urgency of the present crisis. Ethically we are called upon to protect our fellow primates from extinction. To do so we have to protect and, where possible, reinstate the natural habitats of our close relatives. Clearly, if the growing degradation of the Earth's natural environment is not reversed our own species faces extinction as well.

6.   ACKNOWLEDGMENTS

We would like to thank Monica Uddin and Allon Goldberg for insightful discussion and comments. We would also like to thank Michael Mears for discussions about rodent evolution. This research was made possible by grants from the National Science Foundation (NSF) and the National Institutes of Health (NIH).

7.   REFERENCES

Bailey W., Hayasaka K., Skinner C., Kehoe S., Sieu L., Slightom J. and Goodman M. 1992. Reexamination of the African hominoid trichotomy with additional sequences from the primate beta-globin gene cluster. Mol Phylogen Evol. 1, 97-135.

Bailey W.J., Fitch D.H., Tagle D.A., Czelusniak J., Slightom J.L. and Goodman M. 1991. Molecular evolution of the psi eta-globin gene locus: gibbon phylogeny and the hominoid slowdown. Mol Biol Evol. 8, 155-84.

Barroso C.M., Schneider H., Schneider M.P., Sampaio I., Harada M.L., Czelusniak J. and Goodman M. 1997. Update on the phylogenetic systematics of New World monkeys: further DNA evidence for placing the pygmy marmoset (*Cebuella*) within the marmoset genus *Callithrix*. Inter J Primatol. 18, 651-674.

Beard K.C., Qi T., Dawson M.R., Wang B. and Li C. 1994. A diverse new primate fauna from middle Eocene fissure-fillings in southeastern China. Nature 368, 604-609.

Benefit B.R. and McCrossin M.L. 2002. The Victoriapithecidae, Cercopithecoidea. In W.C. Hartwig (ed.): The Primate Fossil Record. Cambridge University Press, Cambridge. pp. 241-253.

Bonhomme F. and Guenet J. 1985. The laboratory mouse and its wild relatives. In PLaK Altman, D.D. (Ed.): Inbred and Genetically Defined Strains of Laboratory Animals. FASEB Press, Bethesda, MD. pp. 1577-1596.

Chaves R., Sampaio I., Schneider M.P., Schneider H., Page S.L. and Goodman M. 1999. The place of Callimico goeldii in the Callitrichine phylogenetic tree: evidence from von Willebrand factor gene intron II sequences. Mol Phylogen Evol. 13, 392-404.

Chen F.C. and Li W.H. 2001. Genomic divergences between humans and other hominoids and the effective population size of the common ancestor of humans and chimpanzees. Amer J Hum Genet. 68, 444-56.

Cortes-Ortiz L., Bermingham E., Rico C., Rodriguez-Luna E., Sampaio I. and Ruiz-Garcia M. 2003. Molecular systematics and biogeography of the Neotropical monkey genus, *Alouatta*. Mol Phylogen Evol. 26, 64-81.

Darwin C. 1859. On the Origin of Species by Means of Natural Selection: or the Preservation of Favoured Races in the Struggle for Life. J. Murray, London.

Darwin C. 1871. The descent of Man, and selection in Relation to Sex. J. Murray, London.

Disotell T.R. 1994. Generic level relationships of the Papionini (Cercopithecoidea). Amer J Phys Anthropol. 94, 47-57.

Disotell T.R., Honeycutt R.L., and Ruvolo M. 1992. Mitochondrial DNA phylogeny of the Old-World monkey tribe Papionini. Mol Biol Evol. 9, 1-13.

Douady C.J., Catzeflis F., Raman J., Springer M.S. and Stanhope M.J. 2003. The Sahara as a vicariant agent, and the role of Miocene climatic events, in the diversification of the mammalian order Macroscelidea (elephant shrews). Proc Natl Acad Sci, USA.

Eggert L.S., Rasner C.A. and Woodruff D.S. 2002. The evolution and phylogeography of the African elephant inferred from mitochondrial DNA sequence and nuclear microsatellite markers. Proc Roy Soc Lond B Biol Sci. 269, 1993-2006.

Fleagle J.G. and Tejedor M.F. 2002. Early platyrrhines of southern South America. In W.C. Hartwig (ed.): The Primate Fossil Record. Cambridge University Press, Cambridge. pp. 161-173.

Fumagalli L., Taberlet P., Stewart D.T., Gielly L., Hausser J. and Vogel P. 1999. Molecular phylogeny and evolution of *Sorex shrews* (Soricidae: Insectivora) inferred from mitochondrial DNA sequence data. Mol Phylogen Evol. 11, 222-35.

Gebo D.L., MacLatchy L., Kityo R., Deino A., Kingston J. and Pilbeam D. 1997. A hominoid genus from the early Miocene of Uganda. Science. 276, 401-404.

Ghiselin M.P. and Jaffe L. 1973. Phylogenetic classification in Darwin's monograph on the subclass Cirrepedia. Syst Zool. 22, 132-140.

Goodman M. 1986. Molecular evidence on the ape subfamily Homininae: Evolutionary Perspectives and the New Genetics. Alan R. Liss, New York, pp. 121-132.

Goodman M. 1999. The genomic record of humankind's evolutionary roots. Amer J Hum Genet. 64, 31-9.

Goodman M., Czelusniak J., Page S.L. and Meireles C.M. 2001. Where DNA sequences place *Homo sapiens* in a phylogenetic classification of primates. In P.V. Tobias, M.A. Raath, J. Moggi-Cecchi and G.A. Doyle (Eds.): Humanity from African Naissance to Coming Millenia. Firenze University Press, Firenze, pp. 279-289.

Goodman M., Page S.L., Meireles C.M. and Czelusniak J. 1999. Primate phylogeny and classification elucidated at the molecular level. In S.P. Wasser (Ed.), Evolutionary Theory and Processes: Modern Perspectives. Papers in Honour of Eviatar Nevo. Kluwer Academic Publishers, Dordrecht, pp. 193-211.

Goodman M., Porter C.A., Czelusniak J., Page S.L., Schneider H., Shoshani J., Gunnell G. and Groves C.P. 1998. Toward a phylogenetic classification of Primates based on DNA evidence complemented by fossil evidence. Mol Phylogenet Evol. 9, 585-98.

Groves C.P. 2001. Primate Taxonomy. Smithsonian Institution Press, Washington, DC.

Gunnell G.F. and Rose K.D. 2002. Tarsiiformes: Evolutionary history and adaptation. In W.C. Hartwig (Ed.): The Primate Fossil Record. Cambridge University Press, Cambridge, pp. 45-82.

Harris E.E. 2000. Molecular systematics of the old world monkey tribe papionini: analysis of the total available genetic sequences. J Hum Evol. 38, 235-56.

Harris E.E. and Disotell T.R. 1998. Nuclear gene trees and the phylogenetic relationships of the mangabeys (Primates: Papionini). Mol Biol Evol. 15, 892-900.

Harrison T. 2002. Late Oligocene to middle Miocene catarrhines from Afro-Arabia. In W.C. Hartwig (Ed.): The Primate Fossil Record. Cambridge: Cambridge University Press, pp. 311-338.

Hartwig W.C. and Meldrum D.J. 2002. Miocene platyrrhines of the northern neotropics. In WC Hartwig (Ed.): The Primate Fossil Record. Cambridge University Press, Cambridge, pp. 175-188.

Hennig W. 1966. Phylogenetic systematics. University of Illinois Press, Urbana.

Jablonski N.G. 2002. Fossil Old World monkeys: the late Neogene radiation. In W.C. Hartwig (Ed.): The Primate Fossil Record. Cambridge University Press, Cambridge, pp. 255-299.

Janecek L.L., Honeycutt R.L., Adkins R.M. and Davis S.K. 1996. Mitochondrial gene sequences and the molecular systematics of the artiodactyl subfamily bovinae. Mol Phylogen Evol. 6, 107-19.

Jeffries R.S.P. 1979. The origin of the chordates - a methodological essay. In M.R. House (Ed.): The Origin of Major Invertebrate Groups. London: Academic Press, pp. 443-477.

Kay R.F., Johnson D. and Meldrum D.J. 1998. A new ptheciin primate from the middle Miocene of Argentina. Amer J Primatol. 45, 317-336.

Kelley J. 2002. The hominoid radiation in Asia. In W.C. Hartwig (Ed.): The Primate Fossil Record. Cambridge University Press, Cambridge, pp. 369-384.

Krajewski C., Wroe S. and Westerman M. 2000. Molecular evidence for the pattern and timing of cladogenesis in dasyurid marsupials. Zool J Linnean Soc. 130, 375-404.

Matthee C.A. and Robinson T.J. 1999. Cytochrome b phylogeny of the family Bovidae: resolution within the Alcelaphini, Antilopini, Neotragini, and Tragelaphini. Mol Phylogen Evol. 12, 31-46.

McKenna M.C. and Bell S.K. 1997. Classification of Mammals Above the Species Level. Columbia University Press, New York.

Meireles C.M., Czelusniak J., Page S.L., Wildman D.E. and Goodman M. 2003. Phylogenetic position of tarsiers within the order Primates: evidence from γ-globin DNA sequences. In P.C. Wright, E.L. Simons and S. Gursky (Eds.): Tarsiers; Past, Present, and Future. Rutgers University Press, New Brunswick, NJ, pp. 145-160.

Meireles C.M., Czelusniak J., Schneider M.P., Muniz J.A., Brigido M.C., Ferreira H.S. and Goodman M. 1999. Molecular phylogeny of dateline New World monkeys (Platyrrhini, Atelinae) based on gamma-globin gene sequences: evidence that Brachyteles is the sister group of Lagothrix. Mol Phylogen Evol. 12, 10-30.

Mercer J.M. and Roth V.L. 2003. The effects of Cenozoic global change on squirrel phylogeny. Science 299, 1568-72.

Norman J.E. and Ashley M.V. 2000. Phylogenetics of *Perissodactyla* and tests of the molecular clock. J Mol Evol. 50, 11-21.

Page S.L., Chiu C. and Goodman M. 1999. Molecular phylogeny of Old World monkeys (Cercopithecidae) as inferred from gamma-globin DNA sequences. Mol Phylogen Evol. 13, 348-59.

Page S.L. and Goodman M. 2001. Catarrhine phylogeny: noncoding DNA evidence for a diphyletic origin of the mangabeys and for a human-chimpanzee clade. Mol Phylogen Evol. 18, 14-25.

Pastorini J., Forstner M.R. and Martin R.D. 2002. Phylogenetic relationships among Lemuridae (Primates): evidence from mtDNA. J Hum Evol. 43, 463-78.

Pastorini J., Martin R.D., Ehresmann P., Zimmermann E. and Forstner M.R. 2001. Molecular phylogeny of the lemur family Cheirogaleidae (Primates) based on mitochondrial DNA sequences. Mol Phylogen Evol. 19, 45-56.

Porter C.A., Czelusniak J., Schneider H., Schneider M.P., Sampaio I. and Goodman M. 1997a. Sequences of the primate epsilon-globin gene: implications for systematics of the marmosets and other New World primates. Gene. 205, 59-71.

Porter C.A., Page S.L., Czelusniak J., Schneider H., Schneider M.P., Sampaio I. and Goodman M. 1997b. Phylogeny and evolution of selected primates as determined by sequences of the ε-globin locus and 5' flanking regions. Inter J Primatol. 18, 261-295.

Roca A.L., Georgiadis N., Pecon-Slattery J. and O'Brien S.J. 2001. Genetic evidence for two species of elephant in Africa. Sci. 293, 1473-7.

Roos C. and Geissmann T. 2001. Molecular phylogeny of the major hylobatid divisions. Mol Phylogen Evol. 19, 486-94.

Rosenberger A.L. 2002. Platyrrhine paleontology and systematics: the paradigm shifts. In W.C. Hartwig (Ed.), The Primate fossil Record. Cambridge University Press, Cambridge, pp. 151-159.

Sarich V.M. and Cronin J.E. 1976. Molecular systematics of the primates. In M. Goodman and R.E. Tashian (Eds.), Molecular Anthropology. Plenum Press, New York, pp. 141-170.

Seiffert E.R., Simons E.L. and Attia Y. 2003. Fossil evidence for an ancient divergence of lorises and galagos. Nature 422: 421-4.

Tosi A.J., Buzzard P.J., Morales J.C, and Melnick D.J. 2002. Y-chromosome data and tribal affiliations of Allenopithecus and Miopithecus. Inter J Primatol. 23, 1287-1299.

Tosi A.J., Disotell T.R., Carlos Morales J. and Melnick D.J. 2003. Cercopithecine Y-chromosome data provide a test of competing morphological evolutionary hypotheses. Mol Phylogen Evol. 27, 510-21.

Tougard C., Delefosse T., Hanni C. and Montgelard C. 2001. Phylogenetic relationships of the five extant *Rhinoceros* species (Rhinocerotidae, Perissodactyla) based on mitochondrial cytochrome b and 12S rRNA genes. Mol Phylogen Evol. 19, 34-44.

Waits L.P., Sullivan J., O'Brien S.J. and Ward R.H. 1999. Rapid radiation events in the family Ursidae indicated by likelihood phylogenetic estimation from multiple fragments of mtDNA. Mol Phylogen Evol. 13, 82-92.

Walsh P.D., Abernethy K.A., Bermejo M., Beyers R., De Wachter P., Akou M.E., Huijbregts B., Mambounga D.I., Toham A.K., Kilbourn A.M., Lahm S.A., Latour S., Maisels F., Mbina C., Mihindou Y., Obiang S.N., Effa E.N., Starkey M.P., Telfer P., Thibault M., Tutin C.E., White L.J. and Wilkie D.S. 2003. Catastrophic ape decline in western equatorial Africa. Nature 422, 611-614.

Wildman D.E., Uddin M., Liu G., Grossman L.I. and Goodman M. 2003. Implications of natural selection in shaping 99.4% nonsynonymous DNA identity between humans and chimpanzees: enlarging genus *Homo*. Proc Natl Acad Sci, USA 100, 7181-7188.

Wood B. and Richmond B.G. 2000. Human evolution: taxonomy and paleobiology. J Anat. 197 (Pt 1), 19-60.

Yamada F., Takaki M. and Suzuki H. 2002. Molecular phylogeny of Japanese Leporidae, the Amami rabbit *Pentalagus furnessi*, the Japanese hare *Lepus brachyurus*, and the mountain hare Lepus timidus, inferred from mitochondrial DNA sequences. Genes Genet Syst. 77, 107-116.

Yoder A.D. 1994 Relative position of the Cheirogaleidae in strepsirhine phylogeny: a comparison of morphological and molecular methods and results. Amer J Phys Anthropol. 94, 25-46.

Yoder A.D., Burns M.M., Zehr S., Delefosse T., Veron G., Goodman S.M. and Flynn J.J. 2003. Single origin of Malagasy Carnivora from an African ancestor. Nature 421, 734-737.

Yoder A.D., Cartmill M., Ruvolo M., Smith K. and Vilgalys R. 1996. Ancient single origin for Malagasy primates. Proc Natl Acad Sci, USA 93, 5122-5126.

Yu N., Zheng C., Zhang Y.P. and Li W.H. 2000. Molecular systematics of pikas (genus *Ochotona*) inferred from mitochondrial DNA sequences. Mol Phylogen Evol. 16, 85-95.

Zehr S. 1999. A Nuclear and Mitochondrial Phylogeny of the Lesser Apes (Primates: genus *Hylobates*). Ph.D. Thesis, Harvard University, Cambridge, MA.

EVOLUTION AND AGING: RELATION TO CELLULAR SENESCENCE AND TELOMERE BIOLOGY

Karl Skorecki and Maty Tzukerman
Rappaport Faculty of Medicine and Research Institute, Technion - Israel Institute of Technology and Rambam Medical Center, Haifa 31096, Israel

Abstract: In most multicellular organisms, which have evolved separate germline and somatic compartments, telomerase and telomere dynamics served key roles in regulating cellular immortality and senescence. Thus, in germline cells persistence of telomerase activity protects both length and stability of chromosome ends maintaining inter-generational viability of the genome. During embryonic and fetal development, a differentiation of somatic cells is associated with suppression of telomerase and attendant attrition of telomere length and integrity with progressive rounds of cell division during the lifetime of the organism. Such attrition of telomere length in the absence of telomerase eventuates in exhaustion of replicative capacity, which corresponds to the cellular phenotype of senescence. In the current manuscript, we focus our attention on the relationship between loss of telomere integrity in the absence of telomerase and the development of cellular senescence with an emphasis on DNA replication and subsequent cell cycle events. In addition, given the ubiquitous finding of downregulation of telomerase activity, which accompanies development of the somatic compartment, the manuscript also reviews recent developments in our understanding telomerase gene regulation. Human embryonic and adult stem cells depend upon telomerase and reservation of telomere integrity for their self-renewal capacity. Therefore, in addition to providing additional basic insights with respect to evolutionary processes, continued research in this area is likely to make an important contribution to regenerative medicine.

S.P. Wasser (ed.), Evolutionary Theory and Processes: Modern Horizons,
Papers in Honour of Eviatar Nevo
K. Skorecki and M. Tzukerman. Evolution and Aging: Relation to Cellular Senescence and Telomere Biology, 313-328.
© 2004 *Kluwer Academic Publishers.*

1.   INTRODUCTION

> "Nothing in biology makes sense except in the light of evo-
> lution" (Dobzhansky, 1973).
> "Aging seems to be the only available way to live a long life"
> (Daniel François Esprit Auber, 1997).

Both of these quotations are eminently appropriate in the context of this monograph dedicated to the 75th birthday anniversary of Eviatar Nevo. For the past half century, and with the help of the Almighty, for many more years to come, Professor Nevo has consolidated the first maxim in diverse life forms ranging from mould to mammals at levels ranging from macro-ecological systems to intricate molecular mechanisms. In terms of the second quotation, at the time of his 75th anniversary, Eibi demonstrates the blessing of maintaining youthful enthusiasm and energy in his continued broad pursuit of scientific excellence and creativity. This brief chapter is dedicated in the hope that Eibi will be blessed with many more years of continued good health and gratifying scientific success.

This chapter will deliberately limit its focus to some very selected molecular aspects relating cellular senescence to the biology of aging. To the extent possible, an evolutionary context will be provided.

First, it is important to distinguish the normal organismal aging process, which involves ubiquitous physiologic changes from age-related increases in disease-susceptibility. The two aspects seem to be related, inasmuch as some of the universal decline in physiologic function with aging may augment specific disease susceptibility. As a nephrologist, I can cite one particular example. There is a well-documented universal decline in renal glomerular filtration rate (approximately $1\,ml/min/1.73m^2$ for each year of life above age 30 (Levey et al., 2002). In and of itself, this expected measurable decline does not affect the day-to-day function of the individual, but in the face of a superimposed nephrotoxic insult, this underlying decline may enhance the propensity to serious renal failure.

Numerous recent excellent reviews have appeared providing a compre-hensive overview of the biology of aging (Toran, 2003 and references therein). These reviews emphasize and rigorously contrast the epidemiology of "lifespan" and "life expectancy", as well as theories regarding the mech-anisms of aging. Of course, the single most striking manifestation of aging is increased mortality with age following maturation. Thus for example, recent epidemiologic data in Europe and USA indicate that in the approximately 40 year interval between completion of human developmental maturation (early adulthood) to retirement age (mid 60's), there is a >20 fold increase in mortality (Arias and Smith, 2003). It is this characteristic of aging which emphasizes the evolutionary and developmental-genetic context of the aging process in complex multicellular organisms. Clearly, the concept of aging and mortality are not applicable to individual unicellular organisms. In contrast, in complex multicellular organisms, *pari passu* with compartmentalization into somatic and germline components, there has developed a process of aging which sets an upper limit to the maximum lifespan of the organism. Numerous theories have been expounded to explain the evolutionary development of this process include, among others, antagonistic pleiotropy, disposable soma, "aging grandmothers" and others (Cole, 1954; Williams, 1957; Hamilton, 1966; Rose and Graves, 1989; Kirkwood and Rose, 1991; Kirkwood, 1996; Hawks, 2003; Trosko, 2003), which will not be discussed herein. In terms of mechanisms of aging, these can be divided into those that do and those that do not postulate a close relationship between the phenomenon of cellular senescence and organismal aging. In the remainder of this paper, I will focus on findings in our laboratory related to cellular senescence as related to telomere biology, in the context of evolution of the aging process.

2. TELOMERE-MEDIATED CELLULAR SENESCENCE AS AN EXAMPLE OF INTRA-INDIVIDUAL DIVERSITY

Diversity is the most striking feature of the biological world, which has comprised the substrate for most of the scholarly research activity of Eviatar Nevo and other evolutionary scientists. Three levels of diversity can be defined for complex multi-cellular organisms including humans. These are inter-individual, inter-cellular, and intra-individual levels or diversity (Fig. 1). Both genetic (mutation and recombination) as well as cultural mechanisms contribute to the wonderful plethora of inter-individual human diversity. This inter-individual diversity also includes differences in rates of aging and disease predisposition. On the other hand, intercellular diversity refers to the differences in gene expression among cell types within an individual, which accounts for differentiated cell function. Thus, at any given time, within a cell type, only a subset of the total estimated repertoire of approximately 30,000 - 40,000 genes are expressed (Baltimore, 2001). It is the particular repertoire of expressed genes that determine the specialized function of any given cell type. Finally, during the development and aging of an individual, durable epigenetic changes occur, but are not necessarily passed on to subsequent generations. These latter "inter-individual" epigenetic changes contribute, in part, to the phenotypic changes characteristic of normal human development and post-maturational aging. Among the best studied of these genome-wide genetic changes are the shortening of chromosome ends or telomeres. Therefore, the current review will focus on selected aspects of the regulation of telomere length in relation to human development and aging.

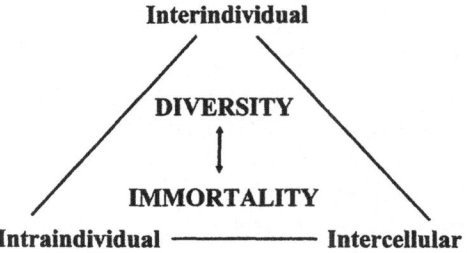

Figure 1. Nature has favored diversity over immortality in formulating the biological world. Genomic mechanisms form the basis for interindividual, intercellular, and intra-individual natural diversity, as described in the text (Tzukerman et al., 2002).

3.   TELOMERES AND TELOMERASE

The unprotected ends of linear chromosomes in eukaryotic species are susceptible to a number of destabilizing processes (Greider and Blackburn, 1985; Blackburn, 2000). Thus, for example, differences in the mechanism of replication for the leading and lagging strands of double-helical DNA give rise to an "end-replication problem" (Blackburn, 2000). Since DNA replication always proceeds in the 5' → 3' direction, the lagging strand is replicated by a process involving a series of Okazaki RNA primers, which are subsequently filled in to generate two complete daughter double helices. Okazaki fragments are unable to prime lagging strand replication at the very end of chromosomes resulting in a situation whereby each round of DNA replication shortens the chromosome ends. In addition, susceptibility to exonuclease digestion and other destabilizing processes mandates the assumption of protective mechanisms. Accordingly, it is not surprising to find that the ends of linear chromosomes in eukaryotic species are charac-terized by a specialized structure. At the core of the structure is a charac-teristic DNA repeat sequence [$(TTAGGG)_n$ in mammals], together with a complex of associated proteins, only some of which have been identified and functionally characterized. One of these is the ribonucleoprotein telomerase, whose reverse transcriptase activity adds telomere repeats to lengthen chromosome ends.

The role of telomeres and telomerase has been most thoroughly studied in terms of its relationship to the phenomenon of cellular senescence in tissue culture. In this context, the term "senescence" refers to the loss of replicative capacity after nontransformed primary cells of human origin are propagated in tissue culture for a given number of population doublings (Olovnikov, 1996). Thus, population doublings are accompanied by attrition of telomere length, and in turn, this is associated with a progressive loss of replicative capacity eventuating in the phenomenon of cellular senescence.

It is evident, that for unicellular organisms as well as for certain cellular subcompartments of complex multicellular organisms, there must exist a mechanism to avert telomere attrition and associated cellular senescence. Thus, for example, in the germline compartment of complex multicellular organisms, in the absence of such a mechanism, each subsequent generation would begin at a compromised starting telomere length, ultimately threatening the viability of that species over one or more generations. Also, in most or all organs, a subpopulation of stem cells is responsible for replenishing the differentiated cells, which turn over during organ renewal. Since such stem cells undergo cell-renewal by repeated rounds of population doublings, at least partial protection against telomere attrition must be in place. In fact, one of the most prominent formulations for organismal aging considers exhaustion of adult cell replicative and renewal capacity to be the primary underlying basis for normal aging (Hawks, 2003). This formulation is supported by experimental studies in telomerase knockout mice (Hayflick, 1961). Average telomere length in laboratory mice is several-fold greater than average telomere length in human cells (somatic cells at birth or germline cells throughout life). Therefore, it was not surprising that for the first several generations the telomerase knockout mice showed no abnormal phenotype, and in particular, no characteristics of premature aging. However, after several generations in the absence of telomerase, progressive deletion of starting telomere length gave rise to mice with interesting panoply of phenotypic abnormalities. Among others, these included characteristics of premature aging such as: graying of coat color and limited capacity to mount a bone marrow recovery response to pharmacological myelosuppression impaired wound-healing, enhanced cancer susceptibility, and other manifestations. The characteristics were consistent with incapacity of adult stem cells to mount an appropriate renewal response to somatic cell loss and genomic instability. In addition, the germline itself was affected by loss of telomere integrity rendering the mice infertile past the sixth generation. Indeed, germline, embryonic, and adult stem cells all maintain a level of telomerase activity appropriate for their required extended proliferative capacity (Wright et al., 1996; Blasco et al., 1997; Ulaner and Giudice, 1997). In contrast, mature somatic cells lose telomerase activity and acquire the susceptibility to telomere attrition and replicative senescence with progressive rounds of cell division. This loss of telomerase activity occurs with cell differentiation during normal embryonic development, and also in the adult when a telomerase positive stem cell undergoes asymmetric cell division giving rise to a differentiated daughter cell. In our laboratory,

we have observed this loss of telomerase activity in human embryonic stem
cells grown in culture under differentiating conditions (Fig. 2). Adult organ
systems may be comprised of several different types of cells in terms of
replicative capacity. Fully differentiated post-mitotic cells (G_0-arrest) carry
out differentiated function and by definition do not divide and are telomerase
negative. Another subset of cells which are not post-mitotic are telomerase
negative and maintain a finite ability to replicate, which is limited by
progressive attrition of telomere length or other associated changes in
telomere function. Approximately 90% of tumors and cancer cell lines also
have an active telomerase (Ulaner and Giudice, 1997; Tzukerman et al.,
2000; Braunstein et al., 2001). It is thought that this expression of telomerase
confers cellular immortality characteristic of cancer growth. Multiple
alternative mechanisms for activation of telomerase with tumorigenesis have
been investigated by numerous laboratories including our own (Autexier and
Greider, 1996; Ulaner and Giudice, 1997). Furthermore, there is intensive
interest in the development of drugs and biological therapies to interfere with
telomerase in the treatment of cancer.

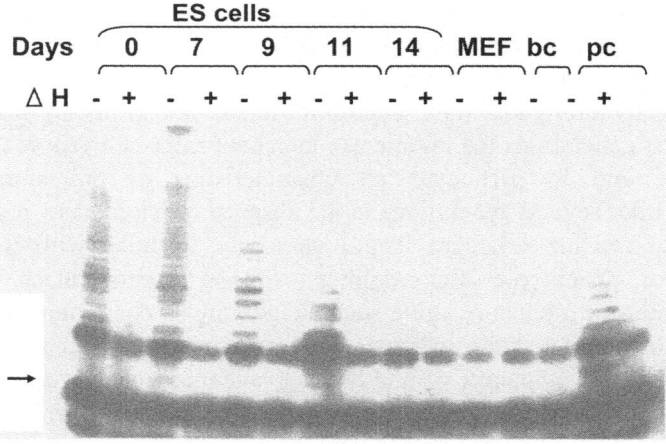

Figure 2. Differentiation of hES cells is associated with downregulation of telomerase
activity. TRAP assay was performed using extracts from hES cells cultured on MEF cells
(day 1) and from hES celss, which were allowed to differentiate from 7, 9, 11 and 14 days. As
controls, TRAP activity was measured in extracts of MEF, buffer (bc), and extracts of
telomerase positive cells supplied in the kit (pc). In preliminary experiments, telomerase
activity was 10 and 50 mg, to ensure that the assay was performed in the linear range, the
results shown here were obtained using 5 mg of each extract without (-) and with (+) heat
inactivation at 85°C for 15 min. A 36-bp internal control (IC) for amplification efficiency and
quantitative analysis was run for each reaction, as indicated by the arrow. The reaction
products were separated on a 10% nondenaturing polyacrylamide gel (Ulaner and Giudice,
1997).

Taken together, the foregoing suggests an overall initial working model for the role of telomeres and telomerase in human development, aging, and tumorigenesis. Some elements of this formulation can be summarized as follows (Fig. 3):

- Telomerase is active in germline cells, which maintain stable telomere length despite multiple rounds of cell division.
- The nuclei of the fertilized egg contain a full complement of chromosomes with an average of telomere length of 12-14kB and active telomerase.
- During early embryonic and fetal development, cellular differentiation is associated with a regulated decrease in telomerase activity, which varies among organ system.
- In the absence of telomerase, telomere length progressively shortens with each round of somatic cell division. In tissue and organ systems in which post-mitotic and differentiated cells are lost by apoptosis or other mechanisms, or which have rapid cell turnover and high proliferative requirements, a subset of adult stem cells (progenitors) maintain a level of telomerase activity which facilitates the replenishment of mature replacement cells. Normal aging may result from a qualitative or quantitative loss in this progenitor population.
- During the process of tumorigenesis, telomerase is usurped for purposes of cancer immortalization in ~ 90% of cancers.
- According to the foregoing formulation, it is evident that the repression of telomerase during embryonic and fetal development represents a developmental genetic program that contributes to organismal aging.

In our laboratory we have focused on the following principal questions:

a) What is the relationship between the unprotected ends of short telomeres in telomerase negative cells following progressive rounds of cell division and replicative senescence?

b) What are the mechanisms for inhibition of telomerase activity with cellular differentiation (and what is the mechanism for reactivation with tumorigenesis)?

c) Can insights gleaned from the foregoing line of research, be exploited to circumvent post-mitotic arrest and replicative senescence in the generation of homogenous populations of derivative differentiated cell types from human embryonic stem cells for therapeutic application in regenerative medicine?

Due to limitations of space, the final section of this chapter will summarize selected aspects of progress in our laboratory in addressing the first two of these questions.

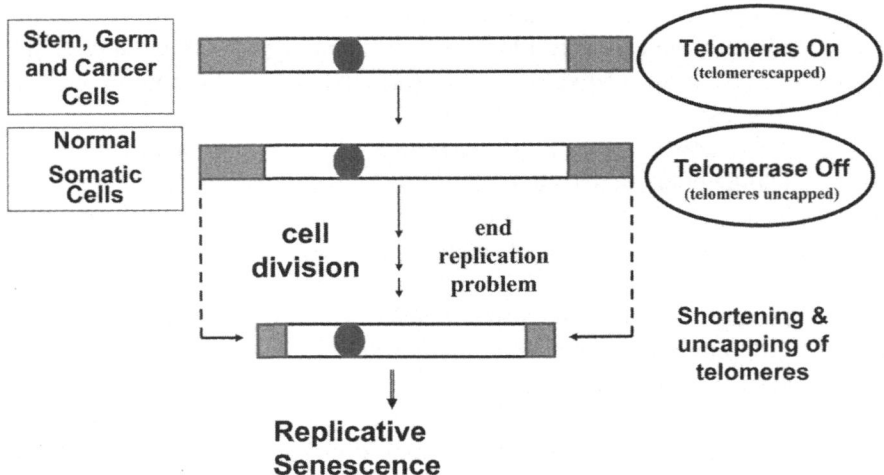

Figure 3. Telomere attrition in the absence of telomere activity. Germ cells maintain telomere length due to telomerase activity. Telomere size is progressively decreased in replicative somatic cells throughout their lifespan to a critical point at which the cells enter replicative senescence (Tzukerman et al., 2002).

4.   THE RELATIONSHIP BETWEEN TELOMERE FUNCTION AND CELLULAR SENESCENCE

Two major categories of mechanisms have been postulated to explain how telomeres might serve as a mitotic clock for population doublings. The first of these relates to the possibility that telomeres and changes in telomere length may serve to alter the expression of adjacent genes. Indeed, in yeast, such a position variegation effect manifested by silencing of genes adjacent to telomeres has been described (Gottschling et al., 1990; Stevenson and Gottschling, 1999). However, in mammalian cells, such an effect has been more difficult to prove definitively although some supporting evidence has emerged (Baur et al., 2001). Instead, we have focused our attention on a second proposed mechanism, which relates to the potential role of telomeres in cell cycle events pertaining to DNA replication and chromosome separation.

Taking advantage of an experiment of nature (Wong et al., 1997), we have reported the effect of an abnormal telomeric structure on the replication timing of an adjacent genomic region (Ofir et al., 1999). In this report, we described delayed replication timing in the genomic region adjacent to a break at chromosome 22q in a patient with a congenital cognitive disability syndrome. Chromosome break syndromes are common causes of cognitive developmental disabilities – but in this case the chromosome end was "repaired" by a building of a stretch of telomeric DNA adjacent to a genomic region, which is not normally juxtaposed to a telomere (Fig. 4). The abnormality was confined to only one of the parental chromosomes, and this enabled comparison of the replication timing of the respective genomic regions between the two parental chromosomes. Replication timing measurements were performed using the fluorescence *in-situ* hybridization (FISH) replication timing assay (Selig et al., 1992; Boggs and Chinault, 1997). Since the other parental chromosome without the break serves as a "built-in" control, this makes the measurement of replication timing differences particularly powerful. Normally, two allelic genomic regions replicate nearly simultaneously during S-phase of the cell cycle. In contrast, in the case of this patient, a consistent pattern was observed, wherein the genomic region abnormally juxtaposed to a telomere replicated later in S-phase compared to the normal parental allelic genomic region. Thus, it appears that the presence of telomeric DNA may modulate the replication timing of adjacent genomic regions. This finding sparked our interest to investigate the potential effect of the normal change in telomere length consequent to progressive rounds of cell division upon later cell cycle events (mitosis). In this regard, in experiments conducted using primary human foreskin fibroblasts, we have found that telomeric regions in pre-senescent cells, on average, replicate later than those in early passage cells. In contrast, in fibroblasts immortalized by constitutive expression of ectopic telomerase, there was no delay in replication timing with progressive rounds of cell division and replicative senescence was averted. In this same study, we went on to examine events in metaphase. It was postulated that if telomeric regions display a pattern of replication timing delay with progressive rounds of cell division, then possibly their behavior in metaphase would also be affected. This required setting up a modified version of the FISH replication assay, which would facilitate the specific genomic region hybridization signals simultaneously with resolution of the state of chromosomes in mitosis. The latter was enabled by simultaneous immuno-fluorescence using antiphosphorylated histone antibodies, and the detailed experimental protocol is provided in reference Ofir et al. (2002). Strikingly, this study and subsequent as yet unpublished experiments pointed to a rather surprising result. In contrast to nontelomeric control regions, for several telomeric regions examined, an excess of singlet hybridization signals were noted, and this phenomenon was more pronounced in pre-senescent compared to early passage normal human fibroblasts. Such singlet signals could represent either DNA that has failed to replicate, or alternatively, failure to separate replicated DNA at adjacent sister chromatids. For various reasons, we favor the latter explanation, and postulate that it may be related to excessive "stickiness" of telomeres. So what does this have to do with replicative arrest? Normal progression through the cell cycle requires appropriate segregation of sister chromatids following completion of mitosis. Incomplete separation of sister chromatids at telomeres, would jeopardize this process, and could serve as a potential mechanism for replicative senescence, as well as genomic instability and enhanced cancer susceptibility. At the moment, this remains a speculative working model, and numerous validation studies need to be carried out to verify or refute all or part of this model.

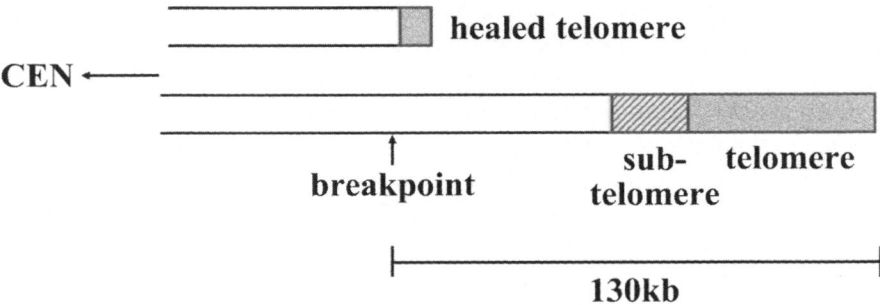

Figure 4. A schematic diagram of the ends of both copies of the q arm of chromosome 22 in patient NT. The gray boxes represent telomeric repeats. The arrow points to the breakpoint. CEN = centromeric direction (Tzukerman et al., 2002).

5.   REGULATION OF TELOMERASE IN DEVELOPMENT AND DIFFERENTIATION

There has been relative paucity of information concerning the role of telomerase and its regulation in embryonic and fetal development, as well as during cellular differentiation. Nevertheless, repression of telomerase activity provides the backdrop upon which cellular senescence and organismal aging can occur. Therefore, we have used human embryonic stem cells in culture as a potentially relevant model system to study the molecular events involved in the regulation of telomerase activity with differentiation. Having shown a clear-cut reduction in telomerase activity with the transition of growth conditions to those which allow hES to differentiate from the pluripotent to the differentiated state, we set about to evaluate the regulation of the promoter for the gene encoding the catalytic subunit of the telomerase hollow enzyme complex. In this regard, it should be recalled that the protein (TERT) and mRNA (TER) components of the telomerase ribonucleoprotein are each encoded at a separate genetic locus and are under independent regulatory control (Feng et al., 1995; Kilian et al., 1997). Although the overall regulation of telomerase activity is exceedingly complex, involving multiple transcriptional, post-transcriptional and post-translational regulatory mechanisms – a number of lines of evidence suggest that transcriptional regulation of hTERT expression is a key limiting regulatory step in determining overall telomerase activity. Thus, for example, there is an excellent correlation between hTERT mRNA and telomerase activity. In contrast, levels of hTR mRNA seem to be dissociated from measure telomerase activity. Furthermore, ectopic expression of hTERT in somatic cells is sufficient to restore telomerase activity and render them immortal (Bodnar et al., 1998; Nakayama et al., 1998; Greider, 1999). Together, these suggest that regulation of the hTERT gene may have a role in determining overall telomerase activity. This is not to discount the multiple additional level of regulation of telomerase activity, which is exceedingly complex involving numerous post-translational modification and protein-protein interactions, representing the convergence of many signaling pathways. However, in view of the key role of transcriptional regulation, our group and others set out to isolate and characterize the 5' flanking region for the gene encoding the catalytic subunit of the human telomerase ribonucleoprotein complex (Takakura et al., 1997; Horikawa et al., 1999; Wick et al., 1999; Tzukerman et al., 2000). This characterization involved promoter-reporter transfection studies in telomerase positive and telomerase negative cell lines, deletional mutation analysis, examination of the effects of signaling pathway inhibitors, electromobility shift and supershift assays, and transcription factor co-transfections and isolation studies, among others. Fig. 5 provides a very schematic rendition of a number of important features. First, it is evident that a core 280 nucleotide segment upstream of the translation initiation site encompassing a first exon and intron, confer a high level of promoter activity in undifferentiated embryonic stem cells and cancer cells and that the activity of this segment is markedly reduced with differentiation. Transcription factor binding to several known and potentially novel elements were identified and shown to be markedly reduced with differentiation. Multiple SP1 elements were identified in this "core" 280nt fragment. As is characteristic of SP1 regulation, one or more of these elements is necessary for baseline promoter activity on the one hand, but on the other hand is also responsible for recruiting histone deacetylases in derepression of the telomerase promoter characteristic of the differentiated state. Derepression in cancer, may or may not involve the same SP1 mediated regulation, and indeed, numerous studies have invoked other regulatory pathways for cancer-mediated activation of telomerase (Shuwen and Jiyue, 2003).

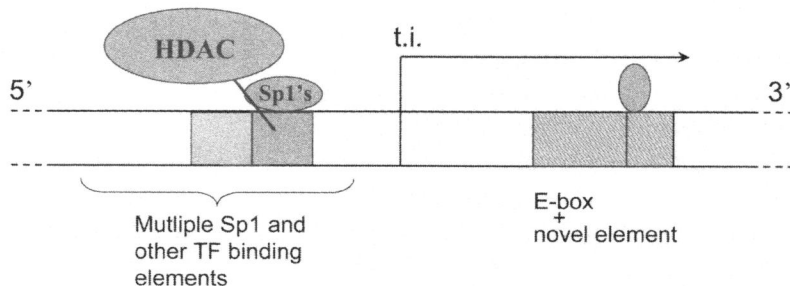

Figure 5. Somatic representation of hTERT promoter core (280nt) promoter activity is high in undifferentiated embryonic stem cells (and cancer cells) and diminishes with differentiation. Transcription factor binding to a "novel element is markedly reduced with differentiation. HDAC: histone deacetylase.

6. POTENTIAL APPLICATION TO AGE-RELATED DEGENERATIVE DISEASE AND STEM CELL THERAPY

Regenerative medicine, is a rapidly evolving field of experimental medicine, where the common theme is to seek ways of replacing cells, which have been irreversibly lost from vital organs leading to organ system failure (e.g., congestive cardiac failure; chronic renal failure leading to end-stage renal disease; auto-immune β-cell destruction in type I diabetes; hepatic cirrhosis neurodegenerative disease and others). It also appears that this process may be accelerated in both congenital and acquired disease states such as diabetes mellitus (Blazer et al., 2002). Since such degenerative organ system diseases are more frequent with aging, this also represents a branch of medical research, which in some senses tries to forestall the major hallmark of aging, namely, increase age associated disease and attendant mortality. One emerging approach is to use pluripotent human embryonic stem cells, as a source for deriving differentiated cell types that re-constitute lost organ system function. The recent interest in this area, was part by the breakthrough report of the generation of human embryonic stem cells, which can be maintained in culture with normal karyotype for hundreds of population doublings (attributed to persistence of telomerase activity), and can also be directed to differentiate into desired specialized cell types (Thomson et al., 1998). However, a limiting factor is the intrinsic paradox that such differentiation into specialized cell types is inherently associated with a progressive loss of replicative capacity. This inherent paradox, naturally limits the potential for scaling up the large numbers of specialized cell types that would be needed for practical clinical application. By combining our interest in human embryonic stem cells (Tzukerman et al., 2000; Assady et al., 2001) with our ongoing studies on the regulation of telomeres and telomerase (Ofir et al., 1999, 2002; Tzukerman et al., 2000, 2002; Braunstein et al., 2001), our laboratory is pursuing a potential strategy to circumvent this limitation. In particular, instead of aiming to produce fully differentiated derivative cell types from pluripotent embryonic stem cells, our current goal is to isolate lineage-restricted progenitor cells, which maintain replicative capacity. As in the adult organism, we postulate that so to during differentiation in culture, these adult stem cells or progenitors will preserve a level of telomerase promoter activity that will sustain there proliferative capacity as a source of cell replacement therapy. Therefore, by stably transfecting undifferentiated human embryonic stem cells in the pluripotent state with an ectopic construct consisting of the core hTERT promoter upstream of a selection or tracking cassette, we hope to select clones that maintain telomerase promoter activity under differentiation conditions. Under such "selection pressure" or "tracking", the desired cells should have the properties of lineage-restricted differentiation on the one hand, but preservation of replicative capacity on the other hand. These strategies are schematized in Fig. 6. Whether or not these strategies will lead to a practical solution, is difficult to determine at this point. However, the experiments involved are already shedding light on important basic biological questions related to the role of telomerase in cellular differentiation and cell cycle regulation.

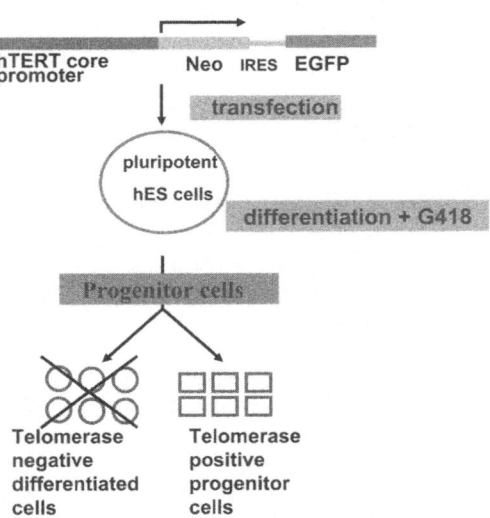

Figure 6. Schematic representation of strategy for selecting progenitor cells from differentiating human embryonic stem cells. Neo: neomycin resistance cassette; IRES: internal ribosomal entry site. See text for details.

7.   CONCLUSION

This chapter focused on a very specific aspect of cellular senescence, namely, the role of telomeres and telomerase. It is clear that numerous other mechanisms are involved, at the level of mitochondria, oxidative changes, and other processes (reviewed in Toran, 2003). We certainly do not wish to oversimplify this complex process of cellular senescence, and leave the reader with the impression that it is solely related to the phenomenon of telomere attrition. Furthermore, the relationship between cellular senescence and organismal aging cannot be considered as a one-to-one relationship. Nevertheless, a common theme in aging research emerges from our studies, which have focused on alternative processes. From all of these studies, it is quite evident that processes, which lead to cellular senescence and organismal aging, seem to have emerged as one evolutionary favored process for many life form phenomena in biology. Therefore, it is evident that the evolutionary context will serve cellular and molecular investigators in unraveling the molecular complexities of cellular senescence and organismal aging.

8.   ACKNOWLEDGMENTS

We appreciate the permission granted by the journal of Molecular Biology of the Cell to reproduce the figures used in this article that appeared in Volume 11 in December 2000.

9.   REFERENCES

Arias E. and Deaths B.: Preliminary Data for 2001. National Vital Statistics Reports 51, 1-45, 2003.

Assady S., Maor G., Amit M., Itzkovitz-Eldor J., Skorecki K.L. and Tzukerman M. 2001. Insulin production by human embryonic stem cells. Diabetes 50, 1691-1697.

Autexier C. and Greider C.W. 1996. Telomerase and cancer: revisiting the telomere hypothesis. Trends Biochem Sci. 21, 387-391.

Baltimore D. 2001. Our genome unveiled. Nature 409, 814-815.

Baur J.A., Zou Y., Shay J.W. and Wright W.E. 2001. Telomere position effect in human cells. Science 292, 2075-2077.

Blackburn E.H. 2000. Telomere states and cell fates. Nature 408, 53-56.

Blasco M.A., Lee H.W., Hande P., Samper E., Landsdorp P., DePinho R. and Greider C.W. 1997. Telomere shortening and tumor formation by mouse cells lacking telomerase RNA. Cell 91, 25-34.

Blazer S., Khankin E., Segev Y., Ofir R., Yalon-Hacohen M., Kra-Oz Z., Gottfried Y., Larisch S. and Skorecki K.L. 2002. High glucose-induced replicative senescence: Point of no return and effect of telomerase. Biochem Biophys Res Comm 296, 93-101.

Bodnar A.G., Ouellette M., Frolkis M., Holt S.E., Chiu C.P., Morin G.B., Harley C.B., Shay J.W., Lichtsteiner S. and Wright W.E. 1998. Extension of life-span by introduction of telomerase into normal human cells. Science 279, 349-352.

Boggs B.A, Chinault A.C. 1997. Analysis of DNA replication by fluorescence *in situ* hybridization. Methods: A Companion to Methods in Enzimology 13, 259-270.

Braunstein I., Cohen-Barak O., Shachaf C., Ravel Y., Yalon-Hacohen M., Mills G.B., Tzukerman M. and Skorecki K.L. 2001. Human telomerase reverse transcriptase promoter regulation in normal and malignant human ovarian epithelial cells. Cancer Res. 61, 5529-5536.

Cole L.C. 1954. The population consequences of life history phenomena. Q Rev Biol. 29, 103-137.

Daniel François Esprit Auber. 1997. In: Bloomsbury Biographic Dictionary of Quotations. Bloomsbury, London (UK).

Dobzhansky T. 1973. Nothing in biology makes sense except in the light of evolution. Amer Biol Teacher 35, 125-129.

Feng J., Funk W.D., Wang S.S., Weinrich S.L., Avilion A.A., Chiu C.P., Adams R.R., Chang E., Allsopp R.C., Yu J., Le S. West M.D., Harley C.B., Andrews W., Greider C.W. and Villponteau B. 1995. The RNA component of human telomerase. Science 2699, 1236-1241.

Gottschling D.E., Aparicio O.M., Billington B.L. and Zakian V.A. 1990. Position effect at S. cerevisiae telomeres: Reversible repression of Pol II transcription. Cell 63, 751-762.

Greider C.W. and Blackburn E.H. 1985. Identification of a specific telomere terminal transferase activity in Tetrahymena extracts. Cell 43: 405-413.

Greider C.W. 1999. Telomerase activation. One step on the road to cancer? Trends Genet 15, 109-112.

Hamilton W.D. 1966. The molding of senescence by natural selection. J Theor Biol. 12, 24-45.

Hawks K. 2003. Grandmothers and the evolution of human longevity. Amer J Human Biol. 15, 380-400.

Hayflick L. 1961. The limited *in vivo* lifetime of human diploid cell strains. Cell Res. 37, 614-636.

Horikawa I., Cable P.L., Afshari C. and Barrett J.C. 1999. Cloning and characterization of the promoter region of human telomerase reverse transcriptase gene. Cancer Res. 59, 826-830.

Kilian A., Bowtell D.D., Abud H.E., Hime G.R., Venter D.J., Keese P.K., Duncan E.L., Reddel R.R. and Jefferson R.A. 1997. Isolation of a candidate human telomerase catalytic subunit gene, which reveals complex splicing patterns in different cell types. Hum. Mol. Genet. 6, 2011-2019.

Kirkwood T.B. and Rose M.R. 1991. Evolution of senescence: late survival sacrificed for reproduction. Phil Trans Roy Soc. Lond B Biol Sci. 332, 15-24.

Kirkwood T.B. 1996. Human senescence. Bioessays 18, 1009-1016.

Levey A.S. et al. 2002. National Kidney Foundation K/DOQI "Clinical Practice Guidelines for Chronic Kidney Disease: Evaluation, Classification and Stratification". Amer J Kid Dis 39, S1-S266 (Supplement 1).

Lin S.Y. and Elledge S.J. 2003. Multiple tumor suppressor pathways negatively regulate telomerase. Cell 113, 881-889.

Meyerson M., Counter C.M., Eaton E.N., Ellisen L.W., Steiner P., Caddle S.D., Ziaugra L., Beijersbergen R.L., Davidoff M.J., Liu Q., Bacchetti S., Haber D.A. and Weinberg R.A. 1997. hEST2, the putative human telomerase catalytic subunit gene, is up-regulated in tumor cells and during immortalization. Cell 90, 785-795.

Nakayama J., Tahara H., Tahara E., Saito M., Ito K., Nakamura H., Nakanishi T., Tahara E., Ide T. and Ishikawa F. 1998. Telomerase activation by hTRT in human normal fibroblasts and hepatocellular carcinomas. Nat Genet. 18, 65-68.

Ofir R., Wong C.C., McDermid H.E., Skorecki K.L. and Selig S. 1999. Position effect of human telomeric repeats on replication timing. Proc Natl Acad Sci, USA 96, 11434-11439.

Ofir R., Yalon-Hacohen M., Segev Y., Schultz A., Skorecki K.L. and Selig S. 2002. Replication and\or separation of some tumor telomeres is delayed beyond S-phase in presenescent cells. Chromosoma 111, 147-155.

Olovnikov A. 1996. Telomeres, telomerase and aging: origin of the theory. Exp Gerontol 31, 443-448.

Rose M.R. and Graves J.L. 1989. What evolutionary biology can do for gerontology. Gerontol 44, B27-B29.

Selig S., Okumura K., Ward D.C. and Cedar H. 1992. Delineation of DNA replication timing zones by fluorescence *in situ* hybridization. EMBO J. 11, 1217-1225.

Shuwen W. and Jiyue Z. 2003. Evidence for a relief of repression mechanism for activation of the human telomerase reverse transcriptase promoter. J Biol Chem. 278, 18842-18850.

Stevenson J.B. and Gottschling D.E. 1999. Telomeric chromatin modulates replication timing near chromosomes ends. Genes Dev. 13, 146-151.

Takakura M., Kyo S., Kanaya T., Hirano H., Takeda J., Yutsudo M. and Inoue M. 1997. Cloning of human telomerase catalytic subunit (hTERT) gene promoter and identification of proximal core promoter sequences essential for transcriptional activation in immortalized and cancer cells. Cancer Res. 59, 551-557.

Thomson J.A., Itskovitz-Eldor J., Shapiro S.S., Waknitz M.A., Swiergiel J.J, Marshall V.S. and Jones J.M. 1998. Embryonic stem cell lines derived from human blastocysts. Science 282, 1145-1147.

Toran B.R. 2003. The Biology of Aging. Mount Sinai J Med. 70, 3-22.

Trosko J.E. 2003. Human Stem cells as targets for the aging and diseases of aging processes. Medi Hypotheses 60, 439-447.

Tzukerman M., Selig S. and Skorecki K. 2002. Telomeres and telomerase in human health and disease. J Pediatr Endocrinol Metab. 15, 229-240.

Tzukerman M., Shachaf C., Ravel Y., Braunstein I., Cohen-Barak O., Yalon-Hacohen M. and Skorecki K.L. 2000. Identification of a novel transcription factor binding element involved in the regulation by differentiation of the human telomerase (hTERT) promoter. Mol Biol Cell 11, 4381-4391.

Ulaner G.A. and Giudice L.C. 1997. Developmental regulation of telomerase activity in human fetal tissues during gestation. Mol Hum Reprod. 3, 769-773.

Wick M., Zubov D. and Hagen G. 1999. Genomic organization and promoter characterization of the gene encoding the human telomerase reverse transcriptase (hTERT). Gene 232, 97-106.

Williams G.C. 1957. Pleiotropy natural selection and the evolution of senescence. Evolution 11348-11411.

Wong A.C., Ning Y., Flint J., Clark K., Dumanski J.P., Ledbetter D.H. and McDermid H.E. 1997. Molecular characterization of a 130-kb terminal microdeletion at 22q in a child with mild mental retardation. Amer J Hum Genet. 60, 113-120.

Wright W.E., Piatyszek M.A., Rainey W.E., Byrd W. and Shay J.W. 1996. Telomerase activity in human germline and embryonic tissues and cells. Dev Genet. 18, 173-179.

THE EVOLUTION OF HUMAN HAIRLESSNESS: CULTURAL ADAPTATIONS AND THE ECTOPARASITE HYPOTHESIS

Mark Pagel[1] and Walter Bodmer[2]

[1]School of Animal and Microbial Sciences, University of Reading, Reading RG6 6AJ, England
[2]Cancer and Immunogenetics Laboratory, Cancer Research UK, Weatherall Institute of Molecular Medicine, John Radcliffe Hospital, Headington, Oxford OX3 9DS, UK

Abstract: Humans lack an outer layer of protective fur or hair, a condition that is unusual among the mammals. We propose that human hairlessness evolved to reduce parasite infection, especially ectoparasites that may carry disease. Unique human cultural adaptations such as the abilities to regulate their environment via fire, shelter, and clothing made hairlessness possible. Clothes and shelters are more flexible than a permanent layer of fur and can be changed or cleaned if infected by parasites. Our hypothesis explains the marked sex differences in body hair and its retention in the pubic regions.

1.   INTRODUCTION

Humans take for granted that they lack a dense layer of hair covering their bodies. And yet, humans are unique among the monkeys and apes in being relatively naked. More generally, hairlessness is rare in the mammals. Of the approximately 3000 extant mammals, the only other effectively hairless species are the elephants, rhinoceroses, hippopotamuses, walruses, pigs, whales, and naked mole rats and at least four of these species are aquatic or semi-aquatic.

S.P. Wasser (ed.), Evolutionary Theory and Processes: Modern Horizons,
       Papers in Honour of Eviatar Nevo
M. Pagel and W. Bodmer. The Evolution of Human Hairlessness: Cultural Adaptations and the
Ectoparasite Hypothesis, 329-335.
© 2004 Kluwer Academic Publishers.

Hairlessness is all the more surprising as it is not without its costs. Humans are more exposed to the sun, and they may suffer greater heat loss when the ambient temperature is low (Newman, 1970; Amaral, 1996). With the exception of the naked mole rats, humans differ from the other hairless mammals in not having a thick or toughened hide for protection. Hairlessness, then, demands some sort of explanation in evolutionary terms.

Before proceeding we must be clear about what we mean by 'hairless'. In fact, humans have about the density of hair follicles expected of an ape of our body size (Schwartz and Rosenblum, 1981). What distinguishes humans is that our hair is so fine and short as to be, for the most part, invisible. For our purposes, then, we use 'hairless' with respect to humans to mean that they lack a dense layer of thick fur.

2.   THE BODY-COOLING HYPOTHESIS

The best-known hypothesis to explain human hairlessness was put forward by Wheeler in a series of papers beginning in 1984. Wheeler argued that the loss of body hair occurred when bipedal hominids first moved to open savanna environments, where they suffered from thermal exposure to the sun. Bipedality may have evolved, in part, to reduce exposure, allowing upright hominids to maintain their body temperature while foraging in the open sun (Wheeler, 1991). Later, Wheeler (1992) suggested that the combination of an upright posture and lack of hair made it easier to radiate heat back into the environment or to lose heat by convective cooling from the wind. However, Wheeler (1992) acknowledges that naked skin increases the rates of both energy gain and loss during periods of too much or too little heat, respectively. This might mean that naked skin is actually a worse solution when the entire day is taken into account: more heat must be dissipated from daytime exposure and at nighttime more heat is lost (Amaral, 1996).

3.   THE AQUATIC APE HYPOTHESIS

Elaine Morgan (1997) has provocatively suggested that between six and eight million years ago, the ancestors to the hominids had a one or two million-year phase of aquatic or semi-aquatic existence. Hairlessness and high levels of body fat evolved in these aquatic apes because fur is not an effective thermal layer under water. This theory then supposes that the aquatic adaptations were retained as ancestral characters throughout at least five million years of subsequent hominid evolution in predominantly terrestrial habitats.

Human fossil remains are often found near bodies of water (Foley, 1987), but evidence for an aquatic phase of proto-hominid existence as yet eludes palaeontologists. Morgan's (1997) theory also fails to explain why features supposedly adaptive to an aquatic lifestyle should have been retained despite several millions of years of substantial evolutionary change in other features of hominids. In a similar vein, the body-cooling hypothesis does not directly explain why hairlessness has been retained despite human populations having occupied colder regions of the Earth for perhaps 100,000 years, and possibly up to 800,000 years for their *Homo erectus* and later *Homo forbearers* (Arsuaga, 2002).

The amount of body hair can change rapidly in evolutionary time as seen from comparing mammoths to extant savannah dwelling and relatively hairless elephants, or domestic pigs and some dog breeds to their closely related and hairy wild cousins. Among modern humans there is variation in the degree of body hair, suggesting substantial genetic variance for this trait. In addition, neither the aquatic-ape nor the body-cooling hypotheses has a ready explanation for the marked difference in body hair between males and females.

4.   HAIRLESSNESS AS AN ADAPTATION TO REDUCE PARASITE LOADS

What is needed are explanations for human hairlessness that link its origin and continued maintenance to advantages that arose and operated throughout hominid evolution, and differentially so between the sexes. Elsewhere (Pagel and Bodmer, 2003), we have put forward the view that hominid cultural adaptations made it possible for hairlessness to evolve in humans as an adaptation to reduce parasite infestation, especially ecto parasites that carry disease. We suggest that hairlessness is maintained by its naturally selected advantages in reducing disease, and by sexually selected effects arising from mate choice for hairless partners. This mechanism could have worked either alone or in concert with other factors that might have favored hairlessness.

Elements of our hypothesis have been around at least since the time of Darwin, although he was convinced that no naturally selected advantage could be adduced for the lack of body hair in humans. Darwin considered the idea - attributed to a Mr. Belt (Naturalist in Nicaragua, 1874, p. 209, cited in Darwin, 1888) - that within the tropics human hairlessness provides a naturally selected advantage for freeing oneself of "the multitude of ticks (acari) and other parasites" (Darwin, 1888, p. 57). He even noted that "as some confirmation of Mr. Belt's view, I may quote the following passage from Sir W. Denison (Varieties of Vice-Regal Life, vol. i, 1870, p. 440): 'it is said to be a practice "with" the Australians, when the vermin get troublesome, to singe themselves'" (Darwin leaves unspecified to which Australians Sir William refers). But he dismissed Belt's idea as unlikely because "none of the many quadrupeds inhabiting the tropics have... acquired any specialized means of relief".

Instead, noting that "in all parts of the world women are less hairy than men", Darwin argued "we may reasonably suspect that this character has been gained through sexual selection" (Darwin, 1888, p. 600). Darwin's theory of natural selection aimed to explain the evolution and maintenance of traits that performed some direct function to the individual such as running faster, gathering food more efficiently, camouflage, and so on. But he also recognized that there was a class of traits that resisted conventional explanation in terms of some sort of mechanical, energetic or physiological function. These are the sexually selected traits, those that, like a peacock's tail, evolve because they are useful to attract or acquire mates but may otherwise actually be disadvantageous to their bearers. Thus, a peacock expends large amounts of energy hauling his heavy tail around and places himself at far greater risk of predation. Darwin's great insight was that these disadvantages are outweighed by the peacock's ability to attract mates.

We suggest, based upon information and ideas not available to Darwin, that the ectoparasite hypothesis is, in concert with sexual selection, the most plausible explanation for hairlessness in humans. There would not have been any real understanding in the late nineteenth century of the important role pathogens play in natural selection. Ectoparasites exact a large toll on the fitness of furry or feathered animals (Lehmann, 1993). Fleas and ticks affect animals directly by biting and causing local irritation, and indirectly by carrying a variety of infectious - including viral - diseases. Animals have specialized muscles for twitching their skin, long tails to swat at flies, and many other anti-parasite morphological and behavioral adaptations (Hart, 1997). Features of beak size and shape in some birds may be adaptations to removing parasites from feathers (Clayton, 1991). Primates devote substantial amounts of time to grooming, increasingly so as group size grows larger, most of which is to remove ectoparasites (Dunbar, 1991). When humans suffer ectoparasite infection it is largely confined to the head and pubic hair suggesting that it is easier to remove ectoparasites from hairless regions.

The fitness effects of parasites are well studied in birds. The colorful feathers and displays of many bird species may be metabolically costly advertisements to prospective mates of the lack of parasites (Hamilton and Zuk, 1982; Read, 1987). Across avian species, those that are more colorful tend to have lower parasite loads, as assessed by a variety of measures (Hamilton and Zuk, 1982; Read, 1987). The idea is that individuals advertise their relative lack of parasites by the quality and sheen of their plumage. Put in this light, a peacock's tail is a billboard, advertising that he has the immune system and the energy to protect him against parasite attack. These may be important considerations to prospective mates. If immune function is heritable, females which mate with colorful mates may have more fit offspring. Three other, more immediate advantages may also accrue. Females may themselves acquire fewer parasites from colorful mates, their nests may, as a result, have fewer parasites to pose risks to the offspring, and finally, colorful mates may be better at provisioning the female and her offspring.

What features of early hominid evolution make hairlessness a plausible response to the toll exacted by parasites? Humans most likely evolved in Africa (Ingman et al., 2000) where biting flies and other ectoparasites are found in abundance. Early humans probably lived in close quarters in hunter-gatherer social groups in which rates of ectoparasite transmission were high. Precisely when humans or their hominid ancestors evolved, hairlessness remained a matter of speculation. What we can say is that having fire and the intelligence to produce clothes and shelter, early humans (and possibly even earlier hominids - *Homo erectus* may have had fire) were well equipped to evolve hairlessness as a means of reducing ectoparasite loads, while avoiding the costs of exposure to sun, cold, and rain. Ectoparasites can and do infest clothing, but clothes, unlike fur, can be changed and cleaned. Infections that do occur on hairless skin can be more easily cleaned than when fur is present. We suggest, then, that a set of cultural adaptations unique to humans made hairlessness a flexible and advantageous naturally selected adaptation.

We do not suggest that the ectoparasite hypothesis explains other mammalian hairlessness, with the possible exception of the naked mole rat. Naked mole rats inhabit arid regions of Kenya, Ethiopia, Somalia, and Israel (Nevo, 1999) where they live underground in large social colonies, rarely coming above ground (Sherman, 2002). Ectoparasite transmission is expected to be high in these colonies, but their climate tends to be regulated within narrow bounds making hairlessness feasible for a species that does not produce clothes or fire. Like humans, they are effectively hairless - having been described as resembling "overcooked sausages with buck teeth" (Sherman, 2002, p. 793) - and like humans, they lack a thick and protective hide.

Sir Ronald Fisher, one of the founders of modern genetically-based Darwinian thinking, emphasized (Fisher, 1930) that sexual selection relies upon a trait having a naturally selected advantage to begin the process of its exaggeration. The ectoparasite hypothesis provides this advantage: initial naturally selected evolution towards reduced amounts of body hair may have been reinforced by sexual selection as hairlessness - by virtue of advertising reduced ectoparasite loads - became a desirable trait in a mate. Unusual among sexually selected traits, reduced body hair would be desirable in both sexes. Greater loss of body hair in females follows from the conventionally stronger sexual selection from male versus female mate choice in humans. Common use of depilatory agents testifies to the continuing attractions of hairlessness, especially in human females.

The retention of hair on the face, head, and pubic regions may also be linked to sexual selection. Head hair may have naturally selected advantages - such as reducing exposure to the sun - that permitted further elaboration by sexual selection. Darwin noted - and contemporary practices attest to - the important role of facial and head hair in attraction and mate choice. The evolutionary retention of pubic hair poses a challenge for the ectoparasite hypothesis, as it provides a warm and humid environment favorable to ectoparasites - and indeed many specialize on these regions. An interesting possibility is that pubic areas may, due to their warmth and humidity, be especially conducive to pheromonal signaling between the sexes. In support of this idea, the density of sweat glands in pubic regions is high (Stoddart, 1990).

The ectoparasite hypothesis makes predictions about human groups. We expect that human groups whose evolutionary history has been in regions of the Earth with higher ectoparasite concentrations will have less body hair. Ectoparasite loads should in general be greater on the hairy parts of our bodies, as anecdotal evidence would already seem to suggest. We should find that apes suffer from higher ectoparasite loads despite having the ability to remove them with their hands. We might also expect that attacks by biting flies are not particularly well defended against by fur - the biting fly simply evolving adaptations to circumvent it.

5.   REFERENCES

Amaral L.Q. do. 1996. Loss of body hair, bipedality and thermoregulation. Comments on recent papers in the J Hum Evol. 30, 357-366.

Arsuaga J.L. 2002. Archaic *Homo sapiens*. In M. Pagel (Ed.), The Encyclopedia of Evolution (Vol. 1, pp. 489-493). Oxford University Press, Oxford.

Clayton, D.H. 1991. Coevolution of avian grooming and ectoparasite avoidance. In J.E. Loyce and M. Zuk (Eds.), Bird-parasite interactions: ecology, evolution, and behaviour (pp 258-289). Oxford University Press, Oxford.

Darwin C. 1888. The descent of man and selection in relation to sex. John Murray, London, 2nd ed.

Dunbar R. 1991. Functional significance of social rooming in primates. Folia Primatol 57, 121.

Fisher R.A. 1930. The genetical theory of natural selection. The Clarendon Press of Oxford University Press, Oxford.

Foley R. 1987. Another unique species: patterns in human evolutionary biology. Harlow, Longman.

Hamilton W.D. and Zuk M. 1982. Heritable true fitness and bright birds: a role for parasites? Science 218, 384.

Hart B.L. 1997. Behavioural defence. In D.H. Clayton, D.H. and J. Moore, J. (Eds.), Host-parasite evolution: general principles and avian models (pp. 59-77). Oxford University Press, Oxford.

Lehmann T. 1993. Ectoparasites – direct impact on host fitness. Parasitol Today, 9, 8.

Ingman M. Kaessmann H., Paabo S. and Gyllensten U. 2000. Mitochondrial genome variation and the origin of modern humans. Nature 408, 708.

Morgan E. 1997. The aquatic ape hypothesis. Souvenir Press, London.

Nevo E. 1999. Mosaic evolution of subterranean mammals. Oxford University Press, Oxford.

Newman R. 1970. Why man is such a sweaty and thirsty naked animal: a speculative review. Hum Biol, 42, 12-27.

Pagel M. and Bodmer W. 2003. A naked ape would have fewer parasites. Biology Letters (in press).

Read A.F. 1987. Comparative evidence supports the Hamilton and Zuk hypothesis on parasites and sexual selection. Nature 328, 68.

Schwartz G.G., and Rosenblum L.A. 1981. Allometry of primate hair density and the evolution of human hairlessness. Amer J Phys Anthropol. 55, 9-12.

Sherman P. Naked mole-rats. In M. Pagel (Ed.), The Encyclopedia of Evolution (Vol. 2, pp 793-795). Oxford University Press, Oxford.

Stoddart D.M. 1990. The Scented Ape: The Biology and Culture of Human Odour. Cambridge University Press, Cambridge.

Wheeler P. 1984. The evolution of bipedality and loss of functional body hair in humans. J Hum Evol. 13, 91-98.

Wheeler P. 1991. The influence of bipedalism on the energy and water budgets of early hominids. J Hum Evol. 21, 117-136.

Wheeler P. 1992. The influence of the loss of functional body hair on hominid energy and water budgets. J Hum Evol. 23, 379-388.

THE EVOLUTION OF THE CULTURAL MEDITERRANEAN LANDSCAPE IN ISRAEL AS AFFECTED BY FIRE, GRAZING, AND HUMAN ACTIVITIES

Zev Naveh and Yohay Carmel

Faculty of Civil and Environmental Engineering, Lowdermilk Department of Agricultural Engineering, Technion, Israel Institute of Technology, Haifa 3200, Israel

Abstract: The early evolution of the cultural Mediterranean landscape in Israel, with special reference to Mt. Carmel, is described with a holistic landscape-ecological systems approach as the coevolution of the paleolithic food gatherer-hunter and his landscapes. In addition to archeological findings and our research on fire ecology and the comparative dynamics of Mediterranean landscapes in Israel and California, we made use of new insights into the self-organization of living systems and landscapes and the theory of nonlinear general evolution. From the Middle Pleistocene onward, this process occurred in two major bifurcations; one in which the pristine forest landscape was converted by human land uses and by natural and intentional set fires into a more open subnatural landscape, and then from the Upper Pleistocene onward into a grass-rich, seminatural, landscape mosaic. The final stage of this coevolution was reached more than 10,000 years ago by the advanced epipaleolithic, pre-agricultural Natufians, whose rich culture and intensive land use have a striking resemblance with those of the pre-European central coastal California Indians. During the third major bifurcation of the Neolithic agricultural revolution, arable seminatural landscapes were converted into agropastoral ones. The coevolutionary symbiotic relationship was replaced by human dominance leading to intensive land uses including burning and grazing. This period is missing from Californian landscapes, 'jumping' almost directly into the agro-industrial age and, therefore, apparently also lacking the great regeneration capacities and adaptive resilience acquired by Mediterranean landscapes.

S.P. Wasser (ed.), Evolutionary Theory and Processes: Modern Horizons,
         Papers in Honour of Eviatar Nevo
Z. Naveh and Y. Carmel. The Evolution of the Cultural Mediterranean Landscape in Israel as Affected by Fire, Grazing, and Human Activities, 337-409.
© 2004 *Kluwer Academic Publishers.*

1.   INTRODUCTION

No bioclimatic region in the world other than the Mediterranean region has endured so long and intensive a period of human-induced perturbations. Nor has any other bioclimatic region suffered so much from the unfortunate combination of a fragile environment and a long history of land abuse and negligence with adverse effects on the land and its people. However, at the same time, no other region has shown, in a more striking way, the resilience and the soil-building capacities of the native vegetation, than the denuded Mediterranean uplands. Probably nowhere else has it been more demonstrated that the people of this region have the power not only to destroy their habitat and deplete their flora and fauna, but they are also able to reclaim them with sufficient motivation and skill, utilizing their biological productivity and preserving their organic and cultural variety (Naveh and Lieberman, 1994; Naveh, 1998a).

Grove and Rackham (2001) rightly stated in their comprehensive and lucid account of the ecological history of European Mediterranean landscapes that the pervasive "ruined landscape theory" in the Mediterranean is far too simplistic and should not be taken literally. It is supportive of this theory that in his review of the environmental history of the Mediterranean mountains, Mc Neill (1992) has called these "skeleton landscapes" in which there is no return from their present human-caused degradation.

Grove and Rackham (2001) have also refuted the deterministic, preconceived theory of a Mediterranean forest "climax", which is automatically degraded by human interference into lower-woody successional stages of maquis, garrigue and batha (or phrygana) and finally into steppe grassland. This follows along the lines of our findings on the dynamics of Mediterranean landscapes in Israel (Naveh, 1971; Naveh and Kutiel, 1990) and in the Mediterranean region in general (Naveh, 1991; Naveh and Lieberman, 1994).

Blondel and Aronson (1999) have attempted to deal with humans as "sculptures of Mediterranean landscapes". They accepted our thesis, by showing many examples of how the long human occupation in the Mediterranean basin had profound consequences on the distribution and dynamics of many organisms and on the current biodiversity. Their report on the results of recent studies showing that organisms may evolve life history traits as a response to human induced habitat changes corroborates our claim on the evolutionary significance of long-term human perturbations. These may have important fitness consequences and can evolve even within few generations if they are submitted to strong selection pressures both by natural and human-induced forces of intensive vegetation management and land uses.

For these issues the Mediterranean landscapes of Israel - and especially those of Mt. Carmel on which we will focus most of our attention - can serve a most suitable example. Here, we have ample archeological evidence of human habitations from the Middle Pleistocene onward. As part of the southern Levant and the "Fertile Crescent" Israel was also one of the first locations for the transition from food collection to food production, marking the beginning of the domestication of plants and animals and the creation of permanent farming communities and settlements during the so-called Neolithic agricultural revolution. As one of the older and better-known human cultural centers, it served also as the cradle for the monotheistic Judeo-Christian religions and Western civilization. Human imprints on the land can be traced back in Israel for longer periods than in any other Mediterranean country.

However, contrary to Blondel and Aronson (1999) mentioning only the evolutionary impact of livestock grazing and burning in the last 10 millennia by pastoralists and agriculturists, we claim that humans started to "sculpture" these landscapes much earlier. They even coevolved together with them, during their biological and cultural evolution in the Pleistocene and together with geological, climatic, and other natural forces and stresses, and especially those caused by fire and foraging of wild herbivores. They converted them gradually from pristine natural landscapes into subnatural, seminatural, agricultural, and rural cultural landscapes. The ensuing closely interwoven natural and cultural processes and patterns contributed much to the great ecological heterogeneity, biological diversity and adaptive resilience of the still remaining nonarable Mediterranean uplands.

As outlined in earlier publications on this subject (Naveh, 1984, 1990; Naveh and Vernet, 1991), we will not restrict ourselves only to those evolutionary and ecological perspectives for which sufficient archeological and geomorphologic evidence is available in the narrow conventional sense of these sciences. Therefore, we will not treat probable ancient human impacts as linear, one-directional cause-effects of human disturbances, but as nonlinear and partly chaotic mutual-causal and reciprocal processes of the coevolution of Mediterranean people and their landscapes.

To support our contentions on the evolutionary significance of the history of human habitation and land use we will refer to the ethno-ecological equivalence in the use of fire for vegetation management by the pre-agricultural Coastal Californian Indians comparable conditions to those of the pre-agricultural Epipaleolithic Natufians of Mt. Carmel. We will also provide indications for the higher resilience and regeneration capacities of the Mediterranean vegetation as compared with its Californian counterparts supporting our claim on the lack of evolutionary convergence between both countries because of the great discrepancies in the duration and intensity of these human impacts.

2.   SOME MAJOR THEORETICAL PREMISES

The holistic and transdisciplinary approach to landscape evolution, on which this essay is based, can be fully comprehended only within the broader context of the present post-modern "scientific revolution". The famous science historian, Kuhn (1970), first coined this term. It takes place when the existing theories no longer adequately explain reality and new paradigms of conceptional schemes have to gradually replace those conventional and well-established paradigms of so-called "normal science". This holistic and transdisciplinary scientific revolution occurred in the last part of the 20th century with the scientific paradigm shift from reductionistic and mechanistic approaches to more holistic and organismic ones. Replacing reliance on exclusively linear and deterministic processes by nonlinear, cybernetic and chaotic processes, this scientific revolution is based on systems thinking of complexity, networks and hierarchic order. It stems from a belief in the objectivity and certainty of the scientific truth towards the recognition of the limits of human knowledge, the need for a contextual view of reality and the need for dealing with uncertainties. It causes the turning away from breaking down, analyzing, and fragmenting wholes into smaller and smaller particles toward wholeness, connectedness, integration, synthesis, and complementarity of ordered complexity, and from mono- and multidisciplinarity to inter- and transdisciplinarity. This scientific revolution is offering a unified worldview that seeks to do justice not only to the physical, biological and the socio-economical, but also to the mental, cultural, and spiritual reality in which we live. It is also leading to profound postmodern cultural transformation, changing many of the ideas dominating Western society since the industrial revolution, and in science and technology - its education, economy, and culture at large.

As described in detail by Naveh and Lieberman (1994) and more recently by Naveh (2000, 2003) and by Carmel and Naveh (2002), we attempted to provide an overarching conceptional framework for a transdisciplinary conception of landscape ecology and its theoretical and practical implications. These concepts are rooted in the General Systems Theory and its recent insights in complex systems and their dynamic self-organization and coevolution in nature and in human societies, enriched by nonequilibrium thermodynamics and chaos theory. Li (2000) has illustrated this holistic landscape paradigm in a more formal way with the help of the mathematical set theory and nonequilibrium thermodynamics. Here, we briefly outline only its most relevant premises.

2.1 A holistic view of landscapes and nature-human relations - the total human landscape of our total human ecosystem

This view implies above all, a paradigm shift from perceiving landscapes as nothing more but large-scale heterogeneous mosaics of physical, chemical and biological landscape elements in repeated patterns of ecosystems, into a holistic view of landscapes as **multifunctional Gestalt systems** in their own right. The German term "Gestalt" has been introduced into psychological Gestalt theory, in which humans are perceived as whole persons, fully embedded in the world, and the world is seen more like a living person than like a nonliving mechanism of separate interacting parts. For studying landscapes in their totality as Gestalt systems, they have to be regarded as a **whole that is more than the sum of its parts.**

As a result, the information about the whole landscape is larger than the sum that can be derived in a mechanistic way from its parts. Therefore the state of the whole must be known to understand the collective parts. This means that from all the natural geophysical, bio-ecological, and cultural landscape components and all other human-made artifacts from its forests, grass- and shrublands, wetlands and rivers, agricultural fields, and from its residential and industrial areas, these patterns and processes contribute to the integral and truly realistic character of each local, regional, and global landscape.

From a hierarchical point of view, all these natural and cultural dimensions are intrinsically related to each other by the general state of the whole and its emergent qualities. The different landscape units and types are closely interlaced into a multilayered, stratified hierarchy of Janus-faced entities or "*holons*" senso Koestler (1968), being both parts of their higher-level supersystem and wholes with regard to their lower-level subsystems. Therefore, instead of a puzzle of separate particles forming a mosaic **in landscapes, we deal with a hierarchically structured interacting network at different multiple nested scales of our global "*Total Human Landscape*"** (THL). Together with increasing spatial, temporal and perceptional scales the complexity of patterns, processes and their resulting functions are increasing. Therefore, a better comprehension of the underlying ecological, historical, and cultural dynamics of representatives of THL is necessary. As will be shown below, this is true also for our prehistorian human influenced, modified, and converted THL.

This holistic landscape conception has to be complemented by a broader holistic view of the role of humans in nature as **integral parts of nature, forming a complex socio-ecological entity with their total environment.** This is the **Total Human Ecosystem** (THE), integrating humans and their total environment at the highest level of the global ecological hierarchy **above** the ecosystem level (Egler, 1964). Landscapes are **the concrete, space/time defined ordered wholes and Gestalt systems of our THE, along different functional, spatial, and perceptional dimensions** providing the spatial and functional matrix for all living organisms, their populations, communities, and ecosystems. Their spatial scales range from the smallest mappable landscape cell or *ecotope* to the *ecosphere* as the largest, global Total Human Landscape.

Whereas the natural landscape elements have evolved and are operating as part of the geosphere and biosphere, their cultural artifacts are a creation of the **noosphere** (from the Greek *noos* = mind). As described lucidly by the great systems thinker and planner, Erwin Jantsch (1980), the noosphere is an additional natural envelope of life in its totality that *Homo sapiens* have acquired throughout the evolution of his neocortex. It is "our mental space" and the domains of our perceptions, knowledge, feeling, volition, and consciousness enabling our self–awareness and cultural symbolization and their linguistic and artistic expression. We claim that the Epipaleolithic food gatherers and hunters at the last phase of their coevolution with their landscapes, in the final stages of the Pleistocene and the end of the (European) Glacier period more than 12,000 years ago, had already attained a high stage of noospheric cultural evolution. Later on, it enabled the development of additional noospheric realms of the info-socio- and psycho-spheres that emerged during the cultural evolution in the Holocene, through which modern man finally became a mighty geological agent with both constructive and destructive powers.

2.2   Autocatalysis and crosscatalysis, autopoiesis and self-organization and their role in evolution

Of great relevance for our discussion are the insights gained on the self-organization of living systems. The spontaneous emergence of new order, creating new structures and new forms of behavior within network patterns of living systems is made possible by their self-regulating feedback loops. Such systems on relatively high-organizational levels, which can renew, repair, and replicate themselves as networks of interrelated component-producing processes in which the network itself is created and recreated in a flow of matter and energy, are called **autopoietic systems** (from the Greek = self-creating or self-renewing). This is true not only for cells (Eigen and Schuster, 1979), organisms, and ecosystems but also for THE landscapes as the spatial and functional matrix of interacting nonhuman and human living systems. This autopoietic process is made possible by **autocatalysis** by which one of the products of the reaction enters a cycle that helps to self-reproduce by creating its own synthesis. In cycles of **crosscatalysis** two or more subsystems are linked, so that they can support each other by catalyzing each other's synthesis and thereby mutually increasing their growth. Such positive feedback loops lead to **hypercycles** of mutually reinforcing processes, typically for systems, i.e., landscapes, which are far from equilibrium, together with the appearance of instabilities leading to new and higher forms of organization.

These nonequilibrium systems are called **dissipative structures** because they maintain continuous entropy production and dissipate accruing entropy, not accumulating in the system, but being part of the continuous energy exchange with their environment. Dissipative structures constitute the simplest case of spontaneous self-organization in evolution. This has opened the way for realizing that evolution toward increasing complexity and organization is the result of structural fluctuations and innovations that can appear suddenly in previously stable systems and drive it subsequently to a new regime at a more complex state (Maturana and Varela, 1975; Prigogine and Stengers, 1984). As will be further explained below, in evolutionary processes these are expressed as *bifurcations.*

Jantsch (1980) has laid the transdisciplinary foundations for a synthetic view of cosmic, geological, biological, ecological, and socio-cultural evolution leading to an all-embracing concept of coevolution and emphasizes our present "Macroshift" from the industrial to the post-industrial information cooperating as the creative player of an entire evolving universe. As a major paradigm shift from the Cartesian and Newtonian view of a mechanistic world it reaches far beyond the post-Darwinian and socio-biological interpretations of evolution. Laszlo (1987, 1994, 2001a,b) has adopted and further developed Jantsch's paradigms of these co-evolutionary patterns of change and transformation in the cosmos, organisms and in modern society, and its far-reaching consequences. This stems from the recognition that realms of evolution in the empirical world do not follow classical disciplinary boundaries although logically aligned with the unfortunate divisions of empirical science between the physical, biological, and social sciences. However, from the perspective of the synthetic evolution, these are not absolute and watertight divisions, but result from the above-mentioned theories based on investigations of systems, leading far from thermal and chemical equilibrium to the formation of dissipative structures.

In this "Grand Synthesis" (Laszlo, 1987, 1994), the evolutionary trajectories are not moving in a continuous and linear progression from the simpler to the more complex type of system, but "leap" by the sudden emergence of successive levels of higher organization. These discontinuous developments of sudden leaps from one kind of stable state occur as the above–mentioned **bifurcations**. In these, abrupt discontinuous changes in system behavior occur as a result of certain parameters crossing an apparent boundary of their domains of attraction in such **metastable systems.** As a result of such subtle "catastrophic" bifurcations, these systems may turn chaotic or disappear or lead to a new state of metastability on a higher level of organization. Their mutually reinforcing auto- and cross-catalytic feedback loops are triggered chiefly by technological innovations. On each level, the amount of cultural information that can be handled by the cycle is greater than that on the lower level, due to a greater diversity and richness of the components and structures. We have adopted this synthetic evolutionary perspective and view the long-term cultural history of human societies and their landscapes in the Mediterranean as proceeding by leaps through such crucial bifurcations at which the past trends broke down, allowing such dynamic systems to emerge on successively higher levels of organization on multiple hierarchical levels. In our case, they led from the primitive food-gathering hunting stage to our present, still chaotic transitional "Macroshift" bifurcation stage. However, in contrast to the human-landscape coevolution shaping the Pleistocene subnatural and seminatural landscapes, the Holocene agropastoral, urban industrial landscapes evolved as the result of human dominance. The introduction of fossil energy during the industrial revolution has caused a crucial bifurcation between the self-organizing autopoietic natural, seminatural, and traditional agropastoral **biosphere landscapes** and the human-created and driven urban-industrial **technosphere landscapes.** The biosphere landscapes are powered solely by solar energy and its conversion through photosynthetic assimilation of autotrophic plants into chemical energy and transmitted in the trophic food chain to heterotrophic herbivores. The rapidly expanding fossil energy powered technosphere landscapes endanger the future of biosphere landscape and their biological evolution and have led to the formation of our disorganized and unsustainable industrial Total Human landscape. Human society has the choice if our present macroshift bifurcation will lead to further biotic degradation and extinction or to a sustainable future for nature and humankind and further biological and cultural evolution (Laszlo, 1994; Naveh, 2003).

Contrary to the homeostatic equilibrium paradigm of the so-called "balance of nature", our studies indicate that Mediterranean woodlands, shrublands and grasslands as seminatural and meta-stable landscapes continue to change among their tree, shrub, herb, and grass layers as long as the same perturbations continue with similar intensities and frequencies. The eminent geneticist Waddington (1975) coined the term **homeorhesis** (from the Greek meaning preserving the flow) for such dynamic flow equilibrium to denote the evolutionary stability of multifactorial systems or the preservation of the flow process of the evolutionary pathway of change through time. While undergoing such short- and long-term cyclic natural and human induced rotations of burning, grazing, browsing, cutting, coppicing, and cultivation superimposed on the seasonal and annual climatic fluctuations, they are apparently driven by positive feedback loops of cross-catalytic hypercycles. Their resulting defoliation pressures were incorporated in the landscape at different spatiotemporal scales. These **human perturbation-dependent** systems have acquired long-term adaptive resilience and evolutionary metastability, which is discussed further in Part II. Their thermodynamic behavior as dissipative structures has clearly pointed out the importance of the re-establishment of this multifactorial homeorhetic flow process by active and dynamic conservation management, furthering the highest attainable multifunctionality of these landscapes (Naveh, 1991, 1994b, 1998a,b; Naveh and Lieberman 1994).

2.3 New approaches to archeology and prehistoric human impacts

Of great significance to our discussion are recent developments in archeology undergoing a similar scientific revolution. According to Runnels (1995), it is shifting from monodisciplinary studies of single sites to inter- and even transdisciplinary studies of the natural and cultural history of whole landscapes and regions in cooperation with scientists from other, relevant disciplines such as geomorphology, biology, ecology, history, anthropology, and arts. These studies are making use of the most advanced methods of these sciences, and other sciences such as geophysics, using seismic refraction and ground penetrating radiation as well as remote sensing and other spatial explicit methods used by landscape ecologists.

An important result of these developments is the attempt to reconstruct the lives, economics, cultures, and beliefs of ancient societies not only on the basis of narrow utilitarian interpretations of the archeological findings of stone tools and other cultural artifacts. Thus, for instance, Ronen (2000) has shown in the example of such broader, transdisciplinary interpretations of pre-agricultural Aceramic Neolithic Cypriotic culture (about 6500 BC) that it could lead to a much better appreciation of the intellectual capacities and of the spiritual world of these people enabling them to utilize their natural resources in a sustainable way for long periods in a peaceful and technological simple way of life. Similar conclusions could be reached also in the case of the Natufians discussed below. Ronen and Adler (2001) have reached similar conclusions on the functions of the parameter walls of Jericho and the Khirokitia walls in Cyprus, regarding these not as part of military fortifications but as ideologically determined "magical defenses, separating at the very least, the built and the unbuilt areas".

Until recently there was a general tendency to deny any significant human influence and to underestimate the importance of human habitation and food gathering and its impacts on their Pleistocene environment and its vegetation. Regarding these as insignificant, they were treated chiefly as hunters. Thus, for instance, Pons and Quezel (1985, p.35) in reviewing human impact in the Mediterranean claimed: "Early man was a hunter and gatherer and had relatively little influence on natural vegetation." McNeill (1992) has accepted uncritically this view, maintaining that until the Neolithic revolution hunters and gatherers played almost no role because of their slender numbers. Although he recognized their early use of fire, he claimed that their activity scarcely affected vegetation complexes. But at the same time he maintained that such prehistoric fires started the "long history of anthropogenic erosion". However, Rackham and Grove (2001) in their review of prehistoric and historic causes of erosion could not find any well-founded evidence that human activities were its major cause, as opposed to a much better correlation with climatic long- and short-term events.

We will present an alternative view of the ecological and evolutionary importance of paleolithic Mediterranean people as food collectors, hunters, and as vegetation manipulators, chiefly with the help of intentional fire. In this context, Leaky and Lewin (1979) stated in their description of early human evolution that it would be foolish to ignore the lessons that contemporary societies of the rapidly vanishing hunters and gatherers could teach us. On the basis of the importance of plants in their diets, economics, time spent on food collecting, and the tools used (mostly made of wood and therefore perishable), they concluded that it would be more accurate to refer to these people as well as to their Paleolithic ancestors, as gatherer-hunters (G-H) and not hunter-gatherers.

In a similar vein, Eisler (1986) critically re-examined the prevailing view regarding the dominating "man the hunter" as the ruler throughout prehistoric times. She offered an alternative of the equilibrium gender between men and women. She stressed the important role of woman in the economy of these early societies and as the religious image of "Mother God" and the symbol of the natural life cycle. This and the deep, spiritual links with nature are reflected also in the Paleolithic cave paintings in the Mediterranean, and as is shown below, also by the Natufians of Mount Carmel. In her opinion, the basis for these misconceptions is the stereotypic view of primitive man as a bloodthirsty militant. This distorted picture is indeed also very different from what we know today about these "primitive" pre-agricultural societies in Africa (Turnbull, 1961) and in New Guinea (Diamond, 2001). Based on his long personal experience, Diamond presents an impressive picture of the high-natural intelligence and environmental awareness of these New Guinean people and their intimate knowledge of plants and animals and their usefulness, which was apparently the result of a long-evolutionary adaptation process of living in intimate dependence on the natural world upon which their survival depended.

These are important insights, which help us in our attempts to reconstruct the cultural evolution of these pre-agricultural people and their still underestimated intellectual capacities. It should lead above all to a re-evaluation of their abilities as efficient "environmental managers" in the use of available natural resources and the important role of women in these activities. They could have even been influenced by their religious beliefs, like the North American Indians, as restraining cultural feedbacks for the overuse of their natural resources.

3. PAST AND PRESENT ENVIRONMENTAL CHARACTERISTICS

3.1 Mediterranean climate and biota

Five regions of the world - the Mediterranean basin, California, central Chile, the Cape of South Africa, and western and southern Australia - share a unique climatic regime of mild, wet winters and dry summers with 90% or more of the annual precipitation falling in the six cool months. These similarities are due to the symmetry of the atmospheric circulation, governed by the positioning and seasonal movement of the subtropical anticyclone on the equatorial side and the positioning of the cyclogenesis in the belt of mid-latitude westerlies on the poleward side. The present, strongly zonal, atmospheric gradient of the polar-to-equator temperature gradients, which is enhanced by the polar ice caps, renders the Earth's glacial history as pertinent to the discussion of the evolution of Mediterranean-type climates and their future (Deacon, 1983).

Thus, the presently observed rapid melting of the arctic ice caps and the rise in temperatures at the lower latitudes could also have far-reaching effects on global climate changes and on the disruption of these Mediterranean climate patterns, accompanied not only by drastic ecological landscape changes but also by social and economic upheavals (Naveh, 1995). As explained in more detail by Allen (2001) in a thorough description of these Mediterranean climates, the bi-seasonal summer drought and winter rainfall pattern already appeared 2-3 million years ago during the mid-late Pliocene chiefly as a result of global cooling during the establishment of permanent ice in the North Atlantic. However, this bi-seasonality has not been consistent and experienced major fluctuations in response to changes in Quaternary ice volume. This is especially true for the Pleistocene, during which our Mediterranean landscapes reached their final geomorphologic structure, converging with major steps of human biological and cultural evolution. It underwent severe glacial periods of colder and wetter conditions and drier, warmer, interglacial periods. The last Penniglacial 7,500-15,000 years ago was a period of considerable climatic and environmental changes in the Levant. On the basis of palynological findings, Weinstein-Evron (1993a) presented a schematic palynological cycling pattern of shifting humid, dry, and intermediate periods. Increased humidity seems always to precede cold conditions in the area. At the time of such humidity peaks, precipitation was higher than today and maquis and forests were probably more expansive in these landscapes. More recent pollen findings by Baruch and Bottema (1999) from Lake Hula in northern Israel show dramatic changes in the last 16,000 years from severe aridity to a highly humid climate. The later part of this period (ca. 14,500-11,000 B.P.) turned out to be the most humid period and the ensuing stage largely coinciding with the Younger Dryas, (ca. 10,500-5,500 B.P.) of renewed aridity to almost full glacial conditions. The early Holocene was marked by a gradual return of humid conditions followed by prolonged stability. We should keep in mind that these pollen patterns can indicate only the general trends of climatic changes, but their resolution is too coarse for detecting the gradual, subtle vegetation changes from eventual human-induced modification of natural to subnatural and to seminatural agropastoral landscape during this crucial cultural and socio-economic transformation stage. By such pollen analysis maquis formations cannot be distinguished from forests with their different oak species and life forms as trees or shrubs caused by different grazing and fire pressures. Only when early agriculturalists carried out large-scale forest clearings and replaced the natural vegetation with cultivars on larger scales by plants whose pollen is wind-transported, such as olives, was this reflected by the occurrence of olive pollen derived from wild olives (Allen, 2001). In fact, according to Baruch and Bottema (1999), around 7,500 B.P. such anthropogenic effects on the vegetation become noticeable, principally marked by a conspicuous rise in the values for olive. These human impacts increased considerably during the second half of the Holocene. Similar phenomena have been noted in pollen diagrams from the Sea of Galilee. As statements on the presence or lack of anthropogenic vegetation changes in the Mediterranean during earlier periods are based solely on such palynological findings spanning hundreds of thousands of years from the Upper Pleistocene to the early Holocene, they may be very misleading. This does not mean, however, that even earlier, specific pollen findings cannot be interpreted as indications of human interferences. As will be shown below, early Mesolithic human activities around 100,000 years ago can be traced back in the Carmel caves from pollen of typical human follower species, taking advantage or even evolving in such human cleared and disturbed camp sites and their "kitchen" waste disposal.

All five Mediterranean climate regions are typical for their evergreen shrublands, dominated by species with evergreen tough and leathery sclerophyllous leaves. In a comprehensive ecological comparison of these regions, Di Castri (1981) showed that these species are generally replaced along environmental gradients of moisture and nutrient availability of other vegetation types. However, these gradients are also greatly influenced by human impacts on these landscapes and their vegetation. A recent overview of both natural and anthropogenic disturbance regimes in these so-called "Mediterranean-Type Ecosystems" has been provided by Rundel (1998).

Blondel and Aronson (1999), Allen (2001), and Grove and Rackham (2001) provided detailed descriptions of the Mediterranean biota. There are great differences in this respect in the different regions and countries. Because of the unique geographic location of the Mediterranean basin between Europe, Asia, and Africa it has served as a meeting point and melting ground for species of varying origins. As the southern most outpost of the eastern Mediterranean, Israel is the most pronounced example of a bridge and corridor for the different biogeographic elements of these regions.

Axelrod (1958) has shown that sclerophylly is an ancient character that long predates the evolution of summer-dry climates. He traced scleropyllous vegetation in North America-Eurasia back to Madro-Tertiary and Mediterranean – tertiary geoflora (the fossil flora with a common geological history), respectively, which had their origin in the southwestern portions of the continents in the late Cretaceous. The Mediterranean geoflora has been derived from Indo-Malesian, Paleo-African, and xerothermic Mesogen stock. Seasonal aridity already appeared sporadically in the Middle Eocene, but the true Mediterranean climate pattern was established only in the Pleistocene. The sclerophyll woody species were apparently best pre-adapted to climatic patterns of increasing drought and lower-winter temperatures that developed during the Pleniglacial in the Levant.

Tchernov (1988) maintained that at the end of the Pliocene and early Pleistocene, dissemination of biota into the southern Levant was only possible for a select number of species. At the onset of the Pleistocene the southern Levant was already divided into several morphotectonic domains that were primarily responsible for landscape formation and the structuring of their fauna and flora into a well-established biogeographic framework. However, during the rest of the Quarternary, tectonic geographic and climatic events continued to play an important role in reshaping these landscapes. These factors resulted in changes of dispersal and abundance of plant and animal groups, sometimes affecting their rate of biotic turnover and extinction. The main route of biotic and hominid dispersal, from Africa to the rest of the world, took place through Israel, as the southern Levantine corridor. The shift toward aridity in the Quarternary became a major causal factor for the extinction of many Afrotropical and Palearctic elements and for the increased separation between tropical Africa and Eurasia. The above-mentioned impact of glacial episodes and the proximity of large desert domains also played a major role in the distribution of Levantine plants, animals, and humans (Horowitz, 1992). According to Pignatti (1978), large numbers of chiefly herbaceous and annual plant taxa, which evolved during the Pleistocene in the Mediterranean basin, were accompanied by further speciation of woody plants and endemism. Presently, in all of these regions, but especially in the Mediterranean basin, with the exception of the rapidly vanishing coastal dunes, wetlands, and marshes the uncultivatable uplands have become the last refuges of nature, which means plants and animals are found spontaneously occurring and reproducing. Wherever these uplands have not yet been converted into dense pine or eucalyptus forests or depleted into scrub and rock deserts, they are distinguished by their great ecological heterogeneity and biological diversity. Although covering only a tiny 1.2% of the Earth's surface, their contribution to species diversity of vascular plants far exceeds their relatively small area of coverage. The largest number of these species can be found in the Mediterranean basin (25,000), making up 20% of the world's total (Cowling et al., 1996). Therefore, it has been recognized as one of the 18 most important biodiversity "hot spots". On the basis of species/area ratio, the Mediterranean territory of Israel is, after Cyprus, by far the richest containing more than 1500 species and a ratio of 0.15 – as compared to 0.01 in Turkey, 0.04 in Greece, and even less in all other Mediterranean landscapes (Naveh and Kutiel, 1990). The Mediterranean flora is especially rich in herbaceous plants including many annuals and exceptional colorful flowering compositae and geophytes with ornamental values. Among these are many rare and endemic plants. Many grasses and legumes are outstanding pasture plants and are cultivated widely in improved pastures, especially in Australia. Among the woody plants, and especially in the Labiatae family, there are many species with great value for pharmaceutical, cosmetic, spice, balsam, and other uses with considerable economic potentials. These plants are now grown, collected, and utilized with increasing intensity for commercial production. The Mediterranean zone of northern Israel is the center of distribution in the Near East "Fertile Crescent" of the wild tetraploid emmer wheat *Triticum dicoccoides*, the progenitor of most tetraploid and hexaploid cultivated wheat, which will play an important role in further wheat improvement. Nevo and his coworkers at the Institute of Evolution of the University of Haifa have studied the rich genetic resources of cultivated wheat since 1971. It served these scientists at the same time as a major model organism for evolution (Nevo, 2001a; Nevo et al., 2003). Its importance for the coevolutionary process of the emergence of agriculture will be discussed below.

A comparison of the floristic and structural diversity and species richness of shrublands and woodlands of northern Israel with those of southern France near Montpellier revealed that regardless of the scale of observation, alpha species richness diversity was higher in Israel across the range of all vegetation structures and land-use histories. It was highest in the semi-arid ecotones of the Mediterranean zone, on Mount Gilboa of Israel in an open shrub community moderately grazed, but chiefly by gazelles. These were dominated by Pistacia lentiscus with many sub-shrubs, reaching 179 species /1000m^2 in comparison with only 79 species in a Quercus coccifera community in France. The same was also true in a comparison in "ecological equivalent" study plots (Naveh, 1969) between Mediterranean-type woodland and oak savanna communities on Mount Carmel and in the Carmel Valley in Central California. The communities of Israel were richer in life forms and had more climbers, geophytes, and woodland legumes. In moderately grazed woodlands in Israel of 1000m^2 we counted 137 species and of these 110 – were mainly annual-herbs, as compared with only 47 species with 41 herbs in California. In both countries, closed shrublands were much poorer than those of Israel and had only 35 species with 17 herbs in Israel and even less in California (Naveh and Whittaker, 1979).

A further comparison between these Mediterranean plant communities with those in central Chile, southwest Australia, and the Cape region in South Africa showed the greater richness in all growth-form categories and higher-total diversity and lower-concentration dominance in Chile, as compared to California. We related the greater richness in woody plants of the Chilean vegetation to a long-term evolutionary history, but the greater richness in herbs - like in Israel - to the longer history of more severe human disturbances by Spanish colonizers, opening the canopies of the Chilean shrub communities about 400 years ago and only 200 years ago in California. However we argued further that because of their much older origin, the Australian and South African Gondwanan shrub communities diverge from the European, Chilean and California ones: They lack in annual species but are extremely rich in woody species, and, therefore, no similar conclusion can be drawn from the length of human habitation and its effect on their structure and diversity (Naveh and Whittaker, 1979).

3.2 The natural environment of Mount Carmel

The chief study site of Mt. Carmel is an isolated mountain ridge, rising from the northern Mediterranean Sea shore of Israel to a height of 450-500 m. It represents a typical example of the rich regional seminatural and rural, cultural landscapes of the eastern Mediterranean and southern Levant, whose great natural and cultural values are threatened presently by the mutually amplifying combination of urban-industrial, agricultural, and recreational pressures.

Fortunately, the unique biological, ecological, geological, archeological and scenic features of Mt. Carmel have been recognized over time by foresighted regional planners like Joseph Bruzkus from the Ministry of the Interior with full support by the Haifa Municipalities and reinforced by public pressure led by the Israel Society for the Protection of Nature. From a total area of 232 km^2, 84 km^2 were declared as the Carmel National Park (CNP) in 1970, incorporating nature reserves of 31 km^2 with about 55 km^2 of densely planted mono-species forests of *Pinus halepensis*.

Used each year by more than two million visitors, the nature reserve is the largest biologically, and culturally the richest, most attractive open landscape, open-door recreational area in the densely populated coastal zone of Israel. The Carmel National Park and its surroundings fulfill important ecological, social, psychotherapeutic, educational and scientific functions. Let us hope that the Israel Nature and Park Authorities will obtain sufficient financial support from the Israel government for continuing to ensure these vital functions for the sake of present and future generations.

Mt. Carmel has a typically mild Mediterranean climate with winter and spring rains ranging from 500-600 mm average annual rainfall in the lower parts to 700-900 mm in the higher parts. According to the UNESCO-FAO (1962) bioclimatic classification, these are the typical drier, warmer, "xerothermic and wetter and slightly cooler" accentuated thermo-Mediterranean bioclimates with a large number of 125-200 pluviothermic dry days in which both the natural and rain-fed agricultural vegetation is highly fire-prone. Therefore, they could also be called **Mediterranean fire bioclimates** (Naveh, 1973). Their great inter- and intraseasonal variability has further increased in the last 10-15 years, most probably because of global climate changes, which has apparently disrupted the typical Mediterranean rainfall and temperature regimes (Naveh, 1995).

Mt. Carmel is distinguished by its great stratigraphic, geomorphologic, and topographic heterogeneity resulting from this dynamic tectonic history and its unique paleogeographic location at the edge of a shallow platform. The dominating rocks consist of dolomites and crystalline limestones, which contributed to the karstic nature and the very scenic appearance of some of these rock formations, especially those facing the Mediterranean. Other rocks are chalks, marls, as well as volcanic tuff and sandstones along the coastal line. The mountain areas are covered chiefly by shallow terra rossa, light and brown rendzina, brown Mediterranean forest soils, and deeper colluvial soils. The soils of the coastal area primarily include deep alluvial soils (Bein and Sass, 1980; Nir 1980; Singer and Ravikovitch, 1980).

Like other mountain landscapes in the Mediterranean, Mt. Carmel is a relatively young geological system, which gained its present geomorphologic form by violent uplift in the Late Tertiary and early Quaternary period as an isolated mountain belt in which Upper Cretaceous (mostly Cenomanian-Turonian) rocks are exposed. Its final shaping was the result of tectonic and volcanic activities during the Pleistocene. These caused intense mountain rising and erosion, sea ingressions and regressions, followed by increasing diversification of local site conditions. Its western boundary, which in places follows ancient reef trends, has been shaped through coastal abrasion of the Mediterranean Sea during the Pleistocene and its changing sea levels. Horowitz (1979) has described these dramatic morphotectonic evolutionary alternations. They created many new habitats such as canyons and gorges with steep slopes and bare rock surfaces, contrasting north-south and east-west exposures in the Carmel uplands and alluvial fans, vernal pools, saline zones playas, bogs, swamps, marshes as well as sand-covered hills and flat in the coastal lowlands. These further increased the edaphic and topographic heterogeneity and microsite diversity and opened many opportunities for plant and animal colonization and speciation. The volcanic activities caused many recurring large and hot wildfires. However, their far-ranging evolutionary and ecological landscape impacts were completely overlooked in these studies.

The most previous thorough palynological study on the Carmel coast near Dor has been carried out to a depth of 10.5 meters reaching the radiocarbon date of 23,400 Y.B.P. at the bottom of a marsh, which was formed during that period (Kadosh, 2002). However, in these anaerobic conditions, only in the upper layer of the early Neolithic at the beginning of the Holocene (between 9,000 to 8,200 Y.B.P.) such palynological pollen data could be obtained. This points to typical Mediterranean vegetation, dominated by *Quercus calliprinos*, with many herbaceous plants, apparently from a more humid and cooler climate than today. Notwithstanding these climatic fluctuations, we also have much earlier evidence from other palynological studies that the Carmel flora - like all other floras of the south Levant - has been mainly of the Mediterranean type from the Pleistocene (Weinstein-Evron, 1991, 1993a,b, 1998). Therefore, we can assume that the major coevolutionary process described below took place in a Mediterranean environment and climate, at least from the botanical point of view.

The bio-climatic variability as well as the unique geographical position of Mt. Carmel near the Mediterranean Sea and its great geomorphologic, lithological, and edaphic landscape heterogeneity has favored the evolution of a rich fauna and flora. The latter is presently comprised of close to 1500 plant species, mostly annual and perennial herbs with several endemic and rare species as well as a great number of ornamental flowering geophytes. For many of the Eu-Mediterranean species, Mt. Carmel is the southernmost limit of their distribution. The latter include *Pinus halepensis,* the only natural occurring conifer tree in Israel. The CNP carries its last larger forest remnants, with a dense woody understory, and a well-developed, multilayered maquis, dominated by *Quercus calliprinos*. The lower belts are covered by more open, park-like woodlands and shrublands dominated by *Ceratonia siliqua, Pistacia lentiscus, Sarcopoterium spinosum,* or *Quercus ithaburensis* with a rich herbaceous understory.

The great macro- and microsite heterogeneity of Mt. Carmel induced the great floristic diversity both on the interspecies level and the intraspecific level. Examples of ecotypic variations have been found on different sites in two of the most abundant woody and herbaceous plants, namely *Pistacia lentiscus* (Swarzboim, 1978) and *Piptatherum miliaceum* (Naveh, 1959). As will be discussed below, it may be also of great evolutionary significance that these species are distinguished by their great regeneration capacities after fire.

The present vegetation of Mt. Carmel consists chiefly of complex mosaics of degrading and regenerating plant communities. Depending on site conditions and past and present land-use pressures, this patchy vegetation ranges from rich, productive open grasslands and woodlands to severely depleted dwarf-shrub communities (batha or phrygana) and denuded rock deserts; from rich multilayered semi-open shrublands and forests to one-to-two-layered, closed, tall shrublands (maquis or mattoral). The latter is composed chiefly by sclerophyllous phanerophytes and dominated in Israel by *Quercus calliprinos*.

Most of these sclerophyll trees and shrubs are distinguished by dual root systems that can spread horizontally and penetrate deeply into rock cracks, vigorously resprouting from their roots after fire, grazing, or cutting. They respond favorably to pruning and coppicing on one stem. If resprouting from suckers is prevented they soon attain the stature of small trees. In this way closed, one-layered, very fire-prone, and unproductive shrub thickets can be converted into rich, multilayered, park-like groves and woodlands. This apparently, was the way sacred oak groves, mistakenly regarded as remnants of "climax" oak communities, have been created.

In a series of extensive multidisciplinary studies carried out since 1991 at Lower Nahal Oren on adjacent slopes with opposite south and north exposures, Nevo and his coworkers investigated the striking differences in their vegetation cover as well as in the biogeographic origin of the flora and fauna. Up until now, they have identified 2000 species including 320 vascular plants displaying qualitative or quantitative divergences between and within slopes. These findings were more recently corroborated in another study in slightly more humid conditions in the western Upper Galilee.

All these studies, ranging from the genome up to the landscape level provide convincing proof that environmental stress - in this case the difference in solar radiation and humidity - act as strong evolutionary forces for natural selection along regional climatic, topoclimatic and microclimatic gradients. Field microscale studies on the relations between molecular evolution and environmental stress on annual grasses abundant on Mt. Carmel, revealed that edaphic (terra rossa vs. basalt soil; rock vs. deep soil) and microclimatic (sun vs. shade; and high vs. low solar radiation) stresses in *Hordeum spontaneum* (wild barley), *Triticum dicoccoides* (wild emmer wheat), and *Aegilops peregrina* point inferentially to natural selection as a major differentiating factor of qualitative and quantitative patterns of genetic diversity at single loci, but primarily at multilocus structures and genome organizations (Finkel et al., 2001; Nevo, 2001b). We should, therefore, not reject the hypothesis that the long-lasting defoliation stress factors induced by foraging wildlife and humans acted as strong natural selection pressures, even if we are not yet in the position to reveal their gene-ecological evolutionary mechanism.

4. THE EVOLUTION OF THE MEDITERRANEAN LANDSCAPE OF MT. CARMEL IN THE PLEISTOCENE AND EARLY HOLOCENE

4.1 Major phases of land use history in Mediterranean landscapes of Israel

Naveh and Dan (1973) and Naveh and Kutiel (1990) distinguished three major periods of human-induced changes in the landscapes and vegetation of Israel:

1. A very long period during the Pleistocene, which marked the major phases of the coevolution of Mediterranean peoples with their landscapes. During this period, the natural landscape was first transformed into a subnatural one, changing the floristic composition, but largely retained the natural vegetation structure and formation. It is the second phase, during which this coevolution reached its peak and intensive vegetation management transformed most of the natural and subnatural landscape into seminatural vegetation that altering their structures took place. In this period the cornerstones were laid for the metastable homeorhetic flow equilibrium on which the vegetation dynamics of these seminatural landscapes are based until present times.

2. A long prehistoric and historic agricultural period in the Holocene, during which the agropastoral cultural landscape was shaped, reached its peak and then gradually declined. Seminatural vegetation

formations were maintained in a metastable stage of homeorhetic flow equilibrium.

3. A recent, short, modern "neotechnological" period in the 20th century of increasing heavy human pressures on these solar energy-powered seminatural biosphere landscapes by fossil-energy powered agro-industrial, rural, and urban industrial technosphere landscapes caused the biological and cultural deterioration of the Mediterranean landscape and the formation of the disorganized Total Industrial Landscape.

In this part of our essay, I will deal chiefly with the Pleistocene and the earlier Holocene periods, marking the major coevolutionary human-landscape stages. For this purpose, I will further broaden the claim made by Di Castri (1981) that thanks to the continuing interaction between natural and anthropogenic features from the Middle Pleistocene onward, human beings coevolved with Mediterranean ecosystems. Coevolutionary features are present in a number of ecological and cultural characteristics in this region. For our coevolutionary theory, I apply the definition of Stebbins (1982) who regarded coevolution as a simultaneous evolution of two genetically independent but ecologically interdependent lines via both biological and cultural templates.

From the Middle Pleistocene onward, the geological and biological landscape evolution coincided with major phases of the biological and cultural evolution of humans. During this long period of more than a million years, the Lower Paleolithic *Homo erectus* was replaced by more advanced food gatherers and hunters such as Middle Paleolithic Neanderthaloids and the first *Homo sapiens* around 100,000 years ago, and subsequently by the intensive food collecting Epipaleolithic *Homo sapiens sapiens* and the food-producing Neolithic *Homo sapiens sapiens* in the early stages of the Holocene, around 12,000 to 10,000 years ago.

We regard these closely coupled evolutionary processes and their mutually beneficial feedback loops of auto- and cross-catalytic networks and their hypercycles as the coevolution of Mediterranean peoples, their landscapes, and vegetation with both natural and human-set fires playing an important role. Naveh and Vernet (1991) described this coevolution in the context of the palaeohistory of the Mediterranean biota. In the caves of Mt. Carmel and especially in those of Nahal Hame'arot ("The River of the Caves") some of its crucial stages have been documented by archeological findings. The significance of this coevolutionary process has been described in detail by Naveh (1984) and will be further updated in this essay.

4.2   Early phases of coevolution from the Middle Pleistocene onwards

According to Axelrod (1958, 1989), the evolution of the Mediterranean in general, including Mt. Carmel, has much in common with California. In both cases, fire and drought played apparently important evolutionary roles. Many wide-ranging wildfires caused by volcanic eruptions as well as by lightening have raged presumably throughout the Pleistocene and on Mt. Carmel also.

Presently, lightening out of a clear sky may occur in rare cases on several days in April-June causing wildfires. However in ancient times, such a fire on a dry day, if not put out immediately, could catch the undisturbed, dense, and highly inflammable woody, herbaceous vegetation and spread rapidly over vast areas. Such fires on wildlands and pastures were mentioned in the Bible in connection with lightening as the "fire of God" and "the heat of summer drought" in the Book of Job 9 (1:6). They also gained special symbolic importance for Mt. Carmel where the prophet Elijah fought the Baal prophets and "the fire of the Lord fell and consumed the burnt sacrifices, and the wood, and the stones, and the dust and licked up the water that was in the trench" (Kings 18, 38).

Such hot fires have probably destroyed most of the woody aboveground vegetation from time to time. Therefore, as explained in more detail elsewhere (Naveh, 1974), only those plants and animals with efficient adaptive resilience could survive such recurring fire stresses as well as increasing aridity. In California, Anderson (1956) showed that even at present times, fire-induced shrub-openings provide ideal opportunities for further speciation, hybridization, and genotype recombination of such woody fire-followers (or "pyrophytes").

The Levant apparently remains the only potential corridor for human migrations out of Africa. These were, as a rule, associated with biotic dispersal events, mainly large mammals. Tchernov et al. (1994) summarized recent findings and the Quarternary chronosequence and faunal remains of the earliest sites that mark the dispersal routes of *Homo erectus* into Eurasia in the Early Acheulian Bone Bearing Beds in northern Israel from Ubeidiya, in central Jordan Valley, near Lake Kinnereth, Gesher Benot Yaakov, south of the Hula Valley on both sides of the Jordan River, and at the Evron Quarry, on the coastal plains of the western Galilee, north of Haifa.

The Ubeidiya formation is among the oldest known stone industries, representing occurrences of early *Homo* outside Africa and the most intensively explored Lower Paleolithic site in Israel and the Levant. Its lithic assembly showing close affinity with Olduvai Upper Bed II dated recently 1.4 myr, in lacustrine and lava flows fluviatile deposits, which accumulated inside central Jordan Valley after the major tectonic activity, forming the

Jordan Rift Valley. Its formation, therefore, most probably marks one of the earliest stations on the route of human dispersal from Africa. The rich mammalian community consists of 45 genera originating from different biogeographically provinces from Africa and Eurasia, exhibiting the transition from early Lower Pleistocene to Middle-Upper Pleistocene fauna. According to palynological evidence for lacustrine sediment (Horowitz, 1979), its paleoenvironment included a hilly area covered with mixed Mediterranean forest with preponderance of sclerophyll phanerophytes such as *Quercus* and *Pistacia,* with indications of a climate that was more humid and cooler than the present semi-arid hot climate prevailing in the Jordan Valley.

The second site of Acheulian occupancy, Gesher Benot Yaakov, is of special significance for our discussion. Here, Stekelis (1960) found burned fibia fractures and within the bifacially worked artifacts, cleavers and hand axes made of basalt, dated younger than 0.800 myr. This is our first archeological evidence of fire not only in Israel, but according to Clark (1960) also of its use by *Homo erectus* in general. The third site of Early Paleolithic activity is at the Evron quarry with Acheulian deposits of flint artifacts and faunal artifacts dated older than 0.800 myr.

The mammalian assemblages, presented by Tchernov et al. (1994) of all these Lower Paleolithic sites, clearly show that already at that time the vegetation was exposed to foraging from a great diversity of ancient herbivores, ranging from elephants and rhinos to camels, gazelles, and gerbille rodents. However, we should not neglect the effect of human habitation and foraging on the surroundings and its vegetation and soil near their camp site and fire places, interacting with fire, grazing, and browsing for game and hunting.

Of special relevance in this respect are the remarks by Carl Sauer (1956, p.53), the eminent regional and cultural geographer, who can be regarded as the first American landscape ecologist (although he was not at all aware of such a term!). In his opening lecture of the now classical Chicago Symposium on "Man's Role in Changing the Face of the Earth" he stated: "The appearance or disappearance, increase or decrease of particular plants and animals may not spell out obligatory climatic change as has been so freely inferred... The intervention of man and animals has also occurred to disturb the balance... Dear thrive on browse; they increase wherever palatable twigs become abundant in brushlands and with young tree growth; ecological factors of disturbance other than climate may determine the food available to them and the numbers found in archeological remains".

The close interaction between the creation and maintenance of anthropogenic environments around human habitations, such as those of the Paleolithic Carmel G-H with their cultural traits, has been emphasized also by Rindos (1984) in his coevolutionary treatment of agriculture. He defined coevolution as an evolutionary process in which the establishment of a symbiotic relationship between organisms, increasing the fitness of all involved, brings about changes in the traits of the organisms.

Considering these interrelations with plants as the first step of "incidental domestication", Rindos (1984, p.137) stated: "the habitual destruction of preservation of species will have major effects on the floristic structure of the region, and eventually on the directions open in plant evolution. Such habitual activities, passed as a cultural trait, are inseparable from human language". Such interactions included destruction and preservation of plant species, opening of gaps in the closed vegetation canopy for food and fuel, the clearing of land for habitation and the trampling of paths, the digging for bulbs and burrowing animals, and the disposing of human kitchen waste. All these intentional and unintentional interferences opened new regeneration niches senso Grubb (1977) creating favorable conditions for herbaceous colonizers

The interferences of these widely scattered and scarce populations with the vegetation canopy, its litter, humus, and upper soil layer was only very slight and very patchy. However, these initial anthropogenic landscape modifications together with accidental forest gaps, opened by wildfires or by other catastrophic events, and later on fires intentionally set by humans, could have been further enhanced by positive feedback loops with fire, foraging and hunting, and their cross-catalytic network relations. Therefore, on the long range they could have carried far-reaching implications on plant evolution and vegetation dynamics and on the landscape as a whole. The richness of the Carmel vegetation in usable plants both for foraging humans and herbivores is clearly indicated in Table 1, showing all vascular species growing near one of the important archeological sites of Mt. Carmel.

Bar-Josef (1984) has already pointed out that tectonics and erosion have obliterated most of the direct archeological evidence in the Pleistocene. No ash deposits and sparse floral remains have been detected in open *in-situ* habitations. Even recent sophisticated flooding methods have not provided large samples of vegetal relics in shallow and eroded Mediterranean upland soils, especially terra rossa in which preservation is very poor. In the specific climatic and edaphic conditions of Mediterranean uplands most of the ashes of forest and brush fires are washed away by the first heavy rains, and remnants become intimately mixed with the thin upper layer of humus-rich terra rossa or rendzina soils (Naveh, 1973). We can, therefore, also hardly expect to find archeological evidence of such fires, especially since the Carmel slopes underwent severe geological erosion and the above described morphogenetic upheavals.

Table 1. Exploitable plant species of the surroundings of Nahal Sefunim determination for human consumption according to Dafni (1984, odified after Naveh, 1984 by Weinstein-Evron, 1998).

WOODY PLANTS					
Trees		**Shrubs**		**Dwarf shrubs**	
Arbutus andrachne	F W	*Calicotome villosa*	Fl Br W	*Cistus salviifolius*	L
Ceratonia siliqua	F! P! Br W	*Genista fasselata*	W	*Cistus creticus*	L
Cercis siliquastrum	Fl F Br W	*Pistacia lentiscus*	Br W	*Coridothymus capitatus*	L
Crataegus aronia	F Br W	*Rhamnus alaternus*	Br W	*Majorana syriaca*	L
Laurus nobilis	L Br W	*Rhamnus palaestina*	F Br W	*Melissa officinalis*	L
Olea europaea	F! Br W!	*Ruscus aculeatus*	L	*Micromeria fruticosa*	L
Phyllirea latifolia	Br W!	*Asparagus aphyllus*	Sh	*Salvia fruticosa*	L
Pinus halepensis	W	*Smilax aspera*	Sh	*Sarcopoterium spinosum*	Br W
Pistacia palaestina	F Br! W	*Tamus communis*	Sh	*Satureja thymbra*	L
Styrax officinalis	W			*Teucrium capitatum*	L
Quercus calliprinos	F Br! W				
HERBACEOUS					
PLANTS		**Legumes**		**Miscellaneous herbs**	
Geophytes	L	*Anthyllis tetraphylla*	P	*Alcea acaulis*	L S
Arisarum vulgare	L	*Coronilla cretica*	P	*Alcea setosa*	L S
Arum dioscioridis	B	*Hippocrepis unisiliquosa*	P	*Anagallis arvensis*	L
Asphodelus ramosus	B	*Hymenocarpos circinnatus*	S P	*Caspella bursa-pastoris*	L
Crocus hyemalis	L Fl	*Lathyrus blepharicarpus*	S P	*Convolvulus caelesyriacus*	P
Cyclamen persicum	B	*Lotus peregrinus*	P	*Daucus carota*	R
Ophrys umbilicata	B	*Medicago orbicularis*	P!	*Erodium gruinum*	P
Ophrys bornmuelleri	B	*Medicago scutellata*	P!	*Erodium moschatum*	P
Ophrys israelitica	B	*Medicago polymorpha*	P!	*Foeniculum vulgare*	L
Ophrys galilaea	B	*Onobrychis squarrosa*	P	*Geranium molle*	P
Ophrys transhyrcana	B	*Pisum elatius*	S L P	*Geranium purpureum*	P
Orchis caspia	B	*Scorpiurus muricatus*	P!	*Geranium rotundifolium*	P
Orchis galilaea	B	*Tetragonolobus paleastinus*	P!	*Isatis lusitanica*	L
Orchis tridentata	B	*Trifolium campestre*	P!	*Kicksia spuria*	P
Serapias levantina	B	*Trifolium clusii*	P	*Mandragora autumnalis*	F
Tulipa agenensis		*Trifolium clypeatum*	P!	*Nigella arvensis*	S
		Trifolium stellatum	P	*Papaver carmeli*	S
Grasses	S P	*Vicia hybrida*	S P!	*Plantago cretica*	L
Aegilops ovata	P			*Plantago afra*	L
Brachypodium	P!	**Asteraceae**		*Salvia hierosolymitana*	L
distachyon	P!	*Calendula arvensis*	L P	*Salvia pinnata*	L
Andropogon distachyus	P	*Carduus argentatus*	L	*Sanguisorba minor*	L P
Avena sterilis	P	*Carlina involucrata*	L	*Sinapsis arvensis*	L
Bromus alopecurus	P	*Catananche lutea*	P		
Bromus syriacus	P!	*Cichorium pumilum*	L S P		
Catapodium rigidum	S B P!	*Gundelia tournefortii*	C		
Dactylis glomerata	S P!	*Hedypnois cretica*	P		
Hordeum bulbosum	P	*Inula viscosa*	L C		
Hordeum spontaneum	P	*Notobasis syriaca*	P		
Hyparrhenia hirta	P	*Rhagadiolus stellatus*	P		
Lopochloa phleoides	P!	*Scorzonera papposa*	B P		
Phleum subulatum	P!	*Senecio vernalis*	P		

Piptatherum miliaceaum	P	Tolpis virgata	P		
Piptatherum blancheanum		Thrincia tuberosa	P		
Stipa bromoides					

F- fruits, S-seeds, B- bulbs, corms, etc., Fl - flowers, L- leaves, Sh- shoots, R- roots, C- capitulum. P- pasture for livestock and browsers (Br). W-wood. (!- high value).

Fortunately, Mt. Carmel is endowed by karstic caves and much of our contentions on this coevolution are supported chiefly by the archeological findings and sequences from the Paleolithic layers of Quaternary pollen spectra in the Tabun cave (Horowitz 1979) and those from the El-Wad cave (Weinstein-Evron, 1996). These and the Jamal cave are neighboring karstic caves, located about 20 km south of Haifa, on the slopes of the south bank of Nahal Me'arot at its outlet to the coastal plain. They have been formed by dissolution of the Cenomanian reef core in the early stages or the Middle Pleistocene during the Mindel and Riss glacial periods in Europe. Thanks to the findings in these caves the coevolutionary process can be traced back here to their Lower Paleolithic layers even if most traces of hearths have been erased by erosion and by changes in sea level followed by sedimentation.

Garrod and Bate (1937) were the first to describe the long and rich archeological sequence in Tabun, extending from the Lower Paleolithic layers to the Middle Paleolithic. In this cave they found direct evidence of human habitation in a nearly complete skeleton of a woman in the younger Mousterian layers (70,000-90,000 years ago) and identified as belonging to the Neanderthaloids of the Middle Pleistocene.

Of special importance our coevolutionary hypothesis and Rindo's are the non-arboreal pollen samples collected by Horowitz (1979) from the lowest Beds F of the Tabun cave. These differed greatly in composition from all the others, but could not be interpreted by possible climatic changes. This could have been, therefore, the first palynological indication of such paleolithic modifications of the natural pristine landscape ecotope into a subnatural one, and, as such, the first coevolutionary step towards the creation of the Lower Paleolithic "cultural" landscape. Many of these herbaceous plants and fire followers evolved during the Pleistocene as opportunistic ruderals, which could best take advantage of the improved light, fertility, and moisture regimes (Dimbleby, 1985). This could have been the case also with some of the plants found in these beds: In one bed, *Scabiosa prolifera* made up 50 percent of the pollen; in another, Compositae were dominant and in the third, Compositae and Graminae were dominant. Similar assemblages of these plants can be found today in close proximity to the Tabun cave and as further discussed below, in other nutrient-rich, ruderalic sites, benefiting also from recent fire events. Further evidence of pollen samples of such ruderalic plants has been provided by Weinstein-Evron (1996) from the epipaleolithic Early Natufian layers to which we will refer below. Most recently Tatskin et al. (1995) investigated wind-blown sedimentary fill of lithic assemblages in the Tabun and Jamal caves. They identified sedimentological and geochemical processes in the formation of sediments of complex sequences from the Lower Paleolithic Acheullean-Yabrudian culture assemblages, exhibiting numerous post-depositional biogeochemical changes, which occurred over a long time span. Some of these can be related to anthropogenic activity, but others could have occurred later, following ground water intrusion into the cave system. With the help of micro-morphological methods they could recognize anthropological sediments of charred organic material, incorporated into microaggregates, scattered pieces of charcoal 0.5-0.9 mm across. In the sediment of layer G of the Tabun cave they detected indications of anthropogenic burning by ferruginized chips of bones. These findings indicate human occupation and the use of fire from the Middle Pleistocene onward in Israel. Ikeya and Poulianos (1970) claimed that the first human "fire cultures" of hearths can be traced even farther back to about one million years in the lowest levels of the Petralona karstic limestone cave in northern Greece, resembling in many aspects our Carmel caves and their surrounding landscape.

We can further assume that more than any other fire feature, as the first extrasomatal energy source, affected not only their environment and shaped their landscapes, but also their life, behavior, and culture. This has already been claimed by Sauer (1956, p.54-55) in the above-mentioned lecture, in which he devoted much attention to fire. He stated:

"Speech, tools and fire are tripod of culture and have been so, we think, from the beginning... About the fireplace, social life took form, and the exchange of idea was fostered. The availability of fuel has been one of the main factors determining the location of clustered habitation".

He mentioned the various sources from woody plants available as fuel and the benefits for human and wildlife consumption from the lush protein rich post fire resprouting woody vegetation and the increase of seeds.

"On burned-over camp sites fire cleared away small and young growth, stimulated annual plants, aided in collecting and became elaborated in time into the fire drive, a formally organized procedure among the cultures of the Upper Paleolithic *"grande chasse"* and of their New World counterpart... Minor element in a natural flora, originally mainly confined to accidentally disturbed and exposed situations, such as windfalls and eroding slopes, have opened to them by recurring burning the chance to spread and multiply. In most cases the shift is from mesophytic to less exacting, more xeric, forms to those that do not acquire ample soil moisture and can tolerate at all times full exposure to sun... In areas controlled by customary burning, a near-ecological equilibrium may have attained, and a biotic recombination maintained by similarly repeated human intervention. This is not destructive exploitation".

Sauer (1961) also pointed out to volcanism in the eastern Mediterranean and the rift valley as the first source of such fires. These contentions by Sauer have been verified by Perles (1977) in her comprehensive review of the prehistoric use of fire, claiming that "Mankind could have involved into *Homo habilis* without fire but it would never have become *Homo sapiens* without fire." This important statement can now be further corroborated by even more substantial proof from both archeological findings on Mt. Carmel and elsewhere. These will be interpreted in light of the insights we gained from our fire ecology studies and the theoretical considerations on synthetic evolution and self-organization.

4.3   Major phases of coevolution in the Upper Pleistocene

4.3.1   The Middle and Upper Paleolithic people and the evolution of the fire-induced subnatural landscapes

According to Bar-Josef (1984) the above–described lithic Acheulian assemblages still show African similarities or even origin. It was only during the Early Upper Pleistocene that the special character of early Mediterranean and Levant Stone cultures emerged, exhibiting special adaptation to these environments. This period also marked the beginning of major coevolution during the gradual intensification of human activities and more sophisticated use of fire by Mousterians and their more sophisticated stone tool and hunting techniques.

The gradual intensification of human interferences was accompanied most probably by a more efficient use of fire by Mousterian gatherers-hunters. According to Perles (1977), mastering fire was adapted about 100000 years ago by these Neanderthaloids who produced lamps to light their caves and torches to carry fire. They could open dense forest and bush thickets to facilitate hunting and food collecting and increase edible food by encouraging the lush regeneration or trees and shrubs, invading grasses, bulbs, and tuberous plants.

That these people had reached a high intellectual level is inferred by their mortuary practices. As reported by Solecki (1977) at the Shanidar cave in the Kurdish mountains, seeds of flowering plants, which grew outside the cave, were found in soil samples taken from around the burials for 'Neanderthaloid IV'. The findings of such plant species, known for their medical and ornamental values can serve as further proof of the modification of the natural vegetation and by paleolithic gatherers and their role in spreading herbaceous plants.

Evidence for the use of fire by Mediterranean Mousterian cultures has been provided also at the Kasitstra caves near Lake Ionina in Greece (Higgs et al., 1967). From this period Vernet (1973) reported findings of charcoal specimens of sclerophylls and phanerophytes such as *Phyllirea* and *Quercus* in southeastern France. From this period onward, the Tabun cave bears provided clear evidence of human use of fire by reddened earth and mixed ashes from hearths. According to Jelinek (1981), the upper Mousterian layers indicate even repeated burning of the whole cave surface, reminiscent of the practices of Australian aborigines and Bushmen for clearing their caves by fire. Ronen (personal communication) speculated that this could have been caused also by accumulation of wind-blown fine ash deposits, originating from the woody vegetation canopy surrounding the cave. It seems, therefore, reasonable to assume that from the early phases of the Late Acheulian and Middle Paleolithic cultures onward, fire became a major driving force in the mutual-causal feedback relations between these Mediterranean ancient people and their environment.

At the same time, we now have a much firmer basis corroborating the claims by Sauer of the cultural importance of human-set fires for food collection and hunting. After observing the beneficial effect of wildfires they could have realized that it could serve not only as the major energy source for heating and cooking but also for opening dense forests and brush thickets. As a result, intentional set spotty fires could have also been used to create more accessible and richer ecotones for food collecting and hunting and to increase edible food by encouraging the lush regeneration of woody plants and invading herbaceous plants. That this has been most probably the case can be inferred from the results of the fire ecology studies in the western Galilee (Naveh, 1960, 1973, 1974) and in the Carmel region (Naveh, 1999), which will be further reported in subchapter VIII.

In addition to other more or less catastrophic natural perturbations, increasing drought and fire caused by volcanic activities and by lightening acted as a strong selection force. As explained in detail by Naveh (1975), those woody and herbaceous genotypes, which developed the most efficient physiological and morphological evolutionary strategies for active and passive vegetative and reproductive regeneration mechanisms, had the best chances to overcome natural and human-induced fire stresses.

The unique combination of natural raging wildfires and human-set fires became, therefore, major landscape shaping factors, converting them gradually from natural, pristine ones to subnatural and seminatural "cultural" landscapes. As will be shown below, this process was further intensified from the Neolithic agricultural revolution onward, in which many of these seminatural biosphere landscapes were converted into agropastoral landscapes. This occurred most probably during the drier, warmer interpluvials in which the Mediterranean climate patterns became established. It created favorable conditions for the germination of light demanding woody plants, such as *Pinus halepensis* and most chamaephytes as well as for the above-mentioned herbaceous plants facilitating their spread over vast areas.

4.3.2   The epipaleolithic Natufians and the evolution of the seminatural landscape of their Total Human Landscape

During the last Pluvial (10-15,000 years ago), the fire-induced landscape modification reached its peak due to prospering and culturally advanced Epipaleolithic economies, such as the Natufians of Mt. Carmel. This period has been described by Braidwood (1967) as the final transitional stage from the more primitive economy of food gathering into the more advanced, better equipped, better organized, more purposeful, and specialized food collecting. As such, a crucial cultural and socio-economic threshold, especially in the later part of the Natufian epipaleolithic culture, has come to be in the last 30 years, one of the major foci of Levantine prehistoric research, with Mt. Carmel as one of its most important centers.

Bar-Josef (1983, 1998) has summarized and discussed comprehensively the major archeological findings and their meaning for the life, culture and economy of the Natufians. Weinstein-Evron (1998) has provided a very thorough account of all the archeological and geophysical studies carried out in the El-Wad cave, in relation to the Early Natufians from the first excavations by Garrod (1929) until her own recent palynological research and her conclusions on their use of the site and environment.

This was apparently the oldest occurrence of Early Natufians in northern Israel, and elsewhere. They occupied this cave and its terrace according to radiocarbon dated charcoal samples from around 12,940 B.P. until around 10,680 BP. This period coincides largely with the European Late Glacial and in spite of the climatic fluctuations between an earlier colder and drier period, followed by more humid conditions and by a colder interval during the young Dryas (ca 1,100-10,000), followed again by a milder climate the typical Mediterranean geographical pattern and the winter rainfall in the Late Pleistocene in the southern Levant was well established. As will be shown below, this has been corroborated for Mt. Carmel and its surroundings also by the palynological studies of Weinstein-Evron and her coworkers.

Bar-Josef (1983, 1998) has presented an impressive picture of the Natufians sophisticated technologies and subsistence strategies for a rather intensive, broad-spectrum utilization of the fauna and flora occurring in typical Mediterranean landscape mosaics in the ecotones of the mountain foothill, coastal plains and river valleys. Living in small, mobile bands and as semi-sedentary groups in hamlets, they became an example of the archeological expression of intensive gathering, hunting and fishing, along with a number of agricultural pre-adaptations, reflected in certain utensils and installations. "The social evolution that occurred in parts of these 'Proto-Mediterranean' populations seems to be an autochthonic change, resulting in differentiation between the human occupations of the Mediterranean and the Irano-Turanian zones and those of the higher mountain and deserts. This differentiation is reinforced in the following millennia and appears to be one of the basic cultural components of the Levant" (Bar-Josef, 1983; 28). Such Proto-Mediterranean populations showed some similarities to Mousterian ancestors, as well as to the scarce remains from the Upper Paleolithic and earlier Epipaleolithic remains. The Natufians were short to medium in stature and their mean age of death was about 30 years, and maximum life span may have been around 50 years. Bar Josef (1983) suggested, that the Natufians lived - like hunter-gatherers of today - **under** the carrying capacity of their environment. If we estimate their population around 500 people along the Carmel coast and its western slopes and wadis, they may have reached densities of about 4 persons/km^2. According to Baumhoff (1963), these are the densities of the coastal California Indians, to which we will refer further below.

Up to now, no other findings outside the Levant have yet posed any indications of a paleolithic entity resembling the Natufians, and probably no other cultural remains from this period have uncovered such a wealth of information providing the basis for a better construction of pre-agricultural living and socio-economic systems. Weinstein-Evron (1998) described in detail all archeological findings in El-Wad cave and the adjacent terrace connected with the organization of their living place. They even constructed houses and thus developed a complex and rich communal, cultural and spiritual life. The layout of this Natufian settlement included a cemetery and a special dumping ground for waste, indicating the high level of social organization achieved already by these early Natufians (Fig. 1).

In this cave they found a rich assemblage of lithics, ground stone and bone implements and numerous human burials, and what Weinstein-Evron and Belfer-Cohen (1993) called "a burst of artistic creativity of artistic objects and decorative items", such as figurines, beads and pendants, as well as decorated sickle hafts. All these findings point also to the advanced hunting, food collecting and preparing technologies, such as the use of archery for hunting and of flint sickles to cut wild grasses and mortars and pestles in the preparation of staple food from roasted cereals and acorns. This roasting of cereals preceded that of the later Neolithic cultures. Their bone industry is far richer in quantity and contains more elaborate, varied morphologies than does any other earlier or later Levantine archeological entity. Many objects bear specific decoration such as the carved shafts from el-Wad and the Kebara caves with young ruminants at the edge.

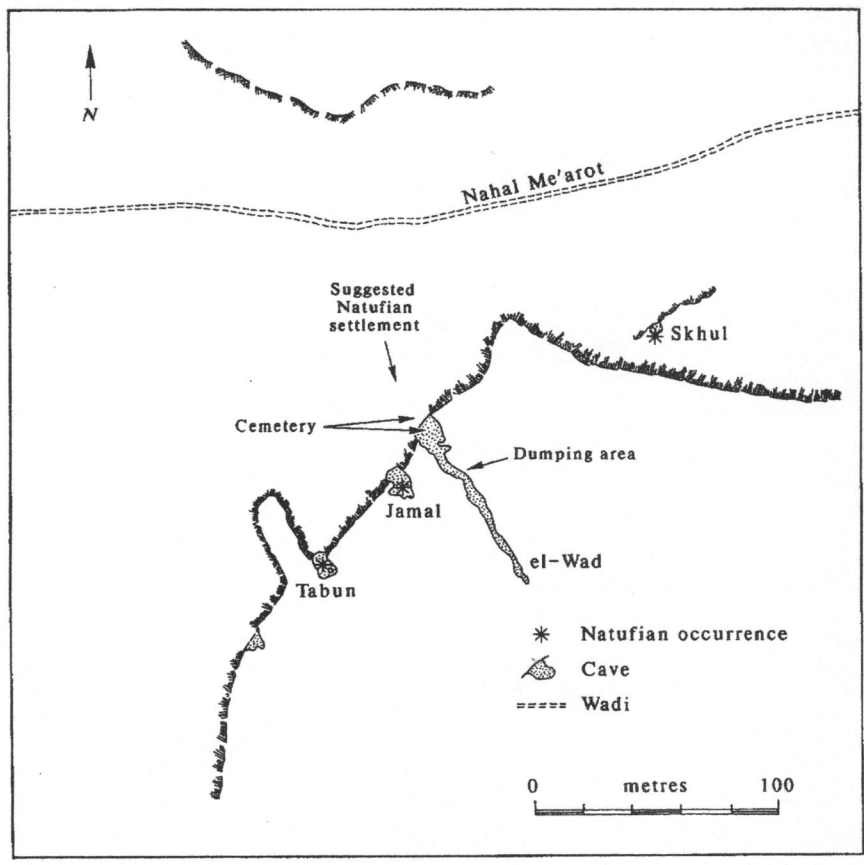

Figure 1. A suggested layout for the Natufian site at El-Wad (Weinstein-Evron, 1998)

Marine shells for jewelry were collected from the shore of the Mediterranean Sea, or more rarely, were brought from the Red Sea. Of special significance for their close symbiotic relations with nature is a horn core with a man's head at one end and a bovid's head at the other end. Red ochre fragments composed of hematite were also found on pestles in this cave, which were used for grinding of both yellow goethite and red hematite rocks. These pigments could have been produced by grinding natural hematite iron oxide extracted from the veins of rock outcrops or alluvium clays or by heating the more common goethite. Such red colors also appear on burned iron oxide, containing limestone and dolomites of volcanic outcrops of Mt. Carmel after hot brush burning with which the Natufians were most probably well acquainted.

The frequency of such volcanic outcrops on Mt. Carmel, shown in Fig. 2, can serve as a good indication for the occurrence of many hot wildfires during volcanic outbreaks in the Pleistocene in addition to those wildfires caused most probably by lightening.

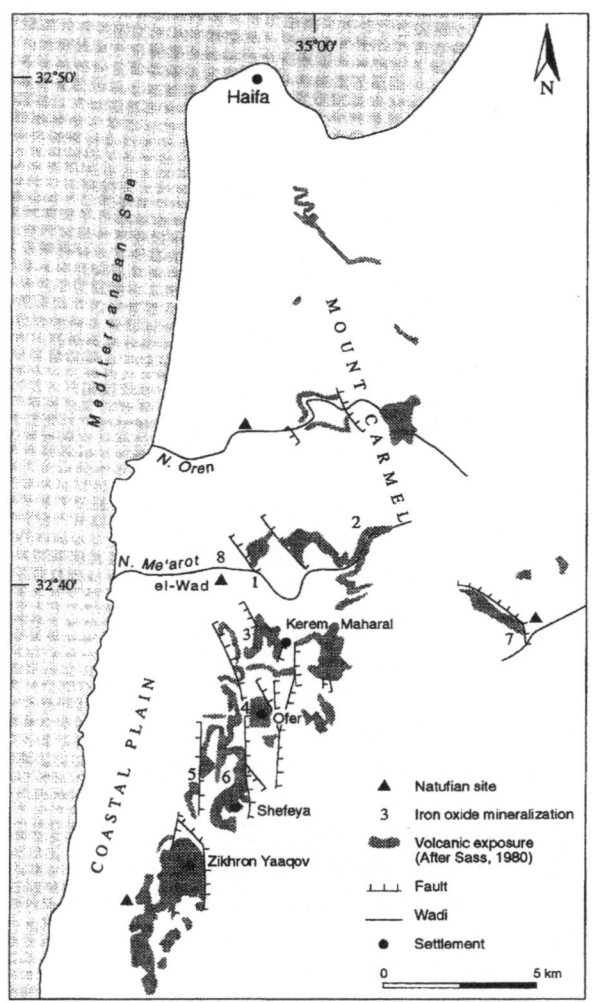

Figure 2. Location map of the volcanic outcrops and iron mineralization sites with Natufian sites of Mt. Carmel (Weinstein-Evron, 1998)

Of great significance for our contentions on the impacts of these Natufians on the vegetation are the palynological findings from el-Wad. As shown in Table 2, these pollen examples are dominated by ruderalic, insect-pollinated and many clusters of pollen, indicating human habitation and activity.

Table 2. The Natufian pollen spectra from El-Wad (Weinstein-Evron, 1998)

Pollen Type	Sample I40c/125	Sample I42b/129	Sample I40a/143
Quercus	10.5	12.8	7.1
Pinus	10.5	12.6	5.5
Olea europaea	-	0.8	2.2
Pistacia	3.5	1.4	4.4
Acer	1.0	-	0.3
Arbutus	0.6	-	1.4
Crataegus	12.1	1.2	10.4
Rhamnus	-	0.6	-
Ceratonia siliqua	7.3	0.4	8.2
Styrax	0.6	-	-
Myrtus	13.7	3.3	8.2
Total AP	**59.8**	**33.1**	**47.7**
Gramineae	3.5	13.0	4.6
Compositae	4.8	3.1	8.7
Centaurea	-	-	0.5
Chenopodiaceae	1.0	0.2	1.1
Umbelliferae	2.9	5.1	8.5
Plantago	0.3	0.4	-
Ephedra	-	0.4	0.3
Malvaceae	-	4.3	-
Polygonaceae	3.5	2.5	4.1
Cruciferae	-	1.4	0.3
Dipsacaceae	0.6	14.0	4.9
Liliaceae	1.0	-	-
Asphodelus microcarpus	-	2.9	-
Papilionaceae	8.9	5.1	7.9
Labiatae	7.1	6.0	5.5
Cucurbitaceae	-	1.0	-
Sarcopoterium spinosum	0.3	1.6	0.3
Caryophyllaceae	-	-	0.3
Rubiaceae	-	-	0.5
Euphorbiaceae	2.6	2.3	1.9
Convolvulaceae	1.0	0.6	0.3
Cistaceae	0.3	2.7	1.4
Primulaceae	1.0	-	-
Ranunculaceae	1.3	-	0.8
Rutaceae	0.3	-	-
Capparis	-	-	0.5
Total Counted	**314**	**486**	**366**
Clusters (No)	**1**	**-**	**76**
Tamarix	-	-	9
Olea europaea	-	-	14
Hydrophilous (No) *Sparganium* Cyperaceae	**-**	**1**	**1**

In this cave, the presence of Mediterranean sclerophylls was also indicated by the earlier findings of Bankroft (1937) of a few pieces of wood charcoal of *Quercus* sp. and *Olea europea* in Upper Paleolithic layers of the El-Wad cave. More recently, Lev-Yadun and Weinstein-Evron (1994) identified 32 pieces, 5-10 cm long of Early Natufian Epipaleolithic wood charcoal of *Quercus calliprinos, Q. ithaburensis, Myrtus communis Cupressus sempervirens, Salix* sp., and *Tamarix* sp.

According to Weinstein-Evron (1998), these findings provide further proof that the Natufians in the Carmel region lived in a typical Mediterranean landscape, representing 3 major habitats: (1) forests and maquis, (2) marshes and other wet land habitats and (3) in disturbed ecotopes, probably in the immediate surroundings, represented by ruderals such as Dipsaceae and Malvaceae. The investigator speculated that the mountains near the cave were densely covered by *Q. calliprinos* dominated maquis, and as exists today, the drier slopes by more open woodland formations of Ceratonia siliqua-Pistacia lentiscus and of the Tabor oak *Q. ithaburensis*. Pods of the carob trees *Ceratonia siliqua* and acorns of the Tabor oak were used as staple food, even until the turn of the last century, whereas *Q. calliprinos* may have been preferred as fuel and for building purposes and the myrtle Myrtus communis may have been used for its therapeutic and aromatic characteristics or even as a ritual plant.

Overrepresentation of gazelles (*Gazella gazelle*) among the faunal assemblages is generally seen as an indication of the exploitation of local game fauna from the more open, savanna woodlands and grassy sites in the coastal plains, but in our opinion also from recently burned wooded landscapes, *Dama mesopotamica, Capreolus capreolus,* and *Cervus elaphus* were probably hunted in the more densely wooded maquis and forests of Mt. Carmel and the coastal plain. In fact, Weinstein-Evron (1998) mentioned that *Microtus guentheri* and *Spalax ehrenbergi* were among the rodents found, suggesting the existence of stands of grasslands around the site, "resulting from human activity in the immediate vicinity of El-Wad".

The high incidence of young gazelles in several Natufian sites on Mt. Carmel may be interpreted as an attempt at a kind of "incipient domestication" although the gazelles' behavior seems not to be suitable for this. However, with the help of repeated burning, they could have created grassy pasture enclosures, surrounded by shrubland or forest to attract gazelles but possibly also to "semi-tame" their kids. Bar-Josef (1983) even speculated whether the sickles were used to cut grass for feeding gazelles.

Tchernov (1975) interpreted the appearance of typical desert rock dweller rodents such as *Acomys russatus, A. cahirinis,* and *Gerbillus dasyuru* as indicators of a principal climatic trend of desiccation during the Late Pleistocene. On the other hand Bottema and van Zeist (1981) suggested that the desiccation from this period onward may not necessarily have been climatically determined, but caused by the increasing human interference with the natural vegetation. The term "desiccations" should be, therefore, only interpreted as a reduction in the woody cover at the expense of the increase of a more open, herbaceous vegetation cover. This also could be a much more plausible explanation for this rodent invasion into the newly created more open and rocky sites, resulting from the increasing burning activities which most probably also encouraged the invasion and colonization of the rock outcrops of the more xeric *Quercus ithaburensis* woodlands by herbaceous plants or their more drought tolerant genotypes – such as those of *Hordeum spontaneum* (Nevo et al., 1986). These assumptions are fully supported by the contentions of the prominent Italian botanist and ecologist, Pignatti (1983) on the important human role not only in the stimulation of the evolution of the flora, but also in the evolution of new habitats in the first seminatural cultural landscapes created in the Mediterranean basin.

In his paleo-ecological interpretations of faunal remains from Mt. Carmel, Tchernov (1984) described these landscapes as "a kind of constant balance between open country and woodland". It is very possible that on the rocky and less fertile terra rossa and light rendzina soils, they were most probably dominated by sclerophyll trees and shrubs with a fire-induced and rapidly spreading light demanding herbaceous understory, maintained by rotational burning. These grasses became dominant in the fertile brown rendzina soils, and others such as *Pragmites* and *Festuca* were dominant in the wetlands. The denser and more productive stands could have been the most preferable habitats for the cutting of grasses and collecting their seeds. The latter could have been accomplished also by collecting their scorched seed dispersal units directly from the ground after burning their dry stands. According to Harlan (1967), these primitive glumed cereals needed to be parched before they could be thrashed and winnowed.

In conclusion, through intensive, vegetation management the Natufians created the first proto-agricultural seminatural Total Human Landscape. This was the most efficient way for channeling high-quality chemical energy of the natural, spontaneous-occurring wild plants and herbivores into human food production on a sustainable basis.

4.3.3   A comparison of the Natufians with Californian Pre-European Indians

One of the major disadvantages in comparing prehistoric G-H with contemporary ones is that the latter cannot be considered true pre- or proto-agriculturalists in their evolutionary development. These are only non-agriculturalists out of constrains, living in marginal environments for farming, such as arid lands or dense tropical forests. Being in contact, one way or another, with modern societies, they are not only exposed to infectious diseases, but they are also culturally "contaminated" by their tools, weapons, and information. This was not the case with the Californian Indians before their closer contacts with the Spanish missioners before 1769. Although living many thousands of years later in time, we can assume that they had reached about the same cultural evolutionary stage as that of the highly developed, sophisticated, pre-agricultural, subsistence economy - the Natufians. Like the Natufians, the Californian Indians had lived in close symbiotic relations with nature making efficient use of resources without destroying them.

Diamond (2001) maintained that the reasons why agriculture (and all following cultural and socio-economic developments) evolved in the Fertile Crescent – including the Levant and Israel some 10,000 years ago - was not because of any inherent intellectual superiority of the inhabitants over the Indians, living in comparable climatic conditions in California. This was chiefly the advantage of a larger biotic diversity and especially of a much higher diversity of wild plants – especially larger seeds of grasses (as well as legumes!) and suitable animal species for domestication.

Thanks to the extensive studies by archeologists, anthropologists, and ethnologists for many years under the leadership of Alfred Kroeber from the University of Berkeley, a plethora of information has become available on the life of these pre-European Indians lending strong support to our contentions. Much of this has been summarized by Heizer and Elsasser (1980) in a very readable and well-illustrated way.

The oldest evidence of human presence in California dates back 9-10,000 years and through time, until the arrival of the first Whites in the sixteenths century, a dense network of more than a hundred Indian tribes and tribuletes spread over the whole state. For our comparison, of greatest interest are those tribes living in ecologically comparable conditions to the Natufians in the ecotones between the coast, the foothills, and the coastal mountain. Such ecological equivalent landscapes stretch from the Carmel to the Santa Barbara regions. These were occupied chiefly by the Esselen, the Salinan, and farther south by the Chumash. As fishermen, gatherers, and hunters those closest to the coast and the coastal foothills and mountains enjoyed the richest and greatest variety of food. Their diet included salmon and many other fishes and sea food, many edible parts of plants with acorns of the seven oak species occurring in Central California, serving as a major staple food and stored in granaries even for two years. They sometimes mixed the acorn meat with edible grass seeds of wild ryes (*Elymus* spp.), and other grasses. In the vicinities of the Franciscan missions they also used wild oats (*Avena fatua*) which was accidentally introduced with wheat seeds from Europe and rapidly invaded the grassland, thanks to its pre-adaptation to similar conditions. Among the game species deer, elk, antelopes, and rabbits as well as waterfowl were also abundant. In this way, a highly versatile and healthy year-round diet was ensured. This helped them to overcome drought periods and famine better than populations of the drier inlands valleys and the South, and enabled them to reach the highest population densities of any North American Indians (Baumhoff, 1963). As reported by Heizer and Elsasser (1980), many more plants were used for medical or other purposes. Their material wealth was reflected in the elaborate round houses with wooden framing of their well-organized villages with several hundred inhabitants, in their spectacular religious cult ceremonies, and the great efforts and skill invested in the ornamentation of their tools and especially their baskets.

According to Heizer and Elsasser (1980), the Californian Indians acted as important agents for the dispersal of certain grasses, berry-producing trees, and shrubs. So far as is known from the archeological record, they never over hunted any animal to the point of extinction, and they did not seem to have affected the overall distribution of game animals and birds. As highly accomplished practical botanists and zoologists, they were also knowledgeable in understanding nature in such a manner as to use it without destroying it – that means in a most sustainable manner. Therefore, as land managers these Indians were in some ways far ahead of us today. Heizer and Elasser (1980) described their elaborate ways of making fire by various types of fire "drills" and many other uses of fire. These included the cutting down of trees, the leaching out of the tannin of acorns with hot water, and the catching of grasshoppers by firing a circular area and then roasting them in a kind of earth oven with hot rocks.

However, most relevant for our comparison is their sophisticated use of intentionally set fires, as described by these authors and in more detail also by Lewis (1973). In order to improve visibility for hunting, the grazing conditions for deer, and to increase the yields of wild grasses, fruits and berries harvest, and to uncover acorns and nuts, they set cooler fires in shorter cycles and on smaller patches than the natural wildfires. Thus, they created heterogeneous mosaics of open forests and woodlands in different regeneration stages, dominated, to a great extent as in the Mediterranean, by sclerophyll phanerophytes with a rich herbaceous understory. It resulted in the establishment of what Lewis described as "a dynamic balance of natural forces". This is comparable to the fire-induced homeorhetic flow equilibrium of the seminatural Epipaleolithic landscapes.

Lewis (1985) carried out extensive anthropological and ecological research on the "pyrotechnics" by native North Americans and Aborigines in North Australia, who still use fire to increase their plant and animal food resources. He concluded that the general aim of such intentional use of fire in some areas, but excluding it from others is to enhance and maintain an overall fire mosaic. This means a complex, more productive, and stable environment than what would have been derived from natural fires in terms of seasonality, frequency, and intensity.

He rightly emphasized that habitat burning was but one component in the total system of foraging adaptation. According to Bean and Lawton (1973), these "semi-agricultural Indians of the central California coast reached a very 'efficient' interlocking of energy extraction processes". They suggested that Lewis' findings could provide new perspectives and ideas for other cultural evolutionists and ecologists showing how high levels of cultural integration and adaptation could have been reached by hunting, fishing, and food gathering communities in intensive proto- and semi-agricultural utilization of their natural resources. This may also be true for the advanced Epipaleolithic Natufians.

Heizer and Elsasser (1980, 220) concluded that we could see in these Indians the "true ecological man who was truly a part of the land and the water and the mountains and the valleys in which they lived. The environmentalists and conservationists of today feel a kinship with the Indians in their respect for nature, a feeling which at times rises to that of the sanctity of the natural world."

4.3.4   Some lessons from this ecological and functional analogy

This comparison lends strong support for the need in replacing the misconception on the passive environmental role of the paleolithic pre-agricultural foragers with the recognition of their active role and its far-reaching implications. The Coastal Californian Indians could also give us some cultural clues on how the pre-agricultural Natufians could reach such a peak in their coevolution with their landscapes. There is now ample archeological evidence that the Natufians developed a complex and rich communal, cultural, and spiritual life, based on an advanced, intensive, broad-spectrum, metastable, proto-agricultural food collecting-hunting-fishing economy. This resembles in many ways that of the Californian Coastal Indians at the time of their first close contacts with Europeans. Presumably, in the gradual conversion of the Carmel slopes and the open coastal plains and hills into a mosaic of seminatural ecotopes, driven by auto- and cross-catalytic feedback loops, both natural and intentionally set fire, operated also as what Renfrew (1979) defined as a "multiplier effect". It induced these mutually beneficial couplings of the Natufians with their landscapes through the conversion of denser forests and maquis into more open, diversified and fine-grained vegetation patterns, richer in herbaceous fire followers. A similar process of landscape conversion has apparently taken place along the Central Coast of California. In both cases their close symbiotic relations with nature ensured their sustainable landscape management during which fire was incorporated as a larger-scale cultural –

that means anthropogenic - perturbation factor, closely coupled with humans and wild animals foraging. In both cases, thanks to longer- as well as shorter-term rotationally shifting defoliation pressures, a homeorhetic flow equilibrium between the woody and herbaceous vegetation layers and their postfire regeneration stages could have been maintained.

4.4 The final phase of coevolution during the Early Neolithic period

Thanks to the extensive studies in Israel by such outstanding geneticists, evolutionists, and paleoecologists as Zohary, Nevo, Kislev, and their assistants a very sound scientific foundation has been provided for the actual evolutionary process of domestication in Israel and the Fertile Crescent. However, the actual causes for this transition from intensive food collection to food production are still a matter of great controversy.

Rejecting exclusively deterministic climatic or demographic explanations, Rindos (1984) has described this as the final evolutionary stage of "specialized domestication of plants and animals" by reciprocal adaptation and coevolution. Broadening the ecological and cultural scales of these symbiotic relationships we can regard them as the culmination of the self-organizing process operating in such human–landscape coevolution. In this process, new evolutionary patterns of greater efficiency in solar energy channeling from selected plants on restricted areas of these biosphere landscapes into human food production and livelihood was favored and, therefore, succeeded in gradually replacing the former patterns of gathering-hunting. This does not exclude the possibility that dramatic ecological changes such as turbulent climate changes and the extinction of big herbivores, together with increasing demographic pressures, could have caused acute crisis situations. Some or all of these factors have apparently induced the breakdown of the economic and cultural patterns, forcing the adaptation of new technologies and new ways of life, for which the above described inventions were most suitable.

Diamond (2001, p.111) has rightly pointed out that the prerequisites for planting cereals as crops were technological innovations and their adoption by the Epipaleolithic Natufians for the exploitation of wild cereals. These included "sickles of flint blades cemented unto wooden or bone handles, for harvesting wild grains; baskets in which to carry the grains home from the hillsides, where they grew; mortars and pestles, so that they could be stored without sprouting; and underground storage pits, some of them plastered to make them waterproof". But it was also preconditioned by the greater abundance of these cereals, for which the earlier technological innovation of controlled-burning, together with rational and judicious vegetation management was at least partly responsible. This, in turn, was the result of the close cognitive and spiritual bonds of the Natufians with their natural surroundings. In this subchapter we will show that fire played a more important role in this Neolithic revolution.

Lewis (1972) has claimed that fire has played an important role in a comprehensive review essay. He also believed that in these earliest phases of the broad spectrum revolution in the oak-pistachio area, the intensification and seasonal extension of man-made fires, coupled with increasing grazing (of either wild or domesticated animals), could have provided the necessary 'shock stimulus' leading to the emergence of agriculture. On the basis of the findings in the Shanidar cave in northern Iraq at the ecotones between the Mediterranean oak-pistachio woodland belt and the semi-arid Assyrian steppe Lewis (1972) speculated that the sudden appearance of numerous larger-seeded Ceralia type grasses at the Proto-Neolithic period, together with pollen grains, which appear to be clearly domesticated, have been caused by increasing human activities and was directly correlated to the reduction of trees and to the findings of extensive, multicolored dry and dusty ash beds in these caves. As mentioned above, according to Ronen such ash deposit findings by Jelinek (1981) in the Tabun cave of Mt. Carmel could have been caused also by accumulation of wind-blown fine ash deposits, originating from the woody vegetation canopy surrounding the cave. Lewis (1972, p.209) reported that "in burns of grasslands and brushlands great amounts of fine ash become airborne which, in the less disturbed air of a cave, would have settled like ordinary house dust".

Kislev (1984, pp.62-63) described this early stage of the emergence of cereal agriculture as "the agrotechnical revolution", initiated by the Natufians with their intensive collection and consumption of seeds including wild diploid tetraploid wheat and barley. He considered the invention of the sickle as a major component of this agrotechnical revolution. It was twice as efficient as manual reaping and allowed the grain to ripen out fully before gathering without a loss of yield. He assumed that in the next stage of the "domestication revolution", the invention of fields by Neolithic people was inspired by the observation of indigenous natural grasslands of annuals, such as wild emmer and barley occurring on open forest belts of deciduous oaks. He further speculated that they may have burned off unwanted grasses and used the cleared space for sowing grain" and then, later on they also cleared dense woods of evergreen sclerophyll low trees and shrubs. This transformation of the landscape from evergreen vegetation to fields with annual winter grasses started in the prepottery Neolithic from 8500-7600 B.C. We have already explained that this transformation was made possible by the use of repeated fires, inducing the post fire grass flush.

However, our fire ecology studies have shown that burning such wild grass stands does not prevent their germination in the following rain season and may even stimulate it (Naveh, 1973), preventing the germination and development of sown grasses; a very different situation is created by the burning of denser tree and shrub stands. The striking effect of their ash seed beds on the increase in grain yield, as a result of the rise in soil fertility (reported below in more detail), could have acted as a major cultural trigger and multiplier effect for initiating the cultivation of these cereals in the early Holocene on Mt. Carmel and elsewhere. As indicated by the demonstrations of Iversen (1971), flint axes could be used for the felling of well-developed oaks and other tall trees in favorable sites, burning them and exploiting their ash seedbeds for the cultivation of these cereals. That this could have been also the case on the deeper and more fertile soils of the broader riverbeds, wadis, and terraces on Mt. Carmel is implied by the findings of the great assemblages of such flint axes in the Sefunim cave by Ronen (1984). Such slash-burn rotations have been repeated several thousand years later by the Neolithic farmers in Europe (Narr, 1956). Archeological evidence for such Neolithic land clearing is indicated also by paleobotanical findings in west Mediterranean *Quercus pubescens* forests by Pons and Quezel (1985). They showed that in southern France, in the Rhone Valley, early slash-and-burn agriculture is suggested by charcoal dated 7350 B.C. There was a simultaneous decrease in the percentage of deciduous oak pollen, and a higher percentage of Labiatae and Leguminosae, *Plantago*, Compositae and other species considered to be weeds of cereal cultivation. Cereal pollen appeared at the same time.

According to Van Andel and Runnels (1995), the first European farmers established themselves in the empty plains of the Thessaly some 9000 years ago. The may have flourished there thanks to the natural irrigation of river and lake flood plains, providing favorable soil conditions for cultivation. After more than a thousand years, they spread to the Balkans and beyond and this may lead to a modified version of the "wave-of-advance model" of the diffusion of agriculture from the Fertile Crescent and the Levant, presumably driven by demographic pressures only.

There is little botanical evidence for major exploitation of the terrace foothills in Greece before the Bronze Age. However, any clearing of the willow and alder cover of these flood plains would have been poorly reflected in the pollen record (Van Andel and Runnels, 1995). Therefore, even if there is no evidence of human occupation from Mesolithic times like on Mt. Carmel and in Israel and in other locations of the Levant; we should not exclude the possibility that also here, such a slash-burn system was introduced as the first step towards crop cultivation together with the first wave of agricultural diffusion.

One possibility is that such cleared and burned forest fields, wherever established, served as "experimental" sites for many species and that those most suitable were further domesticated, while others became facultative weeds, adopting their present dual position as part of both seminatural and agricultural vegetation. Such newly open niches could have favored the emergence of obligatory weeds as an evolutionary "side product" of cultivation and domestication. All these plants together with the post-harvest remnants of the cereals and pulses served as the most valuable pasture plants for the grazing livestock whose domestication was also initiated at the Early Pottery Neolithic. The faunal remains of Mt. Carmel from this period at the Sefunim cave already contained domesticated goats (Ronen, 1984). Goats were apparently much easier to domesticate than gazelles. They are most efficient ungulates for converting the primary production of woody and highly lignified plants, together with the fire-stimulated herbaceous vegetation of the rocky Carmel uplands into animal products.

Figure 3 can give us a clue of the great agro-pastoral potentials of the el-Wad surroundings: In addition to the cultivated fields and the uplands, which could be utilized chiefly in the rain season, also the swamp and marshlands and their fire-stimulated summer-green perennial grasses and legumes could provide highly nutritious fodder. This increase in food production potentials may explain a general increase in population densities by more than five times, as estimated by Hassan (1981) after the completion of domestication around 7,000 BC.

The reduction of the earlier Neolithic mixed, broad-spectrum food collection and production economies into specialist agro-pastoral economies is indicated by late Neolithic and Chalcolithic settlement concentrations, adjacent to their field and lowland pastures on Mt. Carmel (Naveh, 1984).

In conclusion, during the final stages of the coevolution leading to the domestication of agricultural plants and animals, flint axes and prescribed burning have most probably played an important role in the conversion of the natural and seminatural forest and woodland ecotopes into cultivated and grazed agro-ecotopes. It was the beginning of a major step in the evolution of the cultural landscape of Mt. Carmel. However, by the creation of their agro-pastoral Total Human Ecosystem, the early Holocene farmer and livestock breeder moved from the Paleolithic coevolutionary reciprocal relationship with the Carmel landscape towards a more one-sided exploitation and domination of the newly created Agropastoral Human Landscape. He became the chief controlling agent, whose land use practices changed all depending ecosystem variables of vegetation and soil of the state factor equation senso Jenny (1961).

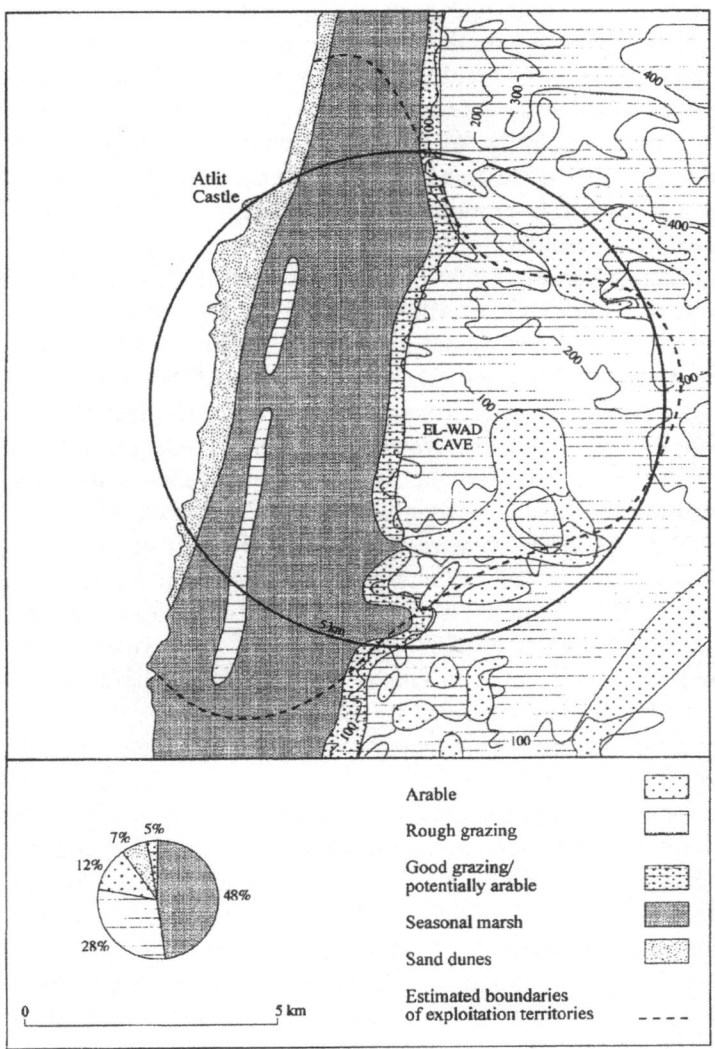

Figure 3. Territorial analysis of El-Wad, within a radius of 5 km (Weinstein-Evron, 1998)

5.   THE AGROPASTORAL TOTAL HUMAN LANDSCAPE OF THE HOLOCENE

In his insightful account of the Neolithic revolution, the German landscape historian Sieferle (1997) rightly emphasized that this was not simply a transformation of a long-lasting stable situation of a pristine natural landscape into a cultural landscape. It disrupted a dynamic process of the gradual formation of seminatural landscapes, described here as a coevolutionary process. Therefore, all conservation and restoration theories and practices aimed at the return of such natural landscapes and its assumed climax vegetation are futile. We can only attempt to conserve and restore the most valuable parts of the cultural agro-pastoral landscapes, which evolved and remained as a side-product of the agricultural economy and life style until the early last century.

According to Sieferle (1996) the Neolithic revolution did not lead to an improvement of the living conditions. On the contrary, farmers had to spend much more time and energy, tilling their fields and caring for and feeding and/or herding their livestock. The nutrient value of the food became much lower and less healthy after it was reduced from the great choice of wild plants and animals into a few crops and into a few energetically expensive animal products. It led from the utilization of extensive areas of natural biomass production to its monopolization on a restricted area of cultivated land in which the crops were exposed to the hazards of great annual and seasonal climatic fluctuations and to the competition by weeds and the attacks by diseases and pests. It was, therefore, a much less stable ecological system in which the coevolution of the passive food collector and hunter (from the point of view of solar energy conversion into biomass production) with his landscapes was replaced by a much more 'active' participation in the food production process of energy conversion, requiring constant care for his crops and livestock. Although he dominated these reciprocal relations, this meant that more chemical and mechanical energy had to be invested and gained from the annual biomass produced than could be converted into human food and other products. The fluctuating annual crop harvested had replaced now long-term, sustainable productivity of the autopoietic production and regeneration capacities of the wild plant and animals of seminatural biosphere landscapes. It became therefore a much more fragile energetic system than that of the Epipaleolithic G-H who lived in a surplus of food, with the probable exception of catastrophic geological and climatic events. However, this was apparently not the case during the Natufian peak of coevolution, especially since he learned to raise the efficiency of energy extraction by the use of fire and vegetation management, like his Coastal Indian cultural and ecological counterpart.

Nevertheless, because of the much more intensive land use, the rapidly expanding Mediterranean agropastoralist populations "defeated" the G-H finally at the end of the Neolithic period. The conversion of the lowlands into agricultural fields almost completely destroyed the natural vegetation. It led chiefly in the less stable soils and in hilly topography to severe erosion. Its traces can still be witnessed in the erosive exposure of calcified paleosols in the central coastal plains of Israel (Dan and Yaalon, 1971).

The final formation of the agropastoral upland and lowland landscapes was initiated several millennia later in the Early Bronze Age, after the domestication of fruit trees. This was achieved by shifting from sexual production of pre-adapted wild populations into vegetative propagation of clones (Zohary and Spiegel-Roy, 1973). It reached its peak only during the Israelite period of the Iron Age from 2100 B.C. onward, with the help of iron tools enabling the uprooting of shrubs and trees and the clearing and terracing of upland fields. Wherever the slopes were too steep or too rocky for arable agriculture, favorable microsites between rock outcroppings were planted with fruit trees such as olives, figs, pomegranates, and vine grapes.

The natural vegetation was apparently left for soil protection along terrace walls and on steeper slopes between the terraces. As described vividly by Feliks (1968) on the basis of biblical sources, planted trees were protected from grazing animals by stone fences, covered with thorny brushes and thistles. All other nonarable uplands served as pastures for goats and other livestock as well as for the wildlife. Wild plants were collected for use as herb spices, medical plants, and sources for fuel and for the construction of tools (including wooden ploughs). Since they were first documented in the Bible, the Talmud and classical literature, these "polyculture" patterns of multiple upland resource utilization lasted with varying intensities on Mt. Carmel and elsewhere throughout the whole history until the middle of the last century. Combined with irrigation, water conservation, crop rotation, manuring, and stubble burning they reached their highest level of agro-and hydro-technological sophistication during Roman times. Such intensive multiple upland use can be regarded as one of the few instances in which agriculture improved the initial ecosystem factors of topography, soil parent material, and moisture regime on a long-term basis, albeit at the expense of the natural flora and fauna.

This total agropastoral landscape transformation was closely connected with the rise of the first proto-urban "technosphere" civilizations and their human-made artifacts. These left many archeological imprints on the Carmel landscape as tells and ruins, chiefly along the coast and in broad riverbeds (Kloner and Olami, 1980).

That this system was not only ecologically, but also economically viable and profitable, can be judged by the fact that in the first century, according to Flavius Josephus, the rocky mountains of the Galilee alone maintained a dense population of 2.5 million people, chiefly farmers, enjoying high incomes from the export of olive oil and wine. The demands of expanding and wealthy populations for timber, fuel, and charcoal lead to extensive forest clearing.

On the basis of many records of fire in both biblical and classical sources (Naveh, 1973, 1974; Liacos, 1974), we can safely assume that natural fires caused by lightening as well as human-set fires continued to be an integral part of the Carmel environment, and also of all other Mediterranean uplands by periodic maquis and shrubland burning to increase herbaceous fodder and to prevent the re-encroachment of dense woods, benefiting at the same time from the lush, woody, post fire regeneration.

We have no information on the timing and intensities of these fires, but is seems very likely, that the natural wildfires, which raged during geological times, were of much higher intensity, more extensive and destructive, especially those connected with volcanic activity. Lightening fires in the cooler and wetter seasons could have been spottier, also causing fires with lower intensity, which therefore had cooler fire temperatures and less far-reaching impacts. The dense woody canopy was opened, at least partly, by the fire-induced vegetation manipulation of the Pleistocene G-H and now may be even more opened by the agropastoralist throughout historical times. For the pastoralists, the desirable fire frequency is dependent chiefly on the rate of the postfire recovery of the woody vegetation that will carry the fire through. It should ensure the complete removal of the woody canopy and encourage the fast development of a dense grass cover for at least a couple of years. For a well developed maquis stand in northern Israel, the interval between such "efficient" burning could have been more than 15 years, but for highly inflammable dwarf shrub communities, or for mixed shrubs and dwarf shrubs, dominated by *Cistus* species, with an upper layer of pine trees, even less than 10 years.

In many cases the unfortunate combination of tree and woodcutting, fire, and grazing certainly caused gradual landscape desiccation, especially in drier regions and on steeper and less fertile slopes. However, as long as the woody vegetation served as the main source for fuel and construction, it was vital for the rural as well as for the emerging urban population to ensure their sustainable supply by the introduction of rotational burning and cutting systems. However, for the reasons mentioned previously, it is impossible to attain reliable pollen samples of the trees and shrubs, which could indicate their true abundance and structure in archeological sites, and, therefore, even less of such rotational management systems.

Therefore, we should be aware of the danger in being misled by the above-mentioned sweeping generalization of the "ruined landscape" theory and the wholesale condemnation of fire, goat, grazing, and woodcutting on Mediterranean uplands. There were great differences between countries and even between adjacent sites and their response to the greatly varying land use pressures. Most shallow and rocky slopes, which were neither terraced nor cultivated, have probably not undergone any extreme change since postglacial and early historic times. The fertile and fine-structured brown rendzina and terra rossa soils have suffered much less from erosion than has been generally assumed, as long as their sclerophyll - woody vegetation canopy has not been uprooted and the upper soil mantle has not been disturbed. Due to the great resilience and recuperative powers acquired during their coevolution, which took place under continued pyric, ungulate, and human pressures, the hardy vegetation provides efficient soil protection, as long as the woody plants can regenerate vegetatively from their extensive rootstocks; the herbaceous perennials from their underground bulbs, rhizomes, and other regenerative tissues; and annuals can draw from sufficient seed reserves (Naveh, 1975, 1994; Naveh and Lieberman, 1994).

Since the downfall of the Byzantine Empire, the historical changes in political regimes and population densities led to great fluctuation in human land uses and their pressures on the landscapes, soils, and vegetation. These induced a series of longer lasting cycles of anthropogenic degradation, re-vegetation and aggradation functions (Naveh and Dan, 1973; Naveh and Kutiel, 1990). Bottema and Van Zeist (1981) and Baruch (1987) found the resulting, rather dramatic changes reflected in the palynological findings in Israel. During periods of very intensive agricultural activities and heavy human population pressures, the natural vegetation receded in the face of the extension of cultivated crops. Ecological deterioration further increased during periods of instability, warfare, and agricultural and population decay such as what occurred in the last Ottoman Empire in the nineteenth century. Baruch (1987) showed that the composition of early postglacial forests and woodlands differed from that of the recent maquis and shrublands and that the abrupt increase in *Sarcopoterium spinosum* dwarf shrub formations in ancient times was connected with the abandonment of olive groves. A similar process could be observed in the first years after the establishment of the State of Israel in olive groves of the Arabic villages, which were abandoned during the Independence War in 1948. From this time onward – and until very recently - the release of the heavy pressure of grazing, burning, cutting, and coppicing on the denuded agro-pastoral upland landscapes of Mt. Carmel and the Galilee experienced a dramatic "regreening" process of rapidly regenerating woody vegetation.

From a hierarchical point of view, the shaping of this Total Human Agro-pastoral Landscape in the Mediterranean has taken place on a larger spatiotemporal scale, superimposed upon the much smaller and shorter local farming and land management practices. The former was dictated by the political rulers of the region and the resulting fate of the people and the land. The several hundred years establishment of a flourishing agro-pastoral landscape in Israel was replaced by a long period of decline, starting after the Muslim conquest (64 A.D.) and lasting for about 1300 years throughout the Arab, Crusaders, and Mameluke and Turkish rules. In many locations, pastoral nomadism gradually replaced settled crop and animal husbandry. The abandonment and neglect of terraces was the main cause for the catastrophic erosion, flooding, and siltation leading to badlands and swamps (Naveh and Dan, 1973). The metastable stage of the homeorhetic flow equilibrium and its thermodynamic consequences of the Pleistocene maintained this dual hierarchical position during the Holocene. On top of both, natural and chiefly climatic longer- and shorter-term annual and seasonal fluctuations were superimposed. The resulting defoliation pressures were, therefore, incorporated in the landscape at these different spatiotemporal scales. In this way these human-perturbation dependent Mediterranean landscapes acquired their long-term adaptive resilience and evolutionary metastability, lacking in all other Mediterranean-climate ecological systems.

6.   THE TOTAL INDUSTRIAL HUMAN LANDSCAPE OF THE LAST CENTURY

In contrast to the gradual transformation from the agricultural to the industrial civilization in Europe, starting already in the 18th century, the industrial revolution was introduced to Israel (or to "Palestine" as it was called then) by the Jewish colonizers, inspired by the Zionist movement, only after World War I. As described in more detail elsewhere (Naveh and Dan, 1973; Naveh and Kutiel, 1990; Naveh and Lieberman, 1994; Naveh, 1998a), the creation of the present Total Industrial Human Landscape started after World War II and the foundation of the State of Israel in 1948. It reached its peak with the combined and synergistic processes of intensification of traditional and modern agricultural land uses and urban-industrial expansion, driven by exponentially growing populations and leading to increasing pressures on the remaining seminatural and agro-pastoral landscapes. Here, like in other industrialized Mediterranean countries, the bifurcation between the solar energy powered autopoietic biosphere landscapes and the fossil energy powered agro- and urban-industrial landscapes resulted in the most adverse effects both on human health and nature chiefly in the densely populated coastal Mediterranean region of central and northern Israel. In the uplands, also, the rapid loss and fragmentation coupled with biological and cultural impoverishment and ecological disruption by accelerated erosion and by soil and water pollution endangers the future of these biosphere landscapes and their organismic evolution. Here the severest repercussions have been caused by the disruption of the agro-pastoral homeorhetic flow equilibrium. This is the result not only from heavy grazing, cutting, and recreational pressures; but also from the complete cessation of traditional agro-pastoral activities and the disruption of the rotational prescribed pastoral brush fire cycles. On Mt. Carmel, these activities have been replaced by the planting of dense monospecies and highly inflammable pine forests on one hand and on the other hand by the noninterference policy in the nature reserve of the Carmel National Park, thereby radically changing the fire parameters. The increase in the amount of fuel, and especially that of highly inflammable pine trees, resulted in higher fire intensities and greater destruction.

This is true for all other Mediterranean countries, wherever the initially vigorous vegetative regeneration of sclerophylls from stunted and almost imperceptible rootstocks was followed by the gradual encroachment of the woody canopy and the almost total suppression of the herbaceous understory in undisturbed and protected maquis. It turned from a blessing more to a curse, when the denser brush thickets became stagnant, and possibly senescent (Naveh, 1971; Naveh and Lieberman, 1994).

Fortunately, the danger of the paradoxically unnatural "nature protection" policy has been recognized by the Israel Nature and Parks Authorities after the disastrous wildfire in the Carmel National Park in 1989, which severely damaged 300 ha of mixed pine and oak forest and maquis including one of the few remaining natural pine forests in Israel. As a result, a great number of studies were carried out after this fire, with very useful management implications for sustainable fire and fuel management, summarized by Neeman et al. (1997). These findings should become part of a dynamic conservation policy aimed at the restoration of the homeorhetic flow equilibrium to ensure the sustainable adaptive resilience and evolutionary metastability, threatened also by the dangers of climate changes facing this region (Naveh, 1995). However, such comprehensive dynamic conservation strategies have not yet been widely adopted neither in Israel or elsewhere. There are many alarming signs that the scenario we predicted 30 years ago seems to have come true. We warned that if these threatening trends of "neotechnological" landscape degradation proceed unhampered, then the few spots of remaining unspoiled open landscapes could be turned into "overcrowded recreational slums" (Naveh and Dan, 1973). Therefore, the need for a diversion of the evolutionary trajectory from breakdown and extinction towards breakthrough and evolution during the present macroshift from the industrial to the post-industrial information society is now more urgent than ever.

7. RESULTS OF POSTFIRE REGENERATION STUDIES AND THEIR EVOLUTIONARY SIGNIFICANCE

To support our contentions on the great postfire regeneration powers, further developed in part 2, some results of the fire ecology studies in the western Galilee and on Mt. Carmel, in the vicinity of paleolithic archeological sites as well as by many other, more recent studies carried out on Mt. Carmel are presented. In several representative sites on Mt. Carmel of biodiversity studies (Naveh and Whittaker, 1979) and fire ecology studies near Nahal Me'arot and Nahal Sefunim (Naveh, 1984, 1994), we counted close to a hundred herbaceous species in gaps of relatively small grassy patches and their ecotones. Of these, about two-thirds had edible parts or were of other human uses, which could have been also of great economic value for the paleolithic food gatherers (see Table 1).

The rapid vegetative regeneration of sclerophyll woody plants and the striking temporary post fire flush of herbaceous plants has been observed in all of our earlier fire ecology studies in the Western Galilee (Naveh, 1960, 1973, 1974) as well as in the most previous one on Mt. Carmel (Naveh, 1999).

This study included a comparison between an open pine forest with a dense shrub understory burned by a hot wildfire and an adjacent unburned site in the summer of 1973.

As shown in Table 3, like in all our other fire ecology studies, we observed the vigorous postfire regeneration of all sclerophyllous trees and shrubs (or phanerophytes) as well as of climbers and of almost all dwarf shrubs (or chamaephytes), geophytes, and hemicryptophytes. We observed the same, almost dramatic increase, in the herbaceous plant abundance and in floristic diversity from less than 7 to 52 species.

The striking rise of the economic value of the burned site, both for humans and wildlife is implied by the fact that out of these 52 herbaceous species 15 have high pasture values and 25 are valuable for human consumption because of their edible bulbs, shoots, leaves, fruits, or seeds. In addition, 5 woody species greatly increased their browsing values because of their lush young leaf growth.

As **obligatory root resprouters** these sclerophyllous woody plants as well as the climbers regained almost one-third of their original dense cover. Their fire-stimulated vegetative regeneration is closely linked with their hydro-ecological behavior as drought-enduring and summer active plants. Since they rely on deep and well-branched root systems, they are capable of starting resprouting immediately after the fire, even in the middle of the summer by the mobilization of stored carbohydrates and possibly also of metabolized water in the roots. Due to their year-round intensive photosynthetic post fire activity, they can recover their former ground cover most approximately 10-15 years later, depending on the site, climate conditions, and prevailing post fire grazing pressures.

Table3. Woody plant cover in unburned and burned 1000 m^2 of open *Pinus halepensis* forest
with a dense shrub and dwarf shrub understory
(Average of 40 plots of 5X5 m; x = no cover value)

	UNBURNED	BURNED
Calycotome villosa	2.00	0.70
Cistus salvifolia	0.20	0
Crategus azarolus	0.19	0
Pinus halepensis	0.20	0
Pistacia palaestina	0.35	0
Rhamnus palaestina	x	0.25
Teucrium creticum	0.40	0
S U B T O T A L	**3.34**	**0.70**
Trees, Shrubs and Dwarf Shrubs 0.5-1m		
Calycotome villosa	6.65	7.55
Cistus salvifolius	55.80	2.35
Cistus villosus	4.60	6.00
Hypericum serpyllifolium	x	0
Olea europea	0	0.25
Osyris alba	0	0.50
Pinus halepensis	0.7	0
Pistacia lentiscus	9.75	6.30
Pistacia palaestina	0.20	0.60
Rhamnus palestina	x	0.25
Salvia judaica	0	0.80
Salvia triloba	1.05	1.10
Sarcopoterium spinosum	0.50	1.00
Satureja thymbra	0	0
Teucrium creticum	2.60	1.35
SUBTOTAL	**82.70**	**28.25**
Shrubs and Dwarf Shrubs 0.5m		
	UNBURNED	BURNED
Ajuga chia	0	x
Calycotome villosa	0	0.90
Cistus salvifolius	0	0.75
Fumana arabica	0.65	x
Helianthemum lavandulifoliu	0	x
Hypericum serpyllifolium	1.35	0.10
Osyris alba	0.02	0
Phagnalon rupestre	0.01	0

Pistacia lentiscus	1.45	3.30
Salvia triloba	0.10	0.50
Sarcopoterium spinosum	2.16	2.45
Satureja thymbra	0.31	0.26
Stachys distan	0	x
Teucrium creticum	0	0.35
SUBTOTAL	x	**0**
Cistus salvifolius	x	3.65
Cistus villosus	x	x
Pinus halepensis	0.10	4.70
Pistacia palaestina	x	0
SUBTOTAL	**0.19**	**8.35**
Climbers		
Asparagus palaestinus	x	x
Hedera helix	x	0
Rubia tenuifolia	0.60	1.0
SUBTOTAL	**0.60**	**1.00**
TOTAL WOODY PLANTS	**92.17**	**46.41**

These factors have to be taken into consideration in the assessment of the evolutionary strategies. Post fire foraging and especially browsing pressures could have acted together with gradually increasing drought periods as additional powerful selective agents. They favored those woody species and biotypes, which very soon after the fire developed hard, thorny or distasteful leaves and twigs together with the highest vegetative regeneration capacities to overcome these defoliation stresses. This recoding of information from fire and grazing may have also pre-adapted them to further defoliation catastrophes from cutting and coppicing. Outstanding examples of such successful evolutionary strategies for maximization of overall survival potentials on Mt. Carmel are the east Mediterranean Kermes oak - *Quercus calliprinos*, and the Eu-Mediterranean Pistachio shrub - *Pistacia lentiscus*. The Kermes oak is distinguished by vigorous resprouting from root crowns and suckers and adventive roots, but *Pistacia lentiscus* branches off laterally from prostrate leafy twigs that send roots and quickly form a dense, compact shrub canopy with high soil- and water-conserving features. As one, if not the most tenacious of the evergreen shrubs, it responds in the first few days after a fire by intensive cambial activity from the root tips and buds (Naveh, 1960). In the drier ecotones of the xero-thermo Mediterranean zone in Israel, on the semi-arid slopes of Mt. Gilboa, an even more drought-enduring ecotype developed than those found on Mt. Carmel (Swarzboim, 1978). Its small sclerophyllous leaves are highly resinous and very soon become distasteful. Therefore, it can withstand even very heavy browsing pressures and because of these high recuperative powers, it has remained on Mt. Carmel and elsewhere as the last woody survivor in frequently burned and heavily grazed pine forests, maquis, and woodlands.

On the other hand, *Calycotome villosa* as well as *Cystus salvofolius, C. villosus,* and all other chamaephytes including *Sarcopoterium spinosum* are **facultative root resprouters.** They can regenerate both from root crowns and from fire-stimulated seed germination, followed by vigorous growth. Most of these species also produce seeds from resprouting plants in the first year after the fire. This explains their relatively large cover values in the burned site. Perennial herbaceous plants, namely, hemicryptophytes and geophytes, have similar dual vegetative and reproductive post fire regeneration mechanisms. Like these chamaephytes, they are typical drought evaders adapting to the dry summers by more restricted physiological activity and especially by reduction of their transpiration surface. They commence resprouting only after the first winter rains from fire-avoiding underground stem bases, bulbs, tubers, and corms as well as from fire-stimulated growth from seeds. These are coupled with morphological and physiological plasticity and aggressiveness in colonization of newly opened, fire-denuded, and mineral-rich patches (Naveh, 1960).

Pinus halepensis is the only coniferous tree found in Israel, with Mt. Carmel as one of its last refuges, occupying a special status among the Mediterranean trees with respect to its response to fire. It relies solely on vigorous seed germination from cones that burst open from the heat of the fire and is, therefore, an **obligatory seed regenerator,** which can undergo natural regeneration after a fire under a dense maquis understory. Like the reseeding chamaephytes, its heliophilous seedlings are capable of establishing themselves on poor, exposed, rocky sites. This is in contrast to the above-described sclerophyll obligatory resprouters requiring more favorable, sheltered, and humus-rich seedbeds, which are not provided by the fire. As a typical pioneering colonizer the lack of resprouting ability is compensated fully by post fire germination of the great number of seeds benefiting from the temporary removal of the competition of the dense maquis understory. This is followed, in general, by a process of natural thinning out, similar to that of most chamaephytes seedlings. It leaves a scattered, rejuvenated stand of pine trees under the regenerating shrub canopy, or in the case of burned planted pine forests, more or less dense, even-aged pine stands. In this study, we found several tall (natural) *Pinus halepensis trees* in the unburned site and even more dead ones in the burned site and as shown in Table 2, we also found numerous seedlings. It can, however, be expected that only a few seedlings growing in suitable regeneration niches will survive and develop into taller trees.

In general, the rate and extent of the post fire herbaceous plant colonization is determined chiefly by the few perennial shade-tolerant herbaceous plants that survived in the dense brush cover as shade-tolerant relics, together with the availability of seeds from their seed banks and other invading plants. As mentioned above, important seed sources for these post fire colonizers in other sites of Mt. Carmel were small, grassy patch openings and edge habitats as well as adjacent fields and waste heaps near human habitations. However, because of the rather heavy grazing pressure by cattle to which this slope was exposed, several geophytes and hemicryptophytes, especially *Carlina* sp., and other low or unpalatable herbs such as *Anthemis melanolepis, Cephalaria joppica,* and *Linum nodiflorum* had a high constancy. Otherwise, we could have encountered a much higher abundance of grasses, especially *Pipthaterum miliaceum,* and other highly palatable perennial and annual grass fire-followers, as in our previous studies of post fire protected maquis shrublands.

The latter are major components of Mediterranean woodlands and grasslands as well as of the semi-arid Steppe grasslands of the xeric Mediterranean ecotones, which are even more fire-prone and may burn year after year and are also the most successful fire-followers. In all these plants, adaptive responses to fire and its avoidance as well as to drought and grazing stresses, are centered naturally around reproductive and growth behavior. Thus, early and prolific seed production, early seed shedding, and distribution by efficient dispersal mechanisms, seed dormancy, and polymorphism – especially in legumes – increase the chances to escape fire and environmental rigor. A good example for the coupling of such survival mechanisms is 'trypanocarpy' – namely, the development of hygroscopic awns, callous tips, and other torsion mechanisms enabling to drill and bury several centimeters deep into the soil the dissimulates of many grasses. Thus, the seeds escape fire, grazing hazards and, at the same time, benefit from more favorable moisture and temperature regimes for germination. Among these grasses are *Hordeum spontaneum, Triticum dicoccoides* and *Avena sterilis*, the progenitor of domesticated barley, wheat, and oats and all endowed with big seeds. In these woodlands and grasslands, perennial grasses and geophytes also demonstrated the most successful strategies for maximizing overall drought, fire, and grazing survival potentials and resilience. *Hordeum bulbosum* and *Poa bulbosa*, which are also very abundant on Mt. Carmel, combine all these reproductive adaptation strategies of the annual grasses with vegetative post fire resprouting from underground bulbs. Dominating even the most heavily grazed and degraded stages of these grasslands, woodlands, and dwarf shrub bathas contribution to the stability and resilience of these grasslands is most significant. It is very unfortunate that these perennial grasses unlike the successful annual grass invaders, have not been introduced to Californian grasslands and woodlands, which are lacking similar resilient perennial pasture plants (Naveh 1967, 1973; Naveh and Dan, 1971).

More recently, on the basis of extensive fossil records in California, Axelrod (1989) also reached similar conclusions for the evolutionary role of lightening and volcanic fires together with the increasingly stressful environments during the late Pleistocene and Quaternary in pre-adapted sclerophyllous taxa and on fire-favored speciation. He rightly maintained that in countries with Mediterranean climates it was the elimination of summer rain that imposed strong selective factors, including increased fire frequency on sclerophyllous vegetation.

In general, this post fire flush of herbaceous species is only temporary and after 3-5 years the woody brush cover will take over again. This temporary post fire flush of herbaceous plants and the fire-stimulated vegetative and reproductive regeneration dynamics, is very similar to that of the California chaparral, where it has been called "*autosuccession*" by Hanes (1971). In northern California, controlled burning of dense chaparral is used as a common practice to raise its carrying capacity for hunting deer as a sport and has resulted in a manifold increase of their carrying capacity (Biswell, 1989). After the hot wildfire on Mt. Carmel in 1986, these ruderalic plants spread from the waste heaps of Kibbutz Beth Oren to several adjacent forest gaps and dominated the herbaceous fire-flush. Here, *Hordeum spontaneum* and *Piptatherum (Oryzopsis) miliaceum* were most abundant. The latter is also the most prolific shade-tolerant fire-follower perennial grass, and is one of the last herbaceous survivors in the maquis thicket regenerating after fires as on Mt. Carmel (Naveh, 1984) and in the western Galilee (Naveh, 1974). Its plentiful millet-like seeds can be baked and used as staple food, in a similar way to its American counterpart *Oryzopsis hymenoides*. The latter is called "Indian Rice" because the Indians of the drier western ranges gathered its seeds in quantities by cutting wild stands. The same practice may have been applied to *Piptatherum miliaceum*, growing in dense and stout bunches, which can be cut shortly before ripening, i.e., annual grasses. In *Hordeum spontaneum,* this has been demonstrated by Harlan (1967) with the help of a flint-sickle of the type found in the Natufian layers of the El-Wad cave on Mt. Carmel. The same may also be true for *Triticum dicoccoides,* the most important progenitor of wheat of which several relict populations can be found on open grassy sites and near rock outcrops on Mt. Carmel. In its major natural habitat on the fire-swept slopes of the eastern Galilee hills facing the Jordan Valley, this grass, together with *Stipa tortilis, Avena sterilis* as well as many others re-colonize after fire year to year. The latter is also the most important grass component of the open Tabor oak savannas and other grasslands whose germination is stimulated by the heat shock of the grass fire (Naveh, 1973).

The above-mentioned fire ecology studies further revealed that these herbaceous fire followers serve as an efficient sink for the follow-up post fire flush of nutrients released in the first winter rain season. In brown rendzina soil, collected two months after a hot wildfire in 1983 on Mt. Carmel, in an open pine forest with a dense and well-developed maquis shrub cover, we found a striking increase in water-soluble nutrients, especially of nitrogen and phosphate in the first winter and spring after the fire in the upper centimeters of the soil (Kutiel and Naveh 1987a,b; Kutiel et al., 1990). This post fire nutrient flush was rather short-termed, but it could be utilized by the herbaceous fire followers for proliferous forage and seed production. These plants, and probably also the rapidly regenerating dwarf shrubs, serve as an important link in the recycling of these nutrients to the soil from which the resprouting, deep-rooted, woody plants can benefit in the following years. In one experiment, pot-grown wheat in the upper 2cm layer of this burned brown rendzina soil, produced 6 times more phytomass and 12 times more seeds. Of special significance for the enhancement of nutrient cycling is the 4.5 times increase in root production, facilitating the manifold increase in nutrient accumulation in the plants. The striking post fire increase in seed production leads strong support of our above-mentioned hypothesis that such favorable ash seedbeds served as triggers for domestication of cereals and their incipient cultivation in slash-burn rotations on Mt. Carmel and elsewhere in the early stages of the agricultural revolution.

This process most probably reached its peak by the Epipaleolithic cultures, and probably had a much greater impact on the vegetation and, therefore, on the landscape as a whole, converting these into seminatural landscapes. Human uses and impacts were further intensified from the Neolithic agricultural revolution onward in which many of these landscapes were converted into agropastoral landscapes and the "humanized" Total Human Landscape was created.

We can, therefore, also safely assume that in addition to other, more or less catastrophic natural perturbations and to increasing drought, as well as human and wildlife foraging, fire originating from volcanic activities and lightening, and from long-lasting human interventions acted as a strong selection force. Those woody and herbaceous genotypes, which developed the most efficient physiological and morphological evolutionary strategies for active and passive vegetative and reproductive regeneration mechanisms had the best chances to overcome the natural and human-induced fire stresses. This occurred most probably during the drier, warmer interpluvials in which the Mediterranean climate patterns became established. It created favorable conditions for the germination of light-demanding woody plants such as *Pinus halepensis* and most chamaephytes as well as for the above-mentioned herbaceous plants facilitating their spread over vast areas.

8.   RESILIENCE TO GRAZING

Because of their adaptation to high light intensities (Langer, 1979) wind pollution, and seed dispersal (Stebbins, 1972), we assumed that Mediterranean grasses have evolved in the drier, more open ecotones, in fire-opened gaps, in more humid forests, maquis, and shrublands. According to Stebbins (1981), they coevolved with grazers and have adapted to being grazed. However, as Belsky (1986) has shown, these adaptations suggest an antagonistic relationship and not with deleterious effects of herbivores. These may have contributed to the evolution of effective structural, chemical, and phenological defense mechanisms. Following the arguments of Axelrod (1959, 1989), both the Mediterranean and the Californian sclerophylls adapted to fire. This may have also pre-adapted them to grazing and browsing. But, as will be discussed below, it has not yet been confirmed that this is also true for the Californian sclerophyll.

On the other hand, in the case of these Mediterranean grasses, there seems to be no doubt that the adaptive defense mechanisms against grazing have also been effective against cutting and fire. As discussed already in part 1 of this essay, in annual grasses as well as in the legumes and other highly valuable pasture plants, these adaptive responses are centered naturally around early and prolific seed production, early seed shedding, distribution by efficient dispersal mechanisms, seed dormancy, and polymorphism – especially in legumes – increasing their chances to escape fire, grazing, and environmental rigor. Of special selective advantage, in this respect, is the capacity to drill and bury their dissimulates several centimeters deep into the soil by trypanocarpy and this may be one of the reasons why *Avena* and *Erodium* species became such successful invaders of California grasslands, in spite of their high palatability.

The same is also true for their ability to compensate fully or partly for lost tissue of biomass production both to grazing and fire with increased tillering, protected ground level meristem, and in perennial plants also with vigorous vegetative reproduction (Stebbins, 1972). *Hordeum bulbosum* and *Poa bulbosa* are the most resistant perennial grasses against heavy grazing pressures in the Mediterranean because they combine all these vegetative and reproductive adaptation strategies, which are apparently lacking in Californian perennial grasses. If they would have been introduced accidentally, like so many annual grasses, they could have contributed much to a better resilience of the Californian grasslands (Naveh, 1967).

At the same time, grazers and browsers together with natural and human-set fires may have helped in the creation and maintenance of open vegetative canopies. Therefore, they may have also played an important role with fire and the paleolithic G-H in the cross-catalytic network of the previously described coevolutionary process.

We have no reliable information on the actual animal husbandry and herding practices not only for prehistoric times, but also for most of the Holocene until the last centuries. Tchernov and Horowitz (1990), in their attempt to fill up this gap, assessed the effect of herding practices by relating fossil remains of goat + sheep (caprovines)/cattle ratios to current pasture carrying capacities in the regions of these bone findings in the last 6000 years. They claimed that the tendency of increasing frequencies of caprovines reflected lowered-carrying capacity associated with "overgrazing".

It is outside the scope of this essay to discuss in detail the results of this 1990 study. However, the major handicaps for the conclusions are: (1) It is impossible to apply modern carrying capacity standards to traditional, and even less to ancient ones, especially since the survey took into consideration the requirements chiefly for modern beef cattle ranching. (2) The results of this survey on carrying capacity in the northern and central Mediterranean parts of Israel, carried out about 50 years ago, were determined to a great deal by the amount of the shrub cover, which is much less valuable for this purpose. The woody-herbaceous vegetation layers relation are highly dynamic and change according to the prevailing homeorhetic flow equilibrium, as described in part 1. (3) "Overgrazing" is a very vague term and can hardly serve as an indicator for the decrease in productivity of these ancient pastures. We can, however, safely assume that these pressures on the open pasturelands increased gradually together with the increasing population densities on one hand and that of expansion of the cultivated land in the uplands, on the other. This happened most probably from the Iron Age onward and after the domestication of fruit trees in the Bronze Age, reaching its first peak in the Hellenistic and Roman periods. But at the same time, stubble grazing of the fields after the harvest, together with other agricultural residues could have partly compensated for these losses. These pressures were most probably considerably reduced during the periods of receding agricultural activities and the abandonment of fields and olive plantations, leading to a new, longer-term cycle of the homeorhetic flow equilibrium. At that time, the neglect of terraces in the uplands and their 'invasion' by grazing livestock and their shepherds could have accelerated their destruction and the resulting erosion events described by Naveh and Dan (1973).

The short-term homeorhetic cycles were maintained chiefly by annual and seasonal climatic fluctuations, by determining the quantity and quality of the available forage and by herding practices. It is very probable that these, like the field crop and horticultural practices, have not been very different from those practiced by the Arab Fellahin and pastoralists until the middle of the last century. The same may be true for most other Mediterranean pastoralists. Contrary to colder temperate regions, where the livestock has to be kept inside and stall-fed in the winter, here they have been grazed all year round without the need for any substantial nutritious supplementary food. Therefore, the availability of fodder during critical periods of drought and dry seasons, before the onset of the heavy winter rains, became a major limiting factor for the number of cattle, sheep, and goats that could survive between years, i.e., wild herbivores. For seminomadic and nomadic pastoralists from the drier regions, the availability of water became an additional severely limiting factor, but on the other hand, they were more flexible in the movements of their herds than the sedentary Mediterranean farmers.

The latter could not allow themselves to keep their livestock numbers for maximum exploitation of the peak spring pasture productivity, but had to adjust these more or less to its lowest levels to get any reasonable economic return from animal products for his family and the market. For the same reason, grazing pressures at the spring peaks were never heavy enough to prevent efficient seed production of most annual pasture plants, on which future productivity depends, especially, since these have developed such efficient autopoietic pathways throughout their long evolutionary history ensuring their high reproductive regeneration capacities.

On the other hand, "overgrazing" leading to irreversible degradation and productivity is typically only for modern livestock husbandry in Mediterranean pastures as well as in California. This additional energy input goes into the trophic pasture-herbivore system by supplemental fodder to overcome the natural limits of biomass production of these seminatural biosphere landscapes. Therefore without special rotational-deferred grazing management, the heavy exploitation during the whole growth and reproduction cycles of the palatable pasture plants can lead even to their complete destruction. The artificial removal of the above-described limiting factor by supplemental food and water supply is also the main cause for "desertification" all over the world.

9. COMPARISON OF VEGETATION REGENERATION POTENTIAL BETWEEN THE MEDITERRANEAN REGION AND CALIFORNIA

Convergent evolution is commonly defined as the expression of a similar set of characteristics among organisms or ecosystems that are phylogenetically unrelated or geographically disjunct when subject to similar agents of natural selection (Cody and Mooney, 1978). The comparative study of Mediterranean-type ecosystems is prominent among studies of convergent evolution (Richardson et al., 2001). Naveh (1967) described similarities between plant species and vegetation formations in California and Israel noting many important differences. Zinke (1973) found resemblance between soil-vegetation relationships in Italy and California. Evidence has been provided for similarity in plant anatomy (Kummerow, 1973) and successional trajectories (Armesto et al., 1995) within Mediterranean regions in Chile and California. Fuentes and Munoz (1995) have argued that convergence between Mediterranean-type eco-systems would have been even stronger if not for the disparate nature of human effects on the landscape in these ecosystems. Reviewing many studies of convergence in Mediterranean-type ecosystems, Cody and Mooney (1978) concluded that much of the similarities can be interpreted in terms of convergence, and much of the biotic dissimilarities can be explained by the many environmental and historical differences between the disparate continents. However, in all these and many other studies, one common conclusion is that dissimilarity among systems is common regardless of the chosen ecological attribute used to examine convergence.

Given this equivocal support for convergence, some studies have claimed that the notion of convergence is imaginary with no basis in reality (Shmida, 1981; Barbour and Minnich, 1990). We believe, however, that the question of convergence cannot be addressed with simple binary (true or false) answers. Rather, it is one of degree, where the strength of convergence is dependent upon the degree of similarity in environmental factors (through evolutionary time) and the degree to which these factors affect the evolution of species and ecosystem traits. Climate is considered the major factor that drives convergence in Mediterranean-type ecosystems (Di Castri and Mooney, 1973). Thus, the amount of convergence between two specific systems is a function of the similarity between their climates and of the degree to which climate determines the character of each trait considered in the assessment of convergence. We hypothesize that strong convergence is expected for traits determined largely by climate (e.g., plant phenology), and nonconvergence is expected for traits determined largely by edaphic or historical factors (e.g., reproductive strategy).

Ecosystem resilience (i.e., ecosystem behavior following disturbances) is an important trait of ecosystems (Likens, 1992). Westman (1986) reviewed concepts and measures of resilience, which he defined as the degree, manner, and pace of change of recovery properties following disturbance.

In particular, woody vegetation regeneration following disturbance can show characteristic rates and spatial patterns within different ecosystems (Glenn-Lewin and van der Maarel, 1992). As a characteristic of ecosystem, resilience is clearly determined by the evolutionary history of the region rather than by its climate. Thus, in different Mediterranean type ecosystems, even with similar vegetation structure, dissimilarity is expected for rates and patterns of vegetation regeneration.

Our chapter has thus far indicated in numerous ways that the evolutionary history of the Mediterranean region has diverged to a new, distinctive path of total human impact during the early Holocene, some 10,000 YBP, while in California, human impact on the land was limited to G-H cultures until 160 years ago (Axelrod, 1977; Mensing, 1998). We envisage that the divergent evolutionary history of the two ecosystem types is reflected in the regeneration potential of the system, both at the landscape level and at the individual level. In order to test if this is indeed the case, we compare the results of available studies of vegetation dynamics in Mediterranean type ecosystems.

We searched the literature for studies that quantified vegetation change following disturbances in California and Mediterranean systems. Quantitative studies of vegetation change at the landscape scale and changes across several decades are not common. The studies we found varied in their spatial and temporal scales and reported results in ways that often differed from our study. Yet, for four studies from California and five studies from the Mediterranean basin we were able to calculate a common system attribute, the average annual change in tree cover as an indicator of vegetation regeneration rate. The studies of Mediterranean ecosystems that provide quantitative data on vegetation change on the landscape scale indicate rates larger than those in California **by a factor of 2 to 20** (Table 4). These results are consistent even when high precipitation areas in California are compared with low precipitation areas in Israel regardless of the type of disturbance from which the area is recovering.

These results strongly suggest divergence rather than convergence in vegetation resilience between the two regions. We interpret these surprisingly profound differences as an indication of the crucial role of evolutionary history in shaping ecosystem characteristics.

A definitive test of our hypothesis may be a controlled experiment in which species from both regions are growing in the same controlled environment under treatments that mimic various natural disturbances. In an effort to address this issue experimentally, we have recently initiated a mutual transplant experiment involving acorns from several Mediterranean and Californian species in order to compare seedling growth rate and responses to various types of experimental disturbance. This study should help determine whether Mediterranean species are inherently capable of more vigorous regeneration than California species in response to various disturbance types.

Table 4. Average annual change in tree cover, calculated using data from different studies of vegetation dynamics in California and in the Mediterranean basin (Carmel and Flather, 2003)

Region	Source	Site name (recovering after)	Precip-itation	Study period (in years)	Average annual change in tree cover
California	Brooks and Merenlender, 2001	Hopland area (clearing)	900	28	0.11%
California	Scheidlinger and Zedler, 1979	San Diego County (not specified)		42	-0.05%
California	Callaway and Davis, 1993	Gaviota State Park, nongrazed (heavy grazing, fire)	600	42	0.43%
California	Callaway and Davis, 1993	Gaviota State Park, grazed (fire)	600	42	0.20%
California	Carmel and Flather (in press)	Hastings Nature Reserve (abandonment of agriculture, fire)	600	56	0.25%
Mediterranean	Paraskevopoulos et al., 1994	Mt. Pilion, Greece (clearing)	475	30	2.03%
Mediterranean	Samocha et al., 1980	Adulam, Israel (heavy grazing)	450	22	1.59%
Mediterranean	Samocha et al., 1980	Bar Giora, Israel (heavy grazing)	550	22	0.92%
Mediterranean	Preiss et al., 1997	Montpelier, France (agricultural practices)	1150	33	0.95%
Mediterranean	Kadmon and Harari-Kremer, 1999	Mt. Carmel, Israel (agricultural practices, heavy grazing)	700	32	1.06%
Mediterranean	Carmel and Kadmon, 1999	Mt. Meron, Israel (agricultural practices, heavy grazing)	900	28	1.3%

10.   DISCUSSION AND CONCLUSIONS

We have presented here the evolution of the cultural Mediterranean landscape in Israel, especially on Mt. Carmel, as an integral part of the evolution of the ecological, social, and economic system of the Total Human Ecosystem (THE). In this THE, the natural landscape elements evolving from the geosphere and biosphere and the cultural artifacts resulting from the evolution of the human noosphere are together forming a hierarchical structured interacting network along multiple nested scales of this Total Human Landscape(THL). I used "cultural" landscapes in the broadest sense, for those landscapes in which human habitation and activities have modified the pristine, natural Pleistocene landscape first into a subnatural, then into a seminatural, and finally into an agropastoral landscape. As a coevolutionary process, it occurred simultaneously with the cultural evolution of the paleolithic G-H, namely, the development of his cognitive and spiritual facilities of thinking, speaking, and acting clearly distinguishing him from his hominid ancestors and relatives.

There seems to be a general consensus that the Neolithic Agricultural Revolution was the first major wave of change in the life of the paleolithic food gatherers and hunters. The Neolithic Revolution has been regarded, also by Laszlo (2001), as the first major bifurcation "that rocked stone-age societies". We do not agree to lump together the hundreds of thousands of years that passed since the arrival of *Homo erectus* in Israel and the Mediterranean basin until the emergence of *Homo sapiens sapiens* into a linear process of cultural and socio-economic evolution. Although we cannot state their exact timing and duration and we know very little of the temporal and spatial dynamics of previous bifurcations, we assume that these occurred as definite phases in the advancement of the G-H and their economies to higher levels of complexity. These are well reflected in the different phases of the evolution of the Pleistocene "cultural" landscapes and proceeded according to Laszlo's (2001) description in several stages. They include innovations in tools and operational devices of greater efficiency that induced changes in social and environmental relations, and brought a higher level of resource production, a faster growth of the population, greater social complexity, and an increased impact on the environment. These new social and environmental conditions catalyzed changes in social organization and economic systems, and in the last stage a new set of values and world-views were introduced to readapt society to new conditions.

In our opinion, such different phases of bifurcations of human cultural evolution more or less characterized these earlier bifurcations of landscape transformations, which led to a more "humanized" seminatural THL. However, contrary to all later bifurcation whose durations have accelerated from thousands to hundreds of years and presently even to less than half a century, they were much slower lasting for very long, geological periods until their impacts left traces in archeological and other findings.

We claim that in this coevolution the use of fire could have been the first technological invention that triggered mutually reinforcing auto-and cross-catalytic feedback loops forming a closely interwoven network with the impacts of human habitation and foraging activities. It led to the expansion of the first small-scale, open, "cultural", habitat patches of ecotopes and of the fire-opened forest, maquis, and shrubland gaps in space and time.

This could have been, therefore, the first major bifurcation changing the life of the primitive paleolithic humans together with the transformation of the Pleistocene pristine landscape into a mosaic of natural and subnatural THL ecotopes. As previously mentioned it was most probably a very slow process, occurring in the Upper Paleolithic, described in subchapter C1. This bifurcation probably led to a new, more dynamic level of self-organization of this evolving subnatural THL. As indicated by the finding of the Neanderthaloids Shanidar cave, this was closely coupled with a higher level of self-organization of these people.

A further major bifurcation, much shorter in time and probably much more extensive in space, occurred during the transition from the Upper Paleolithic to the Epipaleolithic, 20-15,000 years ago. It led to the creation of the intensively managed seminatural landscape of the Natufians. In this case, human impacts caused more far-reaching changes and more complex interacting cross-catalytic loops. In the long-range, human-set fires probably became more important than natural fires and the advanced technologies of food collecting, storing and preparing and for hunting introduced new perturbation factors, which were incorporated in the landscape system and its metastable homeorhetic flow equilibrium. Among these was the selection of preferred grasses for cutting and for seed collection as well as the more intensive grazing by ungulates and especially gazelles, which had replaced the bigger extinct mammals.

The third major bifurcation occurred during the Neolithic Revolution together with the gradual conversion of larger and larger pieces of arable seminatural landscapes into agropastoral ones, simultaneously replacing the coevolutionary symbiotic relationship between people and their landscapes and nature by human dominance. However, the replacement of the productivity of the spontaneous flora and fauna of autopoietic and regenerative biosphere landscapes with the much larger, but less reliable biomass production led to a new, much narrower coevolution between the farmer and his domesticate plants and animals. Although it advanced the efficiency of energy extraction from his crops and animals, it did not ensure more sustainable, natural, renewable resource utilization. On the contrary, it caused many great environmental upheavals, which were not only prevented, but were further aggravated by the most dramatic agrotechnological advances of the modern, intensive, agricultural industry of the last century. This third major bifurcation is missing completely from California that 'jumped' directly into the agro-industrial age almost 10,000 years later.

It is the choice of human society and its leadership to prevent further biotic degradation and extinction and to mobilize all its political, scientific, economic, and spiritual forces to ensure a better sustainable future for nature and humankind. This is also the choice for further biological and cultural evolution (Laszlo, 1994; Naveh, 2003). The hope for a sustainable future of our biosphere landscapes and our post-industrial Total Human Ecosystem as a whole has been expressed lucidly by Laszlo (2000, p.114) in the final sentences of his "Macroshift 2000-2010" book:

> "Endowed with the highest forms of consciousness in our regions of the universe, we are the only species that not only acts, but can also foresee the effects of its actions. As members of a species capable of foresight, we must live up to our responsibility as stewards rather than exploiters of the complex and harmonious web of life on this planet."

11. ACKNOWLEDGMENTS

We hereby gratefully acknowledge the helpful advice for the preparation of this essay by Prof. Mina Weinstein-Evron and Dafna Kadosh of the Zimann Institute of Archeology, University of Haifa. Prof. Weinstein-Evron also allowed us to use figures and tables from her comprehensive book on the Carmel Natufians. This book, as well as all her other studies served as an important source of knowledge and inspiration. Curtis Flather of the Rocky Mountain Research Station, the US Forest Service helped the comparative study of Israel and California in many ways. Hava Lahav of the Israel Society for the Protection of Nature allowed us to make use of her vegetation map, based on her Carmel vegetation survey.

12. REFERENCES

Allen H.D. 2001. Mediterranean Ecogeography. Prentice Hall, Pearson Education Limited, Harlow, England.

Anderson E. 1956. Man as a maker of new plants and new plant communities. In Thomas W.L. (Ed.) Man's Role in Changing the Face of the Earth. University of Chicago Press, Chicago. pp. 763-777.

Armesto J.J., Vidiella P.E. and Jimenez H.E. 1995. Evaluating causes and mechanisms of succession in the Mediterranean regions in Chile and California. In Ecology and biogeography of Mediterranean ecosystems in Chile, California and Australia. Kalin Arroyo M. T., Zedler P.H. and Fox M.D. (Eds.). Springer-Verlag, New York. pp. 418-435.

Aschmann H. 1973. Distribution and peculiarity of mediterranean ecosystems. In Mediterranean-type ecosystems. Di Castri F. and Mooney H.A. (Eds). Springer-Verlag, Berlin. pp. 3-19

Axelrod D.I. 1958. Evolution of the Madro-Tertian geoflora. Bot Rev. 24, 433-509.

Axelrod D.I. 1977. Outline history of California vegetation. In Terrestrial vegetation of California. Barbour M.G. and Major J. (Eds.), John Wiley & Sons, New York. pp. 139-193.

Axelrod D.I. 1989. Age and origin of Chaparral. In The California Chaparral Paradigms Reexamined. Keeley S.C. (Ed.), Nat Hist Mus LA County, Los Angeles. pp. 7-20.

Bankroft H. 1937. Report on charcoal fragments. In The Stone Age of Mount Carmel I. Excavations at the Wadi Mughara. Garrod D.A.E. and Bate D.M.A. (Eds.), Clarendon Press, Oxford. 129 pp.

Bar-Josef O. 1983. The Natufian in the Southern Levant. In The Hilly Flanks and Beyond. Young C.T., Smith P.L.E. and Morhensen P. (Eds.), University of Chicago Press, Chicago. pp. 11-42.

Bar-Josef O. 1984. Near East. In Neue Forschungen zur Altsteinzeit. s. 232-208.

Bar-Josef O. 1998. The Natufian culture in the Levant, Threshold to the origins of agriculture. Evol Anthropol. 6, 159-177.

Barbour M.G. and Minnich R.A. 1990. The myth of chaparral convergence. Isr J Bot. 39, 453-463.

Baruch U. 1987. The Palynology of late Holocene Cores from Lake Kinnereth (Hebrew) University of Jerusalem. Department of Archeology, Jerusalem

Baruch U. and Bottema S. 1999. A new pollen diagram from Lake Hula. Vegetational, climate, and anthropological implications. In Ancient Lakes and Biological Diversity. Kawanabe H., Coulter. G.W. and Roosevelt A.C. (Eds.), Kenobi Productions, Belgium. pp. 75-86.

Baumhoff M.A. 1963. Ecological determinants of aboriginal California populations. In American Archeology and Ethnography. Univ Cal Publ. 49, 144-263.

Bean L.J. and Lawton H.W. 1973. Some explanations into the rise of cultural complexity. In Native California with comments on proto-agriculture. Introduction to Lewis H.T. Pattern of Indian Burning in California; Ecology and Ethnohistory. Anthropological Papers I. Ballena Press Ramona, California.

Bein A. and Sass E. 1980. Geology. In Atlas of Haifa and Mount Carmel. Soffer A. and Kipnis B. (Eds.), Haifa University, Haifa. pp. 14-17.

Belsky A.J. 1986. Does herbivory benefit plants? A review of the evidence. Amer Nat. 127, 870-892.

Biswell H. 1989. Prescribed Burning for Wildland Management in California. University of California Press, Berkeley.

Blondel J. and Aronson J. 1999. Biology and Wildlife of the Mediterranean Region. Oxford University Press, Oxford.

Bottema S. and van Zeist W. 1981. Palynological history of the Near East. In Prehistoire du Levant. CNRS, Paris. pp. 111-123.

Braidwood R.J. 1967. Prehistoric Man. Scott, Foresman and Co, Glenview, Illinois.

Brooks C.N. and Merenlender A. 2001. Determining the pattern of oak woodland regeneration for a cleared watershed in northwest California: A necessary first step for restoration. Restor Ecol. 9, 1-12.

Callaway R.M. and Davis F.W. 1993. Vegetation dynamics, fire, and the physical environment in coastal central California. Ecol. 74, 1567-1578.

Carmel Y. and Kadmon R. 1999. Grazing, topography, and long-term vegetation changes in a Mediterranean ecosystem. Pl Ecol. 145, 239-250.

Carmel Y. and Naveh Z. 2002. The paradigm of landscape and the paradigm of ecosystems – implications for land planning and management in the Mediterranean Region. J Medit Ecol. 3, 24-24.

Carmel Y. and Flather C.H. 2003. Comparing landscape scale vegetation dynamics following recent disturbance in climatically similar sites in California and the Mediterranean basin. Landscape Ecol. (in press).

Castri Di F. 1981. Mediterranean-type shrublands of the World. In. Ecosystems of the World, II- Mediterranean-Type Shrublands. Castri Di F., Goodall D.W. and Specht R.L. (Eds.), Elsevier Sci Publ Comp., Amsterdam. pp. 1-52.

Castri Di F. 1991. An ecological overview of the five regions in the world with a mediterranean climate. In Biogeography of Mediterranean Invasions. Groves R.H. and Castri Di F. (Eds.), Cambridge University Press, Cambridge. pp. 3-17.

Castri Di F. and Mooney H.A. 1973. Mediterranean type ecosystems (origin and structure), 1st ed. Springer-Verlag, Berlin.

Capra F. 1997. The Web of Life. A New Scientific Understanding of Living Systems. Anchor Books Doubleday, New York.

Clark J.D. 1966. Acheulian occupation sites in the Middle East and Africa, a study in cultural variability In Recent Studies in Paleanthropology. Clark J.D. and Howell F.C. (Eds.), Amer Anthropol. 68, 394-412.

Cody M.L. and Mooney H.A. 1978. Convergence versus nonconvergence in Mediterranean-climate ecosystems. Ann Rev Ecol Syst. 9, 265-321.

Cowling R.M., Rundel P.W., Lamont B.P., Arroyo M.K. and Arianoutsou M. 1996. Plant diversity in mediterranean-climate regions. Tr Ecol Evol. 11, 352-360.

Dafni A. 1984. Edible Wild Plants. The Society for the Protection of Nature, Tel Aviv.

Dan J. and Yaalon D.H. 1971. On the Origin and nature of the paleopedological formations in the central coastal fringes areas of Israel. In Paleopedology: Origin, Nature and Dating of Paleosols. Yaalon D.H. (Ed.), Israel Universities Press, Jerusalem. pp. 245-260.

Deacon H.J. 1983. The comparative evolution of mediterranean-type ecosystems. A southern perspective In Mediterranean-Type Ecosystems: The Role of Nutrients. Kruger F.J., Mitchell D.T. and Jarvis J.U.M. (Eds.), Springer-Verlag, Berlin. pp. 3-40.

Diamond J. 2001. Guns, Germs, and Steel. Norton & Company, New York.

Dimpleby G.W. 1985. The Palynology of Archeological Sites. Acad. Press, New York.

Egler F.E. 1964. Pesticides in our ecosystems. Amer Sci. 52, 110- 136.

Eigen M. and Schuster P. 1979. The Hypercycle: A Principle of Natural Self- organization. Springer Verlag, New York.

Eisner R. 1987. The Chalice & the Blade: Our History, Our Future. Harper Collins Publishers.

Feliks J. 1968. Plant World of the Bible. Massada, Tel Aviv (in Hebrew).

Finkel M., Fragman O. and Nevo E. 2001. Biodiversity and interslope convergence of vascular plants caused by sharp microclimate differences at "Evolution Canyon" II, Lower Nahal Keziv, Upper Galilee, Israel. Isr J Pl Sci. 49, 285-29.

Fuentes E.R. and Munoz R. 1995. The human role in changing landscapes in central chile: implications for intercontinental comparisons. In Ecology and biogeography of Mediterranean ecosystems in Chile, California and Australia. Kalin Arroyo M.T., Zedler P. H. and Fox M.D. (Eds.), Springer-Verlag, New York. pp. 401-417.

Garrod D.A.E. 1929. Excavations in the Mugharet El-Wad, near Athlit. April- June 1929. Palest Expl Quart. 61, 220-222.

Garrod D.A.E. and Bate D.M.A. 1937. The Stone Age of Mount Carmel. Vol 1. Excavations at the Wadi Mughara. Clarendon Press, Oxford.

Glenn-Lewin D.C. and van der Maarel E. 1992. Patterns and processes of vegetation dynamics. In Plant Succession. Glenn-Lewin D.C., Peet R.K. and Veblen T.T. (Eds.), Chapman & Hall, London. pp. 11-58.

Grove A.T. and Rackham O. 2001. The Nature of Mediterranean Europe. An Ecological History. Yale University Press, New Haven and London.

Grubb P.J. 1977. The maintenance of species richness in plant communities and the importance of the regeneration niche. Biol Rev. 52, 107-145.

Hanes T.L. 1976. Succession after fire in the chaparral of Southern California. Ecol Monogr. 41, 27-52.

Harlan J.R. 1967. A wild wheat harvest in Turkey. Archeol. 20, 197-201.

Hassan F.A. 1981. Demographic Archeology. Acad. Press, New York.

Heizer R.F. and Elsasser A.B. 1980. The Natural World of the California Indians. University of California Press, Berkeley and Los Angeles.

Henry P.O. 1981. Settlement patterns of the Natufians. In Prehistoire du Levant. CNRS, Paris. pp. 422-431.

Higgs E.S.C., Vita-Finci D., Harris R. and Fagg A.E. 1967. The climate, environment and industries of Stone Age Greece Part III. Prehist Soc. 33, 1- 29.

Horowitz A. 1979. The Quarternary of Israel. Acad. Press, New York.

Horowitz A. 1992. Palynology of Arid Lands. Elsevier, Amsterdam.

Ikeya M. and Poulianos A.N. 1979. ESR Age of the trace of fire in Petralona. Anthropos 6, 44-47.

Iversen J. 1971. Forest clearance in the Stone Age. In Man and the Ecosphere. Ehrlich P.R., Holdren J.P. and Holm R.W. (Eds.), Freeman and Company, San Francisco. pp. 26-31.

Jantsch E. 1980. The Self-Organizing Universe. Scientific and Human Implications of the Emerging Paradigm of Evolution. Pergamon Press, Oxford.

Jelinek A.H. 1981. Middle Paleolithic in the Tabun Cave. Prehistoire du Levant. CNRS, Paris. pp. 265-283.

Jenny H. 1961. Derivation of state factor equations of soils and ecosystems. Soil Sci Proc. pp. 385-388.

Kadmon R. and Harari-Kremer R. 1999. Landscape-scale regeneration dynamics of disturbed Mediterranean maquis. J Veget Sci. 10, 383-402.

Kadosh D. 2002. Palynological Reconstruction of Late Quarternary Climate Changes in the Carmel Coast of Israel. M.A. Thesis, Department of Geography, University of Haifa.

Kislev M.E. 1984. Emergence of wheat agriculture. Paleorient. 10, 61-70.

Kloner A. and Olami Y. 1980. The Early and Middle Canaanite Periods. In Atlas of Haifa and Mount Carmel. Sofer A. and Kipnis B. (Eds.), Haifa University, Haifa. pp.33-36.

Koestler A. 1969. Beyond atomism and holism - the concept of the holon. In Beyond Reductionism: New Perspectives in the Life Sciences. Koestler A. and Smithies J.R. (Eds.), Hutchinson of London, London. pp. 192-216.

Kuhn T.S. 1970. The Structure of the Scientific Revolution. University of Chicago Press, Chicago.

Kummerow J. 1973. Comparative anatomy of sclerophylls of Mediterranean climatic areas. In Mediterranean-type ecosystem: origin and structure. Castri Di F. and Mooney H.A. (Eds.), Springer-Verlag, New York. pp. 157-169.

Kummerow J. 1983. Comparative phenology of Mediterranean-type plant communities. In Mediterranean-type ecosystems - the role of nutrients. Kruger F.J., Mitchell D.T. and Jarvis J.U.M. (Eds.), Springer-Verlag, Berlin. pp. 300-317.

Kutiel P. and Naveh Z. 1987a. Soil properties beneath *Pinus halepensis* on burned and unburned mixed forest on Mt. Carmel, Israel. Forest Ecol Management. 20, 11-24.

Kutiel P and Naveh Z. 1987b. The effect of fire on nutrients in a pine forest soil. Pl Soil 104, 269-274.

Kutiel P., Naveh Z. and Kutiel T. 1990. The effect of wildfire on soil nutrients and vegetation in an Aleppo pine forest on Mt. Carmel, Israel. In Fire in Ecosystem Dynamics, Mediterranean and Northern Perspectives. Goldammer J.G. and Jenkins M.J. (Eds.), SPB Academic Publishing, The Hague. pp. 84-94.

Langer RH.M. 1970. How grasses grow. Stud Biol. 34.

Laszlo E. 1972. Introduction to Systems Philosophy: Toward a New Paradigm of Contemporary Thought. Harper Torchbooks, New York.

Laszlo E. 1994. The Choice: Evolution or Extinction. A Thinking Person's Guide to Global Issues. C. P. Putnam & Sons, New York.

Laszlo E. 2000. Macroshift 2001-20010. Creating the Future in the Early 21st. Century. San Jose, New York.

Laszlo E. 2001a. Macroshift Navigating the Transformation to a Sustainable World. BK Berret-Koehler Publishers, Inc. San Francisco, USA.

Laszlo E. 2001b. Human evolution in the Third Millennium. Futures 33, 649- 658.

Leaky R.E. and Lewin R. 1979. People of the Lake. Avon Books, New York.

Lewis H.T. 1972. The role of fire in the domestication of plants and animals in Southwest Asia: A hypothesis. Man. 7, 198-222.

Lewis H.T. 1973. Pattern of Indian Burning in California; Ecology and Ethnohistory. Anthropological Papers I. Ballena Press Ramona, California.

Lev-Yadun S. and Weinstein-Evron M. 1994. Late Epipaleolithic Wood Remains from El-Wad Cave, Mount Carmel, Israel. New Phytol. 127, 391-396.

Lewis H.T. 1985. Why Indians burned: specific versus general reasons, Symposium and workshop on wilderness fire. Intermountain Forest and Range Experiment Station, U.S.D.A Forest Service, Ogden, Utah. pp. 75-80.

Li B.L. 2000a. Why is the holistic approach becoming so important in landscape ecology? Landscape and Urban Planning. 50, 27-47.

Liacos L.G. 1973. Present studies and history of burning in Greece. Proc. 13th Tall Timbers Fire Ecol. Conf., Tallahassee, Florida, pp. 237-277.

Likens G. E. 1992. The ecosystem approach: its use and abuse. In Excellence in Ecology Vol. 3. Ecology Institute, Oldendorf/Luhe, Germany.

Maturana H. and Varela F. 1980. Autopoiesis and Cognition. Reidel, Dordrecht.

McNeill J.R. 1992. The Mountains of the Mediterranean World. An Environmental History. Cambridge University Press, Cambridge, New York.

Mensing S. 1992. The impact of European settlement on blue oak (Quercus douglasii) regeneration and recuitment in Tehachapi Mountains, California. Madrono 19, 36-46.

Narr K.J. 1956. Early food producing populations. In Man's Role in Changing the Face of the Earth. Thomas W.L. (Ed.), University of Chicago Press, Chicago. pp.134-151.

Naveh Z. 1959. Native types of *Oryzopsis miliacea* in Israel. Bul Res Coun, Sect Bot. 51.

Naveh Z. 1960. Agro-ecological Aspects of Brush Range Improvement in the Maquis Belt of Israel. Ph.D. Thesis, Hebrew University, Jerusalem.

Naveh Z. 1967. Mediterranean ecosystems and vegetation types in California and Israel. Ecol. 48, 445-459.

Naveh Z. 1971. The conservation of ecological diversity of Mediterranean ecosystems through ecological management. In Symposium of British Ecological Society, 11. Duffy E. and Watt A.S. (Eds.), Blackwell Scientific Publications, Oxford. pp. 605-622.

Naveh Z. 1973. The ecology of fire in Israel. Proc. 13th Annual Tall Timbers Fire Ecol. Conf., Tallahassee, Florida. pp. 133-170.

Naveh Z. 1974. Effects of fire in the Mediterranean regions. In: Fire and Ecosystems. Kozlowski T.T. and Ahlgren C.E. (Eds.), Acad. Press, New York. pp. 401-434.

Naveh Z. 1975. The evolutionary significance of fire in the Mediterranean region. Vegetatio. 9, 199-206.

Naveh Z. 1982. The dependence of the productivity of a semi-arid Mediterranean hill pasture ecosystem on climatic fluctuations. Agricul Environm. 7, 47-61.

Naveh Z. 1984. The vegetation of the Carmel and Nahal Sefunim and the evolution of the cultural landscape. In Sefunim Prehistoric Sites in Mount Carmel, Israel. Ronen A. (Ed.), B.A.R. International Series 230. Oxford. pp. 23-63.

Naveh Z. 1990. Ancient man's impact on Mediterranean landscapes in Israel – ecological and evolutionary perspectives. In Man's Role in the Shaping of the East mediterranean Landscapes. Bottema S., Entjest-Nieborg G. and Van Zeist W. (Eds.), AA Balkema, Rotterdam. pp. 43-50.

Naveh Z. 1991. Mediterranean uplands as anthropogenic perturbation dependent systems and their dynamic conservation management. In Terrestrial and Aquatic Ecosystems, Perturbation and Recovery. Ravera O.A. (Eds.), Ellis Horwood, New York. pp. 544-556.

Naveh Z. 1994a. From biodiversity to ecodiversity: a landscape ecology approach to conservation and restoration. Restor Ecol. 2, 180-189.

Naveh Z. 1994b. The role of fire and its management in the conservation of Mediterranean ecosystems and landscapes. In The Role of Fire in Mediterranean-Type Ecosystems. Moreno J.M. and Oechel W.C. (Eds.), Springer, New York. pp. 163-185.

Naveh Z. 1995. Conservation, restoration, and research priorities for Mediterranean uplands threatened by global climate change. In Global Change and Mediterranean-Type Ecosystems. Moreno J.M. and Oechel W.C. (Eds.), Springer Verlag, New York.

Naveh Z. 1998a. From biodiversity to ecodiversity - holistic conservation of biological and cultural diversity of Mediterranean landscapes. In Landscape Disturbance and Biodiversity in Mediterranean-Type Ecosystems. Montenegro G., Jaksic F. and Rundel P.W. (Eds.), Ecological Studies 136. Springer, Berlin-Heidelberg. pp. 23-54.

Naveh Z. 1998b. Culture and landscape conservation - a landscape ecological perspective. In Ecology Today: An Anthology of Contemporary Ecological Research. Gopal B., Pathak P.S. and Saxena K.G. (Eds.), International Scientific Publications, New Delhi. pp.19-48.

Naveh Z. 1999. The role of fire as an evolutionary and ecological factor on landscape and vegetation of Mt. Carmel. J Medit Ecol. 1, 11-26.

Naveh Z. 2000. What is holistic landscape ecology? A conceptual introduction. Landscape and Urban Planning. 50,7-26.

Naveh Z. 2001. Ten major premises for a holistic conception of multifunctional landscapes. Landscape and Urban Planning. 57, 269-284.

Naveh Z. 2002. A transdisciplinary education program for regional sustainable development. Int J Ecol Environ Sci. 28, 167-191.

Naveh Z. 2003. The importance of multifunctional, self-organizing biosphere landscapes for the future of our Total Human Ecosystem – a new paradigm for transdisciplinary landscape ecology. In Multifunctional Landscapes, Volume 1: Theory, Values and History. J. Brandt (Ed.), WIT Press Southampton, UK (in press).

Naveh Z. and Dan J. 1973. The human degradation of Mediterranean landscapes in Israel. In Mediterranean- type Ecosystems. Origin and Structure. Castri Di F. and Mooney HA. (Eds.), Springer Verlag, Heidelberg. pp. 370-390.

Naveh Z. and Whittaker R.H. 1979. Structural and floristic diversity of shrublands and woodlands in Northern Israel and other Mediterranean areas. Vegetatio. 41,171-190.

Naveh Z. and Kutiel P. 1990. Changes in the Mediterranean vegetation in Israel in response to human habitation and land uses. In The Earth in Transition. Patterns and Processes of Biotic Impoverishment. Woodwell G.M. (Ed.), Cambridge University Press, Cambridge. pp. 259-299.

Naveh Z. and Vernet J.L. 1991.The palaeohistory of Mediterranean biota. In Biogeography of Mediterranean Invasions. Groves R.H. and Castri Di F. (Eds.), Cambridge University Press, Cambridge. pp.19-31.

Naveh Z. and Lieberman A.S. 1994. Landscape Ecology Theory and Applications. 2nd ed. Springer, New York.

Narr K.J. 1956. Early food producing populations. In Man's Role in Changing the Face of the Earth. Thomas W.L. (Ed.), University of Chicago Press, Chicago. pp.134-151.

Neeman G., Perevolotsky A. and Schiller G. 1997. The management implications of the Mt. Carmel Research project. Int J Wildland Fire. 7, 343- 330.

Nevo E. 2001a. Evolution of genome-phenome diversity under environmental stress. Proc. Natl Acad Sci, USA 98, 6233-6240.

Nevo E. 2001b. Genetic resources of wild Emmer, *Triticum dicoccoides* for wheat improvement in the third millennium. Isr J Pl Sci. 49, 77-91.

Nevo E., Korol A.B., Beiles A. and Fahima T. 2002. Evolution of Wild Emmer and Wheat Improvement. Springer, Berlin.

Nir D. 1980. Geomorphology. In Atlas of Haifa and Mount Carmel. Soffer A. and Kipnis B. (Eds.), Applied Scientific Research Co. University of Haifa, Israel. pp. 18-19.

Perles C. 1977. Prehistoire du Feu. Masson, Paris.

Paraskevopoulos S.P., Iatrou G.D. and Pantis J.D. 1994. Plant growth strategies in evergreen-sclerophyllous shrublands (Maquis) in central Greece. Vegetatio. 115, 109-114.

Pignatti S. 1978. Evolutionary trends in Mediterranean flora and vegetation. Vegetatio. 37, 175-185.

Pignatti S. 1983. Human impact on the vegetation of the Mediterranean. In Man: Man's Impact on vegetation. Holzer W., Werger M.J.A. and Ikusima I. (Eds.), Dr. Junk, The Hague. pp. 151-162.

Pons S.A. and Quezel M. 1985. The history of the flora and vegetation and past and present human disturbances in the Mediterranean. In Conservation of Mediterranean Plants. Gomez-Campo C. (Ed.), Dr. Junk, The Hague. pp. 25-43.

Preiss E., Martin J.L. and Debussche M. 1997. Rural depopulation and recent landscape changes in a mediterranean region: Consequences to the breeding avifauna. Landscape Ecol. 12, 51-61.

Prigogine I. 1997. The End of Certainty: Time, Chaos, and the New Laws of Nature. The Free Press, New York.

Prigogine I. and Stengers I. 1984. Order out of Chaos. Man's Dialogue with Nature. New Science Library, Shambhala, Boston and London.

Rakham O. and Moody J. 1996. The Making of the Cretan Landscape. Manchester University Press, Manchester.

Raven P. H. 1973. The evolution of Mediterranean floras. In Mediterranean-type ecosystem: origin and structure. Castri Di F. and Mooney H.A. (Eds.), Springer-Verlag, New York. pp. 213-224.

Renfrew C. 1979. Problems in Prehistory. Edinburgh University Press, Edinburgh.

Richardson D.M., Esler K.J. and Cowling R.M. 2001. Mediterranean-type ecosystems - past, present and future. An introduction. J Mediter Ecol. 2, 123-125.

Rindos D. 1984. The Origin of Agriculture. An Evolutionary Perspective. Acad. Press, New York.

Rundel P.W. 1998. Landscape disturbance in Mediterranean-type ecosystems of the world. In Landscape Disturbance and Biodiversity in Mediterranean – Type Ecosystems. Rundel P.W., Montenegro G. and Jaksi F.M. (Eds.), Springer, Berlin. pp. 3-22.

Runnels C.N. 1995. Environmental degradation in ancient Greece. Sci Amer. March: 96-99.

Ronen A. 1979. Guide to the Prehistoric Caves of Mount Carmel. Regional Council of Hof Carmel, Haifa.

Ronen A. 1984. The Sefunim Prehistoric Sites, Mt. Carmel. B.A.R. International Series 230, Oxford.

Ronen A. 2000. Besieged by technology. Scientific American Discovering Archeology. January-February: 92-97.

Ronen A. and Adler D. 2001. The walls of Jericho were magical. Archeol, Ethnol, Anthropol of Eurasia. 2, 97-103.

Samocha Y., Litav M., Fine P. and Vizel Y. 1980. Development rate of woodland trees in the Judea Mountains. Layaaran. 30, 6-15.

Sauer C. 1956. The agency of man on earth. In Man's Role in Changing the Face of the Earth. Thomas W.L. (Ed.), The University of Chicago Press, Chicago. pp. 46-69.

Sauer C.O. 1961. Sedentary and mobile bents in early societies In Social Life of Early Man. Washburn S.L. (Ed.), Viking Fund. Publ. in Anthropology 31, pp. 256-266.

Scheidlinger C.R. and Zedler P.H. 1979. Change in vegetative cover of oak stands in southern San Diego County: 1928-1970. In proceedings of the symposium on the ecology, management and utilization of California oaks. Pacific Southwest Forest and Range Experiment Station. Plumb T.R. (Ed.), Berkeley, California. pp. 81-85.

Sieferle R.P. 1997. Rueckblick auf die Natur: eine Geschichte der Menschheit und seiner Umwelt. In Atlas of Haifa and Mount Carmel. Soffer A. and Kipnis B. (Eds.), Applied Scientific Research Co., University of Haifa, Haifa. Pp. 20-21.

Solecki R.S. 1977. The implications of the Shanidar Cave Neanderthal Flower burial. New York Acad Sci. 293, 114-124.

Stebbins G.L. 1972. The evolution of the grass family. In The biology and utilization of grasses. Younger V.B. and McKell C.M. (Eds.), Academic Press, New York. pp.1-17.

Stebbins G.L. 1982. Darwin to DNA, Molecules to Humanity. W.H. Freeman and Company, San Francisco.

Stekelis M. 1960. The Paleolithic deposits of Jsr Banat Yaqub. Bull. Res. Conc. Israel, Sect G. Geosciences. 9, 346-367.

Swarzboim I. 1978. The Autecology of *Pistacia lentiscus* L. in Israel. Ph. D. Thesis, Technion, Israel Institute of Technology, Haifa (Hebrew with English summary).

Tchernov E. 1975. Rodent faunas and environmental changes in the Pleistocene of Israel. In Rodents in Desert Environments. Pratksh G. and Ghosh P.K. (Eds.), Dr. Junk, The Hague.

Tchernov E. 1985. The fauna of Sefunim cave, Mt. Carmel. In Sefunim Prehistoric Sites in Mount Carmel, Israel. Ronen A. (Ed.), B.A.R. International Series 230, Oxford. pp. 401-422.

Tchernov E. 1988. The paleobiogeographical history of the Southern Levant. In The Zoogeography of Israel. Yom-Tov Y. and Tchernov E. (Eds.), W. Junk, Dordrecht. pp.159- 250.

Tchernov E. and Horwitz L.K. 1990. Herd management in the past and its impact on the landscape of the southern Levant. In Man's Role in the Shaping of the Eastern Mediterranean Landscapes. Bottema S., Entjest-Nieborg G. and Van Zeist W. (Eds.), AA Balkema, Rotterdam. pp. 207-215.

Tchernov E., Horowitz L.K., Ronen A. and Lister A. 1994. The Faunal remains from Evron Quarry in relation to other Lower Paleolithic hominid sites in the southern Levant. Quaternary Res. 42, 328-339.

Tchernov E. and Valla F.R. 1997. Two new dogs, and other human dogs, from the Southern Levant. J Archaeol Sci. 24, 65-94.

Tatskin A., Weinstein-Evron M. and Ronen A. 1995. Weathering and pedogenesis of wind-blown sediments in the Mount Carmel Caves, Israel. Quarternary Proceed. 4, 83-93.

Turnbull C. 1961. The Forest People: A study of the Pigmies of the Congo. Simon and Schuster, New York.

UNESCO-FAO Arid Zone Research Team. 1962. A Bioclimatic Map of the Mediterranean Zone and its Homologues. UNESCO, Paris.

Van Andel T.H. and Runnels C.N. 1995. The earliest farmers in Europe. Antiquity. 69, 481-500.

Vernet J.L. 1975. Les charbons de bois des naiveaux mindelliens de Terra Amata (Nice Alpes-mariitimes). Comptes Rendaus de l'Aca'demie Sciences, Paris 208D. pp. 1535-1537.

Waddington C.H. 1975. A catastrophe theory of evolution. The evolution of an Evolutionist. Cornell University Press, Ithaca. pp.253-266.

Weinstein-Evron M. 1991. New radiocarbon dates from the Early Natufian of El-Wad Cave, Mt. Carmel, Israel. Paleorient. 17, 95-97.

Weinstein-Evron M. 1993a. Paleoecological reconstruction of the Upper Paleolithic in the Levant. Actes of the XIIth International Congress of Prehistoric and Protohistoric Sciences. Bratislava, 1-7 September 1991. Bratislava, pp. 259-270.

Weinstein-Evron M. 1993b. El-Wad Cave. In The New Encyclopedia of Archeological excavations in the Holy Land The Israel Exploration Society Carta, Vol. 4. Stern E., Lewinson-Gilboa A. and Aviram A.J. (Eds.), Jerusalem. pp. 1498-1499.

Weinstein-Evron M. 1998. Early Natufian El-Wad Revisited. ERAUL 77. Service de Pre'histoire, Universite' de Lie'ge.

Weinstein-Evron M. and Belfer-Cohen A. 1993. Natufian figurines from the new excavations of the El-Wad cave, Mt. Carmel, Israel. Rock Art Research. 10, 102-106.

Weinstein-Evron M., Beck A. and Ezersky M. 2003. Geophysical investigations in the service of Mount Carmel (Israel) prehistoric research. J Archaeol Sci. 30, 1331-1341.

Weiss P.A. 1969. The living system: determinism stratified. In Beyond Reductionism: New Perspectives in the Life Sciences. Koestler A. and Smithies J.R. (Eds.), Hutchinson of London, London. pp. 2-55.

Westman W.E. 1986. Resilience: concepts and measures. In Resilience in Mediterranean-type Ecosystems. Dell B., Hopkins A.J.M. and Lamont B.B. (Eds.), Dr. W. Junk Publishers, Dordrecht. Pp. 5-20.

Zinke P.J. 1973. Analogies between the soil and vegetation types of Italy, Greece, and California. In Mediterranean-type ecosystem: origin and structure. Castri Di F. and Mooney H.A. (Eds.), Springer-Verlag, New York. pp. 61-82.

Zohary D. and Spiegel-Roy P. 1975. Beginning of fruit growing in the Old World Science. 187, 319-327.

Zohary M. 1974. Geobotanical Foundations of the Middle East. Gustav Fischer Verlag, Stuttgart.

Zohary M. 1983. Man and vegetation in the Middle East. In: Man's Impact on Vegetation. Holzer W., Werger M.J.A. and Kosmas I. (Eds.), Dr. Junk, The Hague. pp. 287-296.

INDEX